Microfibre Pollution from Textiles

Textile materials shed microfibres during different phases of their life cycle. While much attention has been paid to the impact of microplastics on the environment, there has been less focus on the impact of microfibre pollution, which also poses a serious environmental threat. *Microfibre Pollution from Textiles: Research Advances and Mitigation Strategies* shines a light on the hidden effects of textile microfibre pollution and examines its generation in manufacturing, use including laundering, and disposal. It details advancements and gaps in the quantification and characterization techniques that are emerging in the growing field of study of microfibre pollution in textile technology.

The book

- Examines the contributions of the textile and fashion industries to microfibre pollution, from production to disposal.
- Reviews recently developed methods and technological advancements in the identification and quantification of microfibres from textiles.
- Addresses emerging sustainable mitigation strategies and sustainable textile production methods that can potentially reduce or eliminate microfibre shedding.
- Details the state-of-the-art on existing regulations and standards and provides scope for future research in the area of standard development.

By bridging the gap between environmental and textile studies, this book is aimed at researchers and advanced students in textile and environmental science and engineering.

R. Rathinamoorthy is Associate Professor in the Department of Fashion Technology, PSG College of Technology, Coimbatore, India. He is one of the world's top 2% scientists as reported by Elsevier and Stanford University. His areas of research are biomaterials for textiles and environmental pollution control due to textiles.

S. Raja Balasaraswathi has a Master's degree in Fashion Technology from the National Institute of Fashion Technology, Bengaluru, India. She has conducted extensive research on microfibre pollution since 2019. She has published 17 research papers in international journals, has contributed 3 book chapters, and co-authored a book in the field of microfibre pollution.

Textile Institute Professional Publications

Series Editor: Helen D. Rowe, The Textile Institute

Care and Maintenance of Textile Products Including Apparel and Protective Clothing
Rajkishore Nayak and Saminathan Ratnapandian

Radio Frequency Identification (RFID) Technology and Application in Fashion and Textile Supply Chain
Rajkishore Nayak

Medical Textiles
Holly Morris and Richard Murray

Odour in Textiles
Generation and Control
G. Thilagavathi and R. Rathinamoorthy

Principles of Textile Printing
Asim Kumar Roy Choudhury

Solar Textiles
The Flexible Solution for Solar Power
John Wilson and Robert Mather

Digital Fashion Innovations: Advances in Design, Simulation, and Industry
Abu Sadat Muhammad Sayem

Compression Textiles for Medical, Sports, and Allied Applications
Nimesh Kankariya and René Rossi

Microfibre Pollution from Textiles
Research Advances and Mitigation Strategies
Edited by R. Rathinamoorthy and S. Raja Balasaraswathi

For more information about this series, please visit: www.routledge.com/Textile-Institute-Professional-Publications/book-series/TIPP

Microfibre Pollution from Textiles

Research Advances and Mitigation Strategies

Edited by

R. Rathinamoorthy and

S. Raja Balasaraswathi

CRC Press is an imprint of the
Taylor & Francis Group, an **informa** business

Designed cover image: Shutterstock

First edition published 2024
by CRC Press
2385 NW Executive Center Drive, Suite 320, Boca Raton FL 33431

and by CRC Press
4 Park Square, Milton Park, Abingdon, Oxon, OX14 4RN

CRC Press is an imprint of Taylor & Francis Group, LLC

ISBN: 978-1-032-36276-2 (hbk)
ISBN: 978-1-032-36445-2 (pbk)
ISBN: 978-1-003-33199-5 (ebk)

DOI: 10.1201/9781003331995

Typeset in Times
by codeMantra

Contents

SECTION 4 Potential Mitigation Strategies and Awareness

Preface

Microfibre pollution is becoming a significant issue among emerging contaminants worldwide. Typically, microfibre is a subset of microplastic that is released from the textile material. Due to the increased consumption of synthetic clothing, a great amount of microfibres was reported in various environments. However, it is not restricted only to synthetic textiles; rather, all microfibres have a significant impact on the ecosystem. Moreover, the knowledge about these pollutants is meager among the scientific community, the general public, and governments. Even after much exploration, the existing literature suffers severely due to no uniformity in methods adopted and insufficient standards to regulate the issue.

The book aims to stitch together the various latest research findings of different researchers across the world and provides a better understanding to the readers of the basic know-how. For the convenience of the readers, the contributed chapters in the book are grouped under four categories, namely, "Microfibre Pollution in the Textile and Fashion Industry," "Sources of Microfibre Pollution and Analytical Tools," "Impact of Microfibre Pollution", and "Potential Mitigation Strategies and Awareness."

The first section of the book: Microfibre Pollution in the Textile and Fashion Industry consists of seven chapters as a part of it. Here, the readers will get an insight into fundamentals of the "microfibre pollution." The section also provides a discussion on how fashion and textile industries and products act as a source of such pollution. All the potential release points from textiles, like the manufacturing phase (textile effluents), and user phase (laundry and drying) were addressed. A specific chapter on nonwoven disposable products is also included in this section due to the increased use of masks and nonwoven-based personal care products.

The second section of the book, Sources of Microfibre Pollution and Analytical Tools, delivers information about the gaps in existing analytical methods while elaborating on the latest or upcoming that are being adopted recently by researchers across the world. The section also proposed two new methods based on forensic analysis and an instrumental-based dynamic image analysis for the better quantification of microfibre.

The third section of the book is themed Impact of Microfibre Pollution. Out of the four chapters in this section, the first two chapters analyzed the prevalence of microfibres in various environments and their subsequent impacts. The next two chapters report the transportation mechanism of microfibres from the environment to humans and its health impact as a consequence.

The last part of the book deals with the area of Mitigation Strategies and Regulations, in which the chapters consolidate the existing mitigation and control measures that are discussed in the available literature. The section emphasizes the requirements of interdisciplinary approaches to mitigate the impacts of microfibres. Also, this section throws light on various regulatory norms and standards that have recently been established and yet to be amended. The last chapter of the section outlines the role of different non-government agencies in the control of environmental pollution, where the chapter focuses on "The Microfibre Consortium" and its contributions to this domain, one of its kind organization, specifically instituted to bridge the gap between industry and academia.

We believe that the book will give a new perspective to the readers and research community regarding the role of textiles in microfibre pollution. At the same time, the book will suggest how to eliminate or mitigate such issue and provides possible strategies/methods for industry and the public. The section on the standards will be a highlight of the book which will be very useful to the research community. We would like to mention a special thanks to all contributing authors for volunteering to contribute to this book. Without their effort, the book might have not gotten the better shape as it is now. As authors from different parts of the world had contributed, the word "fibre" was used in both American (fiber) and British English (fibre) in the book, as preferred by the contributed chapter authors.

We look forward to your engagement with the material, your questions, and your thoughts. Feel free to reach out to us with any feedback or reflections you may have. With anticipation and gratitude, we welcome you to delve into the heart of *Microfibre Pollution from Textiles: Research Advances and Mitigation Strategies.*

Dr. R. Rathinamoorthy

Ms. S. Raja Balasaraswathi

Contributors

Fatemeh Mashhadi Abolghasem
Korea Institute of Industrial Technology (KITECH)
Ansan-si, Republic of Korea

S. Raja Balasaraswathi
National Institute of Fashion Technology
Bengaluru, India

Anna Bateman
The Microfibre Consortium
Bristol, United Kingdom

Mansoor Ahmad Bhat
Department of Environmental Engineering, Faculty of Engineering
Eskişehir Technical University
Eskişehir, Türkiye

Elliot Bland
The Microfibre Consortium
Bristol, United Kingdom

Erika Iveth Cedillo-González
University of Modena and Reggio Emilia
Modena, Italy

Carmen K.M. Chan
The Hong Kong Polytechnic University
Hung Hom, Hong Kong

Edith Classen
Hohenstein Institut für Textil Innovation GmbH
Bönnigheim, Germany

Mirjana Čurlin
Faculty of Food Technology and Biotechnology
University of Zagreb
Zagreb, Croatia

Tihana Dekanić
Faculty of Textile Technology
University of Zagreb
Zagreb, Croatia

G.E. De-la-Torre
Universidad San Ignacio de Loyola
La Molina, Perú

Charline Ducas
The Microfibre Consortium
Bristol, United Kingdom

James K.H. Fang
The Hong Kong Polytechnic University
Hung Hom, Hong Kong

Claire Gwinnett
Staffordshire University
Stoke-on-Trent, United Kingdom

Alice Hazlehurst
University of Leeds
Leeds, United Kingdom

Alana M. James
Northumbria University
Newcastle upon Tyne, United Kingdom

Jasmin Jung
Hohenstein Institut für Textil Innovation GmbH
Bönnigheim, Germany

C.W. Kan
The Hong Kong Polytechnic University
Hung Hom, Hong Kong

Kirsten J. Kapp
Central Wyoming College
Wyoming, United States

Disha Katyal, CPHI(C)
Canadian Institute of Public Health Inspectors (CIPHI)
British Columbia, Canada

Juhea Kim
Korea Institute of Industrial Technology (KITECH)
Ansan-si, Republic of Korea

Juran Kim
Korea Institute of Industrial Technology (KITECH)
Ansan-si, Republic of Korea

A.D. Forero López
Instituto Argentino de Oceanografía (IADO)
Bahía Blanca, Argentina

Sophie Mather
The Microfibre Consortium
Bristol, United Kingdom

Rachael Z. Miller
Rozalia Project for a Clean Ocean
Burlington, Vermont, United States of America

Nkumbu Mutambo
Northumbria University
Newcastle upon Tyne, United Kingdom

Satoshi Nakao
Osaka City Research Center of Environmental Science
Osaka, Japan

Tetsuji Okuda
Ryukoku University
Otsu, Japan

Aravin Prince Periyasamy
Aalto University
Espoo, Finland
and
VTT Technical Research Center of Finland
Espoo, Finland

Miranda Prendergast-Miller
Northumbria University
Newcastle upon Tyne, United Kingdom

Tanja Pušić
Faculty of Textile Technology
University of Zagreb
Zagreb, Croatia

R. Rathinamoorthy
PSG College of Technology
Coimbatore, India

G.N. Rimondino
Instituto de Investigaciones en Fisicoquímica de Córdoba (INFIQC)
Córdoba, Argentina

Abbie Rogers
Northumbria University
Newcastle upon Tyne, United Kingdom

Ana Šaravanja
Faculty of Textile Technology
University of Zagreb
Zagreb, Croatia

M.D. Fernández Severini
Instituto Argentino de Oceanografía (IADO)
Bahía Blanca, Argentina

Kelly J. Sheridan
The Microfibre Consortium
Bristol, United Kingdom
and
Northumbria University
Newcastle upon Tyne, United Kingdom

C.V. Spetter
Universidad Nacional del Sur (UNS)
Bahía Blanca, Argentina

Giuseppe Suaria
CNR-ISMAR, Istituto di Scienze Marine
Consiglio Nazionale delle Ricerche
Lerici, Italy

Mark Sumner
University of Leeds
Leeds, United Kingdom

Mark Taylor
Faculty of Mechanical Engineering
University of Leeds
Leeds, United Kingdom

Julija Volmajer Valh
Faculty of Mechanical Engineering
University of Maribor
Maribor, Slovenia

Branka Vojnović
Faculty of Textile Technology
University of Zagreb
Zagreb, Croatia

Songyi Yan
The University of Manchester
Manchester, United Kingdom

Section 1

Microfibre Pollution in the Textile and Fashion Industry

1 An Introduction to Microfiber Pollution

R. Rathinamoorthy
PSG College of Technology

S. Raja Balasaraswathi
National Institute of Fashion Technology

1.1 INTRODUCTION

The textile and fashion industry is one of the fast-growing industries with large business volumes while employing millions. By 2021, the global textile market accounted for 993.6 billion USD, growing at an annual compound rate of 4.0% (Market Analysis Report, 2022). The increased demand for apparel from the fashion industry due to the whirlwind of fast fashion trends drives the industry's growth. Fast fashion is a trend that is easily accessible, reasonably priced, and immediately popularized among the general public after being introduced by catwalk trends (Bick et al., 2018). The fast fashion trend increased the number of collections launched per year. For instance, fast fashion brands such as Zara and H&M launch around 12–24 collections yearly (Centobelli et al., 2022). Fast fashion reduces the buying cycle and lead time to meet customer demand. Maintaining low inventory, quick response to demand, and a shorter supply chain make the fast fashion trend possible (Camargo et al., 2020). While fast fashion helps consumers fulfill their demand for trendy, reasonably priced, and attractive clothing, it also supports 300 million jobs globally and the expansion of global economy (Assoune, 2023a).

The textile sector accounts for about 2% of the worldwide gross domestic product (GDP) (Hanichak, 2021). Despite the industry's enormous volume of business and profit, its influence on sustainability cannot be avoided. A surplus of resources is used up due to the rising demand for fashion products. From 2000 to 2018, textile fiber production increased by a factor of two. When production per capita is examined, it climbed from 7.6 to 13.8 kg/person between 1995 and 2018 (Peters et al., 2021). The environmental impact of fast fashion can be categorized as global, regional, and local environmental impacts, where the global impacts include greenhouse gas emissions during the production and disposal of clothes and ozone depletion due to the use of chemicals in manufacturing (Pejak, 2022). It is estimated that fashion industry will lead to 2.1 billion metric tons of carbon emission by 2030 (Berg et al., 2020). Regional impacts include the utilization of lands, freshwater, and toxic chemicals, which cause environmental pollution. The last category of local environmental impacts includes the local landfills and local air and noise pollution due to the manufacturing of clothes (Pejak, 2022).

DOI: 10.1201/9781003331995-2

Additionally, many natural resources like water, land, and ecosystem get polluted due to the production process of textiles and fashion products. Recent research reported that dyeing and finishing (36%), yarn preparation (28%), and fiber production (15%) are the major pollution-driving sectors of the textile and fashion industry. Out of these, fiber cultivation was reported as the primary consumer of freshwater resources, consuming 31% (Maiti, 2023). The impact of such a process increases daily due to the increasing consumption and lower utilization of fast fashion products (Ellen MacArthur Foundation, 2017). As per World Bank reports, the fashion industry utilized approximately 93 billion cubic meters of water in 2019. Further, the report also mentioned that 20% of the world's wastewater comes from the textile and fashion industry. To the worst, due to the shorter trend cycles, fast fashion products were utilized barely in their lives, and 87% of the total produced garments ended up in a landfill or incinerated before their lifetime (The World Bank, 2019).

1.2 MICROPLASTIC/MICROFIBER ISSUE

In addition to the discussed issues, one of the recently emerged microplastic pollution has a higher association with the fashion industry. Microplastics are often referred to as the plastic particles which are smaller in size, most commonly defined as less than 5 mm in size. The aforementioned fast fashion utilizes more synthetic fibers as a textile product due to their cheaper cost and lower utilization (Young, 2021). The issue of microplastics in the fibrous form arises at all the life stages of these synthetic textiles, from manufacturing, use, and disposal (Raja Balasaraswathi & Rathinamoorthy, 2021). The microplastics released from the synthetic textiles have a higher length-to-diameter ratio than the other shapes, such as spheres, fragments, etc. These fiber-like higher-length particles, generally termed as "microfibers", were initially noted in the wastewater treatment plant sludge and initially termed as micro-sized synthetic fibers (Habib et al., 1998). Past studies demonstrated that the laundry process was the primary source of microfiber release from the textiles. The disposal of laundry wastewater directly emits microfibers into the terrestrial and aquatic environment. Based on their findings, it was estimated that most of the microplastics expected to be deposited in the marine environment are the microfibers released from the textile (Browne et al., 2011). Later researchers confirmed that 35% of the total microplastics in the marine environment are microfibers from synthetic textiles (Boucher & Friot, 2017). Washing synthetic textiles alone can result in the release of 0.28–0.50 million microfibers into the aquatic environment every year (Ellen MacArthur Foundation, 2017). Due to their abundance in the marine environment, marine planktons and other microorganisms sometimes consume them by mistaking them as prey. The microplastics or fibers consumed by the microorganisms were then transferred to other aquatic life and finally can reach humans (Cox et al., 2019). Studies have reported that microfibers and plastics in seafood like fish and edible products like salt are obtained from the sea (Browne et al., 2008; Karami et al., 2017; Kasamesiri & Thaimuangpho, 2020; Lee et al., 2019; Lusher et al., 2013; Van Cauwenberghe et al., 2015a). Though oral admission of microplastics showed several effects on aquatic life forms (Jemec et al., 2016; Kim et al., 2021a), no detailed studies were performed to understand the complete interaction (Qualhato et al., 2023). As far as the human health impact is considered, despite the evidence in several human parts (Çobanoğlu et al., 2021; Huang et al., 2022; Jenner et al., 2022; Pauly et al., 1998; Ragusa et al., 2021), still research gaps are there to be addressed to understand the real impact (Blackburn & Green, 2022). To showcase the other pathways apart from washing, few studies showed evidence of microfibers released from textiles directly into the atmosphere and corresponding pollution in the atmosphere (De Falco et al., 2020). Such pollution directly affects human health as the microfibers are inhaled and adsorbed via breathing and skin, respectively (Bhat et al., 2023; Domenech & Marcos, 2021; Li et al., 2021b; Soltani et al., 2021).

When the terrestrial environments were analyzed for microfiber pollution, the impact of the textile and fashion industry was noted as significant. The release of microfibers was evidenced at all stages of textile materials in both aquatic and terrestrial environments. This chapter provides an introduction to microfiber pollution and the contribution of the textile and fashion industry to this global issue.

1.3 DEFINITION OF MICROFIBERS

When the definition of the term "Microfiber" is analyzed in the context of environmental science and textile engineering, it becomes clear that the former has a distinct connotation. In the textile industry, microfibers are defined as filaments or staple fibers with a linear density between 0.3 and 1 denier. However, in the context of environmental science, microfibers are an emerging contaminant that can be defined as any natural or artificial fibrous materials having thread-like structure whose length-to-diameter ratio is higher than 100 with length ranges between 1 µm and 5 mm (Liu et al., 2019a). To avoid such confusion, researchers have used the term "Fiber fragments" to denote microfibers in the context of environmental science (The Microfibre Consortium, 2023).

The first study that addressed the microfiber abundance in the environment was performed by Thompson et al. (2004). However, correlating the use of textiles and their potential to release microplastic fibers were reported by Browne et al. (2011), and they were the first to apply the term "microplastic fiber" to contaminants found in the environment. Later, other researchers together coined the term "microfiber" for microplastics in a fibrous shape (Van Cauwenberghe et al., 2015b; Woodall et al., 2015).

Also, there is a confusion in differentiating microplastics and microfibers in the environmental context. Microplastics, as it is named, are plastics of smaller sizes. Though several researchers reported different size ranges for microplastics (Table 1.1), plastic particles less than 5 mm are generally considered microplastics. At the same time, microfibers are thread-like structures with a high length-to-diameter ratio that can be natural or synthetic in nature with a size of less than 5 mm. Moreover, synthetic microfibers can fall under the category of microplastics due to their smaller size and synthetic nature. Figure 1.1 represents the various shapes of microplastics reported in the literature. A proper and extensive definition of microfiber pollutants was reported as "Microfibers are any natural or artificial fibrous materials of thread-like structure with a diameter less than 50 µm, length ranging from 1 µm to 5 mm, and length to diameter ratio greater than 100" by Liu et al. (2019a). In this

TABLE 1.1
Summary of Microfiber Definitions as Reported in the Literature

Reference	Definition	Key Characteristics Considered
Karami et al. (2016)	Microplastics are the smaller-sized (ranging between 1 and 1,000 μm) plastic debris that resulted from the fragmentation of large plastic objects.	• Size • Polymer composition
Crawford and Quinn (2017)	Microfiber is a strand or filament of plastic that is in the size range of 1 μm–1 mm in size along its longest dimension.	• Size • Shape
Wang et al. (2020)	Microplastics are plastics less than 5 mm in size obtained out of the breakdown of larger plastics due to physical factors, chemical factors, and biological factors.	• Size • Polymer composition • Source of origin
Moore (2008)	Microplastic debris are marine plastic debris which are less than 5 mm in size.	• Size • Polymer composition
NOAA Marine Debris Program (2009)	Microplastics are plastic particles less than 5 mm in size.	• Size • Polymer composition
Andrady (2003)	Microlitters are fine plastic detritus with a size range by which it can pass through a 500-μm sieve but is retained on one at 63 μm; size in the range of 0.06–0.5 mm in diameter.	• Size • Polymer composition
Liu et al. (2019a)	Microfibers are any natural or artificial fibrous materials having thread-like structures with length to diameter ratio higher than 100 (diameter: less than 50 μm and length: 1 μm–5 mm)	• Size • Shape • Polymer composition
Commission Decision (EU) 2017/1218 (2017)	Microplastics are water-insoluble macromolecular plastic particles of size less than 5 mm.	• Size • Polymer composition
Liu et al. (2022)	Microfiber is also called microplastic fiber, synthetic fiber, or even chemical fiber with a length of less than 5 mm.	• Size • Polymer composition
AATCC TM212 (2021)	Fiber fragment is a short piece (typically $<5\times10^{-3}$ m long) of textile fiber, broken away (or separated) from a textile construction.	• Shape • Size
GESAMP (2015)	The term microplastics refers to small pieces of plastic, in the range of 1 nm to less than 5 mm.	• Size • Polymer composition
The Microfibre Consortium (2023)	Microfiber or fiber fragment refers to a short piece of textile fiber broken from the main textile construction or through its subsequent breakage in the natural environment. Microplastic refers to small pieces of plastic debris measuring 5 mm or less, found in the environment from the disposal or breakdown of consumer products and industrial waste.	• Size • Shape • Polymer composition

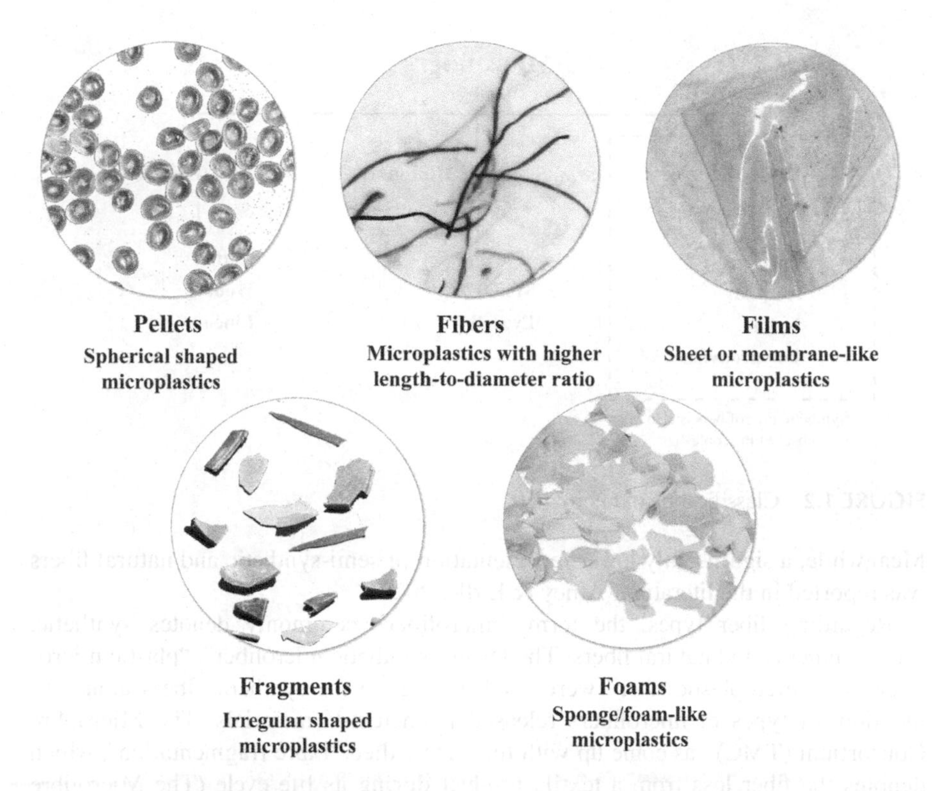

FIGURE 1.1 Different types of microplastics based on their shapes.

book, the term microfiber was used in the environmental context for microplastics with a higher length-to-diameter ratio, as reported in the literature. Table 1.1 summarizes the different definitions for microplastics and microfibers made by considering the size, shape, and polymer composition of the particle by other researchers.

1.4 CLASSIFICATION OF MICROFIBERS

Microfibers can be considered microplastics in the case of synthetic fibers. However, both natural and synthetic microfibers are considered anthropogenic litter. Also, few terms, such as "plastic microfibers" and "microplastic fibers", are also used by researchers to denote microfibers which come under the category of microplastics (Dalla Fontana et al., 2021). Figure 1.2 shows the classification of microfibers. Though the initial studies reported on synthetic microfibers, later researchers found that around 88% of the microfiber found in the marine sediments are natural fibers/cellulosic like cotton, linen, viscose, etc. (Le Guen et al., 2020; Suaria et al., 2020). Approximately 60%–80% of the identified microfibers in the environment are cellulose-based fibers (Suaria et al., 2020). The other category that is often not considered is the semi-synthetic microfiber. Initial studies excluded the natural and semi-synthetic fibers due to their non-plastic nature; however, later, it was also found harmful due to the leaching of chemicals that were applied during manufacturing.

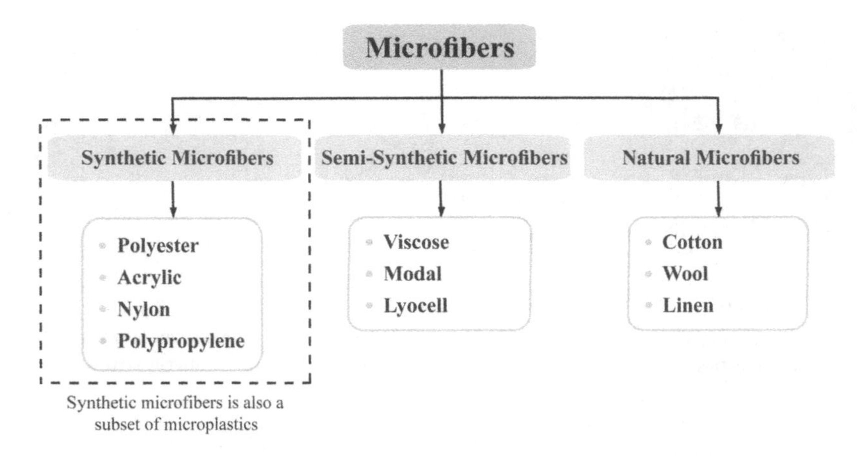

FIGURE 1.2 Classification of microfiber.

Meanwhile, a significantly lower representation of semi-synthetic and natural fibers was reported in the literature (Athey & Erdle, 2022).

Regarding fiber types, the term "microfiber" commonly denotes synthetic, semi-synthetic, and natural fibers. The terms "synthetic microfiber", "plastic microfiber", or "microplastic fiber" were used to refer to the synthetic fibers alone. To mention all types of microfibers released from textile materials, The Microfibre Consortium (TMC) has come up with the term called "Fibre fragmentation", which denotes the fiber loss from a textile product during its life cycle (The Microfibre Consortium, 2023). The detailed definition by TMC can be viewed in chapter 20 of the book.

1.4.1 Microfibers from Natural Fiber-Based Textiles

Natural fibers like cotton, linen, and wool were commonly used in textile manufacturing. Though natural fiber production was reported to consume much water and was reported as unsustainable, initially, there was a belief that all natural fibers are biodegradable and will harm significantly less to the environment when released as microfibers (Zambrano et al., 2019). However, later studies did not agree with such a conclusion because of their abundance in the environment, especially marine basins (Sanchez-Vidal et al., 2018; Zambrano et al., 2020). First of all, while comparing the microfiber emission from textiles during laundry, researchers reported a higher release of microfiber (per unit area of textile) with cellulose-based textiles compared to the synthetic textiles (Cesa et al., 2020; Sillanpää & Sainio, 2017; Vassilenko et al., 2019; Zambrano et al., 2019). This was noted true in the case of blended fabrics also, where more than 80% of microfibers released from a 50/50 polyester/cotton blended fabric was cotton fiber (De Falco et al., 2020). One of the recent studies showcased that the use of denim wear was a significant source of natural cellulose-based microfiber. During the analysis, the study reported that washing a single pair of jeans trousers can release $5.6 \times 10^4 \pm 4.1 \times 10^3$ microfibers per wash cycle (Athey et al., 2020). The study suggested that most of the blue anthropogenic fibers found in the environment are indigo-dyed cotton material. While analyzing the anthropogenic microplastic particles in the Great Lakes and shallow suburban lakes of the Canadian Arctic Archipelago, researchers reported that around 56%–57% of microfibers in both lakes are from indigo-dyed cotton from denim material (Athey et al., 2020). A study performed across the Hudson River in the USA reported that around 50% of the microfibers measured throughout the river are non-plastic in nature (Miller et al., 2017).

Secondly, the primary concerns about natural/cellulosic fiber are the chemicals, finishing agents, and additives used during the different manufacturing stages (Lewandowski et al., 2015). The textile manufacturing process generally uses several toxic chemicals (Ladewig et al., 2015; Xue et al., 2017; Zambrano et al., 2021) and dye substances in its production phase. Such substances leach out from the textiles and adversely affect the environment and aquatic life. Though synthetic and natural fibers are exposed to such chemicals, the leachates from natural fibers soon reach the aquatic environment due to their biodegradable nature (Sharma & Krupadam, 2022). Finishes were most commonly used in textiles to improve their properties on several occasions. Literature shows evidence that finishes like DMDHEU (dimethylol dihydroxy ethylene urea) reduce the biodegradability of the cotton fiber to a level of 6.5%. In contrast, silicone softeners increase the degradability to a level of 5.5% compared to untreated cotton (Li et al., 2010). Research work reported that microfiber emission from cotton fabrics with different finishes like softeners, durable press finishes, and water-repellent finishes showed differences in microfiber release due to changes in the mechanical properties of the fabrics (Zambrano et al., 2021).

Previous studies have reported the capabilities of synthetic microfibers to carry pollutants. For instance, an analysis performed on polyamide and polyester/cotton-blended fabrics showed the potential ability of the fabrics to release 240 and 1,300 micrograms of fluorine per square meter ($\mu g\ F/m^2$) of fabric, respectively. The study reported that European countries alone could emit up to ~0.7 t of fluorotelomer alcohol (6:2 FTOH) per year (Schellenberger et al., 2019). However, research studies were not performed on the effect of these finishes and chemical additives on the biodegradability of natural microfibers and their anthropogenic conversion. Despite a few evaluations performed on synthetic textiles, natural fibers remain untouched. Similarly, the impact of such natural fibers on aquatic lifeforms is unknown, as all the studies reported synthetic microfibers. One of the most recent studies evaluated the impact of polyester and cotton microfibers on the growth rate of juvenile mussels. Exposure to cotton and polyester fibers at 80 fibers per liter concentration showed an 18.7% and 35.6% growth reduction, respectively (Walkinshaw et al., 2023). A similar study reported the presence of cotton microfibers in the commercial fish types obtained from the western Mediterranean Sea (Compa et al., 2018). However, their long-term impact on several other microorganisms must be studied in the future to understand it better.

1.4.2 Semi-synthetic Microfibers

Semi-synthetic fibers are the type of fibers that are made of cellulose but produced artificially, similar to synthetic fibers. Viscose rayon and lyocell are some well-known examples of semi-synthetic fibers. Despite their lower production (10% of total fiber production), several researchers reported rayon as a primary component in their analysis (Assoune, 2023b). A study conducted in 18 different coastal sites in the United States of America reported that 68% of the total microplastics identified are rayon microfibers, and the remaining 24% are polyester. Study results concluded that major cities in those nearby areas are the significant sources of microplastics in the ocean (Yu et al., 2018). While evaluating marine microfiber pollution, other researchers reported that rayon, a semi-synthetic fiber, is equally common in marine environments next to polyester. This information is derived from the literature and apparel production data from countries like India, China, Indonesia, the USA, and Sri Lanka (Mishra et al., 2019). Studies reported that the semi-synthetic fibers significantly contributed to microfiber pollution in regional seashores. Around 55% of the total particles analyzed in the coastal water of Plymouth, United Kingdom, and 63% of the particles in the (Atlantic) southwestern coast of Europe and the western coast of Africa are semi-synthetic fibers (Kanhai et al., 2017; Oldenburg et al., 2021). Coral reefs were collected and analyzed from the coastal region of Belize. The results reported that 50% of the sediments found in the coral system were rayon fibers (Figure 1.3). Due to the higher consumption and use of apparel, the wastewater treatment plant effluent from the manufacturing industry and domestic laundry are the significant sources of semi-synthetic fiber emission (Oldenburg et al., 2021). Despite their cellulosic nature, higher contaminants like plasticizers and other chemicals were identified in the collected sample.

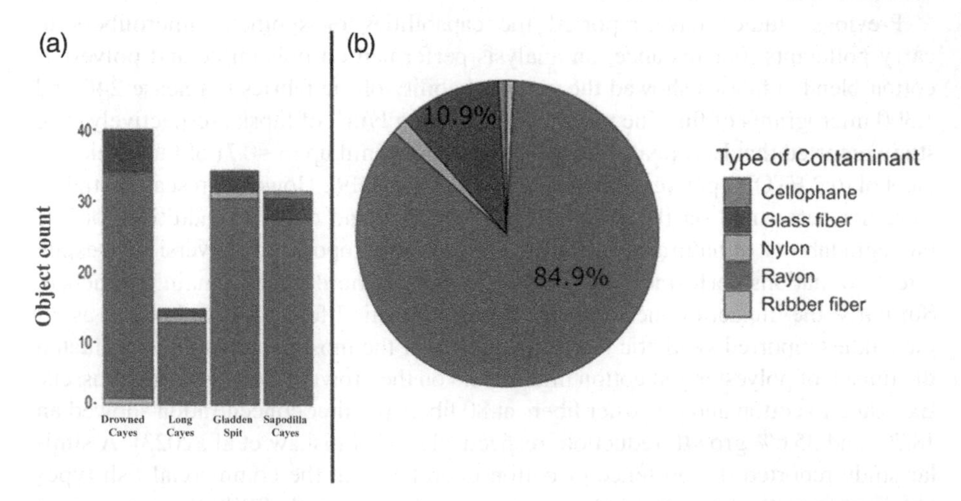

FIGURE 1.3 ATR-FTIR spectroscopy identification of a subset of contaminants within 20 coral samples by Oldenburg et al. (2021). (a) Sample identification by location, and (b) total abundance of five different contaminants identified in our subset of contaminants. (Source: Reprinted with permission.)

Based on this analysis, researchers doubted the real impact of these semi-synthetic fibers on corals and other aquatic animals (Oldenburg et al., 2021). Researchers also reported the dominance of natural and semi-synthetic microfibers in the atmosphere due to their source, physicochemical properties, production process, and applications (Gasperi et al., 2015; Liu et al., 2023).

While analyzing the ingestion of microplastics in aquatic animals, a study evaluated farm fish like *Sparus aurata* and *Cyprinus carpio* at different life stages. The study results showed that 90% of the contaminants noted in the fishes are microfibers, of which 52% are synthetic, 20% are natural, and 28% are semi-synthetic microfibers. Linen, cotton, lyocell, and rayon were reported among the non-plastic microfibers. Of these, rayon is the most frequently identified polymer (21.2%) than other semi-synthetic and natural fibers (Savoca et al., 2021). A similar study analyzed the microfiber contamination in the gastrointestinal tract (GIT) of B. boops specimens from the Tyrrhenian Sea. The results showed that 63.3% of species had ingested microfibers out of the collected samples, with an average of 2.7 fibers per specimen. The characterization studies reported that the ingested fibers are regenerated cellulosic fiber (rayon), one of the primary semi-synthetic fibers (Savoca et al., 2019).

While studying the impact of these anthropogenic microfibers on aquatic biota, other researchers analyzed the impact of polyester and lyocell on the brine shrimp *Artemia franciscana*. The results showed that the lyocell-exposed shrimps had a lower mortality rate than polyester in 48-hour exposure. At the same time, the gut analysis results showed that synthetic and semi-synthetic (lyocell) caused similar amounts of gut damage to the shrimps (Kim et al., 2021). Commercial fish types like Dicentrarchus labrax, Diplodus vulgaris, and Platichthys flesus from the Mondego estuary were reported to have higher (93%) microfiber ingestion than other species. Most of the polymer types reported in these fish types are polyester (31%) and rayon (30%). The author reported that the origin of such fibers is textiles and apparel that are used commercially (Bessa et al., 2018). A similar study was conducted on the freshwater amphipod Gammarus duebeni, exposing it to polyester and man-made cellulose (semi-synthetic) fibers for 96 hours. The results showed that 58.3% of amphipods contained cellulose, and 41.7% contained polyester fibers. Though the 96-hour exposure study did not show any short-term effects on the life of the amphipods, the long-term effects need to be studied yet (Mateos-Cárdenas et al., 2021).

1.4.3 Synthetic Microfibers

Fibers like polyester, acrylic, nylon, and polypropylene are the common synthetic fibers used in apparel and home textile applications. Synthetic textiles were commonly used due to their low cost, easy production, and user-specific properties (Furstenberg, 2021). Synthetic fibers are the most dominant; polyester occupies 60% of world synthetic fiber consumption due to the fast fashion trend (Young, 2021). Studies predicted that around 31% of the total plastic pollution in the marine environment originates from the fashion industry (Brophy, 2020). Research reports predicted that over 22 million tons of microfibers will enter the marine basin between 2015 and

2050 due to the apparel industry (Ellen MacArthur Foundation, 2017). 65.2% of the particles identified in the glaciers are synthetic microfibers, in which the most dominant fiber types are polyester and polypropylene (Ambrosini et al., 2019). Similarly, studies performed in the atmosphere over oceans showed more amount of microplastics with 88.89% of microfibers. Polyester-based microfibers are reported as the most abundant, and the study mentioned that the fibers originated from clothing (Wang et al., 2020). Studies performed worldwide reported more synthetic microfibers in the atmosphere (Gasperi et al., 2015; Liu et al., 2019b; Tunahan Kaya et al., 2018; Wang et al., 2020).

The marine environment and coastal regions are the environments that were highly contaminated due to microplastics, specifically synthetic microfibers. Researchers reported the presence of microfibers in different parts of the sea as sediments and also in the surface water. A study conducted in Yangtze Estuary and the East China Sea by floatation method reported a dominance of 79.1% of microfibers (Zhao et al., 2014). In the arctic polar surface waters and subwaters, more than 95% of microplastics in the shape of microfiber originated from textiles (Sanchez-Vidal et al., 2018). Several regions of Indian subcontinents, such as Mumbai, Tuticorin, and Dhanuskodi beaches, showed an abundance of fibrous microplastics (Tiwari et al., 2019). Furthermore, several researchers evaluated and detailed the aquatic microorganisms and animals that are heavily impacted by the ingestion of microplastics, specifically the impact of synthetic microfibers (Au et al., 2015; Horn et al., 2020; Jemec et al., 2016; Liang et al., 2023; Ziajahromi et al., 2017).

As far as the terrestrial environment is concerned, studies conducted on farmland reported a higher amount of fibrous microplastics, with polyester and polypropylene polymer as significant contributors (Liu et al., 2018). Few studies also reported the presence of higher synthetic microfibers (87.72%) in the gastrointestinal tracts of terrestrial birds (Zhao et al., 2016). The sea foods and other items derived from other animals and insects, such as milk, honey, salts, and drinking water, have been contaminated with microfibers (Diaz-Basantes et al., 2020; Karami et al., 2017; Kutralam-Muniasamy et al., 2020; Paredes et al., 2019; Shen et al., 2021b). Some recent research work also reported synthetic microfibers in the human placenta and guts (Donkers et al., 2022; Zhu et al., 2023). These findings indicate the level of contamination that is being faced with synthetic microfibers.

Though all three different microfibers (natural, synthetic, semi-synthetic) were evidenced in the entire environment and food chain, the real impact of these plastics on humans was unknown. Despite the complete knowledge of the effect of synthetic microfibers on the aquatic biota and other animals, the impact of anthropogenically modified natural and semi-synthetic fibers is yet to be analyzed. Several researchers have already explored the impact of synthetic fibers on the mortality of aquatic organisms and reported their impact (Au et al., 2015; Horn et al., 2020; Jemec et al., 2016; Liang et al., 2023; Ziajahromi et al., 2017). Similarly, analysis of semi-synthetic and natural fibers needs to be performed in the future to understand their actual impact on the aquatic biota. As microplastics reach the human food chain, it is imperative to evaluate the impact of microplastics, specifically the impact of microfibers on human health.

1.5 WHY DO MICROFIBERS NEED SPECIAL ATTENTION?

Though microplastics are of different forms, such as pellets, granules, and fragments, fiber forms are more critical and are extensively addressed as "microfibers", including synthetic, natural, and cellulosic fibers. The main reasons that emphasize the requirement for particular attention are:

i. The abundance of microfiber over other shapes,
ii. Lack of prevention and control methods,
iii. Insufficient knowledge and regulations, and
iv. Unknown environmental impacts.

1.5.1 Abundance of Microfiber over Other Shapes

The first and foremost important reason for requiring special attention is their abundance. Even though the microplastics were initially reported in the literature, later studies confirmed the contribution of microfibers as a significant part of the microplastics. The existence of microfibers was confirmed in the entire environment. To support these findings, we have consolidated literature that shows that microfibers were the most dominating type out of all other shapes reported in the history of microplastics. Figure 1.4 consolidates the contribution of microfiber among all identified shapes of microplastics in different environments by different researchers.

1.5.2 Lack of Prevention and Control Methods

After introducing the term "microfiber" in the research arena, several researchers focused on the products or methods that can be used to reduce the reach of microfibers into the environment. Different approaches were made in the area of controlling microfiber pollution, namely, (i) laundry aids and filters for domestic laundering, (ii) finishing and surface modification of textiles, and (iii) control at manufacturing. The efforts made by the research community on the aforementioned individual sector are detailed below.

1.5.2.1 Laundry Aids and Filters for Domestic Laundering

As the pioneer researchers reported domestic laundry as a significant source (Browne et al., 2011; Napper & Thompson, 2016), many products were developed in the laundry area to control the emission of microfibers from the textiles. Later researchers reported that a single laundry with 6 kg of load could emit up to 700,000 microfibers into the environment (Napper & Thompson, 2016). Though attempts were made to control or evaluate the impact of individual laundry parameters on the microfiber emission, the results were not fruitful. The use of different laundry aids like detergent, softeners, and other surfactants were evaluated (Mermaids, 2015). The interactive effect of the parameters was high, and changes or reduction in one parameter was counter-supported by the others, as reported in Sinner's cycle (Sinner & Verlag, 1960). A detailed review explanation of the effect of different washing parameters in relation to microfiber emission is provided in chapter 3 of this book. Later the

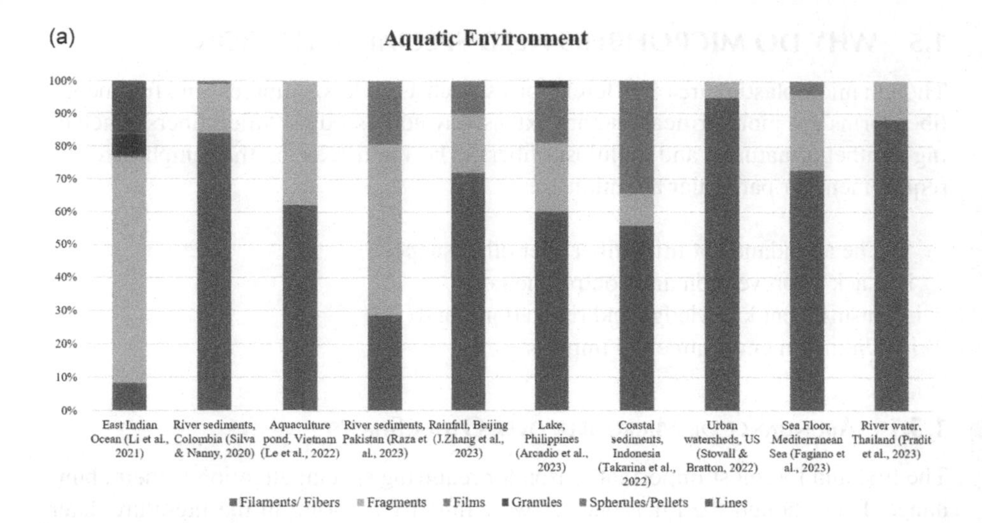

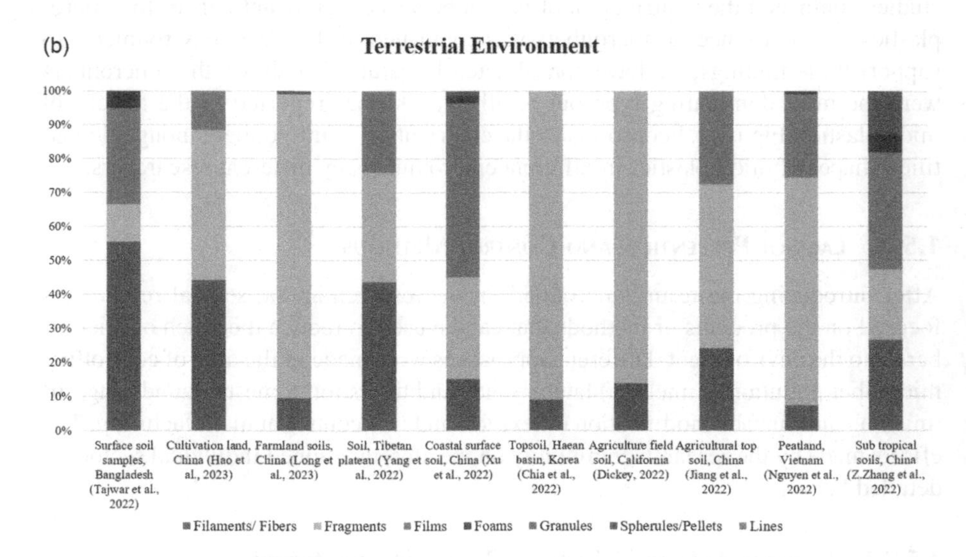

FIGURE 1.4 Domination of microfibers in microplastics in (a) aquatic environment, (b) terrestrial environment, (c) atmosphere, and (d) food items as reported by different researchers (Li et al., 2021; Silva & Nanny, 2020; Le et al., 2022; Raza et al., 2023; Zhang et al., 2023; Arcadio et al., 2023; Takarina et al., 2022; Stovall & Bratton, 2022; Fagiano et al., 2023; Pradit et al., 2023; Tajwar et al., 2022; Hao et al., 2023; Long et al., 2023; Xu et al., 2022; Yang et al., 2022; Chia et al., 2022; Dickey, 2022; Jiang et al., 2022; Nguyen et al., 2022; Zhang et al., 2022; Welsh et al., 2022; Liu et al., 2022b; Perera et al., 2023; Romarate et al., 2023; Uddin et al., 2022; Sarathana & Winijkul, 2022; Perera et al., 2022; Liao et al., 2021; Szewc et al., 2021; Fang et al., 2022; Di Fiore et al., 2023; Makhdoumi et al., 2023; Altunışık, 2023; Zhang et al., 2023b; Tse et al., 2022; Chinglenthoiba & Valiyaveettil, 2023; Crosta et al., 2023; Parvin et al., 2022; Rakib et al., 2021; Ha, 2021).

(*Continued*)

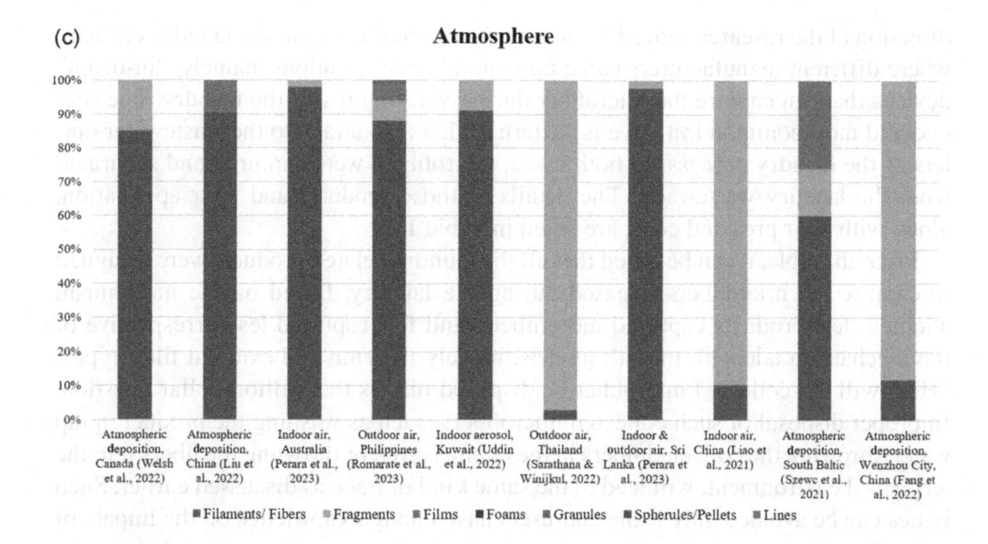

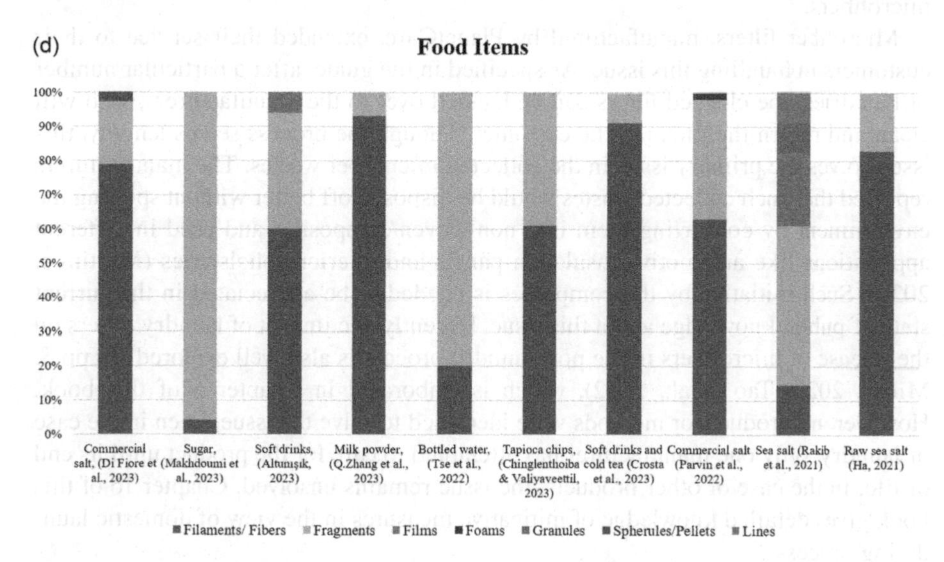

FIGURE 1.4 (*Continued*) Domination of microfibers in microplastics in (a) aquatic environment, (b) terrestrial environment, (c) atmosphere, and (d) food items as reported by different researchers (Li et al., 2021; Silva & Nanny, 2020; Le et al., 2022; Raza et al., 2023; Zhang et al., 2023; Arcadio et al., 2023; Takarina et al., 2022; Stovall & Bratton, 2022; Fagiano et al., 2023; Pradit et al., 2023; Tajwar et al., 2022; Hao et al., 2023; Long et al., 2023; Xu et al., 2022; Yang et al., 2022; Chia et al., 2022; Dickey, 2022; Jiang et al., 2022; Nguyen et al., 2022; Zhang et al., 2022; Welsh et al., 2022; Liu et al., 2022b; Perera et al., 2023; Romarate et al., 2023; Uddin et al., 2022; Sarathana & Winijkul, 2022; Perera et al., 2022; Liao et al., 2021; Szewc et al., 2021; Fang et al., 2022; Di Fiore et al., 2023; Makhdoumi et al., 2023; Altunışık, 2023; Zhang et al., 2023b; Tse et al., 2022; Chinglenthoiba & Valiyaveettil, 2023; Crosta et al., 2023; Parvin et al., 2022; Rakib et al., 2021; Ha, 2021).

direction of the research moved to capture the microfibers from the laundry effluent, where different manufacturers came up with different solutions, namely, "in-drum" devices that can capture the microfiber during washing inside the laundry. The second and most common initiative is "external filters" to attach to the wastewater outlets of the laundry process. In both cases, microfibers were captured and separated from the laundry wastewater. The details of those products and their application, along with their pros and cons, are listed in Table 1.2.

From the table, it can be noted that all the laundry-related products were designed to capture the microfibers released during the laundry. Based on the mechanism adopted, few products captured more fibers, and few captured less. Irrespective of the mechanism adopted, in both models, namely in-drum and external filter types, "How will the collected microfiber be disposed of?" is the million-dollar question. Improper disposal of such collected microfibers, such as washing the product in tap water, throwing the collected fibers in the public sewer, or throwing the fibers into the terrestrial environment, will lead to the same kind of issue as discussed earlier. Such issues can be avoided only if the end-users have enough knowledge on the impact of microfibers.

Microfiber filters, manufactured by PlanetCare, extended their service to their customers in handling this issue. As specified in the guide, after a particular number of laundries, the clogged filters can be handed over to the manufacturers, who will clean and return the filters to the customer. Though the process seems lengthy, this issue solves the primary issue in the collected microfiber wastes. The manufacturers reported that such collected wastes would be disposed off better without spoiling the environment by converting them into non-woven/composites and used in different applications like automotive insulation panels and interior upholsteries (Martinko, 2020). Such initiative by the companies is needed to be appreciated in the current state of public knowledge about this issue. Recently, the impact of laundry dryers on the release of microfibers in the post-laundry process is also well explored (Kapp & Miller, 2020; Tao et al., 2022), which is elaborated in Chapter 6 of this book. However, no products or methods were identified to solve the issue. Even in the case of laundry, only one manufacturer has attempted to care for the product until its end of life; in the case of other products, the issue remains unsolved. Chapter 16 of this book gives detailed knowledge of mitigative measures in the view of domestic laundering process.

1.5.2.2 Surface Finishes and Modification on Textiles

Despite the several attempts made by researchers to control the microfiber release into the environment from laundry, all those measures are not controlling the release. The developed products were just control measures. Hence, recent research focused on proactive solutions for controlling the microfiber release from the point of origin, that is textiles. Attempts were made to finish the textile surfaces with natural polymeric substances like chitosan, sericin, polyvinyl alcohol, etc., using a natural binder or crosslinking agent. The resultant fabrics will have fibers and yarns with complete polymeric coats on their surface. It was expected to provide additional strength and prevent the mechanical damage that usually occurs during the laundry process

TABLE 1.2
Laundry Aids for Microfiber Release Control

Reference	Product	Product Image	Type	Basic Working Principle	Material	Pore Size (μm)	Reduction Efficiency (%)	Price ($)	Cleaning Frequency	Advantages/Disadvantages
Belton (2018), McIlwraith et al. (2019), Napper et al. (2020), *The Coraball* (2023)	Cora Balls		In-drum device	The ball inside the laundry drum captures released fibers on its coral structure.	Recycled & recyclable plastics	–	26%–35%	~42	Clean while tangles of fibers are seen.	**Advantages** • No need for additional space, and • Can be used in all types of washing machine. **Disadvantages** • Difficult to remove tangled fibers, and • Can affect delicate fabrics.
Julia (2017), Napper et al. (2020), Styles (2019), Wilcox (2017)	GuppyFriend Bags		In-drum device	The fibers released from the clothes are collected inside the bag.	Nylon filaments	50 μm	54%–91%	~32	Every wash	**Advantages** • No need for additional space, and • More durable; reduces fiber release in addition to capturing. **Disadvantages** • Temperature sensitive – can be used up to 40°C, and • Bags are susceptible to release fibers.

(Continued)

TABLE 1.2 (*Continued*)
Laundry Aids for Microfiber Release Control

Reference	Product	Product Image	Type	Basic Working Principle	Material	Pore Size (µm)	Reduction Efficiency (%)	Price ($)	Cleaning Frequency	Advantages/ Disadvantages
Brodin et al. (2018), McIlwraith et al. (2019), Napper et al. (2020)	LUV-R Lint Filters		External filter	Fitted into the outlet pipe of washing machine, and fibers are collected in the filter.	Metal mesh	285 µm; 175 µm	29%–87%	~150	10–15 washes	**Disadvantages** • Cannot capture shorter fibers, and • Need additional space.
Napper et al. (2020), PlanetCare (2023), Plastic smart cities (2023)	PlanetCare Filters		External filter	Fitted into the outlet pipe of washing machine, and fibers are collected in the filter.	–	200 µm	25%–80%	~63	20 washes	**Advantages** • Can capture fibers of wide range of lengths, and • Used cartridges can be returned for recycling. **Disadvantages** • Needs additional space, and • Recurring costs on cartridge replacement.

(*Continued*)

TABLE 1.2 (*Continued*)
Laundry Aids for Microfiber Release Control

Reference	Product	Product Image	Type	Basic Working Principle	Material	Pore Size (μm)	Reduction Efficiency (%)	Price ($)	Cleaning Frequency	Advantages/ Disadvantages
Brodin et al. (2018), Napper et al. (2020), Samantha et al. (2019)	XFiltra™		External filter	Fitted into the outlet pipe of washing machine, and fibers are collected in the filter,	–	60 μm	78%–90%	–	–	**Advantages** • Centrifugal separator for dewatering of collected fibers. **Disadvantages** • Needs commercial launch, and • Needs additional space.

(*Continued*)

TABLE 1.2 (*Continued*)
Laundry Aids for Microfiber Release Control

Reference	Product	Product Image	Type	Basic Working Principle	Material	Pore Size (μm)	Reduction Efficiency (%)	Price ($)	Cleaning Frequency	Advantages/Disadvantages
Brodin et al. (2018), Olivia (2020)	Filtrol160™		External filter	Fitted into the outlet pipe of washing machine, and fibers are collected in the filter.	Stainless Steel	100 μm	30%–60%	~160	8–10 washes	**Disadvantages** • Needs additional space, and • Softener usage can clog filters.
Girlfriend Collective (2023)	The Microfiber filter from Girlfriend	ECO365	External filter	Fitted into the outlet pipe of washing machine, and fibers are collected. in the filter	Steel Mesh	200 μm	–	~45	Three washes	**Advantages** • Comparatively cheaper filter. **Disadvantages** • Needs additional space, and • No third-party validation is done.

(De Falco et al., 2018, 2019; Kang et al., 2021; Mermaids, 2015). Upon evaluation, the results were promising in controlling the microfiber release from the textile, but the durability of such finishes was ignored widely. Hence, several research works were still in process to improve the effectiveness of the finishing process. A detailed consolidation of different finishing processes and their impact on controlling microfiber emission are discussed in Chapter 18 of this book.

Some recent research works have shown the potential of surface modification of synthetic textiles as an effective way to control microfiber release. Studies attempted to modify the surface of the polyester using alkaline and enzyme as an agent. The results were promising, and the treatment processes reduced the microfiber release from the polyester up to 80% in the first wash (Rathinamoorthy & Raja Balasaraswathi, 2022a; Ramasamy & Subramanian, 2022; Rathinamoorthy & Raja Balasaraswathi, 2023b). The main advantage claimed in these processes is the durability of such modification. The surface modification process was durable in more than 20 laundries, which is more than sufficient as the microfiber emission is expected to become plateau after 10–15 washes (Cesa et al., 2020; Napper & Thompson, 2016; O'Loughlin, 2018; Rathinamoorthy & Raja Balasaraswathi, 2022). Efforts toward combining surface modification and finishing methods seem unsuccessful for a few types of polymers (Ramasamy & Subramanian, 2023). However, the durability issue persists with the finishes. Despite several attempts, only a few surface modification methods seem to be a potential solution based on their commercial feasibility and the textile industry's knowledge of this process. Still, the research results were laboratory-scale simulations, and real-time attempts at bulk production need to be evaluated regarding economic viability.

1.5.2.3 Control at Manufacturing

As the literature mainly focuses on the laundry process, the impact of the textile manufacturing process on microfiber emission remains unsolved. The first research on the textile industry effluent was performed in the inlet and outlet of textile wastewater treatment plants (WWTPs) (Magnusson & Norén, 2014). Studies reported that more than any other processes, textile industry effluents carry a higher concentration of microfibers per every milliliter of water released despite their higher removal efficiency. Various research works performed on the textile WWTPs showed that the existing mechanisms could remove 80%–95% of the microfibers in the effluent compared to the influent (Murphy et al., 2016; Xu et al., 2018; Zhou et al., 2020). Even after effective removal by WWTPs, the outlet water can still release up to 430 billion microfibers daily (Zhou et al., 2020). Chapter 5 of this book gives a detailed overview of the impact and efficiency of wastewater treatment plants. Based on the quantum of water processed and released per day, their impact on the environment (marine or terrestrial environment-based on where it is released) was noted as significant. Solid waste disposal is the second most crucial issue that needs to be addressed. As we have seen, WWTPs can remove upto 95% of microplastics and fibers; they get sediments into the solid wastes developed in the WWTPs. These wastes, called sludges, were more polluted with higher concentrations of microplastics and fibers than effluents (Alavian Petroody et al., 2021; Zhang et al., 2019).

When discussing preventive measures, researchers reported that the tertiary treatment processes, such as membrane bioreactor and reverse osmosis, are better processes for removing a higher percentage of microplastics from the effluent (Carr et al., 2016; Sun et al., 2019; Talvitie et al., 2017). However, no complete prevention methods were established to control or remove microfibers from the effluent. Similarly, the textile sludges were usually used for the landfills. This process indirectly allows the microfibers and plastics to leach into the surface of the land (Corradini et al., 2019). Specifically, due to the mixing of laundry wastewater in the municipal WWTPs, they also act as a source of microfiber emission. As the municipal WWTPs sludges were used as a fertilizer in the farm lands, they posed a severe threat to the terrestrial environment (Corradini et al., 2019; Zhang et al., 2019). Studies confirmed the release of microfibers and microplastics from farmland to the river due to rain (Miller et al., 2017). There is no proper solution to address this disposal of sludge or solid waste from both textile and municipal WWTPs. A recent research report by Forum of the Future has elucidated the need for the installment of robust wastewater treatment plants which can eliminate microfibers while removing other pollutants (Forum for the Future, 2023). The chemical processing of textile processes like scouring, bleaching, dyeing, and finishing highly consumes water and potential enough to release microfibers. Though researchers addressed the overall outlet water contamination, no studies evaluated the individual process and their contribution, except the recent study on the effluent of the textile screen printing process (Rathinamoorthy & Raja Balasaraswathi, 2023a). A study suggested shifting from highly water-consuming machinery to technologies that use little or no water. Further, none of the textile manufacturing processes were studied for their impact on the environment regarding microfiber release. More focus on production stage factors such as yarn and material type is suggested (Forum for the Future, 2023).

With this meager level of preventive measures and a handful of research data, it is tough to mitigate the environmental (aquatic, terrestrial, and atmospheric) impacts of microfiber pollution. Hence, future studies need to focus on this sector to address the knowledge gaps in mitigation strategies.

1.5.3 Insufficient Knowledge and Regulations

To control and solve any type of problem, understanding the factors influencing the problems and their impact on society is essential. Whereas, in the case of microplastic and microfiber pollution, the knowledge about the issue is very meager among the stakeholders, namely consumers, manufacturers, and the governments of the different nations.

1.5.3.1 Knowledge of Consumer/General Public

The general public is the major consumer of any product; in the case of textiles and apparel, every individual is the consumer. The quantum of knowledge they have about the pollution-related issues of fashion products are very meager than their knowledge of the trend and styles of fashion product (Yan et al., 2020). For instance, due to their culture or celebrity influence, people tend to get influenced, and as a society, they are forced to change or adapt to the current trend. Significantly, people under the age of 30 are highly influenced by the fashion trend. However, the level of awareness about fashion-related pollution and its impact is inadequate. As of now, the knowledge of the use of synthetic dyes and their impact on the environment is known little among the public due to the efforts taken by several researchers, environmentalists, and strict government regulation. Likewise, the issues related to the disposal of dyeing industry wastewater and their environmental impacts are also established among the public to a certain extent. In that case, the significant advantage is that the public can explicitly see the impact of such processes on the environment (Doley et al., 2020).

Whereas emerging issues like microfiber pollution are new to society, and public awareness is absent. This is due to the weird nature of the issues since the problem is not visible to the naked eye of the commoner, and the issue is still not regulated by most governments. Previous studies on estimating the public knowledge of microfibers reported the same results after analysis (Herweyers et al., 2020). The study results reported that the non-traceability or non-visibility of the issue is one of the significant issues in educating people (Herweyers et al., 2020). People see fast fashion products as disposable compared to classic styles with a higher life cycle. Hence, educating the existing consumer and the upcoming generation is essential. It will help the future generations in handling the situation in a better manner than us. The research data should be disseminated among the general public through advertisements, social media platforms, and educational content in schools, as we did for disposable plastics products (Chauhan & Punia, 2022). The role of documentary movies like "True Cost", "Riverblue", "ReDress the Future", etc., are perfect examples of how movies can be effectively used to educate consumers (Rauturier, 2023). Such attempts must be made in the case of microfiber pollution too to educate the public. Studies also suggested that the proper use of a care label or a separate care label or inclusion of the details regarding the percentage of synthetic fibers or the ability of the material to release microfibers in the existing labels can be a viable solution to educate the customer (Salahuddin & Lee, 2021). Other researchers recommended integrating microplastic pollution-related issues in formal education programs like schools and increasing public information campaigns to address the adults (Charitou et al., 2021). Studies reported that reducing psychological distance is a primary way of better communication. Publishing pictures of local foods with the zoom image of associated microplastics and their impact or demonstrating such videos will serve as a better awareness. Researchers believe such attempts will increase understanding and drastically change behavior (Garcia-Vazquez & Garcia-Ael, 2021). Furthermore, studies mentioned that the level of understanding of the issue varies among different groups. Hence, an effective way of communication must be developed to address such a scattered population. Results of the research showed that outcome-driven communication is effective, meaning that the information provided should contain not only the impact of the issue but also the intended solution (Yan et al., 2020).

1.5.3.2 Knowledge of Manufacturers

Though the initial research on microfiber pollution focused on the laundry process, soon after, scientific community realized that microfiber emission occurs at every stage of the textile life cycle. From that, the knowledge of the impact of the manufacturing process on microfiber emission was estimated and realized. However, no complete study has been performed on various textile manufacturing processes and their impact on microfiber release. Very recently, the impact of the cutting and sewing process (Cai et al., 2020) and the influence of the screen printing process on microfiber emission were reported (Rathinamoorthy & Raja Balasaraswathi, 2023a). The remaining studies were mainly focused on the textile wastewater treatment plant (Ramasamy et al., 2022; Xu et al., 2018; Zhou et al., 2020). However, significantly less or no commercial solution was reported to address the microfiber emission at the manufacturing stage. Overall, the level of knowledge that the commercial manufacturers have regarding this microfiber pollution is questionable. As most of the research focused on WWTP effluent to filter the microplastics and fiber, efforts were made to find the solutions in that industry. Studies reported potential modification of the process sequence of the WWTPs with additional process steps to increase the filtration efficiency (Carr et al., 2016; Sun et al., 2019; Talvitie et al., 2017; Wilcox, 2017).

While discussing industrial practices, a few famous brands worldwide have made efforts to control such emissions. Patagonia, a leading activewear manufacturer highly known for its sustainable practices, announced that by 2025, the company would eliminate the use of virgin petroleum fiber in its products with 100% reusable, home-compostable, renewable, or easily recyclable packaging materials. Further to add, the brand aimed to attain a net zero carbon footprint across its entire business line (Patagonia, n.d.). In specific to the microfiber release, Patagonia teamed with Samsung, developed a new wash cycle yielding only fewer microfibers, and incorporated an inbuilt filter system to capture the releasing microfibers. Altogether the new filter and cycle showed a 54% reduction in actual release (Nepori, 2023). However, a 54% reduction will still release many microfibers into the environment. Similarly, leading sportswear brand NIKE has taken initiatives in addressing the microfiber issues, and they are collaborating with the American Association of Textile Chemists and Colorists (AATCC) to develop standards. Further, they have collaborated with Cross Industry Agreement (CIA) to partner with other industries to resolve the issue (Nike, 2019). In association with ZARA Home, Inditex has recently launched a laundry detergent, mainly focusing on microfiber release, claiming to reduce 80% of microfiber emissions during use (Alonso, 2022). H&M Group is another crucial fast fashion group with a significant market share. They have signed in "The Microfibre 2030 Commitment", a global commitment to reduce microfiber emissions to zero. They have funded a research project on "A Management Tool for Microplastics from Textile Production Process" to find a solution to the microfiber issue, in association with the Hong Kong Research Institute of Textiles and Apparel (H&M Groups, 2023). The primary issue that has to be reported after all these efforts is how these industries/brands will adopt these findings or collaboration with their products successfully. Will these initiatives result in fruitful outcomes to solve the prevailing issues? That question needs to be addressed yet. None of those above-mentioned industries reported any of these processes or methods in their process or did not release a product with modified features to release low or no fibers.

In the Netherlands, an infrastructure company has developed a stakeholder network to explore the potential solution for the microfiber issue, explicitly focusing on the textile industry. They have developed standard methods to measure microfibers. The significant advantage of such an initiative is that the forum assembles different stakeholders, namely water companies, infrastructure, and textile-related companies (Campagne & Kentin, 2021; Nguyen, 2023). One of the recent research reports analyzed 55 international fashion brands and reported that 25% of those brands had increased the consumption of synthetics over the last 5 years. In this research, out of the participants, 45% of brands do not have any policy related to microfiber pollution. Thirty-two percent of the surveyed industries did not disclose meaningful information on synthetic fiber consumption (Changing Markets Foundation, 2022). Though few brands have taken initiatives, the majority of the brands found without any inclination toward microfiber-related issues. Whereas, in the manufacturing process case, though a few research works suggested some process changes, none of the methods were commercially evaluated for the bulk production process. Similarly, associated machinery, process cost changes, and other skilled labor requirements were not reported. Future research on such technologies will help transfer laboratory-scale technologies to manufacturing industries.

1.5.3.3 Knowledge of Governments and Other Organizations

Awareness about microplastic and microfiber pollution has increased as of now (2023) compared to a few years ago. Nevertheless, no standard quantification methods were available when discussing the microfiber emission from textiles during laundry, manufacturing, or wastewater treatment plants. The research results obtained so far cannot be directly compared as the authors adopted different methodologies. Hence, the reliability of the quantification was questionable. Chapter 4 of this book elucidates different quantification methods available for microfiber emission study and Chapter 8 of this book gives an overview of standardized quantification methods. These issues became a significant barrier to implementing regulations or strategies on a large scale, for example, a nationwide change. However, AATCC has recently launched weight-based quantification for the microfibers to solve this issue of the unavailability of the standard quantification method (AATCC TM212, 2021). Additionally, the development of ISO and other nationwide regulations (DIN) was in process and expected to be implemented at the end of 2023. The details regarding various standards and regulations developed to quantify microfibers and microplastics were discussed in Chapter 19 of this book. Several governments and nations have implemented laws regulating synthetic textiles and associated microfiber release (Ramasamy & Subramanian, 2021). Such developed regulations were implemented in some places and are yet to be implemented in the rest. Chapter 17 of this book elaborates more on the laws and regulations adopted by nations worldwide. It can be seen that the developed countries have taken initiatives, but the developing countries, a central hub of textile and fashion manufacture, did not have any regulations implemented. This sounds alarming regarding the necessity of formulating and implementing such regulations to protect their environment from the invisible enemy.

Other than this, several private non-profit organizations are also working on the remedial measures that can be used to control these issues. The Microfiber

Consortium (TMC), with more than 70 signatories, including brands, retailers, manufacturers, and researchers, focuses on microfiber control by bridging the gaps between scholarly research and the construction of commercial supply chains (*The Microfiber Consortium*, n.d.). Outdoor Industry Association (OIA) is also one such group that focuses on microfiber issues with its Sustainable Working Group (SWG) (Outdoor Industry Association, 2018). Different organizations, their goals, and strategies are well explored in Chapter 17 of this book.

1.5.4 Unknown Environmental Impacts

There are hundreds of research reports that quantified and predicted the quantity of microfiber and plastic quantities in various environments like marine (Cole et al., 2011; Gago et al., 2018; Mishra et al., 2019), atmosphere (Cai et al., 2017; Finnegan et al., 2022; Roblin et al., 2020; Wang et al., 2020), and terrestrial environments (Ambrosini et al., 2019; Corradini et al., 2019; Liu et al., 2018; Selonen et al., 2020).

1.5.4.1 Aquatic Environment

The marine environment is one of the most polluted places due to these microfibers and plastic pollution. Studies estimated that around 0.28 million tons of microfibers reach aquatic environments yearly through domestic laundry worldwide (Belzagui et al., 2020). Similarly, several lakes and ponds that serve as a drinking water supply were also evaluated and found to be contaminated with microfibers and plastics (Egessa et al., 2020; Felismino et al., 2021; González-Pleiter et al., 2020; Peller et al., 2019). Despite identifying the source and estimating quantity, none of the studies reported any solution to recover or remove the deposited contaminants. Furthermore, the correctness of the estimation was also found to be under "debate", while a few researchers reported that microplastics were over-estimated (Belzagui et al., 2020), and others condemned that it was underestimated and unaccounted (Koutnik et al., 2021). As an immediate impact, studies have reported the presence of microplastics and fibers in the guts and intestines of animals like fishes, plankton, mussels, etc. (Bessa et al., 2018; Kasamesiri & Thaimuangpho, 2020; Lusher et al., 2013; Priya et al., 2023; Ribeiro et al., 2020; Santonicola et al., 2023). This poses a severe threat to aquatic life forms and related food chains. However, their effects on mortality, growth, and reproduction rate were not wholly known with their long-term effects (Mateos-Cárdenas et al., 2021). Studies have also reported that the dyes and toxic chemicals loaded on the fibers and plastic particles will leach into the environment and act as potentially toxic substances (Periyasamy, 2023). A detailed analysis of the sources of microfibers and their impact on the marine environment is discussed in Chapter 11 of this book. Since primitive knowledge of microfiber pollution in the marine environment and its impact on living organisms is a colossal threat ahead of us, how the scientific community will handle this situation before it reaches the state of "beyond the control" is the must-address question.

1.5.4.2 Atmospheric Contamination

The next affected environment due to microfiber pollution is the atmosphere. Studies revealed that wearing synthetic textiles could release many microfibers due to the fabric's self-abrasion and while it abrades against different surfaces. Studies have also reported the presence of microfibers in the atmosphere based on the clothing requirements of different seasons (Tunahan Kaya et al., 2018). Meanwhile, considering the textile manufacturing process, every processing stage can emit microfibers into the atmosphere. Specifically, operations like cutting and sewing processes are capable of microfiber emission. However, no research data are available to date regarding the emission.

Researchers predicted that microfibers from the atmosphere could enter the human body through inhalation and penetration via the skin, depending on the size (Domenech & Marcos, 2021). The other most popular way for microfibers to enter the human body is through food items (Cox et al., 2019). The various mechanisms related to the atmospheric microfiber entering the human body are detailed in Chapters 13 and 14 of this book. Recent studies reported the presence of microplastics in the human lungs, which can be due to the inhalation of atmospheric microplastics (Amato-Lourenço et al., 2021; Gasperi et al., 2018; Jenner et al., 2022; Pauly et al., 1998). Despite their identification in the human body, no studies reported their toxic nature or long-term effects on humans. This area needed a detailed analysis as different modes of interaction will have different parts of the human body. For instance, the inhalation of particles and fibers may directly impact the functions of the lungs, whereas the penetration via skin pores may give a chance to get mixed up with blood flow. Currently, the prevailing knowledge may not serve to find a potential solution, and future research must be performed to get a clear picture.

1.5.4.3 Terrestrial Environment

The disposal of used clothes and WWTP sludge are the two most well-known routes of microfiber reaching the terrestrial environment. Microfibers and plastics in the landscape will directly reach the marine or river via rainwater wash-off. The textiles ended up in landfills can degrade and fragment into microfibers due to exposure to environmental conditions (Laitala et al., 2018). Chapter 2 of this book elaborates on how the aging process could potentially affect microfiber release. The atmospheric pollution of microplastics and microfibers is one of the main reasons for higher terrestrial pollution as the deposit. To confirm this, researchers reported the presence of microfibers in the Italian glaciers (Ambrosini et al., 2019). Higher contamination of both deep and shallow soils in the vegetable fields of Shanghai was reported (Liu et al., 2018), along with farmlands of Wuhan, where the study reported 320–12,560 microplastic particles per kilogram of soil (Chen et al., 2020). Studies have also reported that plastic mulching is also one of the main reasons for such contamination (Huang et al., 2020). Municipal WWTP sludges are commonly used as fertilizer for farmland soil. Research findings showed that the soil samples from the farmland where sludges were applied showed more microplastics (Corradini et al., 2019). The fear of such farmland contamination is due to the possibility of transport of these particles in food items being yielded in those farms.

Recently, terrestrial pollution significantly increased due to COVID-related masks and their improper disposal (Patrício Silva et al., 2021). Polypropylene was the primary fiber used in the mask; lower strength and higher photodegradability of the polypropylene fibers triggered higher emission of microfibers once after their disposal (Chen et al., 2021; Rathinamoorthy & Balasaraswathi, 2022; Shen et al.,

2021a). Chapter 7 of this book details the microfiber pollution in relationship with the COVID-19 pandemic-related PPEs. Despite a few research analyses (Selonen et al., 2020), the long-term effect of such microfibers on soil microorganisms was not explored thoroughly. Likewise, the fertility value of the microfiber-exposed soil has not yet been evaluated. Similar to the other environment, terrestrial environment contamination also aids the transport of microplastics and microfibers in the food chain, as the contamination of terrestrial birds with microfibers has been reported recently by researchers (Zhao et al., 2016). Evidence shows that these particles were already found in vegetables and fruits (Oliveri Conti et al., 2020). Though several identification and quantification analyses were performed, no direct remedial measures or removal technologies have been developed. Hence, it is imperative to take the necessary steps to control and mitigate the issue at the earliest.

1.6 SUMMARY

As the book's first chapter, this chapter introduces the term "microfiber", distinguishing it from the generic term microplastic. Similarly, the chapter also throws light on the differences between "micro denier" fibers commonly used in textile process and the emerging microfiber pollutant. The chapter summarized the definitions of microfiber reported in the literature and introduced the base for differentiating microfibers from other microplastic shapes. The chapter also classifies the different categories of microfibers and their potential risks over other shapes. Apart from synthetic fibers, the chapter includes natural and man-made cellulosic fiber also under this category. This will widen the reader's perspective and showcases the necessity of special attention to microfibers. Additionally, the chapter addresses why microfiber pollution is a big problem. All these problems were detailed, and their connection with the forthcoming chapters of the book was outlined. This will be a guide to the readers to look for specific chapters based on their interests.

REFERENCES

AATCC TM212. (2021). *TM212 fiber fragment release during home laundering*. https://members.aatcc.org/store/tm212/3573/

Alavian Petroody, S. S., Hashemi, S. H., & van Gestel, C. A. M. (2021). Transport and accumulation of microplastics through wastewater treatment sludge processes. *Chemosphere*, *278*. https://doi.org/10.1016/j.chemosphere.2021.130471

Alonso, T. (2022, December 6). *Inditex launches detergent designed to reduce microfibre release*. Fashion Network. https://uk.fashionnetwork.com/news/Inditex-launches-detergent-designed-to-reduce-microfibre-release,1465479.html

Altunışık, A. (2023). Prevalence of microplastics in commercially sold soft drinks and human risk assessment. *Journal of Environmental Management*, *336*. https://doi.org/10.1016/j.jenvman.2023.117720

Amato-Lourenço, L. F., Carvalho-Oliveira, R., Júnior, G. R., dos Santos Galvão, L., Ando, R. A., & Mauad, T. (2021). Presence of airborne microplastics in human lung tissue. *Journal of Hazardous Materials*, *416*. https://doi.org/10.1016/j.jhazmat.2021.126124

Ambrosini, R., Azzoni, R. S., Pittino, F., Diolaiuti, G., Franzetti, A., & Parolini, M. (2019). First evidence of microplastic contamination in the supraglacial debris of an alpine glacier. *Environmental Pollution*, *253*, 297–301. https://doi.org/10.1016/j.envpol.2019.07.005

Andrady, A. L. (2003). *Plastics and the environment*. www.copyright.com.

Arcadio, C. G. L. A., Navarro, C. K. P., Similatan, K. M., Inocente, S. A. T., Ancla, S. M. B., Banda, M. H. T., Capangpangan, R. Y., Torres, A. G., & Bacosa, H. P. (2023). Microplastics in surface water of Laguna de Bay: First documented evidence on the largest lake in the Philippines. *Environmental Science and Pollution Research*, *30*(11). https://doi.org/10.1007/s11356-022-24261-5

Assoune, A. (2023a). *The advantages of fast fashion for consumers*. Panaprium. https://www.panaprium.com/blogs/i/advantages-fast-fashion

Assoune, A. (2023b). *The truth about rayon fabric they are hiding from you*. https://www.panaprium.com/blogs/i/rayon-fabric

Athey, S. N., Adams, J. K., Erdle, L. M., Jantunen, L. M., Helm, P. A., Finkelstein, S. A., & Diamond, M. L. (2020). The widespread environmental footprint of indigo denim microfibers from blue jeans. *Environmental Science and Technology Letters*, *7*(11). https://doi.org/10.1021/acs.estlett.0c00498

Athey, S. N., & Erdle, L. M. (2022). Are we underestimating anthropogenic microfiber pollution? A critical review of occurrence, methods, and reporting. *Environmental Toxicology and Chemistry*, *41*(4). https://doi.org/10.1002/etc.5173

Au, S. Y., Bruce, T. F., Bridges, W. C., & Klaine, S. J. (2015). Responses of Hyalella azteca to acute and chronic microplastic exposures. *Environmental Toxicology and Chemistry*, *34*(11). https://doi.org/10.1002/etc.3093

Belton, P. (2018). *Could these balls help reduce plastic pollution?* https://www.bbc.com/news/business-46137804

Belzagui, F., Gutiérrez-Bouzán, C., Álvarez-Sánchez, A., & Vilaseca, M. (2020). Textile microfibers reaching aquatic environments: A new estimation approach. *Environmental Pollution*, *265*. https://doi.org/10.1016/j.envpol.2020.114889

Berg, A., Granskog, A., Lee, L., & Magnus, K.-H. (2020). *How the fashion industry can urgently act to reduce its greenhouse-gas emissions*. https://www.mckinsey.com/industries/retail/our-insights/fashion-on-climate

Bessa, F., Barría, P., Neto, J. M., Frias, J. P. G. L., Otero, V., Sobral, P., & Marques, J. C. (2018). Occurrence of microplastics in commercial fish from a natural estuarine environment. *Marine Pollution Bulletin*, *128*. https://doi.org/10.1016/j.marpolbul.2018.01.044

Bhat, M. A., Gedik, K., & Gaga, E. O. (2023). Atmospheric micro (nano) plastics: Future growing concerns for human health. *Air Quality, Atmosphere and Health*, *16*(2). https://doi.org/10.1007/s11869-022-01272-2

Bick, R., Halsey, E., & Ekenga, C. C. (2018). The global environmental injustice of fast fashion. *Environmental Health: A Global Access Science Source*, *17*(1). https://doi.org/10.1186/s12940-018-0433-7

Blackburn, K., & Green, D. (2022). The potential effects of microplastics on human health: What is known and what is unknown. *Ambio*, *51*(3). https://doi.org/10.1007/s13280-021-01589-9

Boucher, J., & Friot, D. (2017). *International Union for conservation of nature a global evaluation of sources primary microplastics in the oceans*.

Brodin, M., Norin, H., Hanning, A.-C., & Persson, C. (2018). *Filters for washing machines: Mitigation of microplastic pollution*. https://www.diva-portal.org/smash/get/diva2:1633773/FULLTEXT01.pdf

Brophy, K. (2020, September 18). *FAST FASHION Fast fashion 1 – Why is the fashion industry an environmental problem?* https://blogs.imperial.ac.uk/molecular-science-engineering/2020/09/18/fast-fashion-environmental-problem/

Browne, M. A., Crump, P., Niven, S. J., Teuten, E., Tonkin, A., Galloway, T., & Thompson, R. (2011). Accumulation of microplastic on shorelines woldwide: Sources and sinks. *Environmental Science and Technology*, *45*(21), 9175–9179. https://doi.org/10.1021/es201811s

Browne, M. A., Dissanayake, A., Galloway, T. S., Lowe, D. M., & Thompson, R. C. (2008). Ingested microscopic plastic translocates to the circulatory system of the mussel, Mytilus edulis (L.). *Environmental Science and Technology*, *42*(13). https://doi.org/10.1021/es800249a

Cai, Y., Mitrano, D. M., Heuberger, M., Hufenus, R., & Nowack, B. (2020). The origin of microplastic fiber in polyester textiles: The textile production process matters. *Journal of Cleaner Production*, *267*. https://doi.org/10.1016/j.jclepro.2020.121970

Cai, L., Wang, J., Peng, J., Tan, Z., Zhan, Z., Tan, X., & Chen, Q. (2017). Characteristic of microplastics in the atmospheric fallout from Dongguan city, China: Preliminary research and first evidence. *Environmental Science and Pollution Research*, *24*(32). https://doi.org/10.1007/s11356-017-0116-x

Camargo, L. R., Pereira, S. C. F., & Scarpin, M. R. S. (2020). Fast and ultra-fast fashion supply chain management: An exploratory research. *International Journal of Retail and Distribution Management*, *48*(6). https://doi.org/10.1108/IJRDM-04-2019-0133

Campagne, A., & Kentin, E. (2021, August 27). *How fashion contributes to plastic pollution*. Universiteit Leiden. https://www.leidenlawblog.nl/articles/how-fashion-contributes-to-plastic-pollution

Carr, S. A., Liu, J., & Tesoro, A. G. (2016). Transport and fate of microplastic particles in wastewater treatment plants. *Water Research*, *91*. https://doi.org/10.1016/j.watres.2016.01.002

Centobelli, P., Abbate, S., Nadeem, S. P., & Garza-Reyes, J. A. (2022). Slowing the fast fashion industry: An all-round perspective. *Current Opinion in Green and Sustainable Chemistry*, *38*.

Cesa, F. S., Turra, A., Checon, H. H., Leonardi, B., & Baruque-Ramos, J. (2020). Laundering and textile parameters influence fibers release in household washings. *Environmental Pollution*, *257*. https://doi.org/10.1016/j.envpol.2019.113553

Changing Markets Foundation. (2022). *Synthetics anonymous 2.0: Fashion's persistent plastic problem*. https://changingmarkets.org/wp-content/uploads/2022/12/EX-SUM-ENG-Syntetics-Anonymous-2.0-.pdf

Charitou, A., Naasan Aga-Spyridopoulou, R., Mylona, Z., Beck, R., McLellan, F., & Addamo, A. M. (2021). Investigating the knowledge and attitude of the Greek public towards marine plastic pollution and the EU Single-Use Plastics Directive. *Marine Pollution Bulletin*, *166*. https://doi.org/10.1016/j.marpolbul.2021.112182

Chauhan, N. S., & Punia, A. (2022). Role of education and society in dealing plastic pollution in the future. *Plastic and Microplastic in the Environment*. https://doi.org/10.1002/9781119800897.ch14

Chen, X., Chen, X., Liu, Q., Zhao, Q., Xiong, X., & Wu, C. (2021). Used disposable face masks are significant sources of microplastics to environment. *Environmental Pollution*, *285*. https://doi.org/10.1016/j.envpol.2021.117485

Chen, Y., Leng, Y., Liu, X., & Wang, J. (2020). Microplastic pollution in vegetable farmlands of suburb Wuhan, central China. *Environmental Pollution*, *257*. https://doi.org/10.1016/j.envpol.2019.113449

Chia, R. W., Lee, J. Y., Lee, M., & Lee, S. (2022). Comparison of microplastic characteristics in mulched and greenhouse soils of a major agriculture area, Korea. *Journal of Polymers and the Environment*. https://doi.org/10.1007/s10924-022-02746-1

Chinglenthoiba, C., & Valiyaveettil, S. (2023). *Are we eating chips or microplastics? Isolation, identification and quantification of microplastic in tapioca chips*. https://ssrn.com/abstract=4401456

Çobanoğlu, H., Belivermiş, M., Sıkdokur, E., Kılıç, Ö., & Çayır, A. (2021). Genotoxic and cytotoxic effects of polyethylene microplastics on human peripheral blood lymphocytes. *Chemosphere*, *272*. https://doi.org/10.1016/j.chemosphere.2021.129805

Cole, M., Lindeque, P., Halsband, C., & Galloway, T. S. (2011). Microplastics as contaminants in the marine environment: A review. *Marine Pollution Bulletin*, *62*(12), 2588–2597. https://doi.org/10.1016/j.marpolbul.2011.09.025

Commission Decision (EU) 2017/1218. (2017). *Commission decision (EU) 2017/1218 of* 23 June 2017 *establishing the EU Ecolabel criteria for laundry detergents (notified under document C(2017) 4243)*. https://data.europa.eu/eli/dec/2017/1218/oj

Compa, M., Ventero, A., Iglesias, M., & Deudero, S. (2018). Ingestion of microplastics and natural fibres in Sardina pilchardus (Walbaum, 1792) and Engraulis encrasicolus (Linnaeus, 1758) along the Spanish Mediterranean coast. *Marine Pollution Bulletin*, *128*. https://doi.org/10.1016/j.marpolbul.2018.01.009

Corradini, F., Meza, P., Eguiluz, R., Casado, F., Huerta-Lwanga, E., & Geissen, V. (2019). Evidence of microplastic accumulation in agricultural soils from sewage sludge disposal. *Science of the Total Environment*, *671*. https://doi.org/10.1016/j.scitotenv.2019.03.368

Cox, K. D., Covernton, G. A., Davies, H. L., Dower, J. F., Juanes, F., & Dudas, S. E. (2019). Human consumption of microplastics. *Environmental Science and Technology*, *53*(12), 7068–7074. https://doi.org/10.1021/acs.est.9b01517

Crawford, C. B., & Quinn, B. (2017). *Microplastic Pollutants*. Elsevier.

Crosta, A., Parolini, M., & De Felice, B. (2023). Microplastics contamination in nonalcoholic beverages from the Italian market. *International Journal of Environmental Research and Public Health*, *20*(5). https://doi.org/10.3390/ijerph20054122

Dalla Fontana, G., Mossotti, R., & Montarsolo, A. (2021). Influence of sewing on microplastic release from textiles during washing. *Water, Air, and Soil Pollution*, *232*(2). https://doi.org/10.1007/s11270-021-04995-7

De Falco, F., Cocca, M., Avella, M., & Thompson, R. C. (2020). Microfiber release to water, via laundering, and to air, via everyday use: A comparison between polyester clothing with differing textile parameters. *Environmental Science and Technology*, *54*(6), 3288–3296. https://doi.org/10.1021/acs.est.9b06892

De Falco, F., Cocca, M., Guarino, V., Gentile, G., Ambrogi, V., Ambrosio, L., & Avella, M. (2019). Novel finishing treatments of polyamide fabrics by electrofluidodynamic process to reduce microplastic release during washings. *Polymer Degradation and Stability*, *165*, 110–116. https://doi.org/10.1016/j.polymdegradstab.2019.05.001

De Falco, F., Gentile, G., Avolio, R., Errico, M. E., Di Pace, E., Ambrogi, V., Avella, M., & Cocca, M. (2018). Pectin based finishing to mitigate the impact of microplastics released by polyamide fabrics. *Carbohydrate Polymers*, *198*, 175–180. https://doi.org/10.1016/j.carbpol.2018.06.062

Diaz-Basantes, M. F., Conesa, J. A., & Fullana, A. (2020). Microplastics in honey, beer, milk and refreshments in Ecuador as emerging contaminants. *Sustainability (Switzerland)*, *12*(12). https://doi.org/10.3390/SU12145514

Dickey, V. (2022). *THE DISTRIBUTION OF MICROPLASTICS IN MARSHLANDS SURROUNDED BY AGRICULTURE FIELDS-ELKHORN SLOUGH, CA*.

Di Fiore, C., Sammartino, M. P., Giannattasio, C., Avino, P., & Visco, G. (2023). Microplastic contamination in commercial salt: An issue for their sampling and quantification. *Food Chemistry*, *404*. https://doi.org/10.1016/j.foodchem.2022.134682

Doley, S., Borauh, R. R., & Konwar, M. (2020). Study on awareness of consumers about dye related problems. *International Journal of Current Microbiology and Applied Sciences*, *9*(12). https://doi.org/10.20546/ijcmas.2020.912.142

Domenech, J., & Marcos, R. (2021). Pathways of human exposure to microplastics, and estimation of the total burden. *Current Opinion in Food Science*, *39*. https://doi.org/10.1016/j.cofs.2021.01.004

Donkers, J. M., Höppener, E. M., Grigoriev, I., Will, L., Melgert, B. N., van der Zaan, B., van de Steeg, E., & Kooter, I. M. (2022). Advanced epithelial lung and gut barrier models demonstrate passage of microplastic particles. *Microplastics and Nanoplastics*, *2*(1). https://doi.org/10.1186/s43591-021-00024-w

Egessa, R., Nankabirwa, A., Ocaya, H., & Pabire, W. G. (2020). Microplastic pollution in surface water of Lake Victoria. *Science of the Total Environment*, *741*. https://doi.org/10.1016/j.scitotenv.2020.140201

Ellen MacArthur Foundation. (2017). *A new textiles economy: Redesigning fashion's future*. https://www.ellenmacarthurfoundation.org/publications

Fagiano, V., Compa, M., Alomar, C., Rios-Fuster, B., Morató, M., Capó, X., & Deudero, S. (2023). Breaking the paradigm: Marine sediments hold two-fold microplastics than sea surface waters and are dominated by fibers. *Science of the Total Environment*, *858*. https://doi.org/10.1016/j.scitotenv.2022.159722

Fang, M., Liao, Z., Ji, X., Zhu, X., Wang, Z., Lu, C., Shi, C., Chen, Z., Ge, L., Zhang, M., Dahlgren, R. A., & Shang, X. (2022). Microplastic ingestion from atmospheric deposition during dining/drinking activities. *Journal of Hazardous Materials*, *432*. https://doi.org/10.1016/j.jhazmat.2022.128674

Felismino, M. E. L., Helm, P. A., & Rochman, C. M. (2021). Microplastic and other anthropogenic microparticles in water and sediments of Lake Simcoe. *Journal of Great Lakes Research*, *47*(1). https://doi.org/10.1016/j.jglr.2020.10.007

Finnegan, A. M. D., Süsserott, R., Gabbott, S. E., & Gouramanis, C. (2022). Man-made natural and regenerated cellulosic fibres greatly outnumber microplastic fibres in the atmosphere. *Environmental Pollution*, *310*. https://doi.org/10.1016/j.envpol.2022.119808

Forum for the Future. (2023). *Tackling microfibers at source*. https://www.forumforthefuture.org/investigating-opportunities-to-reduce-microfibre-pollution-from-the-fashion-industry

Furstenberg, D. Von. (2021). *Natural vs. synthetic fiber: What's the difference?* Master Class. https://www.masterclass.com/articles/natural-vs-synthetic-fibers#advantages-of-using-synthetic-fibers

Gago, J., Carretero, O., Filgueiras, A. V., & Viñas, L. (2018). Synthetic microfibers in the marine environment: A review on their occurrence in seawater and sediments. *Marine Pollution Bulletin*, *127*. https://doi.org/10.1016/j.marpolbul.2017.11.070

Garcia-Vazquez, E., & Garcia-Ael, C. (2021). The invisible enemy. Public knowledge of microplastics is needed to face the current microplastics crisis. *Sustainable Production and Consumption*, *28*. https://doi.org/10.1016/j.spc.2021.07.032

Gasperi, J., Dris, R., Mandin, C., & Tassin, B. (2015, September). First overview of microplastics in indoor and outdoor air. *15th EuCheMS International Conference on Chemistry and the Environment*.

Gasperi, J., Wright, S. L., Dris, R., Collard, F., Mandin, C., Guerrouache, M., Langlois, V., Kelly, F. J., & Tassin, B. (2018). Microplastics in air: Are we breathing it in? *Current Opinion in Environmental Science and Health*, *1*. https://doi.org/10.1016/j.coesh.2017.10.002

GESAMP. (2015). *Microplastics in the ocean*.

Girlfriend Collective. (2023). The microfiber filter. In *Girlfriend Collective*. https://www.girlfriend.com/products/water-filter

González-Pleiter, M., Velázquez, D., Edo, C., Carretero, O., Gago, J., Barón-Sola, Á., Hernández, L. E., Yousef, I., Quesada, A., Leganés, F., Rosal, R., & Fernández-Piñas, F. (2020). Fibers spreading worldwide: Microplastics and other anthropogenic litter in an Arctic freshwater lake. *Science of the Total Environment*, *722*. https://doi.org/10.1016/j.scitotenv.2020.137904

Ha, D. T. (2021). MICROPLASTIC CONTAMINATION IN COMMERCIAL SEA SALT OF VIET NAM. *Vietnam Journal of Science and Technology*, *59*(3). https://doi.org/10.15625/2525-2518/59/3/15718

Habib, D., Locke, D. C., & Cannone, L. J. (1998). Synthetic fibers as indicators of municipal sewage sludge, sludge products, and sewage treatment plant effluents. *Water, Air, and Soil Pollution*, *103*(1–4). https://doi.org/10.1023/A:1004908110793

Hanichak, G. (2021). Global textile industry. In *Study*. https://study.com/academy/lesson/global-textile-industry.html

Hao, Y., Sun, H., Zeng, X., Dong, G., Kronzucker, H. J., Min, J., Xia, C., Lam, S. S., & Shi, W. (2023). Smallholder vegetable farming produces more soil microplastics pollution than large-scale farming. *Environmental Pollution*, *317*. https://doi.org/10.1016/j.envpol.2022.120805

Herweyers, L., Carteny, C. C., Scheelen, L., Watts, R., & Bois, E. Du. (2020). Consumers' perceptions and attitudes toward products preventing microfiber pollution in aquatic environments as a result of the domestic washing of synthetic clothes. *Sustainability (Switzerland)*, *12*(6). https://doi.org/10.3390/su12062244

H&M Groups. (2023). *Microfibres*. https://hmgroup.com/sustainability/circularity-and-climate/materials/microfibres/

Horn, D. A., Granek, E. F., & Steele, C. L. (2020). Effects of environmentally relevant concentrations of microplastic fibers on Pacific mole crab (Emerita analoga) mortality and reproduction. *Limnology and Oceanography Letters*, *5*(1). https://doi.org/10.1002/lol2.10137

Huang, S., Huang, X., Bi, R., Guo, Q., Yu, X., Zeng, Q., Huang, Z., Liu, T., Wu, H., Chen, Y., Xu, J., Wu, Y., & Guo, P. (2022). Detection and analysis of microplastics in human sputum. *Environmental Science and Technology*, *56*(4). https://doi.org/10.1021/acs.est.1c03859

Huang, Y., Liu, Q., Jia, W., Yan, C., & Wang, J. (2020). Agricultural plastic mulching as a source of microplastics in the terrestrial environment. *Environmental Pollution*, *260*. https://doi.org/10.1016/j.envpol.2020.114096

Jemec, A., Horvat, P., Kunej, U., Bele, M., & Kržan, A. (2016). Uptake and effects of microplastic textile fibers on freshwater crustacean Daphnia magna. *Environmental Pollution*, *219*, 201–209. https://doi.org/10.1016/j.envpol.2016.10.037

Jenner, L. C., Rotchell, J. M., Bennett, R. T., Cowen, M., Tentzeris, V., & Sadofsky, L. R. (2022). Detection of microplastics in human lung tissue using μFTIR spectroscopy. *Science of the Total Environment*, *831*. https://doi.org/10.1016/j.scitotenv.2022.154907

Jiang, X., Yang, Y., Wang, Q., Liu, N., & Li, M. (2022). Seasonal variations and feedback from microplastics and cadmium on soil organisms in agricultural fields. *Environment International*, *161*. https://doi.org/10.1016/j.envint.2022.107096

Julia. (2017). I tried the Guppyfriend washing bag. https://www.fairforallguide.com/2017/08/22/i-tried-the-guppyfriend-washing-bag/

Kang, H., Park, S., Lee, B., Ahn, J., & Kim, S. (2021). Impact of chitosan pretreatment to reduce microfibers released from synthetic garments during laundering. *Water (Switzerland)*, *13*(18). https://doi.org/10.3390/w13182480

Kanhai, L. D. K., Officer, R., Lyashevska, O., Thompson, R. C., & O'Connor, I. (2017). Microplastic abundance, distribution and composition along a latitudinal gradient in the Atlantic Ocean. *Marine Pollution Bulletin*, *115*(1–2). https://doi.org/10.1016/j.marpolbul.2016.12.025

Kapp, K. J., & Miller, R. Z. (2020). Electric clothes dryers: An underestimated source of microfiber pollution. *PLoS One*, 15(10 October). https://doi.org/10.1371/journal.pone.0239165

Karami, A., Golieskardi, A., Keong Choo, C., Larat, V., Galloway, T. S., & Salamatinia, B. (2017). The presence of microplastics in commercial salts from different countries. *Scientific Reports*, *7*. https://doi.org/10.1038/srep46173

Karami, A., Romano, N., Galloway, T., & Hamzah, H. (2016). Virgin microplastics cause toxicity and modulate the impacts of phenanthrene on biomarker responses in African catfish (Clarias gariepinus). *Environmental Research*, *151*. https://doi.org/10.1016/j.envres.2016.07.024

Kasamesiri, P., & Thaimuangpho, W. (2020). Microplastics ingestion by freshwater fish in the Chi River, Thailand. *International Journal of Geomate*, *18*(67), 114–119. https://doi.org/10.21660/2020.67.9110

Kim, D., Kim, H., & An, Y. J. (2021a). Effects of synthetic and natural microfibers on Daphnia magna-Are they dependent on microfiber type? *Aquatic Toxicology*, *240*. https://doi.org/10.1016/j.aquatox.2021.105968

Kim, L., Kim, S. A., Kim, T. H., Kim, J., & An, Y. J. (2021). Synthetic and natural microfibers induce gut damage in the brine shrimp Artemia franciscana. *Aquatic Toxicology*, *232*. https://doi.org/10.1016/j.aquatox.2021.105748

Koutnik, V. S., Alkidim, S., Leonard, J., Deprima, F., Cao, S., Hoek, E. M. V., & Mohanty, S. K. (2021). Unaccounted microplastics in wastewater sludge: Where do they go? *ACS Environmental Science and Technology Water*, *1*(5). https://doi.org/10.1021/acsestwater.0c00267

Kutralam-Muniasamy, G., Pérez-Guevara, F., Elizalde-Martínez, I., & Shruti, V. C. (2020). Branded milks – Are they immune from microplastics contamination? *Science of the Total Environment*, *714*. https://doi.org/10.1016/j.scitotenv.2020.136823

Ladewig, S. M., Bao, S., & Chow, A. T. (2015). Natural fibers: A missing link to chemical pollution dispersion in aquatic environments. *Environmental Science and Technology*, *49*(21), 12609–12610). American Chemical Society. https://doi.org/10.1021/acs.est.5b04754

Laitala, K., Klepp, I. G., & Henry, B. (2018). Does use matter? Comparison of environmental impacts of clothing based on fiber type. *Sustainability (Switzerland)*, *10*(7). https://doi.org/10.3390/su10072524

Le, N. Da, Hoang, T. T. H., Duong, T. T., Lu, X., Pham, T. M. H., Phung, T. X. B., Le, T. M. H., Duong, T. H., Nguyen, T. D., & Le, T. P. Q. (2022). First observation of microplastics in surface sediment of some aquaculture ponds in Hanoi city, Vietnam. *Journal of Hazardous Materials Advances*, *6*. https://doi.org/10.1016/j.hazadv.2022.100061

Lee, H., Kunz, A., Shim, W. J., & Walther, B. A. (2019). Microplastic contamination of table salts from Taiwan, including a global review. *Scientific Reports*, *9*(1). https://doi.org/10.1038/s41598-019-46417-z

Le Guen, C., Suaria, G., Sherley, R. B., Ryan, P. G., Aliani, S., Boehme, L., & Brierley, A. S. (2020). Microplastic study reveals the presence of natural and synthetic fibres in the diet of King Penguins (Aptenodytes patagonicus) foraging from South Georgia. *Environment International*, *134*. https://doi.org/10.1016/j.envint.2019.105303

Lewandowski, C. M., Co-investigator, N., & Lewandowski, C. M. (2015). Textile chemicals – Environmental data ad facts. In *The Effects of Brief Mindfulness Intervention on Acute Pain Experience: An Examination of Individual Difference* (Vol. 1).

Li, L., Frey, M., & Browning, K. J. (2010). Biodegradability study on cotton and polyester fabrics. *Journal of Engineered Fibers and Fabrics*, *5*(4). https://doi.org/10.1177/155892501000500406

Li, C., Wang, X., Liu, K., Zhu, L., Wei, N., Zong, C., & Li, D. (2021). Pelagic microplastics in surface water of the Eastern Indian Ocean during monsoon transition period: Abundance, distribution, and characteristics. *Science of the Total Environment*, *755*. https://doi.org/10.1016/j.scitotenv.2020.142629

Li, L., Zhao, X., Li, Z., & Song, K. (2021b). COVID-19: Performance study of microplastic inhalation risk posed by wearing masks. *Journal of Hazardous Materials*, *411*. https://doi.org/10.1016/j.jhazmat.2020.124955

Liang, W., Li, B., Jong, M. C., Ma, C., Zuo, C., Chen, Q., & Shi, H. (2023). Process-oriented impacts of microplastic fibers on behavior and histology of fish. *Journal of Hazardous Materials, 448*. https://doi.org/10.1016/j.jhazmat.2023.130856

Liao, Z., Ji, X., Ma, Y., Lv, B., Huang, W., Zhu, X., Fang, M., Wang, Q., Wang, X., Dahlgren, R., & Shang, X. (2021). Airborne microplastics in indoor and outdoor environments of a coastal city in Eastern China. *Journal of Hazardous Materials, 417*. https://doi.org/10.1016/j.jhazmat.2021.126007

Liu, Z., Bai, Y., Ma, T., Liu, X., Wei, H., Meng, H., Fu, Y., Ma, Z., Zhang, L., & Zhao, J. (2022b). Distribution and possible sources of atmospheric microplastic deposition in a valley basin city (Lanzhou, China). *Ecotoxicology and Environmental Safety, 233*. https://doi.org/10.1016/j.ecoenv.2022.113353

Liu, J., Liu, Q., An, L., Wang, M., Yang, Q., Zhu, B., Ding, J., Ye, C., & Xu, Y. (2022). Microfiber pollution in the earth system. *Reviews of Environmental Contamination and Toxicology, 260*(1). Springer. https://doi.org/10.1007/s44169-022-00015-9

Liu, M., Lu, S., Song, Y., Lei, L., Hu, J., Lv, W., Zhou, W., Cao, C., Shi, H., Yang, X., & He, D. (2018). Microplastic and mesoplastic pollution in farmland soils in suburbs of Shanghai, China. *Environmental Pollution, 242*, 855–862. https://doi.org/10.1016/j.envpol.2018.07.051

Liu, K., Wang, X., Fang, T., Xu, P., Zhu, L., & Li, D. (2019b). Source and potential risk assessment of suspended atmospheric microplastics in Shanghai. *Science of the Total Environment, 675*, 462–471. https://doi.org/10.1016/j.scitotenv.2019.04.110

Liu, J., Yang, Y., Ding, J., Zhu, B., & Gao, W. (2019a). Microfibers: A preliminary discussion on their definition and sources. *Environmental Science and Pollution Research, 26*(28), 29497–29501. https://doi.org/10.1007/s11356-019-06265-w

Liu, J., Zhu, B., An, L., Ding, J., & Xu, Y. (2023). Atmospheric microfibers dominated by natural and regenerated cellulosic fibers: Explanations from the textile engineering perspective. *Environmental Pollution, 317*. https://doi.org/10.1016/j.envpol.2022.120771

Long, B., Li, F., Wang, K., Huang, Y., Yang, Y., & Xie, D. (2023). Impact of plastic film mulching on microplastic in farmland soils in Guangdong province, China. *Heliyon, 9*(6), e16587. https://doi.org/10.1016/J.HELIYON.2023.E16587

Lusher, A. L., McHugh, M., & Thompson, R. C. (2013). Occurrence of microplastics in the gastrointestinal tract of pelagic and demersal fish from the English Channel. *Marine Pollution Bulletin, 67*(1–2). https://doi.org/10.1016/j.marpolbul.2012.11.028

Magnusson, K., & Norén, F. (2014). Screening of microplastic particles in and down-stream a wastewater treatment plant. *IVL Swedish Environmental Research Institute, C 55*(C).

Maiti, R. (2023). Fast fashion and its environmental impact. https://earth.org/fast-fashions-detrimental-effect-on-the-environment/

Makhdoumi, P., Pirsaheb, M., Amin, A. A., Kianpour, S., & Hossini, H. (2023). Microplastic pollution in table salt and sugar: Occurrence, qualification and quantification and risk assessment. *Journal of Food Composition and Analysis, 119*, 105261. https://doi.org/10.1016/J.JFCA.2023.105261

Market Analysis Report. (2022). *Textile market size, share & trends analysis report by raw material (cotton, wool, silk, chemical), by product (natural fibers, nylon), by application (technical, fashion), by region, and segment forecasts, 2022–2030*. https://www.grandviewresearch.com/industry-analysis/textile-market

Martinko, K. (2020). *Reusable laundry filter captures 90% of microfibers*. https://www.treehugger.com/planetcare-filter-captures-microfibers-5081906

Mateos-Cárdenas, A., O'Halloran, J., van Pelt, F. N. A. M., & Jansen, M. A. K. (2021). Beyond plastic microbeads – Short-term feeding of cellulose and polyester microfibers to the freshwater amphipod Gammarus duebeni. *Science of the Total Environment, 753*. https://doi.org/10.1016/j.scitotenv.2020.141859

McIlwraith, H. K., Lin, J., Erdle, L. M., Mallos, N., Diamond, M. L., & Rochman, C. M. (2019). Capturing microfibers – Marketed technologies reduce microfiber emissions from washing machines. *Marine Pollution Bulletin*, *139*, 40–45. https://doi.org/10.1016/j.marpolbul.2018.12.012

Mermaids, O. C. W. (2015). *Report on the influence of commercial textile finishing, fabrics geometry and washing conditions on microplastics release.*

Miller, R. Z., Watts, A. J. R., Winslow, B. O., Galloway, T. S., & Barrows, A. P. W. (2017). Mountains to the sea: River study of plastic and non-plastic microfiber pollution in the northeast USA. *Marine Pollution Bulletin*, *124*(1), 245–251. https://doi.org/10.1016/j.marpolbul.2017.07.028

Mishra, S., Rath, C. charan, & Das, A. P. (2019). Marine microfiber pollution: A review on present status and future challenges. *Marine Pollution Bulletin*, *140*, 188–197. Elsevier Ltd. https://doi.org/10.1016/j.marpolbul.2019.01.039

Moore, C. J. (2008). Synthetic polymers in the marine environment: A rapidly increasing, long-term threat. *Environmental Research*, *108*(2). https://doi.org/10.1016/j.envres.2008.07.025

Murphy, F., Ewins, C., Carbonnier, F., & Quinn, B. (2016). Wastewater treatment works (WwTW) as a source of microplastics in the aquatic environment. *Environmental Science and Technology*, *50*(11), 5800–5808. https://doi.org/10.1021/acs.est.5b05416

Napper, I. E., Barrett, A. C., & Thompson, R. C. (2020). The efficiency of devices intended to reduce microfibre release during clothes washing. *Science of the Total Environment*, *738*. https://doi.org/10.1016/j.scitotenv.2020.140412

Napper, I. E., & Thompson, R. C. (2016). Release of synthetic microplastic plastic fibres from domestic washing machines: Effects of fabric type and washing conditions. *Marine Pollution Bulletin*, *112*(1–2), 39–45. https://doi.org/10.1016/j.marpolbul.2016.09.025

Nepori, A. (2023, January 24). *Patagonia launches a sustainable washing machine (with Samsung)*. https://www.domusweb.it/en/sustainable-cities/gallery/2023/01/24/samsung-and-patagonia-team-up-to-make-washing-machines-sustainable.html

Nguyen, A. (2023). *Managing microplastics: Fashions next challenge*. https://www.xerostech.com/coming-soon-managing-microplastics-fashions-next-challenge/

Nguyen, M. K., Lin, C., Hung, N. T. Q., Vo, D. V. N., Nguyen, K. N., Thuy, B. T. P., Hoang, H. G., & Tran, H. T. (2022). Occurrence and distribution of microplastics in peatland areas: A case study in Long An province of the Mekong Delta, Vietnam. *Science of the Total Environment*, *844*, 157066. https://doi.org/10.1016/J.SCITOTENV.2022.157066

Nike. (2019). *Microfibers statement*. https://about.nike.com/en/newsroom/statements/microfibers-statement

NOAA Marine Debris Program. (2009). *Proceedings of the international research workshop on the occurrence, effects, and fate of microplastic marine debris*. www.MarineDebris.noaa.gov

Oldenburg, K. S., Urban-Rich, J., Castillo, K. D., & Baumann, J. H. (2021). Microfiber abundance associated with coral tissue varies geographically on the Belize Mesoamerican Barrier Reef System. *Marine Pollution Bulletin*, *163*. https://doi.org/10.1016/j.marpolbul.2020.111938

Oliveri Conti, G., Ferrante, M., Banni, M., Favara, C., Nicolosi, I., Cristaldi, A., Fiore, M., & Zuccarello, P. (2020). Micro- and nano-plastics in edible fruit and vegetables. The first diet risks assessment for the general population. *Environmental Research*, *187*. https://doi.org/10.1016/j.envres.2020.109677

Olivia. (2020). *How we're reducing microplastic pollution from our washing machine with a Filtrol 160*. https://thiswestcoastmommy.com/microplastic-pollution-washing-machine-filtrol/

O'Loughlin, C. (2018). *Fashion and microplastic pollution investigating microplastics from laundry.*

Outdoor Industry Association. (2018). *Outdoor industry association priority issues brief: Microfibers*. Outdoor Industry Association Priority Issues Brief: Microfibers.

Paredes, M., Castillo, T., Viteri, R., Fuentes, G., & Bodero, E. (2019). Microplastics in the drinking water of the Riobamba city, Ecuador. *Scientific Review Engineering and Environmental Sciences*, *28*(4). https://doi.org/10.22630/PNIKS.2019.28.4.59

Parvin, F., Nath, J., Hannan, T., & Tareq, S. M. (2022). Proliferation of microplastics in commercial sea salts from the world longest sea beach of Bangladesh. *Environmental Advances*, *7*. https://doi.org/10.1016/j.envadv.2022.100173

Patagonia. (n.d.). *Our environmental responsibility programs our environmental responsibility programs*. Retrieved May 28, 2023, from https://www.patagonia.com/our-responsibility-programs.html

Patrício Silva, A. L., Prata, J. C., Duarte, A. C., Barcelò, D., & Rocha-Santos, T. (2021). An urgent call to think globally and act locally on landfill disposable plastics under and after covid-19 pandemic: Pollution prevention and technological (Bio) remediation solutions. *Chemical Engineering Journal*, *426*. https://doi.org/10.1016/j.cej.2021.131201

Pauly, J. L., Stegmeier, S. J., Allaart, H. A., Cheney, R. T., Zhang, P. J., Mayer, A. G., & Streck, R. J. (1998). *Inhaled Cellulosic and Plastic Fibers Found in Human Lung Tissue'* (Vol. 7).

Pejak, N. (2022). *Impact of Fast Fashion Industry on the Environment*. Central European University.

Peller, J. R., Eberhardt, L., Clark, R., Nelson, C., Kostelnik, E., & Iceman, C. (2019). Tracking the distribution of microfiber pollution in a southern Lake Michigan watershed through the analysis of water, sediment and air. *Environmental Science: Processes and Impacts*, *21*(9), 1549–1559. https://doi.org/10.1039/c9em00193j

Perera, K., Ziajahromi, S., Bengtson Nash, S., Manage, P. M., & Leusch, F. D. L. (2022). Airborne microplastics in indoor and outdoor environments of a developing country in South Asia: Abundance, distribution, morphology, and possible sources. *Environmental Science and Technology*, *56*(23). https://doi.org/10.1021/acs.est.2c05885

Perera, K., Ziajahromi, S., Nash, S. B., & Leusch, F. D. L. (2023). Microplastics in Australian indoor air: Abundance, characteristics, and implications for human exposure. *Science of the Total Environment*, *889*, 164292. https://doi.org/10.1016/J.SCITOTENV.2023.164292

Periyasamy, A. P. (2023). Microfiber emissions from functionalized textiles: Potential threat for human health and environmental risks. *Toxics*, *11*(5), 406. https://doi.org/10.3390/toxics11050406

Peters, G., Li, M., & Lenzen, M. (2021). The need to decelerate fast fashion in a hot climate – A global sustainability perspective on the garment industry. *Journal of Cleaner Production*, *295*. https://doi.org/10.1016/j.jclepro.2021.126390

PlanetCare. (2023). *PlanetCare filters, the most efficient washing machine filter ever designed*. https://planetcare.org/

Plastic smart cities. (2023). *PlanetCare*. Plasticsmartcities. https://plasticsmartcities.org/products/planetcare-1

Pradit, S., Noppradit, P., Sengloyluan, K., Suwanno, P., Tanrattanakul, V., Sornplang, K., Nuthammachot, N., Jitkaew, P., & Nitiratsuwan, T. (2023). Occurrence of microplastics in river water in Southern Thailand. *Journal of Marine Science and Engineering*, *11*(1). https://doi.org/10.3390/jmse11010090

Priya, K. K., Thilagam, H., Muthukumar, T., Gopalakrishnan, S., & Govarthanan, M. (2023). Impact of microfiber pollution on aquatic biota: A critical analysis of effects and preventive measures. *Science of the Total Environment*, *887*, 163984. https://doi.org/10.1016/J.SCITOTENV.2023.163984

Qualhato, G., Vieira, L. G., Oliveira, M., & Rocha, T. L. (2023). Plastic microfibers as a risk factor for the health of aquatic organisms: A bibliometric and systematic review of plastic pandemic. *Science of the Total Environment*, *870*. https://doi.org/10.1016/j.scitotenv.2023.161949

Ragusa, A., Svelato, A., Santacroce, C., Catalano, P., Notarstefano, V., Carnevali, O., Papa, F., Rongioletti, M. C. A., Baiocco, F., Draghi, S., D'Amore, E., Rinaldo, D., Matta, M., & Giorgini, E. (2021). Plasticenta: First evidence of microplastics in human placenta. *Environment International*, *146*. https://doi.org/10.1016/j.envint.2020.106274

Raja Balasaraswathi, S., & Rathinamoorthy, R. (2021). *Effect of textile parameters on microfiber shedding properties of textiles*. https://doi.org/10.1007/978-981-16-0297-9_1

Rakib, M. R. J., Al Nahian, S., Alfonso, M. B., Khandaker, M. U., Enyoh, C. E., Hamid, F. S., Alsubaie, A., Almalki, A. S. A., Bradley, D. A., Mohafez, H., & Islam, M. A. (2021). Microplastics pollution in salt pans from the Maheshkhali Channel, Bangladesh. *Scientific Reports*, *11*(1). https://doi.org/10.1038/s41598-021-02457-y

Ramasamy, R., Aragaw, T. A., & Balasaraswathi Subramanian, R. (2022). Wastewater treatment plant effluent and microfiber pollution: Focus on industry-specific wastewater. *Environmental Science and Pollution Research*, *29*(34). https://doi.org/10.1007/s11356-022-20930-7

Ramasamy, R., & Subramanian, R. B. (2021). Synthetic textile and microfiber pollution: A review on mitigation strategies. *Environmental Science and Pollution Research*, *28*(31). https://doi.org/10.1007/s11356-021-14763-z

Ramasamy, R., & Subramanian, R. B. (2022). Enzyme hydrolysis of polyester knitted fabric: A method to control the microfiber shedding from synthetic textile. *Environmental Science and Pollution Research*.

Ramasamy, R., & Subramanian, R. B. (2023). Microfiber mitigation from synthetic textiles - impact of combined surface modification and finishing process. *Environmental Science and Pollution Research*. https://doi.org/10.1007/s11356-023-25611-7

Rathinamoorthy, R., & Balasaraswathi, S. R. (2022). Disposable tri-layer masks and microfiber pollution – An experimental analysis on dry and wet state emission. *Science of the Total Environment*, *816*. https://doi.org/10.1016/j.scitotenv.2021.151562

Rathinamoorthy, R., & Raja Balasaraswathi, S. (2022a). *Effect of surface modification of polyester fabric on microfiber shedding from household laundry*. https://doi.org/https://doi.org/10.1108/IJCST-05-2021-0064

Rathinamoorthy, R., & Raja Balasaraswathi, S. (2022b). Investigations on the interactive effect of laundry parameters on microfiber release from polyester knitted fabric. *Fibers and Polymers*, *23*(7), 2052–2061.

Rathinamoorthy, R., & Raja Balasaraswathi, S. (2023a). Characterization of microfibers originated from the textile screen printing industry. *Science of the Total Environment*, *874*. https://doi.org/10.1016/j.scitotenv.2023.162550

Rathinamoorthy, R., & Raja Balasaraswathi, S. (2023b). Characterization of microfibers released from chemically modified polyester fabrics – A step towards mitigation. *Science of the Total Environment*, *866*. https://doi.org/10.1016/j.scitotenv.2022.161317

Rauturier, S. (2023, January 6). *16 sustainable fashion documentaries you'll be glad you watched*. https://goodonyou.eco/documentaries-environment-sustainability/

Raza, M. H., Jabeen, F., Ikram, S., & Zafar, S. (2023). Characterization and implication of microplastics on riverine population of the River Ravi, Lahore, Pakistan. *Environmental Science and Pollution Research*, *30*(3). https://doi.org/10.1007/s11356-022-22440-y

Ribeiro, F., Okoffo, E. D., O'Brien, J. W., Fraissinet-Tachet, S., O'Brien, S., Gallen, M., Samanipour, S., Kaserzon, S., Mueller, J. F., Galloway, T., & Thomas, K. V. (2020). Quantitative analysis of selected plastics in high-commercial-value Australian seafood by pyrolysis gas chromatography mass spectrometry. *Environmental Science and Technology*, *54*(15). https://doi.org/10.1021/acs.est.0c02337

Roblin, B., Ryan, M., Vreugdenhil, A., & Aherne, J. (2020). Ambient atmospheric deposition of anthropogenic microfibers and microplastics on the western periphery of Europe (Ireland). *Environmental Science and Technology*, *54*(18). https://doi.org/10.1021/acs.est.0c04000

Romarate, R. A., Ancla, S. M. B., Patilan, D. M. M., Inocente, S. A. T., Pacilan, C. J. M., Sinco, A. L., Guihawan, J. Q., Capangpangan, R. Y., Lubguban, A. A., & Bacosa, H. P. (2023). Breathing plastics in Metro Manila, Philippines: Presence of suspended atmospheric microplastics in ambient air. *Environmental Science and Pollution Research*. https://doi.org/10.1007/s11356-023-26117-y

Salahuddin, M., & Lee, Y.-A. (2021). *Critical dialogue on the role of clothing care label for controlling microfiber pollution*. https://doi.org/10.31274/itaa.11859

Samantha, N. A., Diamond, M. L., Erdle, L. M., Lin, J., McIlwraith, H. K., Rochman, C. M., & Sweetnam, D. (2019). *Filters added to washing machines mitigate microfiber pollution*. https://rochmanlab.files.wordpress.com/2019/01/microfiber-policy-brief-2019.pdf

Sanchez-Vidal, A., Thompson, R. C., Canals, M., & De Haan, W. P. (2018). The imprint of microfibres in Southern European deep seas. *PLoS One*, *13*(11). https://doi.org/10.1371/journal.pone.0207033

Santonicola, S., Volgare, M., Di Pace, E., Mercogliano, R., Cocca, M., Raimo, G., & Colavita, G. (2023). Research and characterization of fibrous microplastics and natural microfibers in pelagic and benthic fish species of commercial interest. *Italian Journal of Food Safety*, *12*(1). https://doi.org/10.4081/ijfs.2023.11032

Sarathana, D., & Winijkul, E. (2022). Concentrations of airborne microplastics during the dry season at five locations in Bangkok metropolitan region, Thailand. *Atmosphere*, *14*(1). https://doi.org/10.3390/atmos14010028

Savoca, S., Capillo, G., Mancuso, M., Faggio, C., Panarello, G., Crupi, R., Bonsignore, M., D'Urso, L., Compagnini, G., Neri, F., Fazio, E., Romeo, T., Bottari, T., & Spanò, N. (2019). Detection of artificial cellulose microfibers in Boops from the northern coasts of Sicily (Central Mediterranean). *Science of the Total Environment*, *691*. https://doi.org/10.1016/j.scitotenv.2019.07.148

Savoca, S., Matanović, K., D'Angelo, G., Vetri, V., Anselmo, S., Bottari, T., Mancuso, M., Kužir, S., Spanò, N., Capillo, G., Di Paola, D., Valić, D., & Gjurčević, E. (2021). Ingestion of plastic and non-plastic microfibers by farmed gilthead sea bream (Sparus aurata) and common carp (Cyprinus carpio) at different life stages. *Science of the Total Environment*, *782*. https://doi.org/10.1016/j.scitotenv.2021.146851

Schellenberger, S., Jönsson, C., Mellin, P., Levenstam, O. A., Liagkouridis, I., Ribbenstedt, A., Hanning, A. C., Schultes, L., Plassmann, M. M., Persson, C., Cousins, I. T., & Benskin, J. P. (2019). Release of side-chain fluorinated polymer-containing microplastic fibers from functional textiles during washing and first estimates of perfluoroalkyl acid emissions. *Environmental Science and Technology*, *53*(24). https://doi.org/10.1021/acs.est.9b04165

Selonen, S., Dolar, A., Jemec Kokalj, A., Skalar, T., Parramon Dolcet, L., Hurley, R., & van Gestel, C. A. M. (2020). Exploring the impacts of plastics in soil – The effects of polyester textile fibers on soil invertebrates. *Science of the Total Environment*, *700*. https://doi.org/10.1016/j.scitotenv.2019.134451

Sharma, M. D., & Krupadam, R. J. (2022). Adsorption-desorption dynamics of synthetic and naturally weathered microfibers with toxic heavy metals and their ecological risk in an estuarine ecosystem. *Environmental Research*, *207*. https://doi.org/10.1016/j.envres.2021.112198

Shen, M., Zeng, Z., Song, B., Yi, H., Hu, T., Zhang, Y., Zeng, G., & Xiao, R. (2021a). Neglected microplastics pollution in global COVID-19: Disposable surgical masks. *Science of the Total Environment*, *790*. https://doi.org/10.1016/j.scitotenv.2021.148130

Shen, M., Zeng, Z., Wen, X., Ren, X., Zeng, G., Zhang, Y., & Xiao, R. (2021b). Presence of microplastics in drinking water from freshwater sources: The investigation in Changsha, China. *Environmental Science and Pollution Research*, *28*(31). https://doi.org/10.1007/s11356-021-13769-x

Sillanpää, M., & Sainio, P. (2017). Release of polyester and cotton fibers from textiles in machine washings. *Environmental Science and Pollution Research*, *24*(23). https://doi.org/10.1007/s11356-017-9621-1

Silva, P. M., & Nanny, M. A. (2020). Impact of microplastic fibers from the degradation of nonwoven synthetic textiles to the Magdalena river water column and river sediments by the city of Neiva, Huila (Colombia). *Water (Switzerland)*, *12*(4). https://doi.org/10.3390/W12041210

Sinner, H., & Verlag, H. (1960). Über das Waschen mit Haushaltswaschmaschinen. Hamburg.

Soltani, N. S., Taylor, M. P., & Wilson, S. P. (2021). Quantification and exposure assessment of microplastics in Australian indoor house dust. *Environmental Pollution*, *283*. https://doi.org/10.1016/j.envpol.2021.117064

Stovall, J. K., & Bratton, S. P. (2022). Microplastic pollution in surface waters of urban watersheds in Central Texas, United States: A comparison of sites with and without treated wastewater effluent. *Frontiers in Analytical Science*, *2*. https://doi.org/10.3389/frans.2022.857694

Styles, R. (2019). Guppyfriend washing bag review. https://www.consumer.org.nz/articles/review-guppyfriend-washing-bag

Suaria, G., Achtypi, A., Perold, V., Lee, J. R., Pierucci, A., Bornman, T. G., Aliani, S., & Ryan, P. G. (2020). Microfibers in oceanic surface waters: A global characterization. *Science Advances*, *6*.

Sun, J., Dai, X., Wang, Q., van Loosdrecht, M. C. M., & Ni, B. J. (2019). Microplastics in wastewater treatment plants: Detection, occurrence and removal. *Water Research*, *152*. https://doi.org/10.1016/j.watres.2018.12.050

Szewc, K., Graca, B., & Dołęga, A. (2021). Atmospheric deposition of microplastics in the coastal zone: Characteristics and relationship with meteorological factors. *Science of the Total Environment*, *761*. https://doi.org/10.1016/j.scitotenv.2020.143272

Tajwar, M., Shreya, S. S., Hasan, M., Hossain, M. B., Gazi, M. Y., & Sakib, N. (2022). Assessment of microplastics as contaminants in a coal mining region. *Heliyon*, *8*(11). https://doi.org/10.1016/j.heliyon.2022.e11666

Takarina, N. D., Purwiyanto, A. I. S., Rasud, A. A., Arifin, A. A., & Suteja, Y. (2022). Microplastic abundance and distribution in surface water and sediment collected from the coastal area. *Global Journal of Environmental Science and Management*, *8*(2), 183–196. https://doi.org/10.22034/GJESM.2022.02.03

Talvitie, J., Mikola, A., Koistinen, A., & Setälä, O. (2017). Solutions to microplastic pollution – Removal of microplastics from wastewater effluent with advanced wastewater treatment technologies. *Water Research*, *123*. https://doi.org/10.1016/j.watres.2017.07.005

Tao, D., Zhang, K., Xu, S., Lin, H., Liu, Y., Kang, J., Yim, T., Giesy, J. P., & Leung, K. M. Y. (2022). Microfibers released into the air from a household tumble dryer. *Environmental Science and Technology Letters*, *9*(2). https://doi.org/10.1021/acs.estlett.1c00911

The Coraball. (2023). https://coraball.com/

The Microfiber Consortium. (n.d.). Retrieved November 24, 2021, from https://www.microfibreconsortium.com/

The Microfibre Consortium. (2023). *Technical research report: Recycled polyester within the context of fibre fragmentation.*

The World Bank. (2019, September 23). *How much do our wardrobes cost to the envrionment?* https://www.worldbank.org/en/news/feature/2019/09/23/costo-moda-medio-ambiente

Thompson, R. C., Olson, Y., Mitchell, R. P., Davis, A., Rowland, S. J., John, A. W. G., McGonigle, D., & Russell, A. E. (2004). Lost at sea: Where is all the plastic? *Science*, *304*(5672). https://doi.org/10.1126/science.1094559

Tiwari, M., Rathod, T. D., Ajmal, P. Y., Bhangare, R. C., & Sahu, S. K. (2019). Distribution and characterization of microplastics in beach sand from three different Indian coastal environments. *Marine Pollution Bulletin*, *140*. https://doi.org/10.1016/j.marpolbul.2019.01.055

Tse, Y. T., Chan, S. M. N., & Sze, E. T. P. (2022). Quantitative assessment of full size microplastics in bottled and tap water samples in Hong Kong. *International Journal of Environmental Research and Public Health, 19*(20). https://doi.org/10.3390/ijerph192013432

Tunahan Kaya, A., Yurtsever, M., & Çiftçi Bayraktar, S. (2018). Ubiquitous exposure to microfiber pollution in the air. *European Physical Journal Plus, 133*(11). https://doi.org/10.1140/epjp/i2018-12372-7

Uddin, S., Fowler, S. W., Habibi, N., Sajid, S., Dupont, S., & Behbehani, M. (2022). A preliminary assessment of size-fractionated microplastics in indoor aerosol-Kuwait's baseline. *Toxics, 10*(2). https://doi.org/10.3390/toxics10020071

Van Cauwenberghe, L., Claessens, M., Vandegehuchte, M. B., & Janssen, C. R. (2015a). Microplastics are taken up by mussels (Mytilus edulis) and lugworms (Arenicola marina) living in natural habitats. *Environmental Pollution, 199*, 10–17. https://doi.org/10.1016/j.envpol.2015.01.008

Van Cauwenberghe, L., Devriese, L., Galgani, F., Robbens, J., & Janssen, C. R. (2015b). Microplastics in sediments: A review of techniques, occurrence and effects. *Marine Environmental Research, 111*. https://doi.org/10.1016/j.marenvres.2015.06.007

Vassilenko, K., Watkins, M., Chastain, S., Posacka, A., & Roos, P. S. (2019). *Me, my clothes and the ocean: The role of textiles in microfiber pollution, science feature*. https://assets.ctfassets.net/fsquhe7zbn68/4MQ9y89yx4KeyHv9Svynyq/8434de64585e9d2cfbcd3c46627c7a4a/Research_MicrofibersReport_191004-e.pdf

Walkinshaw, C., Tolhurst, T. J., Lindeque, P. K., Thompson, R. C., & Cole, M. (2023). Impact of polyester and cotton microfibers on growth and sublethal biomarkers in juvenile mussels. *Microplastics and Nanoplastics, 3*(1). https://doi.org/10.1186/s43591-023-00052-8

Wang, X., Li, C., Liu, K., Zhu, L., Song, Z., & Li, D. (2020). Atmospheric microplastic over the South China Sea and East Indian Ocean: Abundance, distribution and source. *Journal of Hazardous Materials, 389*. https://doi.org/10.1016/j.jhazmat.2019.121846

Welsh, B., Aherne, J., Paterson, A. M., Yao, H., & McConnell, C. (2022). Atmospheric deposition of anthropogenic particles and microplastics in south-central Ontario, Canada. *Science of the Total Environment, 835*. https://doi.org/10.1016/j.scitotenv.2022.155426

Wilcox, K. (2017). *Combatting the microfiber pollution problem at the municipality level.*

Woodall, L. C., Gwinnett, C., Packer, M., Thompson, R. C., Robinson, L. F., & Paterson, G. L. J. (2015). Using a forensic science approach to minimize environmental contamination and to identify microfibres in marine sediments. *Marine Pollution Bulletin, 95*(1). https://doi.org/10.1016/j.marpolbul.2015.04.044

Xu, L., Han, L., Li, J., Zhang, H., Jones, K., & Xu, E. G. (2022). Missing relationship between meso- and microplastics in adjacent soils and sediments. *Journal of Hazardous Materials, 424*. https://doi.org/10.1016/j.jhazmat.2021.127234

Xu, X., Hou, Q., Xue, Y., Jian, Y., & Wang, L. P. (2018). Pollution characteristics and fate of microfibers in the wastewater from textile dyeing wastewater treatment plant. *Water Science and Technology, 78*(10), 2046–2054. https://doi.org/10.2166/wst.2018.476

Xue, J., Liu, W., & Kannan, K. (2017). Bisphenols, benzophenones, and bisphenol A diglycidyl ethers in textiles and infant clothing. *Environmental Science and Technology, 51*(9). https://doi.org/10.1021/acs.est.7b00701

Yan, S., Henninger, C. E., Jones, C., & McCormick, H. (2020). Sustainable knowledge from consumer perspective addressing microfibre pollution. *Journal of Fashion Marketing and Management, 24*(3). https://doi.org/10.1108/JFMM-08-2019-0181

Yang, L., Kang, S., Wang, Z., Luo, X., Guo, J., Gao, T., Chen, P., Yang, C., & Zhang, Y. (2022). Microplastic characteristic in the soil across the Tibetan Plateau. *Science of the Total Environment, 828*. https://doi.org/10.1016/j.scitotenv.2022.154518

Young, S. (2021, May 28). *The fabrics with the worst environmental impact revealed, from polyester to fur.* https://www.independent.co.uk/climate-change/sustainable-living/fast-fashion-sustainable-worst-fabrics-b1855935.html

Yu, X., Ladewig, S., Bao, S., Toline, C. A., Whitmire, S., & Chow, A. T. (2018). Occurrence and distribution of microplastics at selected coastal sites along the southeastern United States. *Science of the Total Environment*, 613–614. https://doi.org/10.1016/j.scitotenv.2017.09.100

Zambrano, M. C., Pawlak, J. J., Daystar, J., Ankeny, M., Cheng, J. J., & Venditti, R. A. (2019). Microfibers generated from the laundering of cotton, rayon and polyester based fabrics and their aquatic biodegradation. *Marine Pollution Bulletin*, *142*, 394–407. https://doi.org/10.1016/j.marpolbul.2019.02.062

Zambrano, M. C., Pawlak, J. J., Daystar, J., Ankeny, M., Goller, C. C., & Venditti, R. A. (2020). Aerobic biodegradation in freshwater and marine environments of textile microfibers generated in clothes laundering: Effects of cellulose and polyester-based microfibers on the microbiome. *Marine Pollution Bulletin*, *151*. https://doi.org/10.1016/j.marpolbul.2019.110826

Zambrano, M. C., Pawlak, J. J., Daystar, J., Ankeny, M., & Venditti, R. A. (2021). Impact of dyes and finishes on the microfibers released on the laundering of cotton knitted fabrics. *Environmental Pollution*, *272*. https://doi.org/10.1016/j.envpol.2020.115998

Zhang, J., Ding, W., Zou, G., Wang, X., Zhao, M., Guo, S., & Chen, Y. (2023a). Urban pipeline rainwater runoff is an important pathway for land-based microplastics transport to inland surface water: A case study in Beijing. *Science of the Total Environment*, *861*. https://doi.org/10.1016/j.scitotenv.2022.160619

Zhang, Q., Liu, L., Jiang, Y., Zhang, Y., Fan, Y., Rao, W., & Qian, X. (2023b). Microplastics in infant milk powder. *Environmental Pollution*, *323*. https://doi.org/10.1016/j.envpol.2023.121225

Zhang, Z., Peng, W., Duan, C., Zhu, X., Wu, H., Zhang, X., & Fang, L. (2022). Microplastics pollution from different plastic mulching years accentuate soil microbial nutrient limitations. *Gondwana Research*, *108*. https://doi.org/10.1016/j.gr.2021.07.028

Zhang, J., Wang, L., Halden, R. U., & Kannan, K. (2019). Polyethylene terephthalate and polycarbonate microplastics in sewage sludge collected from the United States. *Environmental Science and Technology Letters*, *6*(11). https://doi.org/10.1021/acs.estlett.9b00601

Zhao, S., Zhu, L., & Li, D. (2016). Microscopic anthropogenic litter in terrestrial birds from Shanghai, China: Not only plastics but also natural fibers. *Science of the Total Environment*, *550*. https://doi.org/10.1016/j.scitotenv.2016.01.112

Zhao, S., Zhu, L., Wang, T., & Li, D. (2014). Suspended microplastics in the surface water of the Yangtze Estuary System, China: First observations on occurrence, distribution. *Marine Pollution Bulletin*, *86*(1–2), 562–568. https://doi.org/10.1016/j.marpolbul.2014.06.032

Zhou, H., Zhou, L., & Ma, K. (2020). Microfiber from textile dyeing and printing wastewater of a typical industrial park in China: Occurrence, removal and release. *Science of the Total Environment*, *739*. https://doi.org/10.1016/j.scitotenv.2020.140329

Zhu, L., Zhu, J., Zuo, R., Xu, Q., Qian, Y., & AN, L. (2023). Identification of microplastics in human placenta using laser direct infrared spectroscopy. *Science of the Total Environment*, *856*, 159060. https://doi.org/10.1016/J.SCITOTENV.2022.159060

Ziajahromi, S., Kumar, A., Neale, P. A., & Leusch, F. D. L. (2017). Impact of microplastic beads and fibers on waterflea (Ceriodaphnia dubia) survival, growth, and reproduction: Implications of single and mixture exposures. *Environmental Science and Technology*, *51*(22). https://doi.org/10.1021/acs.est.7b03574

2 Impact of Artificial Ageing on Microfibre Release from Polyester Textiles

Tihana Dekanić, Ana Šaravanja, and Tanja Pušić
University of Zagreb

Julija Volmajer Valh
University of Maribor

2.1 INTRODUCTION

Globally, the use of plastics is increasing year by year, and figures from 2019 show that the production of plastics has exceeded 368 million tonnes (Yee et al. 2021). The term *plastic* permeates every aspect of human life and the environment. Plastic is composed of natural materials and man-made polymers formed in various chemical processes. Due to its characteristic properties such as lightweight, high durability, corrosion and solar radiation resistance, and plasticity in variants with low processing and production costs, the use of plastics is rapidly increasing both in everyday life and in various applications (Sillanpää and Sainio 2017).

2.1.1 Plastic

Plastic is a synthetic polymer that consists of many repeating monomers, but also contains various additives such as plasticizers, UV blockers, dyes and fillers (Sillanpää and Sainio 2017). The most important processes in plastic production are addition and condensation polymerization, in which polymers are formed into polymer chains (Shrivastava 2018). The process is reversible, and all plastics that undergo various chemical processes can be recycled into new plastics by creating a new form of plastic. The resulting residue cannot be properly disposed of. Over time, the waste is exposed to chemical, biological and atmospheric influences and breaks down into irregular or regular fragments, e.g. microplastics (size <5 mm) and nanoplastics (<0.1 μm).

Due to the decomposition of microplastics into nanoplastics, there is a risk that nanoplastic pollution could dominate the world in the future, as shown in Figure 2.1 (Hernandez et al. 2017, Halle et al. 2017, Gigault et al. 2016). Floating microplastics

DOI: 10.1201/9781003331995-3

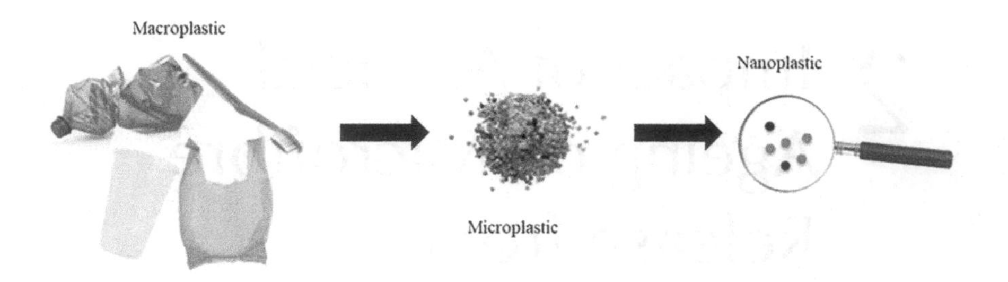

FIGURE 2.1 Illustration of the degradation of plastic to nanoplastic.

are among the most widespread pollutants in the aquatic environment. Due to their extremely small size and density, aquatic organisms consume microplastics. Research shows that nearly 700 aquatic organisms are threatened by microplastic ingestion (Isaac and Balasubramanian 2021, Marn et al. 2020).

Microplastics are of particular concern because their tendency to bioaccumulate increases with decreasing size. Small microplastic particles have a great tendency to adsorb hydrophobic substances (Sillanpää and Sainio 2017). Two forms of microplastics are most commonly cited: *primary* and *secondary.* The primary form consists of microgranules found in cosmetics. The secondary form is created by the decomposition process of larger plastic parts such as polyethylene terephthalate (PET) bottles or by the abrasion of synthetic textiles. Due to human carelessness, microplastics end up in wastewater, sewage, seas and oceans (Isaac and Balasubramanian 2021, Marn et al. 2020).

2.1.2 Textiles

There are several sources of microfibre pollution, and textiles are one of the main sources. Textiles are ubiquitous around us, from clothing, nonwovens, the food industry to the automotive industry. When talking about the estimates of the release of microfibres from textiles, only synthetic textiles are meant. Natural materials such as cotton, linen and silk are usually biodegradable, while non-biodegradable synthetic materials and their blends tend to release microplastics.

Polyesters (PES) belong to a group of fibres composed of macromolecules with a linear structure formed by ester bonds (-CO-O-) linking constitutional units (Rathinamoorthy and Balasaraswathi 2020). PET is the most widely used polymer in the group of PES fibres and is used as thermoplastic polymers in the packaging and textile industries (Hernandez et al. 2017). In 2011, Browne et al. studied the amount of microfibres released from various PES textile products (Browne et al. 2011). The study was based on using different types of laundry in a washing machine without fabric softener or detergent. The results showed that over 1,900 fibres can be removed from clothes per wash cycle. In 2013, Dubais and Liebezeit reported that between 0.033% and 0.039% of fibre mass per laundry weight (w/w) is released from PES materials during the washing process (Dubaish and Liebezeit 2013).

Previous research has studied the negative impact of microplastics on the environment. Certain competent institutions have proposed concrete measures and suggested that in future an appropriate classification of the harmfulness of plastic should be introduced (Browne et al. 2011, Dubaish and Liebezeit 2013, Rochman et al. 2013). Microplastics released from textiles can be in the form of fibres and are called microfibres (Šaravanja et al. 2022). They are present in every step of the production of flat products, from the chemical spinning of the polymer to the method of weaving the threads. Microfibres are released by wearing textiles, during the washing process (the release of the microfibres will decrease as the number of washes increases) and under numerous conditions of use and application. It has been shown that the drying process is one of the main causes of microfibre release, not only from the material inside the machine, but also from microdust particles that have been found in the environment (Rathinamoorthy and Balasaraswathi 2020). For example, the study by Abbasi et al. (2019) showed that a child can absorb more than 900 microfibre particles into their body through dust every year. Textile manufacturing and industrial and household washing are the primary sources of microfibres that are released from textiles. More fibres are released by hydrophilic fibres than by synthetic ones, and their strength can also have an impact on tear (Palacios-Mateo et al. 2021). Zambrano et al. concluded that fabrics with higher abrasion resistance, stronger yarns and lower hairiness are desirable factors for reducing fibre shedding during washing (Zambrano et al. 2019). The amount of microfibres released is affected by the type of construction. The fabric should have a compact weave and as high density as possible (Choi et al. 2021).

2.1.3 Ageing

In a comprehensive study of polymer life, the effects of external factors acting on a polymer must be analysed and quantified (Frigione et al. 2021). The natural ageing of a polymer is the only approach that can guarantee the reliability of the results. In science, natural ageing is a term used when polymer properties change over a period of time. However, to obtain reliable data and conclusions, a long analysis time is required when the polymer is exposed to natural conditions. Therefore, more and more scientists are turning to an alternative – the process of artificial ageing. It aims to make predictions about the longevity of polymers, i.e. to evaluate and predict the long-term properties and characteristics of polymers based on one or more environmental factors that represent the real exposure of polymers under real conditions (Frigione et al. 2021). The real conditions, the so-called weathering, combine UV radiation, humidity and heat and vary based on the country, location, climate zone and season.

Natural and synthetic polymers and materials exhibit their own degree and type of degradation during their lifetime. The result is a change in their thermal, optical, mechanical and functional properties, and these changes reduce or increase their lifetime (Wypych 2018, Singh and Sharma 2008, Liu et al. 2021, Mohammadi et al. 2021, Deshoulles et al. 2021, Kotanen et al. 2021, Asadi et al. 2021). All these changes affect the long-term behaviour of the polymer. The same polymers have been shown to perform differently when exposed to different conditions, whether external or internal. The most important factors that have a direct influence on the properties and characteristics of polymers are those that occur every day under all working conditions. Some of these include water, vapour, oxygen, temperature, UV radiation, solar radiation, chemicals, alkaline or acidic environment, different pollutants and fuels. All polymeric materials have their own specific application or task they have to fulfil (Deshoulles et al. 2021, Kotanen et al. 2021, Asadi et al. 2021). For example, automotive components must be made of thermoplastic polymers. However, in the real atmosphere and under real working conditions, various factors may act simultaneously. Therefore, polymers must be resistant to as many synergistic effects as possible, such as the combination of UV radiation and temperature. When polymers are exposed to the direct effects of high temperatures and UV radiation, they degrade more rapidly (Ishida and Kitagaki 2021). It is known that not all external conditions are constant and predictable and that they depend on various factors related to the place of exposure, such as latitude, altitude and season.

2.1.4 Ageing Simulation

Appropriate environmental chambers and specific equipment are capable of simulating the environmental effects. The conditions in xenon chambers are a rapid simulation of the real long-term environmental conditions to which polymers are exposed in nature, thus mimicking the real conditions. The positive aspect is that such accelerated procedures are increasingly attractive to manufacturers of new smart materials interested in rapid analysis and feedback on them. It should be kept in mind, however, that some of the procedures and combinations of accelerated ageing simulations used may never occur under real-world conditions, so some of these combinations should be viewed with some caution (Frigione et al. 2021).

By varying the time of exposure and combining changes in atmospheric conditions and the strength of exposure, it is easier to obtain and notice information about what is happening to the material and the release of microplastics during the natural or artificial ageing process. From the literature, the ageing process causes the degradation of PES at the molecular level (Lemmi et al. 2021). According to the theory, there are two forms of ageing: chemical and physical. The chemical ageing process occurs as a reaction with external agents such as UV radiation, oxygen, ionizing radiation and water. Physical ageing is the most common form of ageing, in which changes occur in the configuration chain or intermediate molecules that have been modified without changing the chemical structure of the polymer, and changes in composition, such as water absorption, are observed (Lemmi et al. 2021).

The parameters of artificial thermal ageing affect the physical and mechanical properties of the material. From various sources, it can be concluded that the thermal strain and strength are inversely proportional to the exposure temperature in the artificial ageing process. The effects of artificial ageing on materials are increasingly being researched, and ageing affects the degree of crystallinity and other material surface changes associated with the release of microplastics. Defining exposure parameters for artificial ageing is the first step in starting the artificial ageing process (Gandhi et al. 2019). The defined factors should have the task of activating the decomposition mechanism as similar as possible to natural decomposition under normal conditions, in a shorter exposure time (Šaravanja et al. 2022). All accelerated ageing processes use lamps that simulate radiation. These can be mercury, xenon, fluorescent light sources, metal halide, etc. (Tocháček and Vrátníčková 2014). They are the main and responsible light source types for polymer degradation via photochemical decomposition. The best selection, in terms of solar radiation reproducibility, is provided by a xenon light source in the wavelength range of 290–440 nm (Tocháček and Vrátníčková 2014, Kuvshinnikova et al. 2019). The actual principle of the degradation mechanism occurs when the polymer or material is exposed to radiation. The radiation depends on the energy and wavelength of the radiation, but also on the chemical structure of the material or polymer (Pickett et al. 2008). Considering the effects of the water environment and/or rain, deionized water is used in artificial ageing devices (Niemczyk et al. 2019), but polymers are exposed to water (rainwater) in real outdoor conditions, which cannot be compared to the water used in the artificial ageing device (Azuma et al. 2009). To promote polymer degradation, temperature is used to accelerate artificial ageing. To avoid changes that do not occur even under real conditions, it must not exceed or reach the glass transition temperature, the temperature at which the polymer melts (Šaravanja et al. 2022).

It is known that PES sheds microfibres when washed. Accordingly, the focus is on the interaction of PES and water. Below the glass transition temperature, T_g, physical ageing takes place in amorphous parts of the polymers. Penetration of water into the interior of the polymer causes the glass transition temperature to drop. Water reacts with ester groups, so it is a kind of physical ageing with water. It is a reversible process accompanied by irreversible damage that occurs at the fibre interface or in the fibre matrix. Thermal ageing affects the mechanical and physical properties of the material. Strength and thermal elongation are inversely proportional to ageing temperature (Lemmi et al. 2021). Artificial ageing at higher temperatures leads to the weakening of the fibre structure and increased release of microfibres. The aim of this study was to investigate the effects of artificial ageing on a standard PES fabric during exposure to different rain–sun changes and direct sunlight. Given all these facts, the concept of microplastic and its release from PES became the main topic of this research, which is an important environmental issue. Therefore, this research focuses on two theses: how the process of artificial ageing affects the PES structure and what exactly happens to PES when it is exposed to artificial ageing.

2.2 EXPERIMENTAL

Standard PES fabric (PES_N), supplied by Center for TestMaterials, CFT, Netherlands, was used. Fabric structural properties were as follows: plain weave, mass per unit area 156.0 g/m^2, thickness 0.35 mm, warp/weft density 27.7/20.0, warp/weft fineness 30.4/31.9 tex and white colour. The fabric was subjected to artificial ageing simulation over a period of 85 hours in the Xenotest 440 apparatus, SDL Atlas, in accordance with the ISO 105-B04:1994 standard. The fabric is exposed to two different condition simulations of ageing: to direct irradiance without spraying and to simultaneously rain–sun weathering cycle. To define the exposure conditions, the preferred conditions of the normal temperate zone in Europe were used. For this purpose, 1 minute of spraying and 29 minutes of irradiation, with moderate effective humidity of about 40% and an irradiance of 42 ± 2 W/m^2 in the wavelength range 300–400 nm, were chosen. Different exposure conditions and fabric types used in this analysis are provided in Table 2.1.

TABLE 2.1
Legend and Description of the Samples

Label	Time and Exposure Method	Sample
PES_N	0	Standard polyester fabric
PES_85H	85 h, simulation of sunlight	Standard polyester fabric subjected to an ageing process by direct irradiance for 85 hours
PES_K/S_85H	85 hours, simulation of the effects of a combination of rain/sunlight (weathering cycle)	Standard polyester fabric subjected to an 85-hour rain/sun ageing process

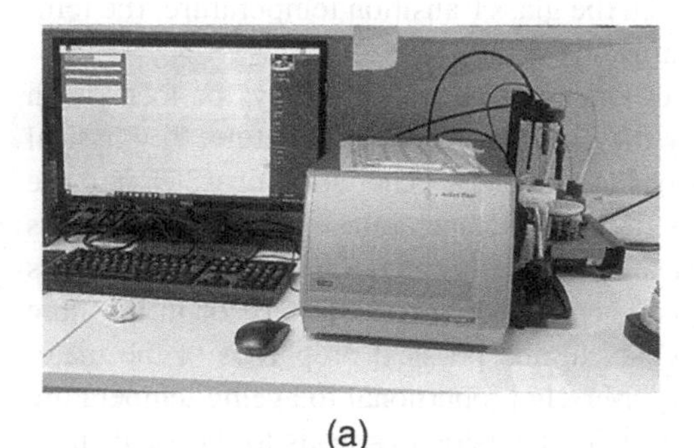

(a)

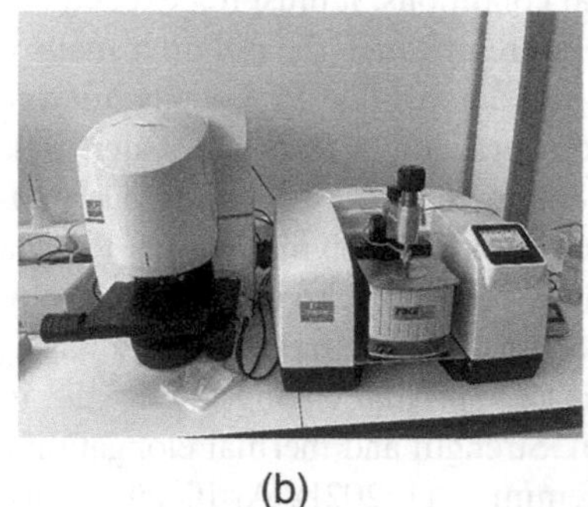

(b)

FIGURE 2.2 Devices for characterizing the samples (a) SurPASS 3, (b) FTIR, Microscope Spotlight 200i.

The characterization of control fabric and aged fabrics was performed on an electrokinetic analyser SurPASS 3, Anton Paar (Figure 2.2a), via zeta potential (ZP, ζ). The ZP of fabrics was measured in the dependence of pH electrolyte (0.001 mol/L KCl), in the pH range from 9 to 2.

The physicochemical properties of control and aged fabrics were analysed using the attenuated total reflection Fourier transform infrared (ATR FTIR) spectroscopy connected to a Spotlight 200i microscope, PerkinElmer (Figure 2.2b). The PerkinElmer Spectrum program IR created a curve in the wavenumber range of $\nu = 4{,}000–650\,cm^{-1}$. It was measured with FTIR and a liquid nitrogen microscope. The observed area was 100×100 µm. It is necessary to adjust the brightness and contrast of the image to obtain the best possible focus of the observed samples.

The physical properties of control and aged fabrics were measured through air permeability, thickness, surface mass, breaking force and elongation. For measuring air permeability, Karl Schoeder KG, Material Testing Machines, D-6940 Weinhein device was used. The working principle was to place a sample on the surface of the device on a circular opening with an area of $20\,cm^2$, lower the holder onto the sample and tighten it. A certain airflow was set, and the fivefold air permeability corresponding to the sample was measured, as shown in Figure 2.3a. The factor (f) taken into

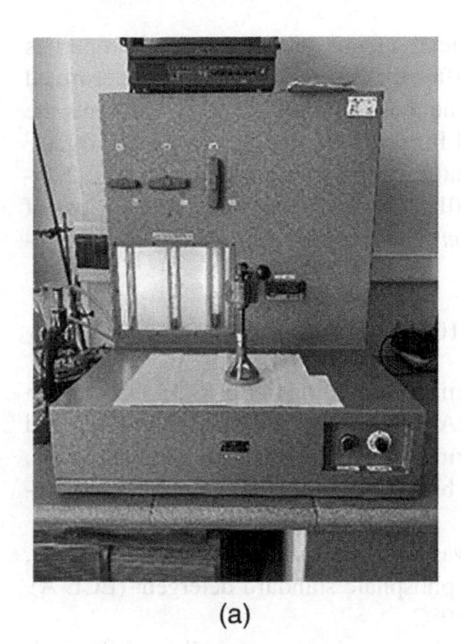

(a)

(b)

FIGURE 2.3 Devices for characterizing the samples (a) air permeability device *Karl Schöder KG, Materialprüfmaschinen, D-6940 Weinhein*, (b) thickness gauge.

account to obtain the permeability value depends on the surface and is 1 for this measuring area of 20 cm². The conditions during the measurement were a pressure of 1,016 mbar and a room temperature of 22°C.

$$VN = f\ VG\sqrt{\frac{PU*TN}{PN*TU}} \tag{2.1}$$

where is test working conditions are as follows:

- Temperature: 22°C (297 K)
- Air pressure in the space: 1,007 mbar
- Read air permeability: 250 L/m² × sp* (*on the instrument)
- PN – reference air pressure 1,013 mbar
- PU – air pressure in the space 1,007 mbar deduction 2 mbar (1,005 mbar)
- TN – reference temperature 20°C
- TU – room temperature 23°C
- VG – read air permeability 250 L/m² × s
- VN – reference air permeability v L/m² × s.

The thickness of the control fabric before and after ageing was determined with a thickness gauge, as shown in Figure 2.3b, according to ISO 5084:1977. The principle of operation was as follows: a sample was placed on the device, a weight of 125 g was applied, and the value was read. The value read was then divided by 10.

In order to analyse the influence of artificial ageing on the structural properties of standard PES, surface mass before and after the ageing process was determined according to the standard ISO 3801:2003. The mass per unit area (Q) represents the mass of one square meter and is expressed in g/m^2. A circular sample of 100 cm^2 area, preconditioned, was weighed on an analytical balance Kern, model ALJ 220–5DNM, with a precision accuracy of ±0.001 g, ensuring that there was no loss of threads. From the mass of weighed sample (m), the mass per unit area (Q) was calculated using the following formula:

$$Q = m \cdot 100 \quad (2.2)$$

Mechanical properties of fabrics were determined according to EN ISO 13934-1:2013 on a Tensolab Strength Tester, Mesdan S.P.A. by determining maximum force and elongation. The measurements were performed on three strips in warp direction, with a distance between clamps of 100 mm, bursting speed of 100 mm/min and pre-tension of 2 N.

All PES fabrics were subjected to ten washing cycles in Rotawash, SDL Atlas, according to EN ISO 6330:2012 using non-phosphate standard detergent (ECE A), with a liquid ratio of 1:8, for 30 minutes at 60°C.

After ten washing cycles, water from washing and rinsing was collected. The collected effluent was subjected to a membrane filtration. The filtrate was obtained by membrane filtration of the effluent, using a 0.2 μm pore diameter polyethersulfone filter. For the purpose of analysing the filter, a paired device FT-IR Microscope from PerkinElmer was used.

2.3 RESULTS AND DISCUSSION

In this study, the influence of different types of exposure on the properties of PES fabrics was analysed. For this purpose, PES fabrics were subjected to 85 hours of artificial ageing simulation, which corresponds to 850 hours under real-world conditions. In order to gain valuable insights, two exposure methods were chosen, namely direct sunlight simulation and rain–sunlight exchange. The effects of weathering and sunlight were observed in comparison with the control PES fabric.

2.3.1 Zeta Potential of Polyester Fabrics

The ZP is interesting from the point of view of surface modification as it indicates the behaviour of the material in a given medium and the surface or structural changes that occur as a result of the application of various treatments and also indicates the charge on the surface of the fabric itself. Most textile materials exhibit a negative charge in aqueous solutions, characterized by ZP, which depends on the number and availability of reactive groups (Luxbacher et al. 2016). The graphical representation shows that the ZP of PES_N is negative in the alkaline pH range and is −23.40 mV, as shown in Figure 2.4. The ZP curve of the tested control fabric shows values slightly higher than the usual values of the ZP of PES, for which previous studies gave a value between −40 and −80 mV

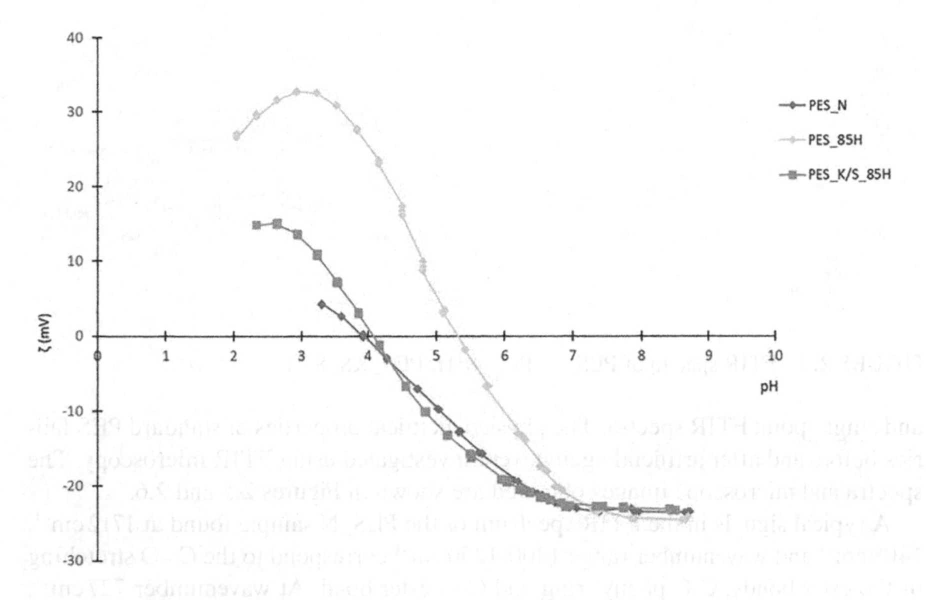

FIGURE 2.4 Zeta potential of polyester fabrics in variation of pH KCl.

(Jacobasch et al. 1985, Pušić et al. 2011, Kaurin et al. 2022). The obtained result indicates that some preparations were present on the fabric and thus prevented the complete dissociation of the -COOH groups from the applied PES fabric (Kaurin et al. 2022). The ZP of PES_85H is −24.08 mV, while it is −23.21 mV for PES_K/S_85H in alkali conditions. The difference between the aged and the original fabric is almost negligible in the alkaline pH range. When the pH of the electrolyte is lowered, ZP differences occur. Already at pH 5.09, PES_85H reaches the isoelectric point and becomes positive, while for PES_K/S_85H, the isoelectric point shifts by a whole unit (pH 3.85). The isoelectric point of PES_N is 3.92, and the curve of ZP change as a function of the pH of the electrolyte is almost the same as that of PES_K/S_85H. The difference in modification can be seen in the shift of the isoelectric point, which confirms the modification of the fabric (85H) compared to the initial one.

The combined exposure to rain/sun leads to a decrease in the ZP compared to PES_85H, as a result of more accessible functional carboxyl groups (-COOH). The reason for this change is the effect of water during the simulation of rain/sun exchange. Namely, during the rain/sun exchange, water washes out the sun-degraded part and hydrolyses the surface of the PES fabric, making it more accessible, the so-called mechanism of surface erosion of the layer, i.e. hydrolysis of the ester bonds of the PES fabric (Woodard and Grunlan 2018).

2.3.2 Fourier Transform Infrared Spectroscopy (FTIR) Microscopy

The physicochemical properties of standard PES fabrics before and after artificial ageing were investigated using ATR FTIR spectroscopy and FTIR microscopy. The ATR FTIR spectra are shown in Figure 2.5. Figure 2.6 shows microscope images

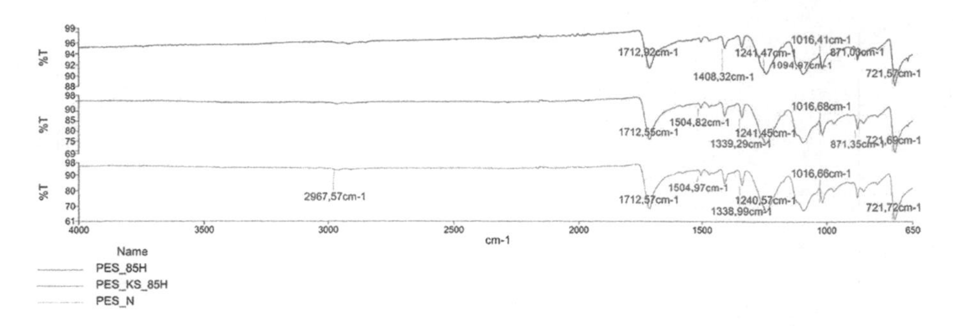

FIGURE 2.5 FTIR spectra of PES_N, PES_85H, PES_KS_85H.

and single-point FTIR spectra. The physicochemical properties of standard PES fabrics before and after artificial ageing were investigated using FTIR microscopy. The spectra and microscope images obtained are shown in Figures 2.5 and 2.6.

A typical signals in the FTIR spectrum of the PES_N sample found at 1712 cm^{-1}, 1408 cm^{-1} and wavenumber range 1000-1250 cm^{-1} correspond to the C=O stretching of the ester bonds, C-C phenyl ring and C-O ester bond. At wavenumber 727 cm^{-1}, a band is seen for the C-H functional group, while at larger wavenumbers, i.e. 2,976 cm^{-1}, it corresponds to the -CH ethyl group (Cecen et al. 2008, Valh et al. 2020, Pereira et al. 2017, Volmajer Valh et al. 2022).

No particular changes were observed in the FTIR spectra of the other two aged samples PES_85H and PES_KS_85H. However, a smaller shift of the C-O ester functional group to larger wavenumbers (1,241 cm^{-1}) and small changes in the region around 2,900 cm^{-1} could be due to the oxidation process.

The differences in pore sizes of PES fabrics are slightly visible under the microscope. PES_N showed larger pores, with an average pore size of 500 μm, than artificially aged fabrics, which showed an average pore size of 300 μm. In addition to the microscope images of the control fabric before and after artificial ageing, the FTIR spectrum was obtained. The signals obtained with the microscope Spotlight 200i are similar to the signals measured with ATR FTIR only.

For the characterization of the filtrate and analysis of the presence of microfibres, a magnification of 5,000 × 5,000 μm was used. Microscopic observation shows microfibres. The presence of microfibres increases with ageing, confirming that ageing has an important influence on the release of microfibres.

2.3.3 Air Permeability and Breaking Force

The results obtained by measuring the air permeability and breaking force are shown in Table 2.2.

The air permeability results indicate that the PES fabric is different before and after artificial ageing. The air permeability of the aged fabric remained the same (~620 J/m^2·s) and decreased compared to the untreated one. PES_N showed an air permeability of 866.19 J/m^2·s. The results can be related to the results of the microscope, where the differences between the pores of the fabrics were observed. In aged fabrics, the pore sizes were smaller compared to PES_N, and the air permeability

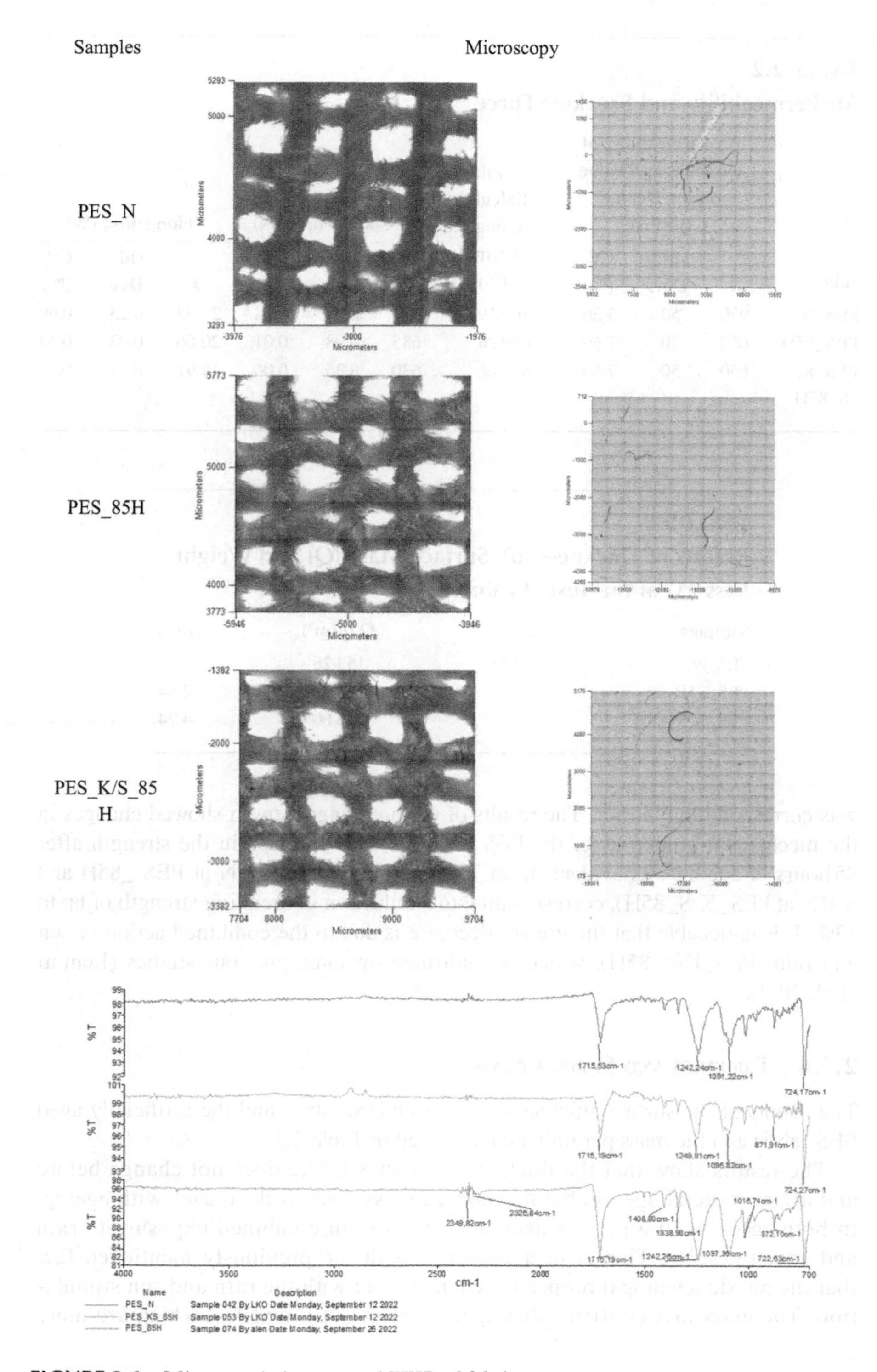

FIGURE 2.6 Microscopic images and FTIR of fabrics.

TABLE 2.2
Air Permeability and Breaking Force

	Air Permeability of the Measured Value on the Device (J/m²·s)			Value Calculated Using the Formula	Breaking Force (N)			Elongation (%)		
Label	X	std. Dev.	CV (%)	(J/m²·s)	X	std. Dev.	CV (%)	X	std. Dev.	CV (%)
PES_N	950	50	5.26	866.19	726	0.26	0.04	23.33	0.23	0.99
PES_85H	650	50	7.69	620.66	685	0.08	0.01	20.06	0.03	0.14
PES_K/S_85H	650	50	7.69	620.66	630	0.35	0.06	19.91	0.11	0.57

TABLE 2.3
Results of Thickness (*d*), Surface Mass (*Q*) and Weight Loss (Δ) of the Tested Fabrics

Samples	*d* (mm)	*Q* (g/m²)	Δ (%)
PES_N	0.34	153.26	–
PES_85H		148.90	2.84
PES_K/S_85H		146.00	4.74

was correspondingly lower. The results of the breaking strength showed changes in the mechanical properties of the PES material. The decrease in the strength after 85 hours of ageing was evident (from 726 N at PES _N to 685 N at PES _85H and 630 N at PES_K/S_85H), corresponding to a reduction in breaking strength of up to 13%. It is noticeable that the greater decrease is due to the combined action of sun and rain (PES_K/S_85H), which is confirmed by some previous studies (Lemmi et al. 2021).

2.3.4 Thickness and Surface Mass

The results of the surface thickness of the standard fabric and the artificially aged PES fabric and the mass per unit area are listed in Table 2.3.

The results show that the thickness of a PES fabric does not change before and after artificial ageing, but the surface mass does. It decreases with ageing, in both cases, with a greater decrease observed in combined exposure to rain and sun (by 40%). This is in accordance with the previously mentioned fact that the physicochemical properties change more with the rain and sun stimulation. The mass loss confirms that ageing affects PES textiles, which are more pronounced due to the combined effect. No changes were observed depending on the artificial ageing method.

2.4 CONCLUSION

Plastic, microplastics and nanoplastics have become a serious problem for human health and the environment. In recent years, this issue has become increasingly important. When evaluating the influence of artificial ageing on the release of microfibres from PES fabrics, the influence of water (moisture) proved to be important. The ZP of the sun-treated standard PES fabric was different from the ZP of the standard PES fabric exposed to the combined effect. Due to the influence of water during sun exposure, more available -COOH groups were formed in the fabric, which is a consequence of the hydrolysis of ester bonds in the PES fabric. The 85-hour sun exposure significantly changed the surface structure compared to the untreated fabric. The difference between the aged and the original fabric was negligible in the alkaline pH range. They were only noticeable in the acidic range, at a pH value below 4. Differences in isoelectric points indicate a difference between aged samples.

Infrared spectroscopy was performed to determine the presence of typical functional groups of PES fabrics before and after artificial ageing. No significant changes in the position of the signals were detected regardless of artificial ageing. A slight change in the shift of the C-O ester functional group to larger wavenumbers and small changes in the region around 2,900 cm^{-1} could be due to the oxidation process.

The focus of this research can be summarized through two aspects: how artificial ageing affects the structure of PES and whether it changes PES textiles and mechanical properties. It can be concluded that some changes occur, but the sun exposure time of 85 hours, which is equivalent to the sun exposure time of 850 hours under real-world conditions (Dekanić et al. 2014), is not enough to accurately detect the changes. The combined effect is stronger than the single effect and confirms that changes occur at the molecular level (Lemmi et al. 2021). At the molecular level, changes such as the cleavage of bonds or the oxidation of long polymer chains can occur, leading to the formation of new molecules, usually with significantly shorter chain lengths.

The characterization of the filter shows the presence of microfibres and other fragments or impurities from the environment. It is believed that this result is a good starting point for further research on how the ageing process affects the release of microfibres.

Considering that in the natural environment it takes a long time for some changes, mechanical or chemical, to occur in polymeric materials, including PES textiles, devices that simulate accelerated ageing are used to obtain data quickly. However, it should be considered that artificial ageing simulation is only a simulation. Not all factors (temperature, UV radiation, rain, etc.) act equally in artificial ageing, so it is necessary to define the factors precisely. They are suitable for any type of polymer and for any climate zone.

ACKNOWLEDGMENT

The work of doctoral student Ana Šaravanja has been supported in part by the "Young researchers' career development project – training of doctoral students" of the Croatian Science Foundation (HRZZ- DOK-2021-02-6750).

The work has been supported by Croatian Science Foundation under the project IP-2020-02-7575, InWaShed-MP, and partly supported by the Slovenian Research Agency (research programme P2-0118).

REFERENCES

Abbasi S et al. 2019. Distribution and potential health impacts of microplastics and microrubbers in air and street dusts from Asaluyeh County, Iran. *Environmental Pollution* 244, 153–164.

Asadi H et al. 2021. Artificial weathering mechanisms of uncoated structural polyethylene terephthalate fabrics with focus on tensile strength degradation. *Materials* 618.

Azuma Y et al. 2009. Outdoor and accelerated weathering tests for polypropylene and polypropylene/talc composites: A comparative study of their weathering behavior. *Polymer Degradation and Stability* 94, 2267–2274.

Browne M A et al. 2011. Accumulation of microplastic on shorelines worldwide: Sources and sinks. *Environmental Science & Technology* 45, 9175–9179.

Cecen V et al. 2008. FTIR and SEM analysis of polyester-and epoxy-based composites manufactured by VARTM process. *Journal of Applied Polymer Science* 108, 2163–2170.

Choi S et al. 2021. Characterization of microplastics released based on polyester fabric construction during washing and drying. *Polymers* 13, 4277.

Dekanić T et al. 2014. Impact of artificial UV-light on optical and protective effects of cotton after washing with detergent containing fluorescent compounds. *Tenside Surfactants Detergents* 51, 451–459

Deshoulles Q et al. 2021. Origin of embrittlement in polyamide 6 induced by chemical degradations: Mechanisms and governing factors. *Polymer Degradation and Stability* 191, 109657.

Dubaish F, Liebezeit G. 2013. Suspended microplastics and black carbon particles in the Jade system, southern North Sea. *Water, Air, & Soil Pollution* 224, 1–8.

Frigione M et al. 2021. Can accelerated aging procedures predict the long term behavior of polymers exposed to different environments? *Polymers* 13, 2688.

Gandhi K et al. 2019. Acceleration parameters for polycarbonate under blue LED photo-thermal aging conditions. *Polymer Degradation and Stability* 164, 69–74.

Gigault J et al. 2016. Marine plastic litter: The unanalyzed nano-fraction. *Environmental Science: Nano* 3, 346–350.

Halle A T et al. 2017. Nanoplastic in the North Atlantic subtropical gyre. *Environmental Science & Technology* 51, 13689–13697.

Hernandez L M et al. 2017. Are there nanoplastics in your personal care products? *Environmental Science & Technology Letters* 4, 280–285.

Isaac M N, Balasubramanian K. 2021. Effect of microplastics in water and aquatic systems. *Environmental Science and Pollution Research* 28, 19544–19562.

Ishida T, Kitagaki R. 2021. Mathematical modeling of outdoor natural weathering of Polycarbonate: Regional characteristics of degradation behaviors. *Polymers* 13, 820.
ISO 5084:1977- Textiles – Determination of thickness of woven and knitted fabrics (other than textile floor coverings)
ISO 3801:1977- Textiles – Woven fabrics – Determination of mass per unit length and mass per unit area.
Jacobasch H-J et al. 1985. Problems and results of zeta-potential measurements on fibers. *Colloid and Polymer Science* 263, 3–24.
Kaurin T et al. 2022. Biopolymer textile structure of chitosan with polyester. *Polymers* 14, 3088.
Kotanen S et al. 2021. Hydrolytic stability of polyurethane/polyhydroxyurethane hybrid adhesives. *International Journal of Adhesion and Adhesives* 110, 102950.
Kuvshinnikova O et al. 2019. Weathering of aromatic engineering thermoplastics: Comparison of outdoor and xenon arc exposures. *Polymer Degradation and Stability* 160, 177–194.
Lemmi T S et al. 2021. Effect of thermal aging on the mechanical properties of high tenacity polyester yarn. *Materials* 14, 1666.
Liu S et al. 2021. Aging mechanisms of filled cross-linked polyethylene (XLPE) cable insulation material exposed to simultaneous thermal and gamma radiation. *Radiation Physics and Chemistry* 185, 109486
Luxbacher T et al. 2016. The zeta potential of textile fabrics: A review. *Tekstil: časopis za tekstilnu i odjevnu tehnologiju* 65, 346–351.
Marn N et al. 2020. Quantifying impacts of plastic debris on marine wildlife identifies ecological breakpoints. *Ecology Letters* 23, 1479–1487.
Mohammadi H et al. 2021. Constitutive modeling of elastomers during photoand thermo-oxidative aging. *Polymer Degradation and Stability* 191, 109663.
Niemczyk A et al. 2019. Accelerated laboratory weathering of polypropylene composites filled with synthetic silicon-based compounds. *Polymer Degradation and Stability* 161, 30–38.
Palacios-Mateo C et al. 2021. Analysis of the polyester clothing value chain to identify key intervention points for sustainability. *Environmental Sciences Europe* 33, 2.
Pereira A P S et al. 2017. Processing and characterization of PET composites reinforced with geopolymer concrete waste. *Materials Research* 20, 411–420.
Pickett J E et al. 2008. Effects of irradiation conditions on the weathering of engineering thermoplastics. *Polymer Degradation and Stability* 93, 1597–1606.
Pušić T et al. 2011. The sorption ability of textile fibres. *Vlákna a textil* 18, 7–15.
Rathinamoorthy R, Balasaraswathi S R. 2020 A review of the current status of microfiber pollution research in textiles. *International Journal of Clothing Science and Technology*.
Rochman C M et al. 2013. Classify plastic waste as hazardous. *Nature* 494, 169–171.
Šaravanja A et al. 2022. Microplastics in wastewater by washing polyester fabrics. *Materials* 15, 2683.
Shrivastava A. 2018. *Introduction to Plastics Engineering*. Elsevier Science.
Sillanpää M, Sainio P. 2017. Release of polyester and cotton fibers from textiles in machine washings. *Environmental Science and Pollution Research* 24, 19313–19321.
Singh B, Sharma N. 2008. Mechanistic implications of plastic degradation. *Polymer Degradation and Stability* 93, 561–584.
Tocháček J, Vrátníčková Z. 2014. Polymer life-time prediction: The role of temperature in UV accelerated ageing of polypropylene and its copolymers. *Polymer Testing* 36, 82–87.
Valh J V et al. 2020. Conversion of polyethylene terephthalate to high-quality terephthalic acid by hydrothermal hydrolysis: The study of process parameters. *Textile Research Journal* 90, 1446–1461.

Volmajer Valh J et al. 2022. Economical chemical recycling of complex PET waste in the form of active packaging material. *Polymers* 14, 3244.

Woodard L N, Grunlan M A. 2018. Hydrolytic degradation and erosion of polyester biomaterials. *ACS Macro Letters* 7, 976–982.

Wypych G. 2018. *Handbook of Material Weathering*, 2nd ed. ChemTec Publishing: Toronto-Scarborough, ON, Canada.

Yee M S et al. 2021. Impact of microplastics and nanoplastics on human health. *Nanomaterials* 11, 496.

Zambrano M C et al. 2019. Microfibers generated from the laundering of cotton, rayon and polyester based fabrics and their aquatic biodegradation. *Marine Pollution Bulletin* 142, 394–407.

3 Factors Influencing Microfiber Emission from Textiles and Methods for Quantifying Microfiber Release during Laundry Processes

Fatemeh Mashhadi Abolghasem, Juhea Kim, and Juran Kim
Korea Institute of Industrial Technology (KITECH)

3.1 INTRODUCTION

Microplastics released into the marine environment are classified as primary and secondary microplastics (Periyasamy & Tehrani-Bagha, 2022). The International Union for Conservation of Nature estimates that among almost 2.5 million tons of shed primary microplastics in the ocean annually, microfibers released due to the laundry process contribute 35% (Boucher & Friot, 2017; Figure 3.1). The type of fabric determines the length, durability, and quantity of the shed microfibers. These fibers are mostly polyesters (PES), notably polyethylene terephthalate (PET) (Ziajahromi et al., 2017). These are the most prevalent fibers in the textile industry, and through washing machine discharge is how they enter wastewater treatment plants (McCormick et al., 2020).

This microfiber release process is influenced by factors such as the fabric age, tightness of the yarn, and chemical composition (Zambrano et al., 2019). Researchers have investigated the influence of various factors, including detergent type, temperature, and washing method. Carney Almroth et al. (2018) used small, specialized simulation machines in which parameters could be set and held constant, while other studies (Kärkkäinen & Sillanpää, 2021) used various types of domestic washers.

Most fiber particles found in water and wastewater originate from household washing machines and abandoned clothing that is dumped in landfills. According to Carney Almroth et al. (2018), microfleece and fleece synthetic textiles released the most fibers (approximately 7,360 fibers/m^2/L) when washed in a laboratory washer, indicating that the fabric design significantly affects the amount of fiber it may shed. Similarly, Napper and Thompson (2016) reported that by using washing machines for

DOI: 10.1201/9781003331995-4

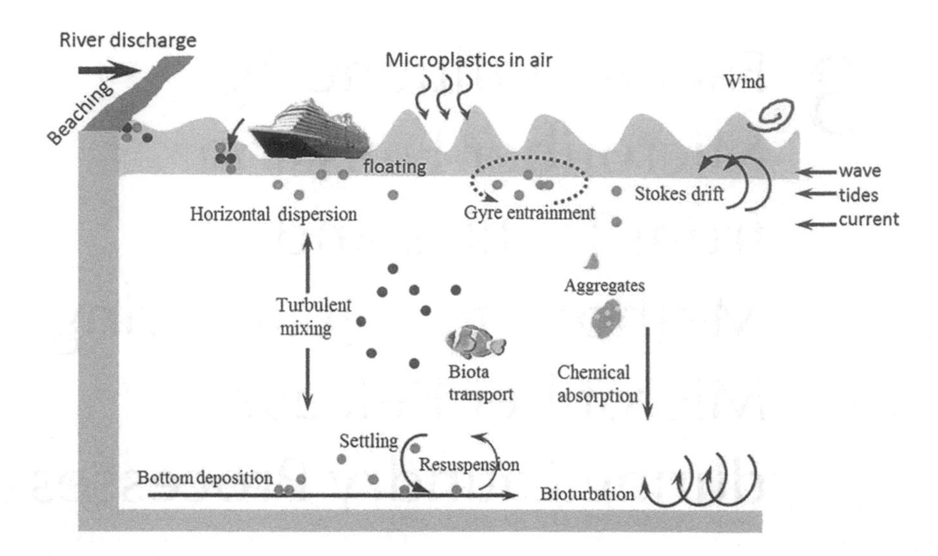

FIGURE 3.1 Scheme depicting how microfibers are generated and their pathways to the marine environment. (*Source*: Adapted with permission from Li et al., 2020, licensed under CC BY 4.0.)

commercial domestic laundering, 6 kg of synthetic textiles (acrylic, PES and cotton mix, and PES) might emit between 140,000 and 700,000 fibers per wash. Browne et al. (2011) claimed that PES apparel can cause the emission of more than 1,900 fibers per laundry cycle. Every garment sheds more than 100 fibers per liter of wastewater, with fleeces producing more than 180% more than normal apparel or blankets (Figure 3.2). According to Hartline et al. (2016), the mass of the microfibers recovered from the tested garments ranged from approximately 0 to 2 g, which is more than 0.3% of the mass of the untreated materials (unwashed garments).

The use of detergents in the laundry process is one of the factors with a significant effect on microfiber emissions, according to Hernandez et al. (2017). Accordingly, using detergents increased the number of microfibers released by approximately 75%, despite the detergent properties (Hernandez et al., 2017). However, Lant et al. (2020) found that the detergent formulation and amount had minor effects on microfiber release. In contrast, De Falco et al. (2018) noted a comparable pattern with woven and knitted PES and polypropylene (PP) materials. Accordingly, the number of microfibers released from garments after laundry has increased as a result of the use of detergents. Moreover, in domestic washing, the powder-type detergent generated more microfiber shedding than the liquid detergent, and overall, the impact of industrial detergents was greater than that of domestic ones (De Falco et al., 2018). Accordingly, employing softeners can reduce the emission of microfibers by 35% (De Falco et al., 2018). Nevertheless, the effect of detergent application has been more persistent and substantial across research than the effect of temperature, as well as using softeners on microfiber emission (De Falco et al., 2018). In another study by O'Brien et al. (2020), using detergents was shown to have no significant effect on the size of the released fibers. However, Cesa et al. (2020) reported that after machine washing, microfiber emissions from synthetic textiles were significantly decreased by detergent usage.

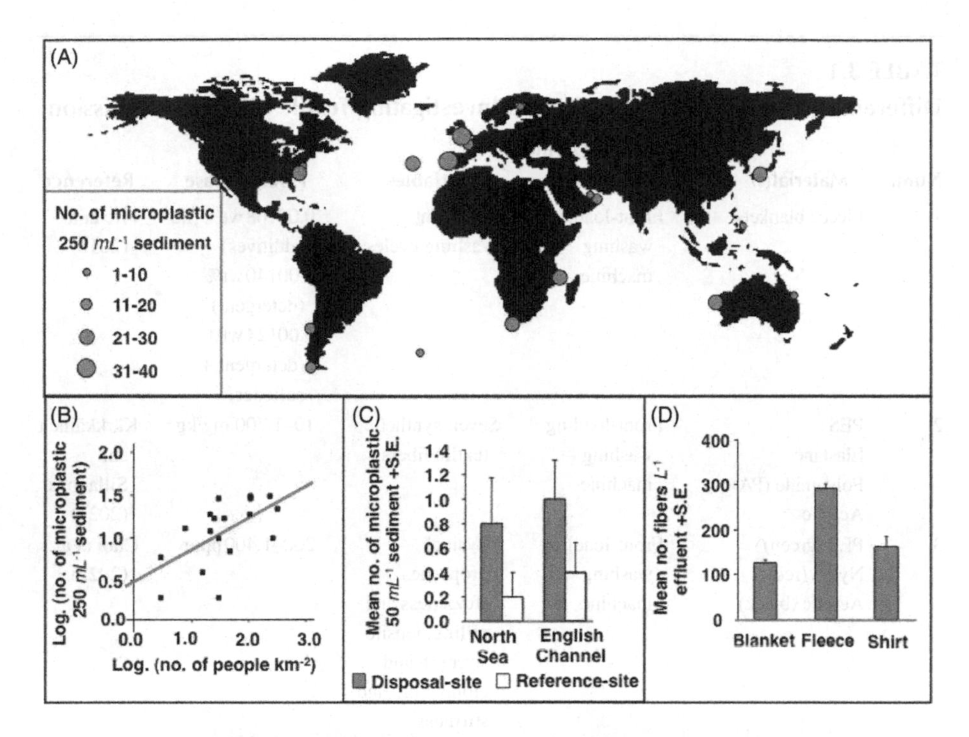

FIGURE 3.2 (a) Global distribution of microplastics identified as plastic by Fourier transform infrared spectrometry in sediments from 18 coastlines. The number of microplastic particles found is represented by the size of the filled circles. (b) Relationship between population size and the number of microplastic particles in shoreline sediment. (c) The number of microplastic particles in sediments from sewage disposal sites and reference sites in the United Kingdom. (d) The amount of PES fibers discharged into wastewater due to washing fleeces, blankets, and shirts in washing machines. (*Source:* Adopted with permission from Browne, M. A., Crump, P., Niven, S. J., Teuten, E., Tonkin, A., Galloway, T., & Thompson, R. (2011). Accumulation of microplastic on shorelines worldwide: sources and sinks. Environmental science & technology, 45(21), 9175–9179. Copyright 2011 American Chemical Society.)

Regarding washing machine models, Hartline et al. (2016) showed that top-loading washers cause more fiber mass emissions than front-loading washers. Commercial washing machines were used in most studies. Moreover, the amount of water used with respect to the number of textiles affects the amount of microfibers released (Kelly et al., 2019). Therefore, the lower water consumption in high-efficiency models could be the reason for the decreased fiber production. The reduced water-to-fabric ratio was associated with a 69.7% drop in microfiber emission in PES fabrics in the study by Lant et al. (2020). The production of microfibers was also increased by washing at higher temperatures (Cotton et al., 2020). Furthermore, according to Hartline et al. (2016), mechanically aged fabrics have higher fiber emission rates than new fabrics. However, in sequential washing trials during laundry with additional washing cycles, the amount of released microfibers frequently decreased until it reached a steady level (Carney Almroth et al., 2018).

Numerous studies have been conducted on the emission of microfibers, with conflicting results that may be related to the application of different materials and various methodologies (Özkan & Gündoğdu, 2021; Table 3.1). Using a Gyrowash

TABLE 3.1
Different Variables Evaluated in Studies Investigating Textile Microfiber Emission

Num.	Material(s)	Washing/Drying Method	Variables	Fiber Release	Reference
1	Fleece blankets	Front-loading washing machine	Detergent Washing cycles	0.00108 wt% (no additives) 0.00140 wt% (detergent) 0.00124 wt% (detergent + softener)	Pirc et al. (2016)
2	PES Elastane Polyamide (PA) Acrylic	Front-loading washing machine	Seven synthetic textile fibers	10–1,700 mg/kg	Kärkkäinen and Sillanpää (2021)
3	PES (green) Nylon (red) Acrylic (black)	Front-loading washing machine	Physical properties (fuzziness and pilling, tensile strength and elongation, and stiffness)	200–1,400 ppm	Choi et al. (2022)
4	Acrylic	Top-loading washing machine	Washing time Drying time Water temperature Washing cycles Detergent	3.64–133.22 mg/kg of fabric due to washing 3.23–35.33 mg/kg of fabric due to drying	Mahbub and Shams (2022)
5	PES	Front-loading washing machine	Textile characteristics (surface, textile geometry, and sewing)	Average decrease in microfiber release of 29.35 mg/kg, corresponding to 43% for PET 1	Dalla Fontana et al. (2021)
6	100% PES 100% acrylic 60% PES/40% cotton blend	Front-loading washing machine	Six devices (internal and external filters)	Control: 0.44 ± 0.04 g per wash XFiltra was the most successful device, reducing microfibers by 78 ± 5%	Napper et al. (2020)
7	100% PES (fleece)	Front-loading washing machine	Air particle characterization	58 ± 60 fibers per 660 g blanket sample	O'Brien et al. (2020)

(Continued)

TABLE 3.1 (*Continued*)
Different Variables Evaluated in Studies Investigating Textile Microfiber Emission

Num.	Material(s)	Washing/Drying Method	Variables	Fiber Release	Reference
8	100% filament PES fleece 100% filament PES double knit with brushed back 100% filament PES single jersey	Gyrowash method	Quantification methods	0.2 mg/g, 16 ± 7 μm diameter	Yang and Nowack (2022)
9	Fleece PES textiles	Gyrowash machine	Characterize nanoplastics and microfibers (microplastic fibers and fibrils)	2.1 × 1,011 nanoplastics, 1.4 × 10*4 microplastic fibers, and 5.3 × 105 fibrils per g	Yang et al. (2021)
10	PES and cotton	Household, electric-vented tumble dryer	Loading mass	7.2–9.4 fibers per kg materials (×10*5)	Tao et al. (2022)
11	Cotton, rayon, PES-cotton blends, and PES	Whirlpool washing machine	Fabric material Detergent Temperature	Cellulose-based fabrics: 0.2–4 mg/g fabric PES: 0.1–1 mg/g fabric	Zambrano et al. (2019)
12	PES, PA, and acetate fabrics	Pulsator laundry machine	Laundry machine and fiber type	Acetate fabric: 74,816 ± 10,656 microfibers/ m^2 per wash	Yang et al. (2019)
13	Knitted fabrics produced from recycled (R-PET) Virgin PES yarns	Gyrowash washing machine	Material type Fiber and yarn type	R-PET knitted fabrics: 4,489.93 fiber/L Virgin PES fabrics: 2,034.26 fiber/L	Özkan and Gündoğdu (2021)
14	100% PES fleece jackets 100% smooth PES t-shirt 100% acrylic sweater	Front-loading washing machine	Different garments Sequential cycles Detergent	220,000–2,820,000 microfibers per kg of textile	Dreillard et al. (2022)
15	100% PES fabrics	Front-loading washing machine	Washing cycles Sewing threads and edge sealing	Average decrease in microfiber release of 29.35 mg/kg	Dalla Fontana et al. (2021)

machine, for instance, Özkan and Gündodu (2021) demonstrated that virgin PES fabrics released 167,436.58 fibers/kg, compared to recycled PET knit fabrics that released 368,094.07 fibers/kg, and with more washing cycles, both fabric types discharged fewer fibers. Fibers can also be released naturally without washing, while wearing garments on a daily basis; according to a previous study by Palacios-Mateo et al. (2021), the amount of microfibers released through laundering is equal to the amount released through rubbing or brushing fabrics (Figure 3.3). However, during washing procedures, unlike during wearing, fibers are released into both air and

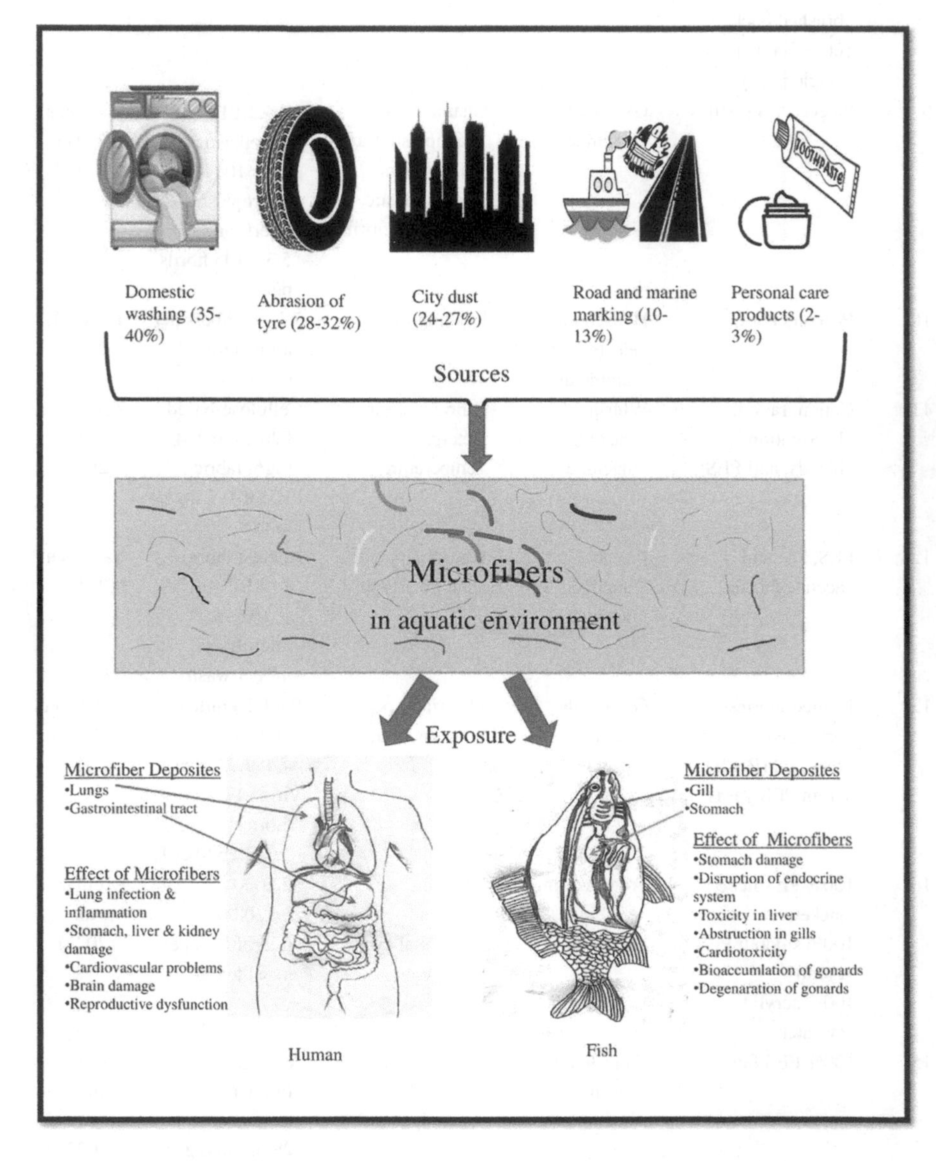

FIGURE 3.3 Sources of synthetic microfiber and the exposure of microfiber toxicity. (*Source:* Adopted with permission from Singh et al., 2020.)

water. However, there has been limited research on the release of synthetic microfibers into the atmosphere. Figure 3.3 represents the various sources of microfibers and microplastics and their impact on aquatic animals and human health.

3.2 EFFECTS OF WASHING FACTORS

Several textile characteristics and washing techniques affect the amount of shed microfibers. Understanding these elements can contribute to mitigation strategies for microfibers of textile origin.

3.2.1 Using Detergents

According to a recent study by Mahbub and Shams (2022), when washing synthetic materials, the use of detergents appears to increase fiber release (162.49 mg/kg) in comparison with not using them (13.32 mg/kg) (Figure 3.4). One explanation for this finding may be that some detergent ingredients, including surfactants, accumulate on the surface of the textile, which reduces friction and stimulates fiber degradation (Mahbub & Shams, 2022), leading to microfiber generation and emission. Moreover, pH, detergent formulation (containing oxidizing substances), and detergent type (powder or liquid) are other factors affecting synthetic microfiber emissions during washing (De Falco et al., 2019). However, neither the detergent type (powder or liquid) nor the detergent amount used had a significant effect on the release of microfibers (Hernandez et al., 2017). However, as highly hydrophobic synthetic fibers are not significantly affected by alkaline detergent solutions when washed at or below the recommended temperatures, the washing temperature appears to be the primary determinant of microfiber release (Hernandez et al., 2017).

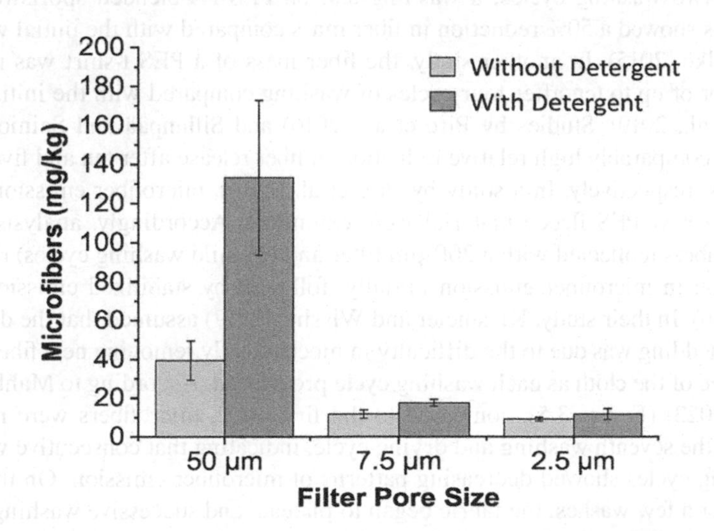

FIGURE 3.4 Microfibers released during washing due to the addition of detergent; statistically, no significant differences between parameters are indicated by * ($p > 0.05$). Error bars represent the standard deviation for three experimental replicates. (*Source:* Adopted with permission from Mahbub & Shams, 2022.)

There was also a relationship between the amount of shedding and the quantity of detergent used. Since they are the primary active ingredients in detergents, surfactants are believed to be responsible for the majority of fiber shedding (Hernandez et al., 2017). When surfactants were added for about 0.75 g/L (the suggested quantity by detergent companies), the amount of fibers increased; however, when the dose was raised to 1.5 or 2.25 g/L, no further emission was noted. However, Pirc et al. (2016) found no evidence of the effect of detergents (with and without liquid detergent 0.00108 vs. 0.00140% w/w, respectively) on microfiber emission from PES blankets.

Furthermore, the presence of softeners reduced microfiber emissions compared to detergents, which was related to the softeners' reduced friction within fibers during laundering (De Falco et al., 2018). However, Pirc et al. (2016) did not observe any remarkable decrease in fiber shedding due to the use of softeners. Without more specific information, it may be challenging to generalize; the results can be different because of the various ingredients and concentrations of the tested detergents and softeners, as well as the various washing procedures used.

3.2.2 Washing Cycles and Duration

While emissions into the environment were not a problem before, the deterioration of textiles was evaluated to identify the limits of useful fabric life. Browne et al. (2011) conducted the first study revealing the laundry process as a cause of synthetic microfiber pollution. They claimed that "all fabrics produced over 100 fibers per each liter of wastewater" and that "a single garment can shed more than 1900 fibers per wash." However, the experimental technique and substantial details regarding textiles have not been provided (Browne et al., 2011). In general, more washings cause fewer microfibers to be shed during each wash (Hernandez et al., 2017). Following four and two washing cycles, a washing test on PES-PA-blended sportswear and PES shirts showed a 50% reduction in fiber mass compared with the initial washing cycle (Folkö, 2015). In another study, the fiber mass of a PES t-shirt was reduced by a factor of up to ten after four cycles of washing compared with the initial wash (Kelly et al., 2019). Studies by Pirc et al. (2016) and Sillanpää and Sainio (2017) showed a comparably high relative reduction in fiber release after ten and five washing cycles, respectively. In a study by Pirc et al. (2016), microfiber emissions from washing a new PES fleece material were examined. Accordingly, analysis of the released fibers (collected with a 200-μm filter and ten mild washing cycles) revealed a reduction in microfiber emission initially, followed by stabilized emission (Pirc et al., 2016). In their study, Klasmeier and Wissing (2017) assumed that the decrease in fiber shedding was due to the difficulty in mechanically removing new fibers from the surface of the cloth as each washing cycle progressed. According to Mahbub and Shams (2022) (Figure 3.5), compared to the first cycle, microfibers were released less after the seventh washing and drying cycle, indicating that consecutive washing and drying cycles showed decreasing patterns of microfiber emission. On the other hand, after a few washes, the fabric began to plateau, and successive washing cycles revealed a declining trend in microfiber release (Mahbub & Shams, 2022).

In contrast, during the course of five successive washes (40 rpm, 40°C, 45 minutes), a steady fiber emission rate with 0.025 mg of fiber per gram of fabric on average

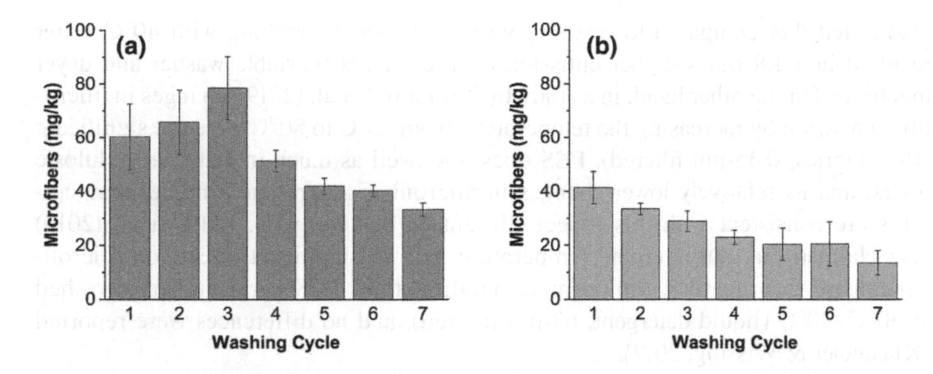

FIGURE 3.5 Total microfibers released due to subsequent washing. Error bars represent the standard deviation for three experimental replicates. (*Source:* Adopted with permission from Mahbub & Shams, 2022.)

(PES fiber, 0.45-μm filter) was reported (Hernandez et al., 2017). The same authors investigated the potential impact of the washing time on fiber shedding (washing times: 1, 2, 4, and 8 hours), but no significant alterations in the quantity of fibers were observed (Hernandez et al., 2017). Therefore, the researchers concluded that frequent washing did not significantly influence the gradual release of additional fibers over time (Hernandez et al., 2017).

As a result of prolonged rotational forces and physical tension on the clothing, Mahbub and Shams (2022) study revealed that when washing and drying periods were increased from 30 to 60 minutes, microfibers were released in larger quantities by 1.4 and 2 times, respectively. Moreover, the longer duration of the laundry process, the addition of a detergent, and the use of hot water increased the release of microfibers. Kelly et al. (2019), who developed a novel approach to determining the effects of temperature, agitation, water levels, and washing time on microfiber emissions, similarly showed that lengthening the washing process causes more microfibers to be released. Moreover, high microfiber release has been reported with increased washing time and physical tension (Kumar & Mariappan, 2022).

The physical forces in the laundry process (simulated washing vs. regular washing), fabric structure (interlock knitting vs. fleece), and various analytical methods (amount of separation during filtration) may be the major causes of the conflicting findings regarding fiber release among studies on the relationship between laundry duration and washing cycles.

3.2.3 Washing Temperature

According to De Falco et al. (2018), microfiber emissions increased by approximately 500 fibers per gram of fabric (PES, with detergent, and 5-μm filtered) by increasing the laundry temperature to 60°C. The possible cause of this phenomenon might be the greater surface degradation of the PES fabric in the presence of alkaline detergent and at higher temperatures (De Falco et al., 2018). Mahbub and Shams (2022) also noted that compared to washing with 20°C water, washing with 40°C water resulted in a 1.8 times higher emission of microfibers (portable washer and dryer machine). On the other hand, in a study by Zambrano et al. (2019), changes in microfiber emission by increasing the temperature from 25°C to 80°C were not significant (PES fabrics, 0.45-µm filtered). PES does not swell as much in water as cellulosic fibers, and its relatively lower changes in microfiber generation at higher temperatures are consistent with this aspect (Bryant & Walter, 1959). Kelly et al. (2019) also claimed that duration and temperature had no significant effects on microfiber release from textiles. Similarly, in another study, PES garments were washed at 30°C–40°C (liquid detergent, 63-µm filtered), and no differences were reported (Klasmeier & Wissing, 2017).

Therefore, the exact effect of increasing the temperature is not yet completely clear. To obtain better results for this purpose, the impact of using a wider range of fabrics with regard to materials, surface patterns, and washing temperatures may have to be considered when analyzing.

3.2.4 Loading Mass and Washing Machine Type

In a washing process, Volgare et al. (2021) analyzed the release of microfibers, focusing on the washing load, and reported that it gradually decreased as the washing load increased (Figure 3.6). The results of the study showed that, owing to a synergistic action between the water-volume-to-fabric ratio and mechanical stress during washing, reducing the washing capacity increased the number of shed microfibers by nearly five times (Volgare et al., 2021). Furthermore, because of the higher physical tension that the fabric is subjected to in the case of a low washing load, it is also challenging to differentiate the effects of other washing parameters, such as detergent type, on the release of the fabric (Volgare et al., 2021). The highest discharge of microfibers as a result of increased water volume was previously reported by Kelly et al. (2019). Therefore, the authors recommended that users decrease microfiber release by avoiding delicate cycles (using a high water volume), utilizing full wash loads, and converting to low-water-volume appliances (North American high-efficiency washing machines).

Washing PES jackets in a top-loading machine caused significantly higher fiber emissions than in a front-loading machine (20-µm filtered, 419–526 vs. 25–122 mg, respectively) (Hartline et al., 2016). This outcome was believed to be caused by the strong stirring mode of top-load washers as opposed to front-load washers (Hartline et al., 2016). In addition, rinsing in a quicker spin-dry cycle promotes microfiber release. Therefore, the speed and intensity of agitation during washing may enhance fiber emission if their rates are sufficiently high. Kelly et al. (2019) used Lab Color Space and tergotometers to determine the quantity of fibers released in relation to water level and agitation. Accordingly, the amount of fibers released can be calculated using a regression correlation with color variations (such as luminance) captured by the Lab Color Space (Kelly et al., 2019). They noticed that agitation made less of a difference than a high water-to-fabric ratio, and washers with less water were assumed to help limit microfiber emissions (Kelly et al., 2019).

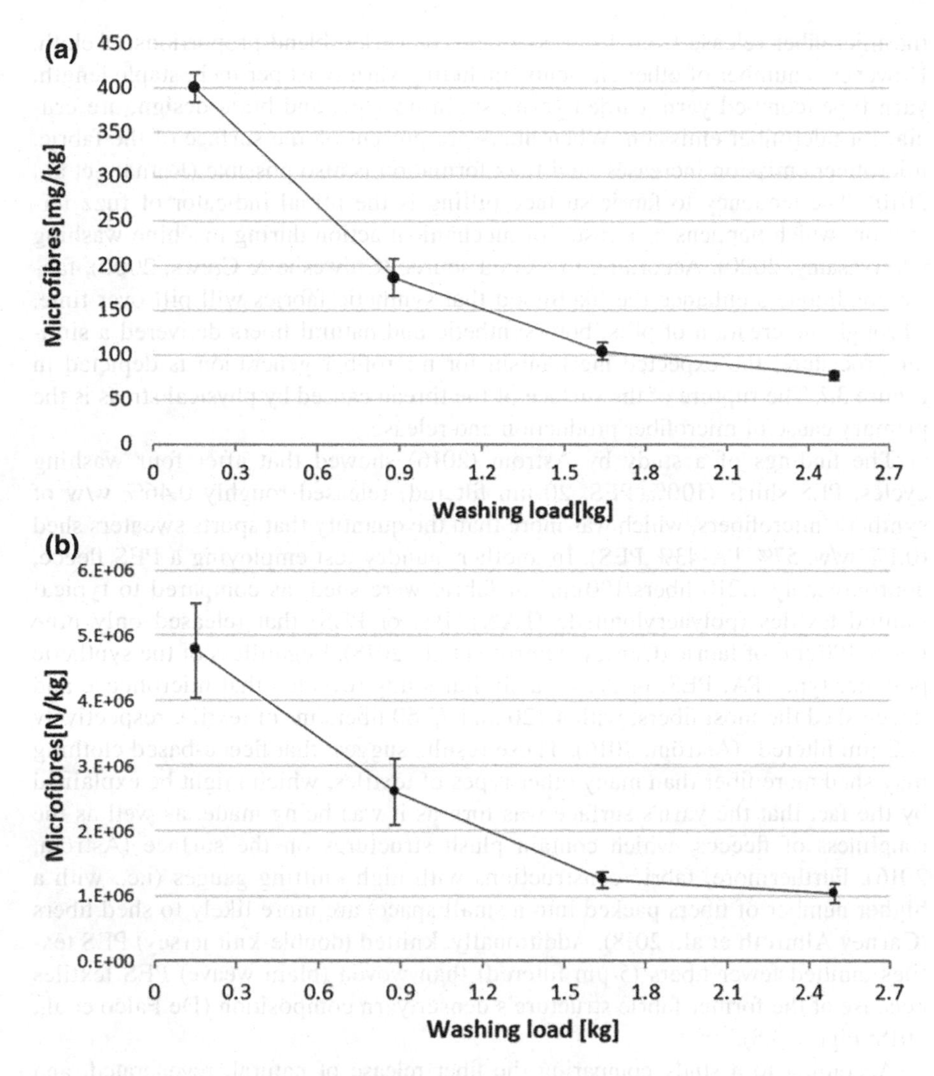

FIGURE 3.6 (a) Microfibers released per kilogram of washed fabric (expressed in mean mg/kg ± SD, $n = 2$) during the washing process performed with 0.15, 0.88, 1.64, and 2.50 kg. (b) Number of microfibers released per kilogram of washed fabric (expressed in mean N/kg ± SD, $n = 2$) during washing processes performed with 0.15, 0.88, 1.64, and 2.50 kg. (*Source:* Adopted with permission from Volgare et al., 2021.)

3.2.5 Fiber Type

The types of yarn (staple or filament yarns) and fabric (designs and structures) both play important roles in the amount of microfibers released during the washing process. Furthermore, the blend proportion is largely responsible for the emission of microfibers (Napper & Thompson, 2016). Regarding microfiber generation and their mix percentage ratios, findings from earlier studies (Napper & Thompson, 2016) are more broadly applicable. Nevertheless, it is impossible to generalize the microfiber release from 100% synthetic or varied blend proportions of cloth. However, a number of other elements, including yarn twist per inch, staple length, yarn type (combed yarn, carded yarn), spinning type, and braid design, are crucial for microfiber emission. When fibers are present on the surface of the fabric, microfiber emission increases, and fuzz formation is also possible (Ratnam et al., 2010). The tendency to fabric surface pilling is the initial indicator of fuzz formation, which happens as a result of mechanical action during machine washing (Periyasamy, 2020). According to a cited source (Chiweshe & Crews, 2000), fabric conditioners enhance the likelihood that synthetic fabrics will pill over time. Through the creation of pills, both synthetic and natural fibers delivered a similar procedure; the expected mechanism for microfiber generation is depicted in Figure 3.7. The rupture of the surface of the thread caused by physical stress is the primary cause of microfiber production and release.

The findings of a study by Åström (2016) showed that after four washing cycles, PES shirts (100% PES, 20-μm filtered) released roughly 0.46% w/w of synthetic microfibers, which was more than the quantity that sports sweaters shed (0.1% w/w, 57% PA-43% PES). In another laundry test employing a PES fleece, approximately 1,210 fibers/100 cm^2 of fabric were shed, as compared to typical knitted textiles (polyacrylonitrile (PAN), PA, or PES) that released only nine fibers/100 cm^2 of fabric (Carney Almroth et al., 2018). Regardless of the synthetic polymer type (PA, PES, or PAN), a similar study revealed that microfleece and fleece shed the most fibers, with 4,120 and 7,360 fibers/m^2 of textile, respectively (1.2-μm filtered) (Åström, 2016). These results suggest that fleece-based clothing may shed more fiber than many other types of textiles, which might be explained by the fact that the yarn's surface was torn as it was being made, as well as the roughness of fleeces, which contain plush structures on the surface (Åström, 2016). Furthermore, fabric constructions with high knitting gauges (i.e., with a higher number of fibers packed into a small space) are more likely to shed fibers (Carney Almroth et al., 2018). Additionally, knitted (double-knit jersey) PES textiles emitted fewer fibers (5-μm filtered) than woven (plain weave) PES textiles because of the former fabric structure's denser yarn composition (De Falco et al., 2018; Figure 3.8).

According to a study comparing the fiber release of natural, regenerated, and synthetic fabrics with a framed (knitted) structure, regenerated and natural textiles may release more fibers than PESs throughout the laundry process (Zambrano et al., 2019). The significant weight loss of the textiles examined with a Martindale Abrasion Tester was assumed to indicate that cotton and viscose had lower fabric abrasion resistance than synthetic fibers (Zambrano et al., 2019). As was the case with fleeces earlier, this point of view emphasizes the significance of the fabric surface shape in fiber shedding. Moreover, PP and PES fabrics produced from staple yarns (50 and 35 mm lengths, respectively) shed more fibers than fabrics made of filaments (with an average length of 340 m) (De Falco et al., 2018). Because staple yarns consist of shorter fibers in contrast to filament yarns, they are easier to remove from fabrics (De Falco et al., 2018).

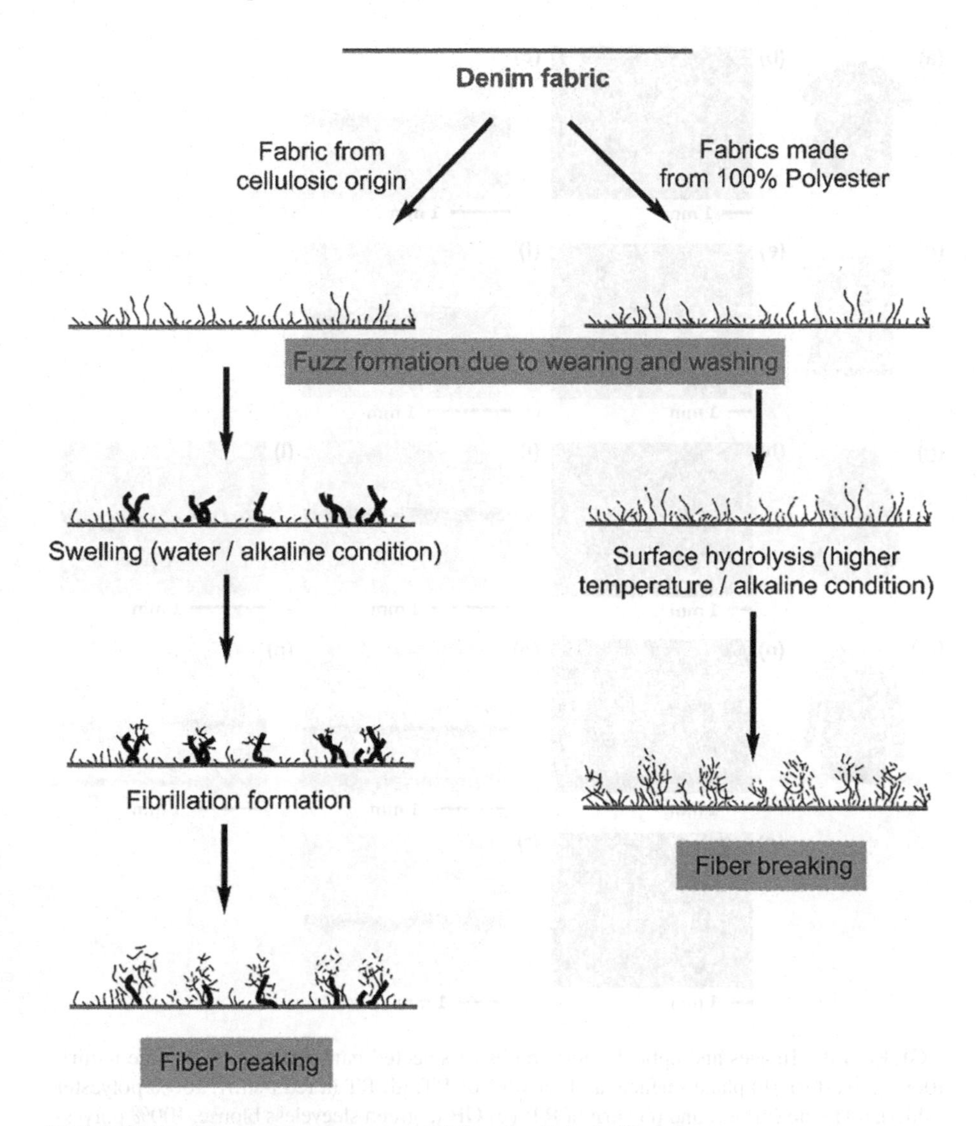

FIGURE 3.7 Scheme illustrating how PET and cellulosic fibers in jeans can produce microfibers. (*Source:* Adopted with permission from Periyasamy, 2020.)

3.2.6 Textile Characteristics

Another important factor that affects microfiber emissions during washing is the structure of the fabric. According to the study by De Falco et al. (2019), t-shirts made of 100% PES released a comparable amount of microfibers during the laundry process (125.0 ± 32.1 and 124.1 ± 12.4 mg/kg of microfibers, respectively). This similar behavior during washing tests is due to the same fabric structure and yarn characteristics of both t-shirts. The amount of microfibers emitted varies depending on the type of

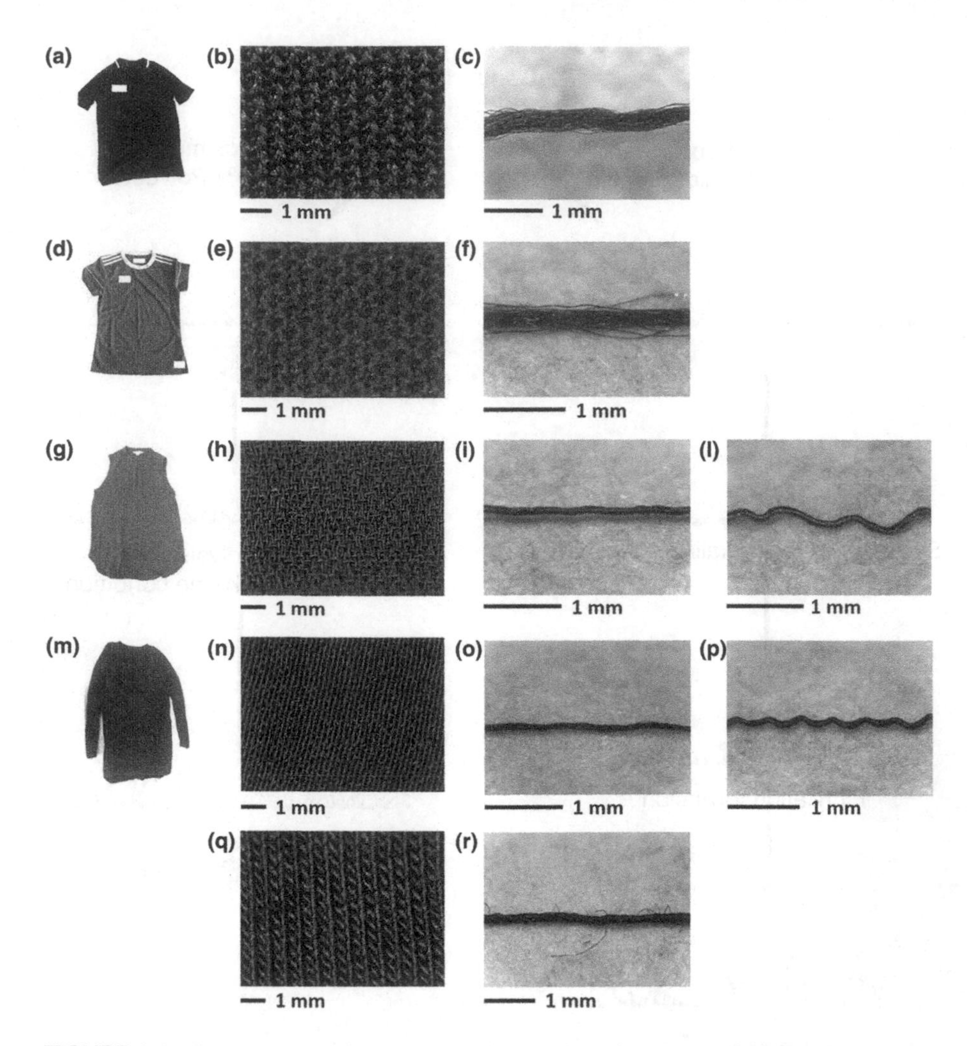

FIGURE 3.8 Images and optical micrographs of selected garments: (a) BT (a blue t-shirt, 100% polyester), (b) plane surface, and (c) yarn of BT; (d) RT (a red t-shirt, 100% polyester t-shirt), (e) plane surface, and (f) yarn of RT; (g) GB (a green sleeveless blouse, 100% polyester blouse of which 65% is recycled polyester), (h) plane surface, (i) warp and (j) weft yarns of GB; (k) GT (a green long-sleeved top, a top whose front is made of 100% polyester and whose back is made of a blend of 50% cotton and 50% modal), (l) plane surface, (m) warp and (n) weft yarns of GT front polyester part, (o) plane surface, and (p) yarn of GT back modal/cotton part. (*Source:* Adopted with permission from De Falco et al., 2019.)

fabric used. For instance, according to De Falco et al. (2018), because the structure of knitted fabric is less compact than that of woven cloth, it may release more microfibers. Furthermore, fabrics constructed with short staple fibers emitted more microfibers than those made with continuous filaments. According to the results of a study by Choi et al. (2022), acrylic released the most microfibers amount during laundry, followed by PES and nylon. This research used three distinct types of fabrics with various physical

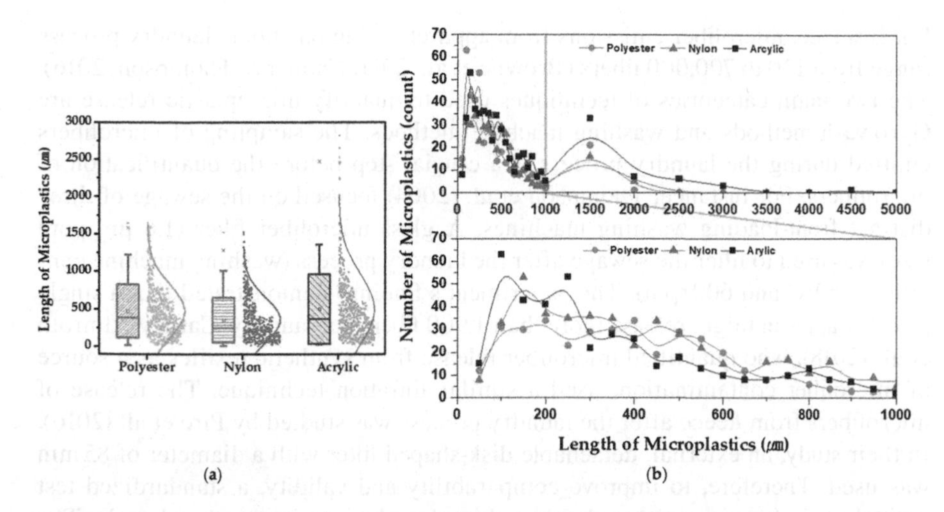

FIGURE 3.9 Distribution of the lengths and numbers of microplastic fiber. (a) The frequency distribution diagram of microfibers length for each fabric type. (b) The number of microplastic fibers of various lengths by fabric type. (*Source:* Adopted with permission from Choi et al., 2022, licensed under CC BY 4.0.)

and chemical compositions (Figure 3.9). The quantity of microfibers produced after the laundry process varied depending on the type of fabric polymer, which was explained by variations in the mechanical qualities of the textiles during washing. When subjected to the mechanical action of washing, fabrics with better abrasion resistance, flexural stiffness, and yarn breaking strength are less likely to develop fuzziness or release microfibers (Choi et al., 2022). Furthermore, according to Napper and Thompson (2016), mechanical factors during the yarn fabrication process, such as the dyeing process and air-textured yarn, might also affect microfiber emissions.

The impact of edge sealing and sewing threads on microfiber emissions was examined by Fontana et al. in 2021. They found a notable difference between the samples (double heat-sealing (PET1) vs. overlock machine (PET2)). The study's major finding indicates that PET2 discharged more fibers following washing than PET1 did (Dalla Fontana et al., 2021). The type of thread and sewing are likely to mask the effects of other characteristics of the textile. Therefore, a fabric with distinct textile properties, including heat sealing, may be less prone to the release of microfibers.

Overall, it can be concluded that the edges, type of sewing, and textile geometry all have a significant impact on microfiber emissions, as well as important data for creating the eco-design of the product.

3.3 QUANTIFICATION METHODS AND ANALYTICS OF TEXTILES MICROFIBERS

Despite the fact that various studies have examined the quantification of microplastic pollution from laundry, it is challenging to draw accurate conclusions regarding the magnitude of effects due to the multiplicity and diversity of approaches.

For instance, microfiber emissions from apparel during one home laundry process range from 120 to 700,000 fibers (Browne et al., 2011; Napper & Thompson, 2016). The two main categories of techniques used to quantify microplastic release are Gyrowash methods and washing machine methods. The sampling of microfibers emitted during the laundry process is a crucial step before the quantification of microfibers. For instance, Thompson et al. (2004) focused on the sewage of three distinct front-loading washing machines. A glass microfiber filter (1.6 μm pore size) was used to filter the sewage after the laundry process (washing machine conditions: 60°C and 600 rpm). The experiment's findings demonstrated that a single piece of apparel might release more than 1,900 fibers per laundry. Carney Almroth et al. (2018), who quantified microfiber release from synthetic textiles as a source of microfiber contamination, used a similar filtration technique. The release of microfibers from fleece after the laundry process was studied by Pirc et al. (2016). In their study, an external, detachable disk-shaped filter with a diameter of 85 mm was used. Therefore, to improve comparability and validity, a standardized test procedure and test apparatus should be developed on an international scale. The subjects covered by the standardized test procedures could include washing technique, fabric structure, and washing cycle.

On the other hand, owing to the complexity of the matrix and diversity of microfibers, it can be difficult to detect, identify, and quantify microfibers in process effluents and environmental matrices. Scanning electron microscopy (SEM) and optical microscopy have generally been employed to quantify the amount of microfibers in environmental samples (Rocha-Santos & Duarte, 2015; De Falco et al., 2018). Electron and optical microscopy have been used to determine the shape, size, quantity, and surface morphology of the microfibers (Pirc et al., 2016; Carney Almroth et al., 2018; Pirc et al., 2016). However, the fundamental problem is that optical microscopy cannot be used to obtain exact information about the material and additive composition. Therefore, it is necessary to develop additional techniques for locating and measuring microfibers.

The R method is as follows: PE granules measured in toluene-d8 at 60°C. PET fibers measured in CDCl3/TFA 4:1 at 25°C. PS beads measured in CDCl3 at 25°C (Peez et al., 2019; Olesen et al., 2017; Gü et al., 2016). The microfiber quantity, size, number, surface shape, and material types were determined using Fourier transform infrared (FTIR) and Raman spectroscopy. The particle detection limit for 1–10 μm for Raman spectroscopy and FTIR was 10–50 μm (Braun et al., 2018). Therefore, it is typical to perform combined spectroscopic and microscopic measurements. For example, the polymeric compositions of microfibers extracted from washing wastewater were determined using SEM combined with attenuated total reflectance (ATR), FTIR spectroscopy, and Raman measurement techniques (Mahon et al., 2017; Pirc et al., 2016). Furthermore, FTIR demonstrated differential capabilities in identifying naturally occurring and artificially created cellulosic fibers when used with various spectral acquisition tools, such as single-reflection ATR spectroscopy, ATR microscopy, and transmission microscopy (Comnea-Stancu et al., 2017). However, owing to misidentification caused by spectral changes, spectroscopic methods have disadvantages in large-scale measurements, the identification of additives, and the determination of adsorbed materials, which may limit the characterization of microfibers released from garments that have undergone multiple finishes (Braun et al., 2018).

To identify polymer fragments, thermal methodologies such as thermogravimetric analysis and differential scanning calorimetry are used, which are often associated with chromatography–mass spectrometry (Shim et al., 2017). These techniques provide additional details in this area. Thermal desorption – gas chromatography – mass spectrometry (TDS-GC/MS) (combined with thermogravimetry) and pyrolysis–gas chromatography–mass spectrometry enable the determination of mass balancing, specific polymer types, and additive detection (Braun et al., 2018; Nuelle et al., 2014; Dümichen et al., 2015). However, these techniques do not provide details regarding the size, shape, quantity, or morphology of the fibers.

A variety of structured quantification methods are required to better understand the outcome and influence of microfibers. This is because of the wide variety of microfiber polymer types, sizes (diameters and lengths), and the shortage of reliable identification methods for microfibers in various samples. With less uncertainty and better evaluation of microfiber emission from textiles, accumulation in wastewater treatment systems, and wide environmental distribution, researchers will be able to compare microfiber size ranges, varieties, and abundances.

3.4 CONCLUSION

The effect of synthetic fibers on microfiber pollution in the environment is covered in this chapter. The chapter's findings demonstrated that different textile properties and washing methods had an impact on the quantity of shed microfibers. Moreover, analysis methods for the synthetic microfibers emitted are shown to be essential for determining potential causes. According to previous studies, several methods have been developed to measure the release of synthetic microfibers during laundry cycles. In general, methods for measuring the amount of released synthetic microfibers include counting the fibers using a microscope, calculating the mass of the microfibers, and evaluating the results with a Raman spectrometer, fourier transform infrared (FTIR) detector, or inductively coupled plasma mass spectrometry (ICP-MS) detector. Understanding the involved elements can contribute to mitigation strategies for microfibers of textile origin. In terms of the microfiber emission process, a suitable source of yarn, type, fabric structure, and structural compactness could be influential. Most of the previous studies are based on the ingredients and characteristics of laundry. Very few studies have been done on textile finishes to date, and the use of environmentally technological finishes has not yet been assessed in terms of microfiber emission. Thus, it is recommended that increasing the knowledge of this environmental pollution is the most significant factor for pollution prevention or offering a sustainable lifestyle.

It is essential to modify international standards and guidelines for controlling microfiber pollution to reduce their usage through making consumers aware of the harmful effects of synthetic textiles. The expense of the product in comparison with the current synthetic textile will be the main obstacle to the adoption of this approach. Furthermore, despite the fact that research on laundry process and laundry products is gaining interest, they do not control the emission of microfibers. The fundamental area of focus should be devoted to the textile material's quality, which is currently lacking considerably. Besides, in the future, identifying key variables that impact the release of synthetic microfibers from synthetic clothes during washing and drying cycles may contribute to the development of strategies to decrease environmental pollution caused by such microfibers.

ACKNOWLEDGMENT

This research was supported by the National Research Council of Science & Technology (NST) grant by the Korean government (MSIT) (No. CAP20022-000).

REFERENCES

Åström, L. (2016). *Shedding of Synthetic Microfibers from Textiles*. Gothenburg, Sweden: Göteborgs Universitet.

Boucher, J., & Friot, D. (2017). *Primary Microplastics in the Oceans: A Global Evaluation of Sources* (Vol. 10). Gland, Switzerland: IUCN.

Braun, U., Jekel, M., Gerdts, G., Ivleva, N., & Reiber, J. (2018). Microplastics analytics: sampling, preparation and detection methods. In *Discussion Paper, BMBF Research Focus "Plastics in the Environment"*.

Browne, M. A., Crump, P., Niven, S. J., Teuten, E., Tonkin, A., Galloway, T., & Thompson, R. (2011). Accumulation of microplastic on shorelines woldwide: Sources and sinks. *Environmental Science & Technology*, *45*(21), 9175–9179.

Bryant, G. M., & Walter, A. T. (1959). Stiffness and resiliency of wet and dry fibers as a functi of temperature. *Textile Research Journal*, *29*(3), 211–219.

Carney Almroth, B. M., Åström, L., Roslund, S., Petersson, H., Johansson, M., & Persson, N. K. (2018). Quantifying shedding of synthetic fibers from textiles; a source of microplastics released into the environment. *Environmental Science and Pollution Research*, *25*(2), 1191–1199.

Cesa, F. S., Turra, A., Checon, H. H., Leonardi, B., & Baruque-Ramos, J. (2020). Laundering and textile parameters influence fibers release in household washings. *Environmental Pollution*, *257*, 113553.

Chiweshe, A., & Crews, P. C. (2000). Influence of household fabric softeners and laundry enzymes on pilling and breaking strength. *Textile Chemist and Colorist and American Dyestuff Reporter*, *32*(9), 41–47.

Choi, S., Kim, J., & Kwon, M. (2022). The effect of the physical and chemical properties of synthetic fabrics on the release of microplastics during washing and drying. *Polymers*, *14*(16), 3384.

Comnea-Stancu, I. R., Wieland, K., Ramer, G., Schwaighofer, A., & Lendl, B. (2017). On the identification of rayon/viscose as a major fraction of microplastics in the marine environment: Discrimination between natural and manmade cellulosic fibers using Fourier transform infrared spectroscopy. *Applied Spectroscopy*, *71*(5), 939–950.

Cotton, L., Hayward, A. S., Lant, N. J., & Blackburn, R. S. (2020). Improved garment Longevity and reduced microfiber release are important sustainability benefits of laundering in colder and quicker washing machine cycles. *Dyes and Pigments*, *177*, 108120.

Dalla Fontana, G., Mossotti, R., & Montarsolo, A. (2021). Influence of sewing on microplastic release from textiles during washing. *Water, Air, & Soil Pollution*, *232*(2), 1–9.

De Falco, F., Di Pace, E., Cocca, M., & Avella, M. (2019). The contribution of washing processes of synthetic clothes to microplastic pollution. *Scientific Reports*, *9*(1), 1–11.

De Falco, F., Gentile, G., Avolio, R., Errico, M. E., Di Pace, E., Ambrogi, V., Avella, M., & Cocca, M. (2018). Pectin based finishing to mitigate the impact of microplastics released by polyamide fabrics. *Carbohydrate Polymers, 198*, 175–180.

Dreillard, M., Barros, C. D. F., Rouchon, V., Emonnot, C., Lefebvre, V., Moreaud, M., Guillame, D., Rimbault, F., & Pagerey, F. (2022). Quantification and morphological characterization of microfibers emitted from textile washing. *Science of the Total Environment, 832*, 154973.

Dümichen, E., Barthel, A. K., Braun, U., Bannick, C. G., Brand, K., Jekel, M., & Senz, R. (2015). Analysis of polyethylene microplastics in environmental samples, using a thermal decomposition method. *Water Research, 85*, 451–457.

Folkö, A. (2015). *Quantification and Characterization of Fibers Emitted from Common Synthetic Materials during Washing Microplastic Fibers Discharged from a Fleece Shirt during Washing*. Stockholms, Sweden: Stockholm University.

Löder, M. G. J., Kuczera, M., Mintenig, S., Lorenz, C., & Gerdts, G. (2015). Focal plane array detector-based micro-Fourier-transform infrared imaging for the analysis of microplastics in environmental samples. *Environmental Chemistry, 12*(5), 563–581.

Hartline, N. L., Bruce, N. J., Karba, S. N., Ruff, E. O., Sonar, S. U., & Holden, P. A. (2016). Microfiber masses recovered from conventional machine washing of new or aged garments. *Environmental Science & Technology, 50*(21), 11532–11538.

Hernandez, E., Nowack, B., & Mitrano, D. M. (2017). Polyester textiles as a source of microplastics from households: A mechanistic study to understand microfiber release during washing. *Environmental Science & Technology, 51*(12), 7036–7046.

Kärkkäinen, N., & Sillanpää, M. (2021). Quantification of different microplastic fibres discharged from textiles in machine wash and tumble drying. *Environmental Science and Pollution Research, 28*(13), 16253–16263.

Kelly, M. R., Lant, N. J., Kurr, M., & Burgess, J. G. (2019). Importance of water-volume on the release of microplastic fibers from laundry. *Environmental Science & Technology, 53*(20), 11735–11744.

Klasmeier, J., & Wissing, M. (2017). Waschmaschinenablauf als mögliche Eintragsquelle von Textilfasern (Mikroplastik) in Gewässer.

Kumar, A., & Mariappan, G. (2022). Effect of laundry parameters on micro fiber loss during washing and its correlation with carbon footprint. *Journal of Natural Fibers, 19*(16), 14744–14754.

Lant, N. J., Hayward, A. S., Peththawadu, M. M., Sheridan, K. J., & Dean, J. R. (2020). Microfiber release from real soiled consumer laundry and the impact of fabric care products and washing conditions. *PLoS One, 15*(6), e0233332.

Li, Y., Zhang, H., & Tang, C. (2020). A review of possible pathways of marine microplastics transport in the ocean. *Anthropocene Coasts, 3*(1), 6–13.

Mahbub, M. S., & Shams, M. (2022). Acrylic fabrics as a source of microplastics from portable washer and dryer: Impact of washing and drying parameters. *Science of the Total Environment, 834*, 155429.

Mahon, A. M., O'Connell, B., Healy, M. G., O'Connor, I., Officer, R., Nash, R., & Morrison, L. (2017). Microplastics in sewage sludge: Effects of treatment. *Environmental Science & Technology, 51*(2), 810–818.

McCormick, M. I., Chivers, D. P., Ferrari, M. C., Blandford, M. I., Nanninga, G. B., Richardson, C., Fakan, E. P., Vamvounis, G., Gulizia, A. M., & Allan, B. J. (2020). Microplastic exposure interacts with habitat degradation to affect behaviour and survival of juvenile fish in the field. *Proceedings of the Royal Society B, 287*(1937), 20201947.

Napper, I. E., Barrett, A. C., & Thompson, R. C. (2020). The efficiency of devices intended to reduce microfiber release during clothes washing. *Science of the Total Environment, 738*, 140412.

Napper, I. E., & Thompson, R. C. (2016). Release of synthetic microplastic plastic fibres from domestic washing machines: Effects of fabric type and washing conditions. *Marine Pollution Bulletin*, *112*(1–2), 39–45.

Nuelle, M. T., Dekiff, J. H., Remy, D., & Fries, E. (2014). A new analytical approach for monitoring microplastics in marine sediments. *Environmental Pollution*, *184*, 161–169.

O'Brien, S., Okoffo, E. D., O'Brien, J. W., Ribeiro, F., Wang, X., Wright, S. L., samanipour, S., Rauert, C., Toapanta, T. Y., Albarracin, R., & Thomas, K. V. (2020). Airborne emissions of microplastic fibres from domestic laundry dryers. *Science of the Total Environment*, *747*, 141175.

Olesen, K. B., van Alst, N., Simon, M., Vianello, A., Liu, F., & Vollertsen, J. (2017). Analysis of microplastics using FTIR imaging: application note. *Agilent Application Note Environment*.

Özkan, İ., & Gündoğdu, S. (2021). Investigation on the microfiber release under controlled washings from the knitted fabrics produced by recycled and virgin polyester yarns. *The Journal of the Textile Institute*, *112*(2), 264–272.

Peez, N., Janiska, M. C., & Imhof, W. (2019). The first application of quantitative 1 H NMR spectroscopy as a simple and fast method of identification and quantification of microplastic particles (PE, PET, and PS). *Analytical and Bioanalytical Chemistry*, *411*, 823–833.

Palacios-Mateo, C., van der Meer, Y., & Seide, G. (2021). Analysis of the polyester clothing value chain to identify key intervention points for sustainability. *Environmental Sciences Europe*, *33*(1), 1–25.

Periyasamy, A. P. (2020). Effects of alkali pretreatment on lyocell woven fabric and its influence on pilling properties. *The Journal of the Textile Institute*, *111*(6), 846–854.

Periyasamy, A. P., & Tehrani-Bagha, A. (2022). A review of microplastic emission from textile materials and its reduction techniques. *Polymer Degradation and Stability*, 109901.

Pirc, U., Vidmar, M., Mozer, A., & Kržan, A. (2016). Emissions of microplastic fibers from microfiber fleece during domestic washing. *Environmental Science and Pollution Research*, *23*(21), 22206–22211.

Ratnam, T. V., Rajamanickam, R., Chellamani, K. P., Shanmuganandam, D., Doraiswamy, I., & Basu, A. (2010). *SITRA Norms for Spinning Mills*. Coimbatore, India: South India Textile Research Association.

Rocha-Santos, T., & Duarte, A. C. (2015). A critical overview of the analytical approaches to the occurrence, the fate and the behavior of microplastics in the environment. *TrAC Trends in Analytical Chemistry*, *65*, 47–53.

Shim, W. J., Hong, S. H., & Eo, S. E. (2017). Identification methods in microplastic analysis: a review. *Analytical Methods*, *9*(9), 1384–1391.

Sillanpää, M., & Sainio, P. (2017). Release of polyester and cotton fibers from textiles in machine washings. *Environmental Science and Pollution Research*, *24*(23), 19313–19321.

Singh, R. P., Mishra, S., & Das, A. P. (2020). Synthetic microfibers: Pollution toxicity and remediation. *Chemosphere*, *257*, 127199.

Tao, D., Zhang, K., Xu, S., Lin, H., Liu, Y., Kang, J., Yim, T., Giesy, J. P., & Leung, K. M. (2022). Microfibers released into the air from a household tumble dryer. *Environmental Science & Technology Letters*, *9*(2), 120–126.

Thompson, R. C., Olsen, Y., Mitchell, R. P., Davis, A., Rowland, S. J., John, A. W., McGonigle, D., & Russell, A. E. (2004). Lost at sea: Where is all the plastic? *Science*, *304*(5672), 838–838.

Volgare, M., De Falco, F., Avolio, R., Castaldo, R., Errico, M. E., Gentile, G., Ambrogi, V., & Cocca, M. (2021). Washing load influences the microplastic release from polyester fabrics by affecting wettability and mechanical stress. *Scientific Reports*, *11*(1), 1–12.

Yang, L., Qiao, F., Lei, K., Li, H., Kang, Y., Cui, S., & An, L. (2019). Microfiber release from different fabrics during washing. *Environmental Pollution*, *249*, 136–143.

Yang, T., Luo, J., & Nowack, B. (2021). Characterization of nanoplastics, fibrils, and microplastics released during washing and abrasion of polyester textiles. *Environmental Science & Technology*, *55*(23), 15873–15881.

Yang, T., & Nowack, B. (2022). Reply to comment on "characterization of nanoplastics, fibrils, and microplastics released during washing and abrasion of polyester textiles." *Environmental Science & Technology*, *56*(14), 10545–10546.

Zambrano, M. C., Pawlak, J. J., Daystar, J., Ankeny, M., Cheng, J. J., & Venditti, R. A. (2019). Microfibers generated from the laundering of cotton, rayon and polyester based fabrics and their aquatic biodegradation. *Marine Pollution Bulletin*, *142*, 394–407.

Ziajahromi, S., Neale, P. A., Rintoul, L., & Leusch, F. D. (2017). Wastewater treatment plants as a pathway for microplastics: Development of a new approach to sample wastewater-based microplastics. *Water Research*, *112*, 93–99.

4 Characterization and Identification of Microfiber Particles Released in the Washing Process

Mirjana Čurlin, Tanja Pušić, and Branka Vojnović
University of Zagreb

4.1 INTRODUCTION

The washing process is a complex multiphase and multiscale process that affects mass transfer in porous textiles exposed to numerous physicochemical influences (Liu et al. 2019). Mechanical action, chemical action, duration, and temperature are four factors in the washing process called Sinner's cycle, which was used by Dr. Herbert Sinner with the intention of explaining the basic principles of cleaning (Sinner 1960, Beringer and Kurz 2011). The proportion of the factors is theoretically equal, while in real systems they are compensated by a proportional increase or decrease of certain factors.

The fundamental differences in the observation of the washing process are primary effects (Lambert et al. 2016), secondary effects (Fijan et al. 2008, Fijan and Turk 2012, Bockmühl et al. 2019, Reynolds et al. 2022), and the environmental impact, considering that household washing has been identified as a major cause of microfiber particle pollution (Cesa et al. 2019, De Falco et al. 2019, Henry et al. 2019, Hernandez et al. 2017, Kang et al. 2021, Kelly et al. 2019, Napper and Thomson 2016, Pirc et al. 2016, Schöpel and Stamminger 2019, Yang et al. 2021). Accordingly, numerous publications, global guidelines, and legislation focus on the issue of qualifying and quantifying the microfibers released in the washing process of synthetic textiles. In general, the washing effect depends on the specific interactions between textiles, soiling, water quality, and detergent composition, which are superimposed by mechanical agitation. Table 4.1 shows the basic partners with subvariables employed in a washing process. The variability of all these parameters or partners in the washing process complicates the interpretation of the washing effects.

DOI: 10.1201/9781003331995-5

TABLE 4.1
Partners with the Most Important Subvariables in the Washing Process

S.No	Partner	Subvariables
1.	Water	Water level/liquor, hardness, heavy metal contamination, pH, conductivity, microbiological properties, biofilm
2.	Textiles	Composition (natural, synthetic, blends), structural characteristics, dimensions, weight, finish (waterproof, oil and water resistant, color, anti-crease, antibacterial, flame retardant, etc.), pilling, durability, dry and wet properties (tear and wear resistance)
3.	Soils	Degree of contamination (low, medium, high), different types of stain (protein, starch, sugar, pigments, …), aging, heat
4.	Washing machine *Horizontal-axis* *Vertical-axis*	Temperature, time, mechanical agitation (rotation speed, running and standing position and diameter of the drum, washing ratio, liquor quantity, dimensions, and weight of the textiles according to composition)
5.	Washing agent *(liquid, powder, gel)*	*Inorganic ingredients* (alkali, phosphates, phosphonates, aluminosilicates, bleaching agents, silicates) *Organic ingredients* (surfactants, citrates, functional polymers, fluorescent whitening agents, enzymes, fragrances, solvents, precursors)
5a.	After treatment aids *(Softeners, stiffeners)*	Chemical composition, concentration

4.2 DETERGENCY

The washing process begins with wetting and penetration of the bath into the textiles. During the process in the washing machine, several phenomena take place: (i) removal of dirt from the structure of the textiles, (ii) suspension and prevention of redeposition of removed dirt, (iii) discoloration of stains remaining in the textiles, (iv) deposition of shading dyes, and (v) deposition of fluorescent whitening agents (FWAs) (Bueno et al. 2018).

During the mixing, turbulence, and heating of the water medium during the washing process, momentum and heat transport phenomena occur, which affect the detachment of the dirt particles from the fiber and their transfer to the bath. Calcium and magnesium ions from the water tend to precipitate with the detergent components and form residues on the textiles and parts of the washing machine (Young and Matijević 1977). The presence of heavy metal ions such as copper, iron, or manganese in trace amounts negatively affects the course of the washing process, e.g., catalytic decomposition of oxygen bleaches and quenching of fluorescence of FWA and UV (ultraviolet) absorbers (Dekanić et al. 2021, Mainali et al. 2013). Therefore, the detergent contains builders as ingredients that prevent scaling and harmful effects in the washing process through chelation or ion exchange mechanisms (Li et al. 2019).

Therefore, the circular laundry chart (Sinner's cycle) also includes an inner cycle, Figure 4.1 which implies that the effectiveness of chemistry, temperature, time, and mechanical agitation is a consequence of water as the medium that connects them (Smulders 2002, Stamminger 2010).

Water transfers heat, which facilitates the removal of substances such as waxes and greases, which change to a liquid phase at as low as 40°C (Smulders 2002). Polar soils and water-soluble detergent components are dissolved in water, while insoluble detergent components and soils are finely dispersed. Dirt is composed of a variety of substances such as fats, proteins, pigments, and other colored particles that have quite different adhesive properties on the fabric (Smulders 2002). To remove the dirt, the Van der Waals attractive forces between the fiber and the dirt particle must be overcome by electrostatic repulsion interactions, which are assisted by builders and alkalis in detergents (Dienner 1983, Mercadé-Prieto and Bakalis 2014).

Finely dispersed particles can form agglomerates, which increases the tendency to redeposit stains (Višić et al. 2021). Insoluble stains, e.g., proteins, starch, and mannans, are hydrolytically degraded under the influence of enzymes and

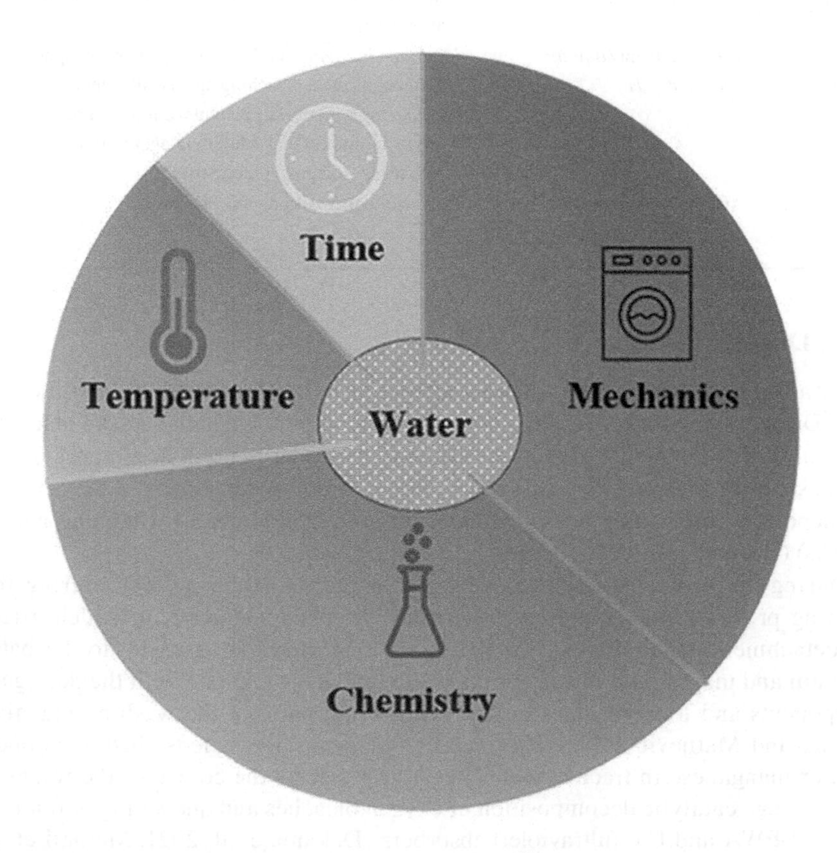

FIGURE 4.1 Sinner's cycle. (*Source:* Designed by M. Čurlin.)

converted into soluble products. Colored pigments are degraded by oxidation and transformed into soluble or colorless compounds. Textiles can be contaminated by pathogenic microorganisms from various sources (infected textiles, biofilm), which cause an unpleasant odor. In modern washing, the process is controlled by innovative conditions (wash water temperature, wash cycle duration, sensors, etc.), design (front-loading vs. top-loading), changes in the chemical composition of detergents and types of laundry additives, and drying methods (electric dryers vs. air drying) (Bockmühl et al. 2019, Bloomfield et al. 2011, Jacksch et al. 2020).

The physicochemical effects of washing are enhanced by the amplification of hydrodynamic effects, the importance of which for soil removal depends on the properties of the particles and other variables listed in Table 4.1 (Soljačić and Pušić 2005). Mechanical actions during washing, classified as position-dependent (stretching, bending, or twisting of textiles), enhance the mass transfer of detergents and stains (Mercadé-Prieto and Bakalis 2014). Mechanical effects can be considered as small-scale and large-scale effects. Small-scale effects include mass transfer, mechanical removal of microscopic soils trapped between fibers, and dissolution of solid particles on the surface of textiles. The large-scale effects include the movement of the textiles and the hydrodynamic flow of the detergent solution, where the textiles are in direct contact with the free bath (Park et al. 2013). The mechanical and chemical stresses to which the textiles are subjected during the process in the washing machine result in the detachment of microfibers from the textiles.

4.3 INFLUENCE OF WASHING PROCESS VARIABLES ON PARTICLES RELEASED FROM TEXTILES

The potential for fibril release from textiles depends on the composition (plain or blended), structure (woven, knitted, or nonwoven), texture (loose or dense), type of yarn (number of threads, textured, hairy, regular, open-end), and type of fibers (natural, synthetic, cut or filament), and the finish type that is applied on the textiles (durable, semidurable, nondurable) (Miraftab and Horrocks 2007, Carney Almroth et al. 2018). In general, textile fibers are spun into yarns that are twisted in various ways along the fiber axis. The fibers that make up yarns can be staple fibers (short length) and filaments (indefinite length).

Yarns are structurally arranged in woven fabrics (made by interlacing the warp in the longitudinal direction and the weft in the width direction) and in knitted fabrics (made by the interlooping of yarns). Nonwoven fabrics consist of fibers joined by bonding agents, i.e., mechanical or thermal bonding methods (Chakraborty 2012). The Poisson's coefficient in woven fabrics results from the interaction between warp and weft and can be expressed in terms of the structural and mechanical parameters of the system (Postle et al. 1988). Poisson's ratio is a fundamental mechanical property of the material that affects the properties of the material and helps to better interpret the behavior of textiles (Penava et al. 2014).

The synergy of factors in the Sinner cycle affects textile properties (shrinkage, deformation, damage, handle, porosity, roughness, color, etc.) (Toshikj et al.

2019). The changes in washing according to Bishop (1995) can be attributed to the following:

i. The physical effects of water, temperature, and mechanics on the different fiber types with respect to fabric construction and drying method
ii. The chemical effects of the individual components of detergents on different fibers, in terms of the usual processing of textiles
iii. Combined physicochemical effects of the washing protocol and detergent components on the durability and color transfer of common dyes to fibers.

Textile composition plays an important role in wrinkle formation and elastic recovery. Hydrophilic textiles swell when they absorb water, the arrangement of molecular chains changes, and the rotation along the fiber axis also increases due to the absorption and desorption of moisture, so that the chains can achieve higher mobility, especially in the amorphous region. Hydrophobic fibers, which absorb little or no water, show only an insignificant effect (Yun and Park 2016).

It is necessary to consider the path of modification of textiles through stability, resistance, and durability —the migration potential of finishing agents under dry and wet conditions, possible interactions, and ecological characteristics (Roshan 2014). Modification of the material surface is a very effective way to generate considerable added value: e.g., protection of textiles from interaction with liquids, the ability to allow them to bead off or possibly be easily removed, and to obtain clean and unmodified textiles (McCarthy 2011). For example, fibrils released from textiles can be carriers of functional substances, e.g., chemical pollutants such as per- and polyfluoroalkyl substances (PFAS), brominated flame retardants (BFR), and organophosphate esters (OPE), micro/nanoparticles, dyes, pigments, surfactants, other contaminants, and washing aids which additionally impact environmental systems (Saini et al. 2016, Schellenberger et al. 2019, Shabana et al. 2015).

The challenges in modern washing technology are not only to reduce energy and water consumption but also to extend the life cycle of textiles. Textile parameters (fiber type, yarn characteristics, structural parameters, age) and external parameters (type of washing machine, BR, speed, time, temperature, drying) impact MF release during the washing cycles (Palacios-Mateo et al. 2021). Mechanical forces, e.g., rubbing during the usage of textiles cause the ends of fibers to form pills or fuzz. It is more expressive with synthetic fibers under the influence of mechanical stress in dry and wet conditions at elevated temperatures (Pušić et al. 2021). Pilling of polyester textiles may be associated with a slight migration of fibers that form a ball. Regardless of the mechanism, textiles with more loose ends per unit area, create more fuzz and shed more MFs (Carney Almroth et al. 2018). The main external parameter relevant for the detachment of formed pills from the textile surface is BR (a higher amount of MFs was released in a higher BR wash). The size and mass of MFs released from synthetic (polyester) textiles during simulated household washing under controlled laboratory conditions depends on their structure, and washing conditions (use of detergents, temperature, duration of washing, successive cycles) (Volgare et al. 2021). Kelly et al. presented the amount of particles released from a textile source with defined properties under the influence of washing process variables (type of machine, wash duration, mechanics —number of metal balls, bath ratio—BR).

This variability raises numerous issues that require standardization of washing process protocols and quantification methods. The American Association of Textile Chemists and Colorists has published TM212–2021, Test Method for Fiber Fragment Release during Home Laundering (AATCC TM212-2021). The test method determines the mass of fiber fragments in an accelerated washing machine, while degradability, chemical toxicity, and physical hazards are other important elements of textile release that require additional test methods (Belzagui et al. 2019).

The composition and properties of wastewater from the washing process vary widely due to the diversity of textiles, washing method, and factors associated with washing such as temperature, hydrodynamics, detergent, and water volume (Cesa et al. 2019, De Falco et al. 2019, Henry et al. 2019, Hernandez et al. 2017, Kang et al. 2021, Kelly et al. 2019, Napper and Thomson 2016, Pirc et al. 2016, Schöpel and Stamminger 2019, Yang et al. 2021). Yarns made of staple fibers shed more MFs than those made of filament. Scissor-cut samples shed 30 times more MF than textiles cut by laser (Palacios-Mateo et al. 2021).

Due to all the parameters discussed in this section, the wastewater after the washing process has large differences in the concentration of pollutants, fibers, and solids, resulting in turbidity of the water, high pH (alkaline), and the presence of organic compounds (Pušić et al. 2022). The discharge of the wastewater causes major problems due to the high amount of pollutants and high pH, which negatively affects the biological processes of water purification (Višić et al. 2015).

4.4 MICROFIBERS AND OTHER PARTICLES IN WASTEWATER—DISPERSION SYSTEM

In general, wastewater contains various organic and inorganic constituents with different physicochemical properties. The pollutants or types of pollutants contained in wastewater depend on the sources of pollution. Different types of pollutants in the form of particles are present in wastewater or are formed during the treatment process, especially in wastewater from the textile industry. Particles can be categorized based on their size, i.e., dissolved (<0.001 μm), colloidal (0.001–1 μm), supra-colloidal (1–102 μm), and settleable (>102 μm) (Dulekgurgen et al. 2006).

During the washing process, the water-soluble dirt dissolves in the water, and the insoluble components of the detergent and the insoluble dirt are finely dispersed, forming a dispersion system. Nowadays, research pays great attention to particles of textile origin in washing water, mainly containing fiber fragments or filaments. These fragments, when washing synthetic materials, are microplastic particles known as microfibers (MF), which are considered a major problem of environmental pollution, especially wastewater. All plastic particles < with a size of 5 mm are considered microplastics. They can appear in different forms: filaments or fibers, fragments, spheres, and fluffy forms, while depending on their origin they can be divided into primary and secondary (He et al. 2018, Choobar et al. 2019). Microplastic particles in the form of filaments and fragments left over from textiles are referred to as microfibers (MF) in the literature. It is important to note that MF in the technical language (AATCC TM212–2021), this term refers to fibers with a linear density less than 1 denier or 1 dtex (den, dtex, and tex are units for the linear density of yarn; *tex* is the mass in grams of 1000 km of yarn; *den* is the mass in grams of 9.000 m of yarn).

According to data in the literature (De Falco et al. 2020, Napper and Thompson 2016, Pirc et al. 2016) about 30% of microfibers from washing machines are released into the environment as a byproduct of washing synthetic fiber fabrics. Data at the European Union (EU) level show that daily discharges through a conventional wastewater treatment plant are about 10–60 g microfibers/day, depending on the total volume of the plant (Gatidou et al. 2019). During the wastewater treatment process (WWTP), the percentage of fiber removal is high, but microscopic particles and fiber fragments remain in the treated wastewater. Due to the small size of the fibers, a significant portion of the microfibers pass through the screens used for the pretreatment of wastewater (usually coarse and fine screens, 1.5–6 mm) (Napper and Thompson 2016).

The predominant MF particles in the effluent included textile fibers and some fragments and films. Fragments and films were mainly composed of polypropylene (PP) while fibers had a much broader chemical composition, including PP, polyethylene terephthalate (PET), polyamide (PA), and polyacrylonitrile (PAN). Commonly used fibers from textiles are PET, PA, and PAN fibers are often found in wastewater from household washing machines and professional laundries. However, the content of PP fibers is elevated at the output of WWTP, which is because PET and PAN fibers with higher density are more likely to be deposited during the wastewater treatment process (Zhang et al. 2021). Although MFs lag behind, they contribute <0.1% of the microplastics that pollute the environment, depending on the operation of the WWTP. Although the removal of microplastics from treated wastewater in WWTP reaches 69%–99%, it is important to emphasize that this removal refers to the phase transfer of microplastics from liquid to sludge (Reddy and Nair 2022, Gatidou et al. 2019, Li et al. 2018).

There are numerous methods for the analysis of textile washing wastewater as a dispersion system based on optical or laser technology. For the separation of solids, mainly physical methods are used, usually filtration or a combination of different separation techniques.

4.5 PARTICLE ANALYSIS METHODS IN WASHING WASTEWATER

There is a non-uniformity of the microplastic (MP) particle analysis methods in wastewaters, especially in wastewaters from textile washing results from the already mentioned complexity of the entire washing system and the synergistic effects that strongly influence the quantity, size, and type of particles. Variations in the results presented are numerous for almost all techniques used, and this is especially true for systems where fibers are present.

4.5.1 Microscopic and Spectroscopic Methods

Microscopic methods are usually the first step in microplastic analysis, as they allow visual identification of particles and provide data on number, size, color, and morphology. These analyses are usually followed by spectroscopic analyses to confirm the nature of the particulate material. These methods can be used for the analysis of wastewater samples as well as for the analysis of residuals after particle separation. Although characterized by the simplicity of performing the microscopy method, results are inconsistent depending on the size and shape of the particles, highlighting very poor results (greater variation in replicates) for fibers and for white and transparent particles, as well as numerous false positives related to non-plastic particles (Kotar et al. 2022). To increase the accuracy and reproducibility of the MP particle analysis results of this valuable but still limited method, professional experience as well as continuous training of the analyst is of particular importance.

4.5.2 Particle Size Distribution (PSD) Methods

Particle size distribution methods in wastewater can provide data on the type and size of suspended contaminants, which is very important when the problem of particle release is observed during the washing process. These methods are based on laser technology and are mainly used for laboratory analysis of particle size and are not suitable for the determination of particle morphology. The laser diffraction technique (ISO 13320:2020) is based on the passage of a laser beam through a sample containing suspended particles. Depending on their size, the particles diffract or scatter light at specific angles. Larger particles diffract light at small angles relative to the laser beam, while smaller particles diffract light at large angles. The particle size is calculated using the Mie or Fraunhofer theory of light scattering (Ryzak and Bieganowski 2011, Sochan et al. 2014). Due to this evaluation of the physical property of the particle, which is related to its size, the method chosen to determine the size of the particle affects the results obtained. Particle size distribution in the form of a histogram or in the form of a function obtained by normalization, dividing the value of each interval by the total number of counted particles, represents a simplified description of the distribution of particles present in the sample, which differ from each other in terms of size and shape. Particles have random, complex, and irregular shapes, which are most often reduced to "spheres" when recalculating, so we speak of "sphere-equivalent diameter".

The mentioned approximation introduces a significant error in the calculation of the diameter if fiber fragments or filaments that are released from textile materials in the washing process are determined. Various auxiliary substances that are an integral part of the washing system (undissolved detergent ingredients, impurities) can affect the precision of the measurement and can be confused with released fiber fragments. The results of methods for detecting particles in water from washing based on numerical representation, such as microscopes, show the distribution as a distribution of numbers. The results of laser diffraction present their original result as a volume distribution. Conversion from numbers to volume or vice versa is possible, but conversion of volume results from laser diffraction to a numerical basis can lead to undefined errors and is recommended only for comparison with results from microscopy. The results of this conversion show large differences in the mean sizes of the particles present in the system. Laser diffraction is used mainly for laboratory analysis and is not suitable for online measurements and particle morphology determination.

The mentioned problems in the analysis of fibrous particles in textile washing water can be solved by the application of digitization, i.e., by connecting computers with equipment and applying computer analysis methods in real time. Accordingly, the field of analysis methods known as digital image analysis (DIA) is evolving (Yu et al. 2009), encompassing the microscopic techniques previously described. To address the problem of online analysis within this group of analyses, flow cells are being developed for online measurements such as fluid imaging flow cytometry (Hyeon et al. 2023).

4.5.3 Digital Image Analysis —Fluid Imaging Flow Cytometry

In fluid imaging flow cytometry, particles are analyzed from images acquired from flowing streams of sample fluids to measure the total number of particles and record the extended morphological parameters of individual particles. This non-destructive method allows larger samples to be processed more quickly for a wider range of particle sizes (0.9–1,000 μm). The online particle imaging analysis system with permanent digital recording of samples can quickly acquire multiple measurements of microplastics and reduce the time and effort required to identify microplastics. Flow cytometry allows quantification of microplastic fibers by capturing images and extracting shape parameters that indicate the influence of laundry load and chemical additives (i.e., detergents and fabric softeners) as important factors in the release of microplastic fibers in wastewater. Fluid imaging flow cytometry eliminates all the shortcomings of the previously mentioned methods and allows the adaptation of analytical methods for digitization.

4.6 SEPARATION OF PARTICULATE MATTER

Removal of microplastics from various water matrices is a major challenge today. The removal of released fibers from the washing water of synthetic textiles can be done inside the washing machine with different types of drum devices and external filters. Research shows that special filters for washing machines can remove up to 78%–80% of microplastic particles with fibrous structures on average (Erdle et al. 2021). Since textiles can carry chemical contaminants into wastewater, these filters can also reduce emissions of chemical pollutants. After installing filters for washing machines, an average 41% reduction in fibrous microplastic particles in wastewater was observed. In this way, the inflow of wastewater containing textile fragments to the wastewater treatment plant or discharge to the receiving water is prevented. Wastewater treatment plants are considered the last line of defense, providing a barrier between microplastic particles and the environment. In addition, laws and regulations must control the use and uncontrolled release of microplastics into the environment.

4.6.1 Separation for Removal Particles - WWTP Scale

The first step in wastewater treatment refers to the separation of solids from wastewater, with some degree of mechanical treatment required in the pre/primary or even secondary treatment of municipal and industrial (textile) wastewater. In this stage of processing, floating, settled, and suspended solids are completely separated, i.e., solid particles are removed from the flow to increase the efficiency of contaminant removal in the further stages of processing. Depending on the size of the pollution particles from the various wastewaters, optimal separation is achieved by using the appropriate range of grids (>0.5 mm) or perforations (>3 mm) and various sizes of screens (microsieves), multilayer filters, centrifugal separators, clarifiers, or lagoons. The addition of a settling and coagulating agent, which promotes the agglomeration of substances, further increases the efficiency of mechanical separation. Thus, the dissolved and very fine particles from the wastewater are converted into agglomerates that can be easily separated to allow the reduction of solids. As a result, a significant reduction of pollutants such as COD/BOD by 65% and phosphorus by 60% can be expected. The removal efficiency of microplastics after successful primary, secondary, and tertiary treatment stages in wastewater treatment plants is 65%, 0.2%–14%, and 0.2%–2%, respectively (Murphy et al. 2016, Talvitie et al. 2017). For the removal of microplastics in the size range of 20–190 µm, advanced treatment methods such as membrane bioreactors (MBR), rapid sand filtration, electrocoagulation, and photocatalytic degradation have proven to be effective, and these methods help to increase the removal efficiency to >99%. Krishnan et al. (2023) emphasize that smaller particles MP < 100 µm are not efficiently removed in conventional wastewater treatment plants. Granules and fragments are easier to remove than fibers and pellets. Advanced membrane technologies such as MBR which combine the bioreactor aerobic/anaerobic/anoxic conditions and membrane filtration have a high-efficiency removal rate (99.9%) for all forms of MP. In particular, the higher retention efficiency of MBR processes combined with a nominal pore size of 0.4 µm for polymeric microfiltration membranes in large-scale wastewater treatment plants can be highlighted. Ultrafiltration membranes are also effective in removing microplastic particles. They enable the removal of particles in the size range of 0.001–0.1 µm. The restriction on the use of polymeric membranes refers to secondary contamination with microplastics resulting from damage to the membrane due to chemical cleaning or high-pressure backwashing performed for the purpose of unclogging (Talvitie et al. 2017). The authors Cai et al. (2022) studied an integrated MBR and reverse osmosis system (RO) to remove MPs with a high removal efficiency of 98%. They also mentioned that small-sized fibers (<200 µm) could break through the RO system.

The separation of media with different phases is carried out in hydrocyclones, which can also be used to separate fine microplastic particles of different densities in a single-stage and serial multi-stage process. Mini hydrocyclones can be successfully used to remove >80% of microplastics >20 µm with well-defined process parameters in terms of separation ratio, loading pressure, loading flow rate, and solids concentration (Liu et al. 2022). Hydrocyclone with injected air microbubbles into the water can achieve high separation efficiency and concentration rate at the same time. Cyclonic air flotation technology can be applied to small-scale water treatment, such as separating microplastics in a washing machine (Yuan et al. 2022).

4.6.2 Separation for Identification and/or Quantification—Lab Scale

In order to perform identification and quantification of microplastic particles, the microplastic must first be separated or extracted from the source matrix, e.g., wastewater from washing. Laboratory membrane filtration processes are mostly used to separate microplastic particles from wastewater. Membrane filters made of different materials and with different pore sizes and dimensions are used.

Regarding the type of retention of particles, membrane filters are divided according to the depth and width of the pores (Cai et al. 2020). *Pore depth* structure type of filters includes stainless steel mesh, nylon, glass fiber/cellulose fiber, and nitrocellulose/mixed cellulose filters with deep and curvy pore canals. *Pore width* structure type of filters includes polycarbonate membrane with circular pores on filters and shallow and straight canals. The pore size (i.e., the diameter of a circular pore) measured using a microscope is the actual pore size of the membrane filter. The predominant forms of microplastics in wash water are fibers and fragments, as well as insoluble detergent particles, which differ significantly in shape from the spherical particles based on which pore size is determined. The pore size of the filter is important for quantifying MP particles, as long fibers and small particles can pass through large pores and not be quantified, while small pores can lead to clogging of the pores and subsequent long-term filtration. The authors Cai et al. (2020) emphasize that the results of filtering water matrices are strongly influenced by the structure and size of the filter pores. By using a different filter structure, MP achieves particle retention values of 50%–100%. The sizes of fibers and fragments found on the filters did not correspond to the pore size of the filter, e.g., fragments with a length of 37.2 μm were found on filters with a pore size of 50 μm. All this suggests that special attention should be paid to the analysis of the number and shape of MP fragments and fibrils, considering the complexity of the washing system.

4.6.3 Remarks on Identification and Separation Methods

Any of the above methods can be used with more or less success to identify and separate the released particles in the washing process. Depending on the dispersive system and its complexity, it is very important to know the advantages and limitations of each method. Tables 4.2 and 4.3 show the main features of selected identification and separation methods.

From all that has been presented, it is clear that certain methods of particle identification and separation are being successfully applied, while others are only in the testing phase.

Considering the complexity of the washing system and the influence of numerous factors, the identification methods and especially the separation methods at the wastewater treatment plant scale and laboratory scale require appropriate equipment

TABLE 4.2
Advantages and Limitations of Selected Identification Methods

Methods	Advantages	Limitations
Microscopy (Stereo, fluorescence, transmission electron, scanning electron, atomic force)	• Non-destructive • Quick and easy • High resolution • Identification of shape, size, and color • High accuracy	• Lack of data for transparent and small particles • Offline methods • Low reproducibility • No result on polymer composition
Spectroscopy (Raman, FTIR)	• Non-destructive • No false positive or negative data • High accuracy	• Expensive measurement equipment • Time – consuming • *Offline* methods • Low reproducibility
Particle size distribution (PSD)	• Non-destructive • Quick and easy	• *Off line* methods • No result on polymer composition • No determination of particle morphology
Fluid flow cytometry	• Non-destructive • *Online* methods • Larger samples processed • Quick methods • A wider range of particle sizes	• Extensive knowledge of image analysis • Undefined instruments

TABLE 4.3
Advantages and Limitations of Selected Separation Methods on WWTP and Lab Scale

Methods of Separation	Advantages	Limitations
WWTP scale Grids, screens (microsieves), multilayer filters, centrifugal separators, clarifiers, or lagoons Membrane filtration, membrane bioreactor Flotation technology	• Non-destructive • Quick and easy • Standard equipment	• Particle leakage • Fouling of membrane • Contamination with polymeric membranes
Lab scale Membrane filtration	• Non -destructive • Quickly methods • A wider range of particle sizes • Standard filtration equipment	• Interference of sample and membrane materials • Particle identification

selection, additional professional experience, and continuous training of the analyst in order to successfully separate the particles and thus protect the environment from microplastic pollution.

4.7 ANALYTICAL PROCEDURE FOR QUANTIFICATION AND IDENTIFICATION

Sample preparation and processing are followed by the identification and quantification of microplastic particles from the sample. The first step in the identification and quantification of microplastics is mainly a visual inspection of the samples with a microscope, usually a stereomicroscope. Visual inspection allows the classification of microplastic particles based on their physical properties such as color, shape, size, etc.

In order to obtain accurate and precise data on the amount of microplastic particles released during the washing of textiles made of synthetic textile materials, it is necessary to check all the steps of the analytical measurement procedure, which can be schematically represented in Figure 4.2.

Pyrolysis gas chromatography —mass spectrometry (Py-GC/MS) is a method for the thermal analysis and identification of polymers and organic macromolecules and is therefore widely used for the identification and quantification of microplastics of MP particles (Corall et al. 2022, Gomiero et al. 2021). The principle of the method is the pyrolysis of MP particles under inert conditions, after which the pyrolyzed products are separated on a chromatographic column and analyzed using GC-MS. The results of the analysis are pyrograms, based on which the chemical composition of the particles is determined. The qualifications and identifications of the peaks in the pyrograms can be confirmed by comparing the mass spectrum of each peak with those of the F Search all in one (MS 08) (Frontier Laboratories Ltd., Japan) and NIST / EPA / NIH (NIST 05) data search libraries (Dimitrov et al. 2022).

Non-destructive spectroscopic methods used in microplastics research to determine the chemical identity of microplastics in various matrices are Fourier transform

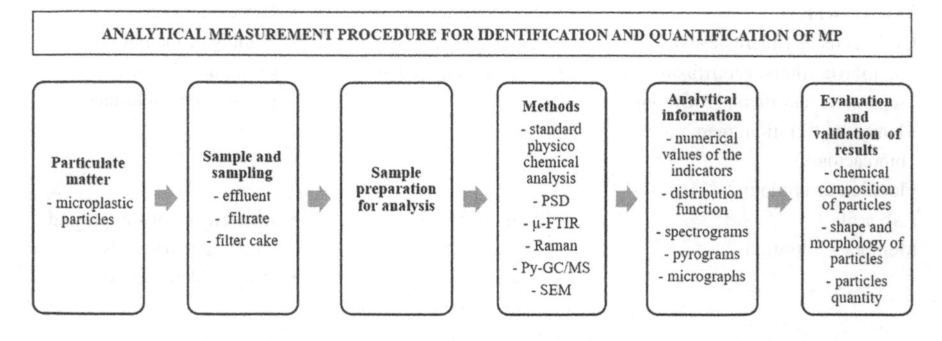

FIGURE 4.2 Scheme of analytical measurement procedure for identification and quantification of MP.

infrared (FTIR) and Raman spectroscopy on a macroscale and microscale, μ-FTIR and μ-Raman (Li et al. 2018).

The FTIR method is based on the effect of infrared radiation, which excites vibrations depending on the composition and molecular structure of the substance. The results of the analysis are IR spectra specific to polymeric materials, which can be displayed as transmittance, reflectance, and attenuated total reflectance (ATR) modes. A fast and quite reliable method that can handle large sample volumes. Samples must be IR-active, sample preparation requires careful drying to remove IR-active water, and measurement problems can occur with samples smaller than 5 μm. FTIR microspectroscopy creates a fingerprint of the material under study by measuring the change in the dipole moment of the molecule excited by infrared radiation to produce a spectrum that can be used to identify materials.

The interaction between the laser and the atoms/molecules of the samples leads to differences in the frequency of the backscattered radiation, and so-called Raman shift detection occurs, and Raman spectra specific to individual substances are generated. This method is suitable for the analysis of particles larger than 1 μm and is also the only available method for the analysis of microplastic particles in the range of 1–20 μm. Raman spectrometry allows the analysis of colored and dark particles, and relatively fast chemical mapping can be performed, resulting in rapid and automated data acquisition and processing. There are strong fluorescence interferences due to biological, organic, and inorganic contaminants (e.g., biofilm) that complicate the identification of microplastic particles. Therefore, sample pretreatment (extraction and isolation of MP particles) is required to perform this analysis. Raman spectrometry is a time-consuming method. It is necessary to repeat the procedure for each individual particle since only one particle of a given weight can be evaluated in a test series. The base of polymer spectra is limited and not all polymer compounds can be detected. Automatic mapping of μ-Raman is still in the development phase.

Scientific research has shown that for reliable and effective identification and quantification of microplastic particles, it is not sufficient to use only one analytical method. For the results to be as meaningful and reliable as possible, several different analytical approaches must be combined, following certain guidelines to prevent additional contamination of the samples with particles known to be present under the test conditions (fibers, airborne plastic particles, accessories, etc.).

The pyrogram and spectrogram dataset and all online results coupled with multivariate data analysis algorithms; successfully extract new significant information for the evaluation of an MF as a risk pollutant of washing wastewater (Dimitrov et al. 2022).

4.8 CONCLUSION

Despite numerous studies on the problem of particles in wastewater, there is still insufficient knowledge about the actual amount and size of microfibers released during the washing process. The contribution of this chapter is to provide an overview of the complexity of this problem, which includes textiles as donors of MFs, the washing process as a trigger of the release, and the heterogeneity of the effluent as a factor affecting the pollution problem. Therefore, the complexity of the observed dispersive system requires a multidisciplinary approach to optimize the parameters related to the structural properties of the textiles, the importance of hydrodynamics, the temperature, and the composition of the washing baths in order to reduce wastewater pollution.

Moreover, it is necessary to define and adapt each step of the analytical measurement procedure. In the era of modernization, as well as ubiquitous digitalization, which is considered a necessity, it is necessary to perform online monitoring of dispersive systems in combination with advanced analytical methods and statistical techniques. From all this, the area of future research can be deduced, which will cover the scientific, professional, and legal levels with elements of the circular economy.

ACKNOWLEDGMENTS

The work has been supported by the Croatian Science Foundation under the project IP-2020–02–7575, Assessment of microplastic shedding from polyester textiles in the washing process, InWaShed-MP.

REFERENCES

AATCC TM212-2021. Test method for fiber fragment release during home laundering.

Belzagui F, Crispi M, Olivarez A et al. 2019. Microplastics' emissions: Microfibers' detachment from textile garments. *Environmental Pollution*, 248, 1028–1035.

Beringer J and Kurz J. 2011. Hospital laundries and their role in medical textiles. In *Handbook of Medical Textiles*, ed. V Bartels, 385–386. Oxford: Woodhead Publishing.

Bishop D. 1995. Physical and chemical effects of domestic laundering process. In *Chemistry of the Textile Industry*, ed. CM Carr, 125–171. Glasgow: Springer.

Bloomfield SF, Exner M, Signorelli C et al. 2011, April. The infection risks associated with clothing and household linens in home and everyday life settings, and the role of laundry. *International Scientific Forum on Home Hygiene*.

Bockmühl DP, Schages J and Rehberg L. 2019. Laundry and textile hygiene in healthcare and beyond. *Microb Cell*, 6, 299–306.

Bueno L, Amador C and Bakalis S. 2018. Modeling the deposition of fluorescent whitening agents on cotton fabrics. *AIChE Journal*, 64, 1305–1316.

Cai H, Chen M, Chen Q et al. 2020. Microplastic quantification affected by structure and pore size of filters. *Chemosphere*, 257, 127198.

Cai Y, Wu J, Lu J et al. 2022. Fate of microplastics in a coastal wastewater treatment plant: Microfibers could partially break through the integrated membrane system. *Frontiers of Environmental Science & Engineering*, 16, 96.

Carney Almroth, BM, Åström, L, Roslund S et al. 2018. Quantifying shedding of synthetic fibers from textiles; a source of microplastics released into the environment. *Environmental Science and Pollution Research*, 25, 1191–1199.

Cesa FS, Turra A, Checon HH et al. 2019. Laundering and textile parameters influence fibers release in household washings. *Environmental Pollution*, 113553.

Chakraborty JN. 2012. Strength properties of fabrics: Understanding, testing and enhancing fabric strength. In *Understanding and Improving the Durability of Textiles*, ed. P Annis, 31–56. Oxford: Woodhead Publishing Limited.

Choobar BG, Shahmirzadi MAA, Kargari A et al. 2019. Fouling mechanism idetification and analysis in microfiltration of laundry wastewater. *Journal of Environmental Chemical Engineering*, 7, 2, 103030.

Corall I, Giorgi V, Vassura I et al. 2022. Seconary reactions in the analysis of microplastics by analytical pyrolysis. *Journal of Analytical and Applied Pyrolysis*, 161, 105377.

De Falco F, Cocca M, Avella M, Thompson CR. 2020. Microfiber release to water, via laundering, and to air, via everyday use: A comparison between polyester clothing with differing textile parameters. *Environmental Science & Technology*, 54(6), 3288–3296.

De Falco F, Di Pace E, Cocca M et al. 2019. The contribution of washing processes of synthetic clothes to microplastic pollution. *Scientific Reports*, 9, 6633.

Dekanić T, Pušić T, Soljačić I et al. 2021. The influence of iron ions on optical brighteners and their application to cotton fabrics. *Materials*, 14, 4995.

Dienner H. 1983. *Fleckenfernung-aber richtig!* Leipzig: VEB Fachbuchverlag.

Dimitrov N, Čurlin M, Pušić T et al. 2022. Application of GC/MS pyrolysis for assessment residues of textile composites after filtration of washing and rinsing effluent. *Separations*, 9, 10, 92.

Dulekgurgen E, Doğruel S, Karahan Ö et al. 2006. Size distribution of wastewater COD fractions as an index for biodegradability. *Water Research*, 40, 273–282.

Erdle L, Sweetnam D, Rochman CM et al. 2021. Washing machine filters reduce microfiber emissions: Evidence from a community-scale pilot in Parry Sound, Ontario. *Frontiers in Marine Science*, 8(1–9), 777865

Fijan S, Fijan R, Šostar Turk S. 2008. Implementing sustainable laundering procedure for textile in a commercial laundry and thus decreasing wastewater burden. *Journal of Cleaner Production*, 16, 258–263.

Fijan S, Šostar Turk S. 2012. Hospital textiles, are they a possible vehicle for healthcare-associated infections? *International Journal of Environmental Research and Public Health*, 9, 3330–3343.

Gatidou G, Arvaniti OS, Stasinakis AS. 2019. Review on the occurrence and fate of microplastics in sewage treatment plants. *Journal of Hazardous Materials*, 367, 504–512.

Gomiero A, Øysæd KB, Palmas L et al. 2021. Application of GCMS-pyrolysis to estimate the levels of microplastics in a drinking water supply system. *Journal of Hazardous Materials*, 416, 1–7.

He D, Luo Y, Lu S et al. 2018. Microplastics in soils: Analytical methods, pollution characteristics and ecological risks. *Trends in Analytical Chemistry*, 109, 163–172.

Henry B, Laitala K, Klepp IG. 2019. Microfibres from apparel and home textiles: Prospects for including microplastics in environmental sustainability assessment. *Science of the Total Environment*, 652, 483–494.

Hernandez E, Nowack B, Mitrano MD. 2017. Polyester textiles as a source of microplastics from households: A mechanistic study to understand microfiber release during washing. *Environmental Science & Technology*, 51(12), 7036–7046.

Hyeon Y, Kim S, Ok E, Park C. 2023. A fluid imaging flow cytometry for rapid characterization and realistic evaluation of microplastic fiber transport in ceramic membranes for laundry wastewater treatment. *Chemical Engineering Journal*, 454(1), 140028. https://doi.org/10.1016/j.cej.2022.140028

Jacksch S, Kaiser D, Weis S et al. 2020. Influence of sampling site and other environmental factors on the bacterial community composition of domestic washing machines. *Microorganisms*, 8, 30.

Kang H, Park S, Lee B. 2021. Impact of chitosan pretreatment to reduce microfibers released from synthetic garments during laundering. *Water*, 13, 2480.

Kelly MR, Lant NJ, Kurr M et al. 2019. Importance of water-volume on the release of microplastic fibers from laundry. *Environmental Science & Technology*, 53, 11735–11744.

Kotar S, Mc Neish R, Murphy-Hagan C et al. 2022. Quantitative assessment of visual microscopy as a tool for microplastic research: Recommendations for improving methods and reporting. *Chemosphere*, 308(3), 136449.

Krishnan RY, Manikandan S, Subbaiya R et al. 2023. Recent approach and advanced wastewater treatment technology for mitigation emerging microplastics contamination – A critical review. *Science of the Total Environment*, 858(1), 159681.

Lambert E, Maitra W, Scheid F et al. 2016. Differentiated evaluation of washing performance in washing machines of test stain strips as a function of temperature, washing duration and load size. *Tenside Surfactants Detergents*, 53, 424–437.

Li X, Chen L, Mei Q, et al. 2018. Microplastics in sewage sludge from the wastewater treatment plants in China. *Water Research*, 142, 75–85.

Li C, Wang L, Lijie Y et al. 2019. A new route for indirect mineralization of carbon dioxide-sodium oxalate as a detergent builder. *Scientific Reports*, 9, 12852.

Liu H, Gong R, Xu P et al. 2019. The mechanism of wrinkling of cotton fabric in a front loading washer: The effect of mechanical action. *Textile Research Journal*, 89, 3802–3810.

Liu L, Sun Y, Kleinmeyer Z, et al. 2022. Microplastics separation using stainless steel mini-hydrocyclones fabricated with additive manufacturing. *Science of the Total Environment*, 840, 156697

Mainali B, Pham TT, Ngo HH et al. 2013. Maximum allowable values of the heavy metals in recycled water for household laundry. *Science of the Total Environment*, 1, 452–453:427–324.

McCarthy JB. 2011. *Textiles for Hygiene and Infection Control/Surface*. Cornwall: Woodhead Publishing.

Mercadé-Prieto R, Bakalis S. 2014. Methodological study on the removal of solid oil and fat stains from cotton fabrics using abrasion. *Textile Research Journal*, 84, 52–65.

Miraftab M, Horrocks AR. 2007. *Ecotextiles – The Way Forward for Sustainable Development in Textiles*. Cambridge: Woodhead Publishing in association with the Textile Institute.

Murphy F, Ewins C, Carbonnier F, et al. 2016. Wastewater treatment works (WwTW) as a source of microplastics in the aquatic environment. *Environmental Science & Technology*, 50(11), 5800–5808.

Napper IE, Thompson RC. 2016. Release of synthetic microplastic plastic fibres from domestic washing machines: Effects of fabric type and washing conditions. *Marine Pollution Bulletin*, 112, 39–45.

Palacios-Mateo C, Meer Y, Seide G. 2021. Analysis of the polyester clothing value chain to identify key intervention points for sustainability. *Environmental Sciences Europe*, 33, 251–255.

Park S, Yun C, Kim J et al. 2013. The effects of the fabric properties on fabric movement and the prediction of the fabric movements in a front-loading washer. *Textile Research Journal*, 83, 1201–1212.

Penava Ž, Penava D, Nakić M. 2014. Istraživanje utjecaja utkanja osnove i potke na Poissonov koeficijent tkanine. *Tekstil*, 63, 217–227.

Pirc U, Vidmar M, Mozer A et al. 2016. Emissions of microplastic fibers from microfiber fleece during domestic washing. *Environmental Science and Pollution Research*, 23, 22206–22211.

Postle R, Carnaby A, Jong S. 1988. *The Mechanics of Wool Structure*. Chichester: John Wiley & Sons.

Pušić T, Vojnović B, Čurlin M et al. 2022. Assessment of polyester fabrics, effluents and filtrates after standard and innovative washing processes. *Microplastics*, 1, 494–504.

Pušić T, Vojnović B, Dimitrov N et al. 2021. Generiranje i otpuštanje fibrila i funkcionalnih čestica u procesu pranja. In *Book of Proceedings 2nd International Conference The Holistic Approach to Environment*, eds. A Štrkalj, Z Glavaš, 512-520. Sisak: Association for Promotion of Holistic Approach to Environment.

Reddy AS, Nair AT. 2022. The fate of microplastics in wastewater treatment plants: An overview of source and remediation technologies. *Environmental Technology & Innovation*, 28, 102815.

Reynolds KA, Verhougstraete MP, Mena KD et al. 2022. Quantifying pathogen infection risks from household laundry practices. *Journal of Applied Microbiology*, 132, 1435–1448.

Roshan P. 2014. *Functional Finishes for Textiles Improving Comfort, Performance and Protection*, 1st Edition. Amsterdam: Woodhead Publishing.

Ryzak M, Bieganowski A. 2011. Methodological aspects of determining soil particle-size distribution using the laser diffraction method. *Journal of Plant Nutrition and Soil Science*, 174, 624–633.

Saini A, Thaysen C, Jantunen L et al. 2016. From clothing to laundry water: Investigating the fate of phthalates, brominated flame retardants, and organophosphate esters. *Environmental Science & Technology*, 50, 9289–9297.

Schellenberger S, Jönsson C, Mellin P et al. 2019. Release of side-chain fluorinated polymer-containing microplastic fibers from functional textiles during washing and first estimates of perfluoroalkyl acid emissions. *Environmental Science & Technology*, 53, 14329–14338.

Schöpel B, Stamminger R. 2019. A comprehensive literature study on microfibres from washing machine. *Tenside Surfactants Detergents*, 56, 94–101.

Shabana R, Shahnaz PK, Tanveer H. et al. 2015. Colour fastness properties of polyester/cotton fabrics treated with pigment orange and various functional finishes. *Asian Journal of Chemistry*, 27, 4568–4574.

Sinner H. 1960. *Über das Waschen mit Haushaltwaschmaschinen: in welchem Umfange erleichtern Haushaltwaschmaschinen und -geräte das Wäschehaben im Haushalt?* Hamburg: Haus und Heim-Verlag.

Smulders E. 2002. *Laundry Detergents*. Weinheim: Wiley-VCH.

Sochan A, Polakowski C, Łagód G. 2014. Impact of optical indices on particle size distribution of activated sludge measured by laser diffraction method. *Ecological Chemistry and Engineering Science*, 21, 137–145.

Soljačić I, Pušić T. 2005. Njega tekstila: Čišćenje u vodenom mediju, Sveučilište u Zagrebu, Tekstilno-tehnološki fakultet.

Stamminger R. 2010. Reiningen. In *Lebensmittelverarbeitung im Haushalt*, ed. U Gomm, 28. Bon: Aid infodienst.

Talvitie J, Mikola A, Koistinen A et al. 2017. Solutions to microplastic pollution – Removal of microplastics from wastewater effluent with advanced wastewater treatment technologies. *Water Research*, 123, 401–407.

Toshikj E, Demboski G, Jordanov I et al. 2019. Functional properties and seam puckering on cotton. Shirt influenced by laundering. *Tekstilec*, 62, 4–11.

Višić K, Pušić T, Čurlin M. 2021. Carboxymethyl cellulose and carboxymethyl starch as surface modifiers and greying inhibitors in washing of cotton fabrics. *Polymers*, 13, 1174.

Višić K, Vojnović B, Pušić T. 2015. Problematika zbrinjavanja i pročišćavanja otpadnih voda - zakonski propisi. *Tekstil*, 64, 109–121.

Volgare M, De Falco F, Avolio R. et al. 2021. Washing load influences the microplastic release from polyester fabrics by affecting wettability and mechanical stress. *Scientific Reports*, 11, 19479.

Yang H, Chen G, Wang J. 2021. Microplastics in the marine environment: Sources, fates, impacts and microbial degradation. *Toxics*, 9, 41. https://doi.org/10.3390/toxics9020041.

Young SL, Matijević E. 1977. Precipitation phenomena of heavy metal soaps in aqueous solutions, III. Metal laurates. *Journal of Colloid and Interface Science*, 61, 287–301.

Yu RF, Chen HW, Cheng WP et al. 2009. Simultaneously monitoring the particle size distribution, morphology and suspended solids concentration in wastewater applying digital image analysis (DIA). *Environmental Monitoring and Assessment*, 148, 19–26.

Yuan F, Li X, Yu W et al. 2022. A high-efficiency mini-hydrocyclone for microplastic separation from water via air flotation. *Journal of Water Process Engineering*, 49, 103084.

Yun C, Park CH. 2016. The effect of fabric movement on washing performance in a front-loading washer III: Focus on the optimized movement algorithm. *Textile Research Journal*, 86, 563–572.

Zhang L, Liu J, Xie Y et al. 2021. Occurrence and removal of microplastics from wastewater treatment plants in a typical tourist city in China. *Journal of Cleaner Production*, 291, 125968.

5 Microfibres from Textile Industry Effluents

Carmen K.M. Chan, James K.H. Fang and C.W. Kan
The Hong Kong Polytechnic University

5.1 INTRODUCTION

A report from McFall-Johnsen (2020) claimed that the textile industry uses one-quarter of the world's chemicals and is responsible for 20% of global water pollution, making it the second biggest polluter. Apart from pollution, since most textile materials production uses scalable amounts of water, the scale of textile effluents becomes enormous. The textile industry is one of the world's highest water-use industries (Rather et al., 2019), and the Ellen MacArthur Foundation (2017) estimated that textile production represented 4% of the global freshwater withdrawal. The vast amount of water use is linked to the release of microfibres (MFs) in textile wastewater effluents during production and usage. Microfibers in textile wastewater have become a pressing environmental concern, as they are one of the major sinks and sources of global microplastic pollution (Woodall et al., 2014; Belzagui et al., 2019; Acharya et al., 2021).

Microplastics (MPs) are synthetic polymers often defined as plastic particles smaller than 5 mm (5,000 μm) (Thompson et al., 2004; Arthur et al., 2009; Costa et al., 2010), which includes particles in the nano-size range (1 nm) (GESAMP, 2015). However, Frias and Nash (2019) proposed a more comprehensive definition of MPs: "any synthetic solid particle or polymeric matrix, with regular or irregular shape and with size ranging from 1 μm to 5 μm, of either primary or secondary manufacturing origin, which is insoluble in water". Whereas, microplastic fibres (MPFs) are derived from petrochemicals originating from synthetic-based textiles and are considered a subset of MPs (Henry et al., 2018; Xu et al., 2018). MPFs are also defined as fibrous or threadlike pieces of plastic with a length between 100 μm and 5 mm and a width of at least 1.5 orders of magnitude shorter (Zhao et al., 2014; Fischer et al., 2016; Barrows et al., 2017).

Water is used as a medium for the wet processing of textiles, including sizing, desizing, bleaching, mercerisation, dyeing, printing, finishing and washing, to achieve the aesthetic and performance required (Madhav et al., 2018). These processes result in fibre fragmentation caused by mechanical abrasive actions between the fibres, yarns and fabric surfaces. Furthermore, adding more chemicals and additives may weaken the bonding between larger molecules and break them down into smaller molecules with natural decomposition. The chemical reaction results in a molecular

DOI: 10.1201/9781003331995-6

breakdown that not only makes the fibre degrade into smaller fragments physically but also releases chemicals potentially hazardous in the wastewater effluents, which discharge to different water bodies. Concerns are consequently raised regarding the effects of these effluents on the environment, marine life and human health.

Synthetic textiles have the most prominent global textile fibre production share, taking up 64% in 2021. The global fibre production demand has significantly grown from 8.4 kg per person in 1975 to 14.3 kg per person in 2021, with less than 0.6% recycled from pre-consumer or post-consumer textiles. Nonetheless, the trajectory will continue to grow for another 34% increase from 2020 to 2030 if the business is as usual (Textile Exchange, 2022). MFs are widely distributed and found in diverse environments such as marine, freshwater and air. Browne et al. (2011) first estimated that the accumulation of MPs was associated with shoreline population density worldwide, showing that 85% of the MPs were MPFs. According to Boucher and Friot (2017), 35% of MPs released into the ocean are from synthetic textiles, with 25% originating from wastewater. By combining the releases of MPFs shown by existing domestic washing studies, the Ellen MacArthur Foundation (2017) predicted that an additional 22 million tonnes of MPFs will go into the sea between 2015 and 2050 if demand continues to follow the current pattern. Since textile production will grow remarkably, the MF pollution issue will be magnified and more challenging to resolve.

As evidence grows, it is certain that the release of MFs by industrial sources is currently underestimated and a significant part of the pollution problem is caused by MFs from this source (Xu et al., 2018; Wang et al., 2020; Zhou et al., 2020; Chan et al., 2021). In view of these factors, addressing MF pollution and fibre fragmentation in wet processing units which are discharged through wastewater treatment plants (WWTPs) is of paramount importance. This chapter describes the release of MF from industrial sources and its pathways in detail. The effectiveness of wastewater treatment in reducing MF pollution and methods of detection and quantification of MF in wastewater effluents are also discussed. Finally, owing to the lack of legal requirements on industrial MF release, some potential control and means of mitigation that can be used to tackle the issue without delay are explored.

5.2 INDUSTRIAL TEXTILE EFFLUENTS AS A SOURCE OF MICROFIBRE POLLUTION

Different textile wet processing will eventually generate industrial textile effluents or sludges containing MFs because wastewater treatment is not 100% efficient in retaining them (Mintenig et al., 2017). Like MPs, MPFs are nondegradable and take hundreds of years to decompose, thus inevitably accumulating in the environment. MPFs are emerging as the most prevalent type of secondary MP debris in the aquatic environment (Browne et al., 2011; Woodall et al., 2014; Talvitie et al., 2015; Zhao et al., 2018; Barrows et al., 2018; Zhu et al., 2019).

Even though WWTPs are reported to be 95%–99% effective, these plants are not explicitly designed for MF retention as they can bypass the WWTPs (Magnusson & Norén, 2014; Murphy et al., 2016; Ziajahromi et al., 2017; Lares et al., 2018). Therefore, due to the enormous discharge volumes, there is strong evidence that WWTPs are significant sinks for MP pollution (Carr et al., 2016; Mintenig et al., 2017; Lares et al., 2018; Lv et al., 2019). Other than discharge wastewater, solid waste or sludge is generated for disposal from WWTPs. Sludge is a by-product of WWTPs commonly applied to agricultural land as fertilisers. A number of studies have reported long MFs found in soil, including Zubris and Richards (2005) as well as more recent studies (Selonen et al., 2020; Yuan et al., 2022).

The earliest extensive study was conducted by Browne et al. (2011), who stated that the accumulation of MPs has links to population density at shoreline sites around the globe and that MPFs contribute to 85% of the MPs identified from washing clothes. As a result, domestic laundry was considered the most common source of MF pollution and is widely researched as fibres shed from textiles at wash. Nevertheless, textiles also release fibres into the environment during production; there is a lack of studies examining the loss of MFs in the production phase. Fibre loss and release from this source are still not fully understood. Roos et al. (2017) suggested that the microfibre shedding risk of garments is reduced if the MPFs are removed from the fabric production stage. Zhou et al. (2020) and Chan et al. (2021) claimed that MFs released from industrial sources could be as significant and make no less impact than domestic laundry.

During usage and production, fabrics are subjected to mechanical and chemical stresses during domestic washing or industrial wet processing of fabrics (Shafiq et al., 2016). Raised fabrics, such as fleece fabrics, are made by abrasive actions to make them softer and more appealing. Loops are loosely knitted and brushed on the fabric surfaces releasing loose fibres through wet processing when they go through dyeing as well as air when they are cut open and brushed. The wastewater from wet processing mills is usually treated in WWTPs before they discharge into the environment. However, MFs are still unavoidably released into the atmosphere, effluent and sludge, which are still abundantly found everywhere because of their size, these processes are not designed and cannot retain them.

Figure 5.1 demonstrates different wet processing operations that can be the potential causes of MF release during the production process. Wet processing is the most renowned production phase of MF release because these processes use water and discharge effluent directly or indirectly via mill WWTPs or Centralised Effluent Treatment Plants (CETPs) to the water bodies. According to the input material form and operation types, they can be divided into three major groups of wet processing. Textile refers to fabric materials made of synthetics, semi-synthetics and natural fibres. Most natural plant fibres like cotton are cellulosic, which usually requires extra pre-treatment. Broadly speaking, the rest of the fibre types including semi-synthetics or synthetics commonly go through bleaching, dyeing, printing and finishing. Synthetic leather is a popular material used in the textile industry to simulate real leather. The wet method of production uses polyurethane (PU). Synthetic textile fabric is used as backing. The textile base is made wet by dipping it into the water and coating it with PU fluid. This is followed by solidifying in N,N-Dimethylformamide (DMF) sink and washing off before drying. Sometimes, products such as garments are finished after being sewn in their final state. This route explicitly applies to garment-dyed and washed products, such as denim. MFs fragmented from production can be released in wastewater at the industrial wet processing units and become the primary

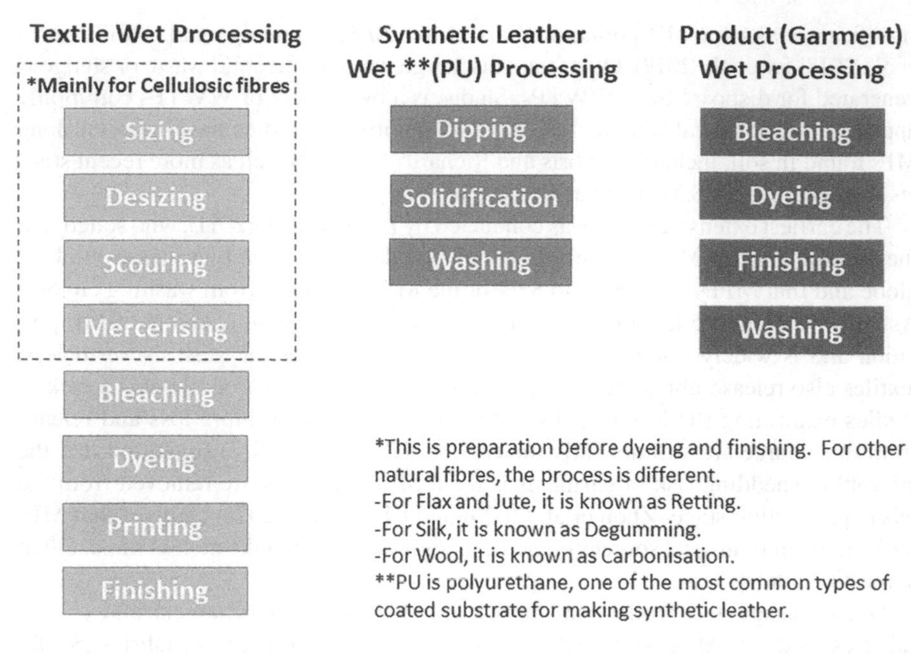

FIGURE 5.1 Common wet processes in textile, synthetic leather and products. (*Source:* Modified from Madhav et al., 2018 and Zhu et al., 2021.)

sources of industrial effluents. Although limited MF studies focus on the release from textile industrial effluents, those identified in the Web of Science until 2022 are summarised in Table 5.1.

5.2.1 Textile Mill Effluents

This section focused on effluents that are directly released from textile mills. The first study was from Swerea IVF (2018); effluents were collected from five mills in Sweden, number of MFs released was projected at 100–1,450 MFs per litre. However, a coarse mesh size of 100 μm and a comparatively small volume of 0.85 mL were used. There was a potential for underestimation, as much later research claimed that filter pore size at 100 μm ignored a significant proportion of MF (Zhou et al., 2020; Akyildiz et al., 2022), and even the most considerable number of MFs identified was below this length (Chan et al., 2021). Despite whether these mills are indirect or direct discharge types, the high end of 1,450 MFs/L identified was about three times more than the average of later studies.

Zhou et al. (2020) analysed the number of MFs/L of wastewater from three distinct types of mills in China and reported a significant difference. They varied in terms of fibre type as well as treatment method. The best-performing mill had just 5 MFs/L in its effluent because it is a tertiary treatment plant with a membrane bioreactor (MBR) and reverse osmosis (RO). Tertiary treatment is probably the key reason for explaining the low level of MFs found compared to other mills. This result was echoed by Lares et al. (2018a), who found that MBR was about 2.5 times more efficient than conventional active sludge (CAS) technology.

TABLE 5.1
Summary of Textile Industrial Effluents

Reference Source/ Type	No of Mills/ (No of Receiving Mills)	Discharge Type	Major Fibre Type	Treatment Processes	MP Retention/ Removal Efficiency	Sample Region	Adjusted Discharged as Effluent to the Aquatic Environment	Effluent MF Length	The First Filtration Mesh Size (μm)	Finest Filter Size (μm)	Sample Volume (L)	Major Characterisation Method	F/L from Direct Effluent
Textile Mill Effluents													
Swerea IVF (2018)	5	Direct	Polyester, polyamide, cotton, viscose	n.a.	n.a.	Sweden	100–1,450 MFs/L	50–500 μm (75%)	n.a.	100	0.85	FTIR	
Zhou et al. (2020)	3	Indirect	Viscose, cotton, synthetics	Primary, secondary and tertiary	84.7%–99.5% (secondary removal 44.3%–96%)	China (Hangzhou)	0.05–90[a] MFs/L	<1,000 μm (>90%) 100–300 μm highest %	n.a.	0.45	0.75	Stereo microscopy	5–1800 MFs/L
(Chan et al., 2021)	1	Indirect	Polyester	Primary, Secondary	n.a.	China (Changzhou)	18.1[a] MPFs/L	451 μm (<1000 μm 92%) <100 μm highest%	10	1.5	5	Raman	361.6 MFs/L
(Akyildiz et al., 2022)	1	Indirect	Wool, Cotton, Acrylic, Polyamide, Polyester, Polypropylene, Viscose	Primary	54%	Turkey	15.5–120.2[a] MFs/L	<1000 μm (82%) 100–500 μm highest %	n.a.	0.7	1	FTIR	310–2404 MFs/L

(*Continued*)

TABLE 5.1 (*Continued*)
Summary of Textile Industrial Effluents

Reference Source/ Type	No of Mills/ (No of Receiving Mills)	Discharge Type	Major Fibre Type	Treatment Processes	MP Retention/ Removal Efficiency	Sample Region	Adjusted Discharged as Effluent to the Aquatic Environment	Effluent MF Length	The First Filtration Mesh Size (μm)	Finest Filter Size (μm)	Sample Volume (L)	Major Characterisation Method	F/L from Direct Effluent
Laundries Textile Effluents													
Brodin et al. (2018b)	Five laundries (×3 WWTPs)	Indirect (without WWTP)/ direct)	Polyester, cotton, polycotton, nylon, rubber	Primary Primary, secondary	Chemical×2 (65%–96%) Biological×1 97%	Sweden	15–78 MFs/L (worst scenario estimate) 1–5.2 MFs/L (best scenario estimate)	<400 μm are dominant	n.a.	0.65	0.078–0.5	FTIR/SEM-EDS	500–375,000 (MPFs/L only)
Alipour et al. (2021)	2	Direct	Acrylic, polyester	Nil	n.a.	Iran	48–81 MPFs/L	<500 (81.6%–85.6%)	37	25	10	Stereo microscopy	
Industrial Wastewater Treatment Effluents (Mainly Received from Textile Mills)													
Xu et al. (2018)	1 (33)	Direct	Polyester, viscose, natural fibres (cotton, linen, wool), polypropylene	Primary, secondary	95.1% (primary 76%, secondary 83.7%, tertiary 95.1%)	China	16.3 MFs/L	30–1,000 μm (76.7%)	10	5	12	FTIR	

(*Continued*)

TABLE 5.1 (*Continued*)
Summary of Textile Industrial Effluents

Reference Source/ Type	No of Mills/ (No of Receiving Mills)	Discharge Type	Major Fibre Type	Treatment Processes	MP Retention/ Removal Efficiency	Sample Region	Adjusted Discharged as Effluent to the Aquatic Environment	Effluent MF Length	The First Filtration Mesh Size (μm)	Finest Filter Size (μm)	Sample Volume (L)	Major Characterisation Method	F/L from Direct Effluent
Xu et al. (2019)	5 (unknown)	Direct	Polyester, viscose, polyethylene, polystyrene	n.a.	89.17%–97.15%	China	7.7–9.48 MPs/L	<1,000 μm (>70%)	25	5	15	FTIR	
Wang et al. (2020)	2 (unknown)	Direct	Polyester, polyamide, polypropylene, polystyrene	Primary, secondary	Unknown	China (Changzhou)	8–23 MPFs/L	<500 μm (89%)	13	0.45	1	Raman	
Zhou et al. (2020)	2 (130)	Direct	n.a.	Primary, secondary and tertiary	92.8%–97.4%	China (Hangzhou)	537.5 MFs/L	<3,000 μm (>70%) <1,000 μm (100%)	n.a.	0.45	0.75	Stereo microscopy	

FTIR, Fourier transfer infrared spectroscopy; SEM, scanning electron microscope; EDS, energy dispersive X-ray spectroscopy)

[a] 95% efficiency is used for the next wastewater treatment before discharge to the waterbodies.

The China mill reported by Chan et al. (2021) was a homogenous production house of polyester fleece fabrics all year round. The data could easily benchmark MF release from this type of fabrication at 361.6 MFs/L from a typical wet processing mill with its secondary treatment plant. Compared with that of Akyildiz et al. (2022) in collected wastewater samples over 5 months in Turkey, there was a considerable variation in MF release on different days of production. It would have been contributed by the varied materials processed in these periods. However, the percentage of incoming fibre materials for processing and their relationship to release were not compared. It was hard to conclude whether it was solely owing to the difference in material type processed and their textile (fibre, yarn and fabric) parameters, as these were reported in domestic laundry studies, which have a noticeable impact on the quantity of MFs released (Rathinamoorthy & Raja Balasaraswathi, 2021; Palacios-Marín et al., 2022). MFs of cotton and other fibres were numerous in both inflow and outflow wastewater samples. In contrast, viscose was not found in the outflow samples, suggesting it was retained in sludge because of its higher density (Akyildiz et al., 2022).

According to Chan et al. (2021), almost all the textile mills' discharge mainly showed fibre lengths smaller than 500 μm. The shortest length group of less than 100 μm is the most dominant, implying that earlier studies using 300 μm filter pore size potentially underestimated the MF release. The study was the first to propose 5 L as the optimal sampling volume, giving a more consistent result. Although it is essential to balance the time used for filtration, the reliability of using different sampling volumes must be further validated (Chan et al., 2021).

5.2.2 Commercial Textile Laundry Effluents

Only two studies were categorised as relevant to commercial textile laundries. This area may go unnoticed because some post-consumer textile products are washed at commercial laundries on a smaller scale than industrial textile mills. Although similar mitigation solutions from domestic washing may apply to commercial laundries, their dominance should not be neglected with the scale of water used and effluent discharged. Their MPFs released were as high as textile mills or industrial effluents which is illustrated in Table 5.1.

Five commercial laundries were selected with different sources of products, from workwear, hotels, mats and hospitals by Brodin et al. (2018b). These products were made from varied materials: cotton, polyester, polycotton, nylon and rubber. The detection of MFs was more challenging, as they also contained other MPs. Approximately 27%–46% of the total MP samples were MFs, making them the most abundant type. Substantial amounts of debris may be exposed to these products, and an extra separation procedure may be needed for analysis.

Beyond the release of MFs in effluent from textile production facilities, the laundries are expected to release MFs and microparticles in the environment (Brodin et al., 2018). Since the laundry process removes loose particles (such as dirt), the treatment of the textiles can be harsh. Many other factors can also influence the severity of MF shedding, including if the fabric is entirely new in production versus aged textiles (Hartline et al., 2016). Studies by Cesa et al. (2020) and Vassilenko et al.

(2021) demonstrated that MFs release more in the first few washes, and, therefore, are likely to be similar when washed at the laundries. Small particles between 5 and 15 μm were dominant in the Brodin et al. (2018b) study, regardless of the types of textiles washed or whether the laundry had a wastewater treatment facility. From the microscopic, FTIR and SEM analysis, it could be concluded that microplastics were not dominant in this size range. Most of the particles (in the 5–15 μm range) were of other materials (for example, minerals, metal fragments, silica, aluminium silicate, yeast and starch). The release of MP particles varied significantly between the different laundries. The wastewater treatment type was greatly impacted by reducing the number of particles.

According to the results obtained from the treatment of the mats (chemical) and two kinds of workwear (chemical and biological) laundries, the amount of fibre-shaped particles released from the treated effluent decreased by 65%, 96% and 97%, respectively. Thus, it shows that wastewater treatment at the laundries can efficiently reduce the levels of particles released to the WWTPs. A specific study of carpet laundries by Alipour et al. (2021) reported Iran's fast-growing carpet cleaning services, which pointed to the MPF release from commercial laundries in Ahwaz and Sari. Laundries in both areas directly discharge their effluents to the Karun River and absorption wells with relatively high emissions of 48–81 MPFs/L. The amount of wastewater generated by washing one square metre of carpet in Ahwaz and Sari was 81 and 48 MFs/L, respectively. Sari WWTP received water was found to contain 4.9 and 12 MPFs/L microfibres during the spring and winter. Therefore, carpet-washing workshops released significantly high concentrations of MPFs. Further findings showed that short fibres with 37–300 μm were found at approximately 57% as the most dominant group; for <500 μm, it was above 80%. The severe erosion of the high-speed spinning and drying actions in these laundries further amplified the quantity of potential MFs entering the environment.

5.2.3 Industrial Wastewater Treatment Effluents

Xu et al. (2018) first reported MFs in textile industrial WWTP effluents. The plant, at 30,000 tonnes of daily capacity, receives wastewater from 33 textile printing and dyeing mills in a textile industry park in China. The average abundance was reduced from 334.1 to 16.3 MFs/L after treatment with 95.1% efficiency, incorporating synthetic and non-synthetic origins. MF abundances were also related to suspended solids but not COD, pH or nitrogen content.

Another study by Xu et al. (2019) reviewed five textiles out of eleven industrial WWTPs in China and found that the average abundance was 7.7–9.48 MPFs/L with 89.17%–97.15% retention efficiency. The most frequent length identified is 100–500 μm. Wang et al. (2020) investigated another five industrial WWTPs, but only two were linked to the textile industries. These WWTPs contributed to 8–23 MPs/L; the removal efficiency was 92.8%–97.4%. This finding was comparable to the earlier studies that focused on MPFs. Nevertheless, the outcome concluded that MPs from domestic, industrial, agricultural and aquacultural sources were insignificant. Polymer types were not examined from each WWTP, making it difficult to correlate MPs and MFs.

The Zhou et al. (2020) study focused on MFs, which were from industrial WWTPs receiving influents from 130 textile mills in China. In these WWTPs, removal rates ranged from 92.8% to 97.4%, and MF pollution levels were 537.5 MFs/L, significantly higher than the rest of the research but comparable to the surface waste contraction at 600 MFs/L in nearby areas as these industrial WWTPs directly discharge effluents into surface water.

5.3 PATHWAYS OF MICROFIBRES FROM TEXTILE INDUSTRIAL EFFLUENTS

Discussions on pathways have focused mainly on municipal WWTPs so far. In their first quantitative assessment of the world's ocean, Boucher and Friot (2017) calculated the immediate releases of primary MPs of 1.5 Mt/year, regionally exceeding the weight of those secondary MPs from mismanaged waste. In this study, the main pathways of MPs from the source to the ocean were road runoff (66%), wastewater (25%) and wind transfer (7%). It was estimated that 34.8% of the release was from the domestic laundry of synthetic textiles, followed by 28.3% from erosion of tyres while driving. Domestic laundry is the most significant pathway, which explains why it receives the most attention (Hernandez et al., 2017). Studies on industrial effluent are limited and rarely discussed, with fewer estimates.

Figure 5.2 is a flow diagram created to show the potential pathways and releases of MFs from industrial sources. Black arrows illustrate pathways for MFs to the aquatic environment. The sources are classified into pre- and post-consumer origins. Industrial wet processing mills are typically produced for pre-consumer use. Many mills have sophisticated wastewater treatment systems; some may discharge directly into the natural environment. Direct discharge refers to wastewater effluent

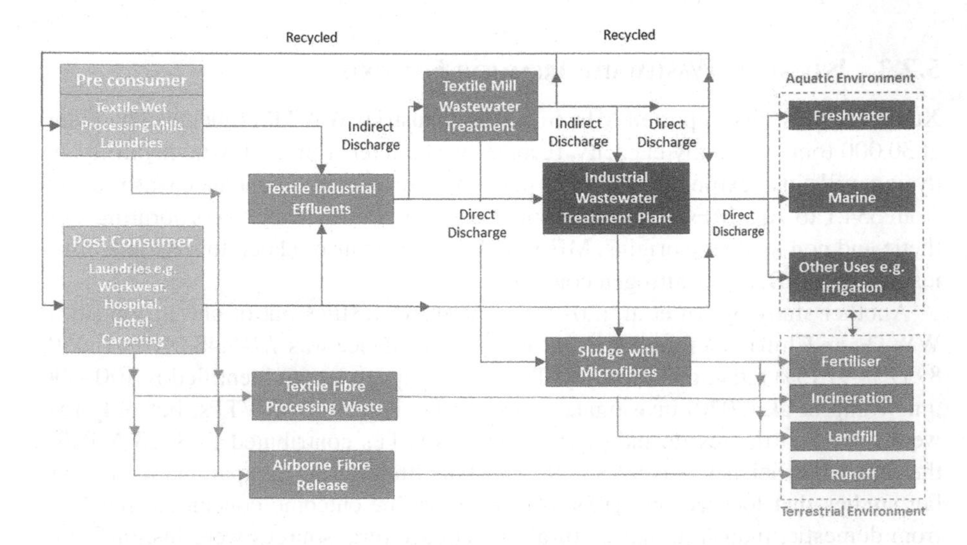

FIGURE 5.2 Sources and pathways of microfibres from industrial wet processing of textiles. (*Source:* Modified from Madhav et al., 2018 and Zhu et al., 2021.)

generated in the first place, released directly without further treatment. Some of the mills and laundries may not have treatment facilities or effluent quality that meet the direct discharge standards required by local laws and will be directed to CETPs before subsequently being discharged to the aquatic environment. These are indirect discharge pathways. It is worth noting that sometimes effluents will be recycled to reduce water use.

Referring to the terrestrial pathway in Figure 5.2 (Grey arrows), WWTPs produce sludge as a by-product; depending on their content, some can still be used as fertilisers. Practically, most of them are incinerated, which is more appropriate for industrial sludge because of the prominent level of toxicity. Airborne MFs are released into the atmosphere during production, and textile MF waste intentionally disposed of for incineration or landfill follow the grey arrow paths. MFs that cannot be captured might blow by wind current as runoff. Aquatic and terrestrial pathways can interact through storms and wind (Liu et al., 2019; Wright et al., 2020).

5.4 EFFECTIVENESS OF WASTEWATER TREATMENT IN REDUCING MICROFIBRE POLLUTION

5.4.1 General Performance

Mintenig et al. (2017) proposed that WWTPs serve not only as sinks but also as sources of MPs and thus were critically indispensable in MP pollution. WWTPs as pathways for MP release have drawn more attention recently, with an exponentially growing number of related publications in the last few years (Sun et al., 2019). There is sound evidence that MPs, MPFs and MFs can easily bypass WWTP filtration and other solid separation processes, as they are not designed for such a purpose (Murphy et al., 2016; Ziajahromi et al., 2017; Conley et al., 2019).

Magnusson and Norén (2014) and Talvitie et al. (2017a) demonstrated that the supply of MPs from WWTP effluents to the aquatic environment might be substantial because of the enormous daily discharge volumes. Still, their relative importance concerning other sources/entrance routes is difficult to estimate due to the lack of quantitative studies. The reviewer found that there is still a lack of research in this area, and it is poorly understood (Hu et al., 2019; Xu et al., 2021).

In the study by Conley et al. (2019), MPF removal efficiency was reported at 80.2%–97.2%, which was relatively high but still significantly less than the total MPs. Talvitie et al. (2017) suggested that advanced wastewater treatment (e.g., a membrane bioreactor) is needed to improve the removal efficiency of small-sized MPs (<100 μm). This efficiency level was echoed by Ziajahromi et al. (2017), who reported that fibres were the dominant type of MPs detected in most effluent samples and were not completely removed even after some advanced treatment processes. A UNEP (2018) report estimated the MPF retention efficiencies from four main types of wastewater treatment options, as shown in Table 5.2.

WWTPs in developed countries are generally believed to be more efficient. For example, Mintenig et al. (2017) discovered that a German WWTP removed 98% of MFs after advanced filtration. In new economies, WWTPs are usually of a lower standard because of inadequate sewage infrastructure. In 2014, China alone

TABLE 5.2
The Microplastic Fibre Removal Efficiency for Different Wastewater Treatment Options

Wastewater Treatment Options	Microplastic Fibres Removal Efficiency (%)
Preliminary	58.0
Primary	87.0
Secondary	92.2
Tertiary	96.5

Source: Modified from UNEP (2018).

accounted for 69% of all polyester fibre production globally, with the combined output of China, India and Southeast Asia representing over 80% of the global total (Henry et al., 2018). Nevertheless, developing countries produce and consume more synthetic textile materials at 62.7% compared to 48.2% in developed countries (Boucher & Friot, 2017). These regions are still developing and will tend not to have a tertiary treatment standard commonly available, as illustrated in Table 5.3 and are of more desperate concern.

Figure 5.3 illustrates a typical MP flow in various stages of primary, secondary and tertiary WWTPs, which should also apply to MPFs. For non-synthetic MFs, there is no independent study focusing on this group of fibres alone on MF retention. Therefore, most data are aggregated as MPs are similar to MPFs in chemical structure, and MPFs are like non-synthetic MFs in their fibrous shape. The assumption must be made until there is more data available.

Pre-treatment or primary treatment has an immense impact on the size distribution and removal of sizable MPs via skimming on primary clarifiers and settling heavy MPs trapped in solid flocs during drift removal and gravity separation. The efficiency at this stage was reported at 35%–59%. Sun et al. (2019) concluded that secondary treatment, typically comprising biological and clarification, could increase retention by an additional 0.2%–14%. At this stage, suspended matter aggregates together, forming a floc that will consequently be removed at the settling stage (Carr et al., 2016). However, there were more fragment particles removed relative to the fibres. Despite an absence of a larger than 500 μm size that was believed to be effectively removed in this secondary treatment process (Mintenig et al., 2017; Talvitie et al., 2017; Ziajahromi et al., 2017; Conley et al., 2019). Figure 5.3 shows that the overall MP reduction in the tertiary treatment was estimated to decrease further by 0.2%–2%. Talvitie et al. (2017) suggested that advanced wastewater treatment could improve the removal efficiency of small-sized MPs (<100 μm). Different tertiary technologies were compared, MBR was considered the most effective at 99.9%, followed by rapid sand filtration (RSF) at 97% and dissolved air flotation (DAF) at 95%. Disc filter (DF) tended to vary between 40% and 98.5%.

TABLE 5.3
The Distribution of Wastewater Treatment Options in Wastewater Treatment Plants of Different Regions

	NAFTA (Incl. Rest of North America)	Western Europe	Japan	Central Europe & CIS	Asia (Excl. Japan, India and China)	Africa	Latin America & Caribbean	Oceania	India	China	Middle East
Share of population covered in (OECD stat. 2023)	100%	93%	100%	9%	4%	0%	14%	12%	0%	0%	2%
Share going to preliminary treatment	0%	0%	0%	0%	0%	0%	0%	0%	0%	0%	0%
Share going to primary treatment	17%	8%	13%	5%	65%	65%	53%	18%	65%	3%	30%
Share going to secondary treatment	46%	21%	57%	20%	35%	35%	28%	32%	35%	97%	39%
Share going to tertiary treatment	37%	72%	30%	75%	0%	0%	18%	50%	0%	0%	31%
Reference	OCED stat (2023)	OCED stat (2023)	OCED stat (2023)	OCED stat (2023)	Due to lack of better data, the treatment share was assumed to be the same as in India.		OCED stat (2023)	OCED stat (2023	Kalbar, stat (2018)	Zhang et al. stat (2016)	OCED stat (2023)

Source: UNEP (2018).

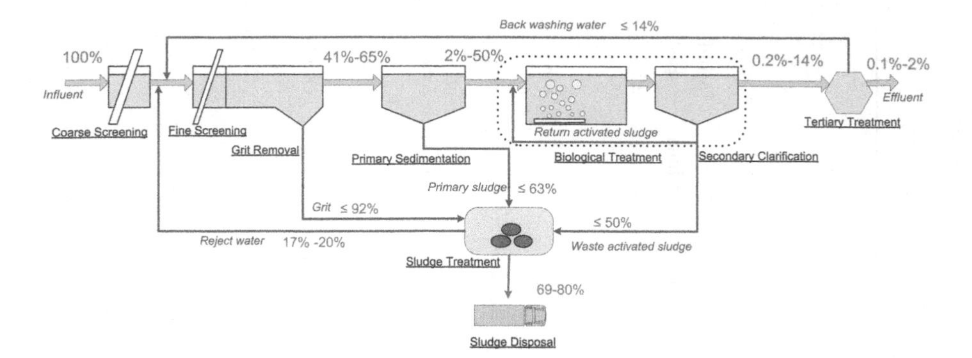

FIGURE 5.3 Estimated microplastic flow in wastewater treatment plants with primary, secondary and tertiary treatment processes. The sludge phase was calculated according to the particle balance. (*Source:* Adapted from Sun et al., 2019.)

5.4.2 Performance Specific to Industrial Textile Effluents

Referring to Table 5.1, the Industrial WWTPs have a better MF retention performance at 90% or above from all reported sources (Xu et al., 2018, 2019; Zhou et al., 2020). The textile mill has a lower retention rate at 84.7% even though it has MBR tertiary treatment processes, whereas, it was just at 54% for the Turkish mill, which only has primary treatment (Akyildiz et al., 2022). These textile mills will discharge their effluent to industrial WWTPs. This pathway with an additional treatment process should have lowered the MF release significantly and become less risky than those mills or laundries that discharge directly to the aquatic environment. However, even after advanced treatment, most effluent samples were not completely removed (Ziajahromi et al., 2017). In contrast to other shapes, the fibres retained were more effective, and a higher reduction was found after pre-treatment (Talvitie et al., 2015; Ziajahromi et al., 2017).

Xu et al. (2018) found that most textile fibres attached to the gravel and flocs were removed in the initial primary sedimentation stage at 76% and then 83.7% after secondary treatment. Therefore, the additional treatment as a second procedure in CETP potentially contributes to the higher efficiency of overall MF removal. MFs larger than 1,000 μm were most effectively removed at preliminary treatment than those between 500 and 1,000 μm. There was no selectivity in smaller MFs.

Only a handful of research has investigated industrial textile wastewater. Indeed, the reported figures between studies were relatively difficult to compare, as there were large variations in the sampling methodologies conducted across various dimensions. For example, the finest filter sizes vary as much as 153 times (i.e., from as small as 0.65 μm to as big as 100 μm). Furthermore, according to the industrial textile effluent described in Swerea IVF (2018), the production parameters, such as capacity and materials processed, were not captured. This missing detail made analysis difficult, as these parameters might have a significant role in understanding the true scenario of MF fragmentation. For instance, some textile processes might use smaller volumes of water, which could cause higher concentrations of MFs in their effluent and potentially misinterpret the result. Nonetheless, industrial effluents likely have a dramatically higher concentration of MFs than municipal. For example, a comparison of results from Xu et al. (2018) with the municipal effluents from Yang et al. (2019) and Lv et al. (2019) in China revealed that there was 28–1,310 times more concentrated effluent of MFs discharged directly to aquatic environments from industrial sources than from municipal WWTPs.

5.5 DETECTION AND QUANTIFICATION OF MICROFIBRES IN TEXTILE EFFLUENTS

Although there are some achievements lately in standardising the test method for MFs from domestic laundry AATCC TM212 (2021), ISO/DIS 4484 (2022) and TMC method(2019) (The Microfibre Consortium, 2019), there is neither an established nor international standard for the detection and quantification of MFs in textile effluents. Some collaboration has been initiated by Cowger et al. (2020) to improve the reporting of MPs, making research results more reproducible, comparable and transparent. These can be a good reference in preparing the development of applicable standards for industrial effluents.

Due to the variations of procedures in conducting MPs or MFs analysis, the key challenge is more on the analytical processes. In Figure 5.4, Kang et al. (2020) have summarised the most widely used procedures, from sample collection and pre-treatment to characterisation. However, it should be noted that depending on the purpose and nature of the samples, not all the methods are mandatory and can be adjusted accordingly.

5.5.1 SAMPLING

Eunomia et al. (2017) stated that potentially the best methods currently devised for wastewater effluent MP measurement use large sample volumes (≥1 m³), a small mesh size for filtration (<10 μm) and automated material analyses. However,

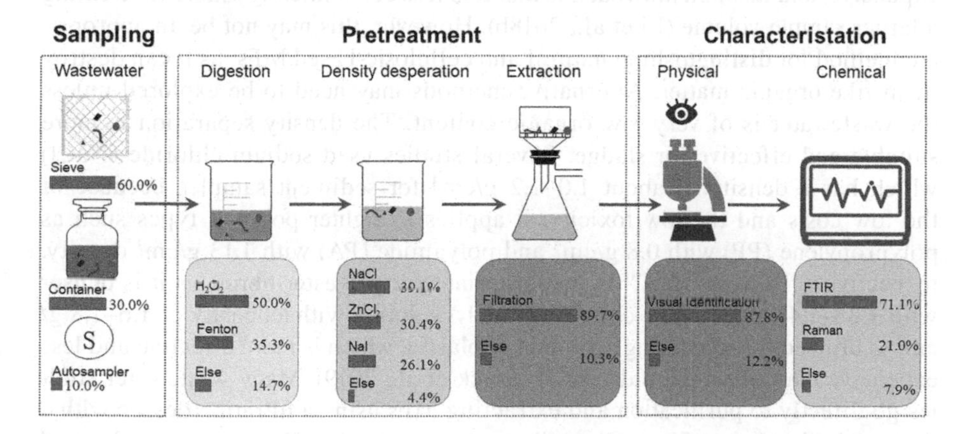

FIGURE 5.4 Summary of microplastic detection methodologies and analysis techniques from wastewater. (*Source:* Kang et al., 2020.)

examining studies conducted to date have rarely found the use of large sample volumes reaching 1 m^3. Four out of ten studies shown in Table 5.1 have used larger volumes. A larger volume can have higher accuracy; however, it means a longer time to complete with more resources which might not be justified economically. Some studies also tried taking a sample over different days of production to report an average. These results may not be easy to analyse without understanding the relationship between processed effluents and the production parameters, as the composition was expected to be highly heterogeneous. Collecting samples a few times on a commercial scale is not cost-effective. Chan et al. (2021) collected the samples over 6 hours, covering three production cycles. Referencing wastewater guidelines from the ZDHC Foundation (2016) are more feasible to analyse as production over a shorter period within a day stabilises, and results are more conclusive. The outcome was the proposal of using 5 L as the optimal volume. By all means, further studies are needed to cover a broader type of textile production process. Indeed, other researchers have used finer mesh sizes with a positive relationship to potentially retaining finer and shorter MFs (Ziajahromi et al., 2017). In Table 5.1, three out of ten have used a mesh size of over 10 μm in 13, 25 and 35 μm, which might cause underestimation as textile MFs are between 10 and 30 μm in diameters (Xu et al., 2018; Schmiedgruber et al., 2019; Dreillard et al., 2022).

5.5.2 Pre-treatment

Most organic matter in the sample can be destroyed by active digestion by 30% (v/v) H_2O_2 (Li et al., 2018), and this step can significantly reduce major interferences in the subsequent spectroscopic analysis (Dyachenko et al., 2017). Alternatively, other chemicals, such as Fenton reagents or enzymes, are sometimes used for other advantages, such as reaction time and temperature and more rapid organic removal; however, H_2O_2 remains the most popular option to remove organic matter in the sample (Li et al., 2018; Sun et al., 2019). H_2O_2 is relatively expensive, and its main drawback is that it is less economically viable for treating a larger sample volume (Li et al., 2018b). However, this may not be an appropriate method for distinguishing natural and cellulosic-based MFs, as it can destroy them like organic matter. Alternative methods may need to be explored unless the wastewater is of very low organic content. The density separation is more suitable and effective for sludge. Several studies used sodium chloride (NaCl) which has a density of about 1.0–1.2 g/cm^3 for sediment samples because of the low costs and the low toxicity, it applies to lighter polymer types such as polypropylene (PP) with 0.8 g/cm^3 and polyamide (PA) with 1.13 g/cm^3 density, respectively. However, NaCl is inappropriate for polyester fibres, as it is denser with 1.38–1.41 g/cm^3. Zinc chloride ($ZnCl_2$) solution with a density of 1.6–1.8 g/cm^3 is utilised for separating all kinds of plastic, which is more effective and less expensive than sodium iodide (NaI) (Stock et al., 2019). Many wastewater studies go directly to purification and extracting MFs using a filtration device with a fine mesh filter below 5 μm. Glass fibre, paper and nylon filters were mostly used (Ryan et al., 2020).

5.5.3 Characterisation

For many years, the characterisation of polymers has relied on optical spectroscopy methods to provide information on polymeric materials' identity and chemical compositions. One significant finding was the inaccurate ability to identify MP or MF solely by microscopic visual inspection. Instead, each particle needs to be verified as plastic using a spectroscopic technique that can handle tiny particle sizes, such as micro-Fourier-transform infrared microscopy (μ-FTIR) or micro-Raman spectroscopy (Dyachenko et al., 2017). These techniques are the most common and are highly recommended by Prata et al. (2019) for characterising MFs with similar performance. Raman spectroscopy is a laser-based method that provides better resolution than infrared spectroscopy. It is well suited when the process requires focusing on small regions of a sample. It can also address the identification of MFs as small as 1 μm. The Raman spectrum yields similar but complementary information to that found with FTIR. Optical microscopy can be used with both instruments to establish a standardised method for the qualitative analysis and characterisation of MFs (Lares et al., 2019). FTIR was considered the most popular method in general MP and MF studies. Scanning Electronic Microscope (SEM) was used only by one study in Table 5.1.

5.6 POTENTIAL CONTROL AND MITIGATION FOR TEXTILE INDUSTRIAL WASTEWATER EFFLUENT

5.6.1 Source Control

Notably, industrial effluents carry a significantly higher concentration of MFs than municipal effluents when discharging into the environment, without a doubt (Xu et al., 2018; Zhou et al., 2020; Chan et al., 2021). In addition, pathways such as commercial laundries (Alipour et al., 2021) were not treated by standard WWTPs before they were discharged into the environment. The higher concentration of MFs in industrial effluent was alarming, as little data is available. Therefore, identifying these sources and finding the best options to divert these effluents is crucial. Likewise, some direct discharge industrial WWTPs may also deliver a higher concentration of MFs, as sedimentation is the most effective process; those pathways having a second treatment in CETP should also lower the MF concentration in the second process.

Eunomia et al. (2017) and Brodin et al.'s (2018) studies proposed that there were more loose fibres in the first wash of a garment, as high as three times that of subsequent washes. Therefore, MFs shed during the manufacturing process are incompletely washed off. As a result, reducing MF loss in production may be a more effective means to tackle the problem than actions during the usage phase.

5.6.2 Technology Advancement

Advanced treatment technologies, such as the Membrane Bioreactor (MBR) and Zero Liquid Discharge (ZLD), are available in the industry to reduce water stress and MF pollution, which have proven effective in reducing MP discharge by 97% (Mintenig et al., 2017). MBR may be the most efficient method among the common wastewater treatment technologies to eliminate MFs from wastewater. However, the appropriateness of selecting a technology that fits MFs requires more study as the efficiency depends on the material, morphology, size and density of input MPs (Xu et al., 2021). Zhou et al. (2020) suggested that dissolved air floatation (DAF) is a preferred technology for separating MFs, as they are of comparatively lower densities than effluent. Since industrial textile effluent, with its vast volume, still releases massive quantities of MFs, it suggested milder processing conditions and cleaner chemical materials can ease the release and reduce its impact. In addition, DAF can be a more economical option than reverse osmosis and ZLD. The drawback of advanced technology is the comparatively high investment costs, sometimes high energy demand and even higher cost of operation. Hence, their availability in industrial WWTPs is still limited. Furthermore, there is no specific role in MF removal in membrane technology (Westphalen & Abdelrasoul, 2018). For instance, the removal of MFs from textile wastewater using membrane technology has not been reported (Xu et al., 2018). Therefore, further research and development are required to achieve a more affordable and effective means of preventing MFs from reaching the environment.

5.6.3 Sludge Management

Since WWTPs are the major source and sink of MFs, a higher retention rate using advanced treatment technology should considerably influence reducing release. However, proper sludge handling must go in parallel to ensure the problem is not being transferred from aquatic to terrestrial. Precautions are required to ensure that the higher concentrations of MFs in sludge do not pose a problem in terrestrial environments. MFs found in terrestrials have the potential to overshoot aquatic emissions. Suppose the retention rate of WWTPs continues to improve. In that case, the concentration of MFs in sludge will increase and pollute the terrestrial ecosystem, and with airborne MFs, it was suggested it might overshoot aquatic (Gavigan et al., 2020). Therefore, improving the technology in WWTPs is not a complete solution to stop MF release from polluting waterways. In order not to transfer the issue to the soil ecosystem, a robust solid waste management system must be established. Textile sludge using the incineration technique at 800°C was reported by Iqbal et al. (2014) as an effective method of removing heavy metals and reducing their volume. This temperature is much higher than the melting point of all fibres, so it should probably decompose all textile fibres. The ash can also be used for block preparation in the construction industry.

5.6.4 Legislation and Policy

Increasing recycled water within the WWTPs is considered a quicker fix. Lack of knowledge and test methodology, inability to include MPF requirements in sourcing policy from retailers and limited financial incentives to invest are the most significant barriers to delaying industrial transformation. Innovation in dry finishing technologies, such as using carbon dioxide as a dye carrier, is cleaner than wet processing. More sustainable technology or green production is a new trend that can help resolve

MFs and water pollution in one basket. Combining these benefits can multiply the effects on accelerating investment decisions.

5.7 FUTURE PRIORITIES

5.7.1 Understand the Complete Picture and Urgency to Address Microfibre Pollution

Research from the Ellen MacArthur Foundation (2017) and Rudenko (2018) revealed that 30% of garments produced are unsold, and 25% of those sold are never worn. Therefore, the estimated figures for MF pollution should be adjusted to reflect the true scenario when considering these factors. Furthermore, owing to the fact that most of the environmental pollution estimates were derived from previous domestic laundry studies, it has demystified the significance of industrial effluent as a source of MF emissions persistently.

Existing estimates calculate MFs shed or lost from domestic laundry related to different textile compositions, fabric types, wash conditions and methodologies (Bruce et al., 2017; Lant et al., 2020; Cai et al., 2020; Tiffin et al., 2021). These figures vary widely and were difficult to compare by significant orders of magnitude. Only limited studies estimate global MF loss based on domestic laundry (Belzagui et al., 2020). In addition, these figures are not updated according to the latest findings with most of the length <1,000 μm and finer linear density; therefore, previous studies using larger mesh sizes up to 300 μm have underestimated the abundance. So, recalculation must be made to compare to the earlier estimations meaningfully.

MFs are tiny and can be characterised by various densities, resulting in their dispersal through different layers in the water column (Wright et al., 2013). Consequently, removing all MFs that have accumulated and scattered in the oceans is less feasible. Besides, the rate at which MPs enter the environment exceeded the removal rate (Auta et al., 2017). Although a few studies identified several bacteria species capable of degrading plastic polymers, including PET, in soil (Asmita et al., 2015), such microbes were not used for mass cleanup. They potentially pose risks not yet fully identified concerning bacteria released into the environment. Even though there is a solution to clean plastic debris in the ocean (Davidson, 2019); however, the size is significantly longer than 1,000 μm, which is still a long way to go for MFs to be effectively cleaned up, with the majority found to be shorter than this. Therefore, a precautionary approach is more viable to tackle the problem, being more realistic for reducing the number of MFs generated and preventing them from entering the aquatic environment.

5.7.2 Raising Stakeholder Awareness

Raising awareness among all stakeholders in the supply chain of MF release is vital to develop mitigation measures. Every stakeholder should understand the issue to play a role and be a part of the solution. For industry, the mitigation measures may include varying the choice of materials for design and production. In addition, consumers can communicate concerns about fibre fragmentation from domestic laundry to retailers and manufacturers so that they can improve fabric performance. Similar to domestic washing studies, consumer awareness campaigns are more popular than those of industry stakeholders. However, awareness should extend to the industrial value chain, with each stakeholder having a role in developing, monitoring and implementing new solutions to reduce fibre fragmentation. As a result, this will create long-term and sustainable solutions to the problem of MF pollution.

5.7.3 Establishing a Standard Test Method for Measurement and Reporting

Controlling and monitoring the MF release is currently impractical since measurements are not comparable due to the lack of a standardised test for evaluating MFs released from WWTPs. Thus, establishing a standard testing method and reporting is a recognised priority. Analysis methods employed in the identification of MPFs and MFs vary considerably. The non-standardised sampling approach represents a massive challenge in research on MFs or wider MPs. For this reason, findings are hardly comparable across research groups. A group of researchers have released a guideline to standardise reporting and make data more reproducible and comparable (Cowger et al., 2020). Therefore, it is expected that later studies shall be improved. Nevertheless, consistent and standardised sample processing and analysis procedures are essential to estimate the abundance of MFs and MPs and identify their migrating pathways.

The test method must be suitable to be employed in the broader industry. Academia or the industry alone cannot bridge this knowledge gap. There is a need to collaborate and develop diversified and innovative solutions that can be scalable for implementation. For example, The Microfibre Consortium (The Microfibre Consortium, 2021) plays a key role and drives to pull together industry expertise, academia and government to develop solutions. At this moment, a preliminary control guideline on MFs in wastewater was released by TMC (The Microfibre Consortium, 2022). A standard test method is expected to be developed by 2023 and widely adopted by the industry.

Making the new standard practical, a collaborative effort in creating an imaging database of confirmed MFs could enhance method robustness, thereby aiding the use of existing equipment in commercial laboratories, such as high-resolution light microscopes. Besides, this enables the method to minimise the need for costly micro-FTIR or micro-Raman spectroscopes to confirm the characteristics of all MFs, representing an economic and logistic barrier to implementation.

5.7.4 Better Product Design and Manufacturing Processes

Product design improvement is recommended by adopting critical parameters that substantially impact MFs released during washing (MERMAIDS Consortium, 2017). These parameters include longer fibre length, higher yarn twist, coarser yarn count, higher fabric density and fewer textile auxiliaries that can reduce MF shedding. Transparent production parameters imply significant steps that must be taken during the design phase to create an impact. Concurrently, further support is necessary for decision-makers to motivate designers and manufacturers to materialise these parameters. In particular, it will be critical to enable the capability to measure the effectiveness of these parameters through unbiased testing.

The recommended specifications of manufacturing processes are yet to be explored. Future studies can encompass an extensive scope for evaluating different production parameters, including fibre type, fabric constructions, machine settings, chemical agents and wastewater treatment technologies. These parameters are building blocks for standard implementations, making improvements scalable for performance.

Shedding reduction may also be implemented by strengthening the cohesive bonds between fibres. However, research in this area is insufficient, and the data gathered is inconclusive about its effectiveness. Alternatively, there are suggestions on the efficiency of pre-wash, air filtration and exhaustion at sites (Almroth et al., 2018). However, since industrial facilities have more efficient controls over releases than domestic laundries, these proposals must be applied cautiously, and there must be appropriate disposals without going further into the environment.

5.7.5 Responsible Consumption

There are several changes that need to happen at the consumer level. These include reducing consumption, extending garment life and disposing of them properly to enhance circularity. Consumers need to be recognised as part of the solution. Reduced consumption and manufacturing would reduce the quantity of MFs released into the environment with less material use. However, a significant challenge involves the simultaneous and systematic change in consumers, multi-stakeholders and governments' efforts. According to the Ellen MacArthur Foundation (2017), less than 1% of clothing is recycled into apparel. A more sustainable approach should be to slow down the overall consumption of resources and support circularity using innovative business models.

5.8 CONCLUSION

The release of MFs from industrial wastewater treatment plants is as substantial as that from domestic laundry. Existing research has already suggested that the impact of MF release from this source may even be more severe. This is because MFs in industrial effluents are more abundant in wastewater treatment plants, while some pathways, such as direct, are still understudied. Additionally, as a large volume is discharged daily into aquatic environments, the impact may be more significant than previously imagined.

Brodin et al. (2018) emphasised that MFs shed from the initial wash are three times greater than those from subsequent washes of the same garment. The evidence strongly suggested that the creation of loose fibres happens during production. Therefore, further opportunities exist to address this issue during the early stages before products are marketed and to prevent MFs from being dispersed into the environment. As garments made are more than it is sold and worn, we can estimate that the impact of industrial MF release can be much higher than previously predicted. Therefore, common sense dictates that addressing this issue upstream would be more effective.

Studies focusing on MF release during production are still in their infancy; many unanswered questions exist. Adopting a consistent protocol will allow for comparisons across studies, which are crucial when identifying areas of high impact. It is necessary to raise awareness and encourage industry and academia to conduct further investigations. The ability to measure MFs in WWTPs can support industry and regulatory bodies in defining the maximum threshold for releases of MFs, thus enabling the evaluation of mitigation measures and improving industrial wastewater treatment practices.

Once MFs can be quantified consistently, control policies are feasible to implement in the value chain. Subsequently, this will drive additional efforts to take remedial measures further upstream from designing and developing materials in fibre, yarn or fabric form. Generally, the most effective way of addressing any production issue is to fix it as early as possible before it is too late or too costly to fix later. Although the advancement in WWTP technologies can effectively stop MF release as effluents, it should still be considered equally important unless dry processes in production can replace these conventional wet processes.

REFERENCES

AATCC. (2021). *Test method for fiber fragment release during home laundering*. https://members.aatcc.org/store/tm212/3573/

Acharya, S., Rumi, S. S., Hu, Y., & Abidi, N. (2021). Microfibers from synthetic textiles as a major source of microplastics in the environment: A review. *Textile Research Journal*, *91*(17–18), 2136–2156. https://doi.org/10.1177/0040517521991244

Arthur, C., Baker, J., & Bamford, H. (2009). Proceedings of the International Research Workshop on the Occurrence , Effects , and Fate of Microplastic Marine Debris. *NOAA Marine Debris Program National, January*, 530. https://marinedebris.noaa.gov/proceedings-international-research-workshop-microplastic-marine-debris

Akyildiz, S. H., Bellopede, R., Sezgin, H., Yalcin-Enis, I., Yalcin, B. & Fiore, S. (2022). Detection and analysis of microfibers and microplastics in wastewater from textile company. *Microplastics*, *1*(4), 572–586. https://doi.org/10.3390/microplastics1040040

Alipour, S., Hashemi, S. H. & Alavian Petroody, S. S. (2021). Release of microplastic fibers from carpet-washing workshops wastewater. *Journal of Water and Wastewater*, *31*(6), 2021. https://doi.org/10.22093/wwj.2020.216237.

Asmita, K., Shubhamsingh, T., Tejashree, S., Road, D. W. & Road, D. W. (2015). Isolation of plastic degrading micro-organisms from soil samples collected at various locations in Mumbai, India. *International Research Journal of Environment Sciences*, *4*(3), 77–85. https://pdfs.semanticscholar.org/5f34/8903e9b2eef3588bd3753692d3ff949a3210.pdf

Auta, H. S. S., Emenike, C. U. & Fauziah, S. H. (2017). Distribution and importance of microplastics in the marine environment: A review of the sources, fate, effects, and potential solutions. *Environment International*, *102*, 165–176. https://doi.org/10.1016/j.envint.2017.02.013

Barrows, A. P.W., Cathey, S. E. & Petersen, C. W. (2018). Marine environment microfiber contamination: Global patterns and the diversity of microparticle origins. *Environmental Pollution*, *237*, 275–284. https://doi.org/10.1016/j.envpol.2018.02.062

Barrows, A. P.W., Neumann, C. A., Berger, M. L. & Shaw, S. D. (2017). Grab: Vs. neuston tow net: A microplastic sampling performance comparison and possible advances in the field. *Analytical Methods*, *9*(9), 1446–1453. https://doi.org/10.1039/c6ay02387h

Belzagui, F., Crespi, M., Álvarez, A., Gutiérrez-Bouzán, C. & Vilaseca, M. (2019). Microplastics' emissions: Microfibers' detachment from textile garments. *Environmental Pollution, 248*, 1028–1035. https://doi.org/10.1016/j.envpol.2019.02.059

Belzagui, F., Gutiérrez-Bouzán, C., Álvarez-Sánchez, A., & Vilaseca, M. (2020). Textile microfibers reaching aquatic environments: A new estimation approach. *Environmental Pollution, 265*, 114889. https://doi.org/10.1016/j.envpol.2020.114889

Boucher, J. & Friot, D. (2017). Primary microplastics in the oceans: A global evaluation of sources. *Primary Microplastics in the Oceans: A Global Evaluation of Sources*, 43. https://doi.org/10.2305/iucn.ch.2017.01.en

Brodin, M., Norin, H., Hanning, A.-C. & Persson, C. (2018a). *Filters for washing machines Mitigation of microplastic pollution.* https://urn.kb.se/resolve?urn=urn:nbn:se:ri:diva–58472

Brodin, M., Norin, H., Hanning, A.-C., Persson, C. & Okcabol, S. (2018b). *Microplastics from industrial laundries – A laboratory study of laundry effluents.* https://urn.kb.se/resolve?urn=urn:nbn:se:ri:diva–58473

Browne, M. A., Crump, P., Niven, S. J., Teuten, E., Tonkin, A., Galloway, T. & Thompson, R. (2011). Accumulation of microplastic on shorelines woldwide: Sources and sinks. *Environmental Science and Technology, 45*(21), 9175–9179. https://doi.org/10.1021/es201811s

Bruce, N., Hartline, N., Karba, S., Ruff, B., & Sonar, S. (2017). *Patagonia Microfiber Pollution and the Apparel Industry, Bren School of Environmental Science & Management.* 1–98. http://www.esm.ucsb.edu/research/2016Group_Projects/documents/PataPlastFinalReport.pdf

Cai, Y., Yang, T., Mitrano, D. M., Heuberger, M., Hufenus, R., & Nowack, B. (2020). Systematic Study of Microplastic Fiber Release from 12 Different Polyester Textiles during Washing. *Environmental Science and Technology, 54*, 4847–4855. https://doi.org/10.1021/acs.est.9b07395

Carney Almroth, B. M., Åström, L., Roslund, S., Petersson, H., Johansson, M. & Persson, N. K. (2018). Quantifying shedding of synthetic fibers from textiles; a source of microplastics released into the environment. *Environmental Science and Pollution Research, 25*(2), 1191–1199. https://doi.org/10.1007/s11356-017-0528-7

Carr, S. A. (2017). Sources and dispersive modes of micro-fibers in the environment. *Integrated Environmental Assessment and Management, 13*(3), 466–469. https://doi.org/10.1002/ieam.1916

Carr, S. A., Liu, J. & Tesoro, A. G. (2016). Transport and fate of microplastic particles in wastewater treatment plants. *Water Research, 91*, 174–182. https://doi.org/10.1016/j.watres.2016.01.002

Cesa, F. S., Turra, A., Checon, H. H., Leonardi, B., & Baruque-Ramos, J. (2020). Laundering and textile parameters influence fibers release in household washings. *Environmental Pollution, 257*, 113553. https://doi.org/10.1016/j.envpol.2019.113553

Chan, C. K. M., Park, C., Chan, K. M., Mak, D. C. W., Fang, J. K. H. & Mitrano, D. M. (2021, January 14). Microplastic fibre releases from industrial wastewater effluent: A textile wet-processing mill in China. *Environmental Chemistry, 18*(3), 93–100. https://doi.org/10.1071/en20143

Changing Markets Foundation. (2018). *The false promise of certification.*

Conley, K., Clum, A., Deepe, J., Lane, H. & Beckingham, B. (2019). Wastewater treatment plants as a source of microplastics to an urban estuary: Removal efficiencies and loading per capita over one year. *Water Research X, 3*, 100030. https://doi.org/10.1016/j.wroa.2019.100030

Costa, M. F., Ivar Do Sul, J. A., Silva-Cavalcanti, J. S., Christina, M., Araújo, B., Spengler, Â., Tourinho, P. S., Costa, M. F., Ivar Do Sul, J. A., Silva-Cavalcanti, J. .S, Araújo, M. C. B., Spengler, Â. & Tourinho, P. S. (2010). On the importance of size of plastic fragments and pellets on the strandline: A snapshot of a Brazilian beach. *Environmental Monitoring and Assessment, 168*, 299–304. https://doi.org/10.1007/s10661-009-1113-4

Cowger, W., Booth, A. M., Hamilton, B. M., Thaysen, C., Primpke, S., Munno, K., Lusher, A. L., Dehaut, A., Vaz, V. P., Liboiron, M., Devriese, L. I., Hermabessiere, L., Rochman, C., Athey, S. N., Lynch, J. M., Frond, H. De, Gray, A., Jones, O. A. H. H., Brander, S., … Nel, H. (2020). Reporting guidelines to increase the reproducibility and comparability of research on microplastics. *Applied Spectroscopy*, *74*(9), 1066–1077. https://doi.org/10.1177/0003702820930292

Davidson, G. (2019). *Ocean cleaning device succeeds in removing plastic for the first time.* EcoWatch. https://www.ecowatch.com/indigenous-peoples-day-abandoning-columbus-day-2640950160.html

Dreillard, M., Barros, C. D. F., Rouchon, V., Emonnot, C., Lefebvre, V., Moreaud, M., Guillaume, D., Rimbault, F. & Pagerey, F. (2022). Quantification and morphological characterization of microfibers emitted from textile washing. *Science of the Total Environment*, *832*(March), 154973. https://doi.org/10.1016/j.scitotenv.2022.154973

Dyachenko, A., Mitchell, J. & Arsem, N. (2017). Extraction and identification of microplastic particles from secondary wastewater treatment plant (WWTP) effluent. *Analytical Methods*, *9*(9), 1412–1418. https://doi.org/10.1039/c6ay02397e

Ellen MacArthur Foundation. (2017). A new textiles economy: Redesigning fashion's future. In *Ellen MacArthur Foundation*. https://www.ellenmacarthurfoundation.org/assets/downloads/publications/A-New-Textiles-Economy_Full-Report_Updated_1-12-17.pdf

Fischer, E. K., Paglialonga, L., Czech, E. & Tamminga, M. (2016). Microplastic pollution in lakes and lake shoreline sediments e A case study on Lake Bolsena and Lake Chiusi (central Italy) *. *Environmental Pollution*, *213*, 648–657. https://doi.org/10.1016/j.envpol.2016.03.012

Frias, J. P. G. L., & Nash, R. (2019). Microplastics: Finding a consensus on the definition. *Marine Pollution Bulletin*, *138*(November 2018), 145–147. https://doi.org/10.1016/j.marpolbul.2018.11.022

Gavigan, J., Kefela, T., Macadam-Somer, I., Suh, S. & Geyer, R. (2020). Synthetic microfiber emissions to land rival those to waterbodies and are growing. *PLoS One*, 15(9 September), 1–13. https://doi.org/10.1371/journal.pone.0237839

GESAMP Joint Group of Experts on the Scientific Aspects of Marine Environmental Protection. (2015). Sources, fate and effects of microplastics in the marine environment: A global assessment. *Reports and Studies GESAMP*, *90*(August), 96. https://doi.org/10.13140/RG.2.1.3803.7925

Hann, S. Sherrington, C. Jamieson, O. Hickman, M. Kershaw, P. Bapasola, & A. Cole, G., Hann, S., Sherrington, C., Jamieson, O., Hickman, M., Kershaw, P., Bapasola, A., Cole, G., Andrady, A. L. L., Hann, S., Papadopoulou, L., Braddock, M., Sherrington, C., Jamieson, O., Hickman, M., Kershaw, P., Bapasola, A., Cole, G., Andrady, A. L. L., … Eunomia. (2017). *Investigating options for reducing releases in the aquatic environment of microplastics emitted by products* - Interim Report. *Report for DG Env EC, Vol. 62, N*(February), 1596–1605. https://doi.org/10.1002/lsm.22016

Hann, S., Sherrington, C., Jamieson, O., Hickman, M. & Bapasola, A. (2018, February 23). *Investigating options for reducing releases in the aquatic environment of microplastics emitted by products*. In Eunomia. Retrieved March 23, 2023, from https://www.eunomia.co.uk/reports-tools/investigating-options-for-reducing-releases-in-theaquatic-environment-of-microplastics-emitted-by-products/

Hartline, N. L., Bruce, N. J., Karba, S. N., Ruff, E. O., Sonar, S. U. & Holden, P. A. (2016, October 13). Microfiber masses recovered from conventional machine washing of new or aged garments. *Environmental Science & Technology, 50*(21), 11532–11538. https://doi.org/10.1021/acs.est.6b03045

Henry, B., Laitala, K. & Klepp, I. G. (2018). *Microplastic Pollution from Textiles : A Literature Review* (Issue 1). Consumption Research Norway – SIFO, Oslo and Akershus University College of Applied Sciences.

Hernandez, E., Nowack, B. & Mitrano, D. M. (2017). Polyester textiles as a source of microplastics from households: A mechanistic study to understand microfiber release during washing. *Environmental Science and Technology*, *51*(12), 7036–7046. https://doi.org/10.1021/acs.est.7b01750

Hu, Y., Gong, M., Wang, J. & Bassi, A. (2019). Current research trends on microplastic pollution from wastewater systems: A critical review. *Reviews in Environmental Science and Biotechnology*, *18*(2), 207–230. https://doi.org/10.1007/s11157-019-09498-w

Iqbal, S. A., Mahmud, I. & Quader, A. K. M. A. M. A. (2014). Textile sludge management by incineration technique. *Procedia Engineering*, *90*, 686–691. https://doi.org/10.1016/j.proeng.2014.11.795

ISO. (2022). *ISO – ISO/DIS 4484-2 – Textiles and Textile Products – Microplastics from Textile Sources – Part 2: Qualitative and Quantitative Evaluation of Microplastics*. ISO. https://www.iso.org/standard/80011.html

Kalbar, P. P., Muñoz, I., & Birkved, M. (2018). WW LCI v2: A second-generation life cycle inventory model for chemicals discharged to wastewater systems. *Science of the Total Environment*, *622–623*, 1649–1657. https://doi.org/10.1016/J.SCITOTENV.2017.10.051

Kang, P., Ji, B., Zhao, Y. & Wei, T. (2020). How can we trace microplastics in wastewater treatment plants: A review of the current knowledge on their analysis approaches. *Science of the Total Environment*, *745*, 140943. https://doi.org/10.1016/j.scitotenv.2020.140943

Lant, N. J., Hayward, A. S., Peththawadu, M. M. D. D., Sheridan, K. J., & Dean, J. R. (2020). Microfiber release from real soiled consumer laundry and the impact of fabric care products and washing conditions. *PLoS ONE*, *15*(6), e0233332–e0233332. https://doi.org/10.1371/journal.pone.0233332

Lares, M., Ncibi, M. C., Sillanpää, M. & Sillanpää, M. (2018). Occurrence, identification and removal of microplastic particles and fibers in conventional activated sludge process and advanced MBR technology. *Water Research*, *133*, 236–246. https://doi.org/10.1016/j.watres.2018.01.049

Lares, M., Ncibi, M. C., Sillanpää, M. & Sillanpää, M. (2019). Intercomparison study on commonly used methods to determine microplastics in wastewater and sludge samples. *Environmental Science and Pollution Research*, *26*(12), 12109–12122. https://doi.org/10.1007/s11356-019-04584-6

Li, X., Chen, L., Mei, Q., Dong, B., Dai, X., Ding, G. & Zeng, E. Y. (2018). Microplastics in sewage sludge from the wastewater treatment plants in China. *Water Research*, *142*, 75–85. https://doi.org/10.1016/j.watres.2018.05.034

Li, J., Liu, H., & Paul Chen, J. (2018). Microplastics in freshwater systems: A review on occurrence, environmental effects, and methods for microplastics detection. *Water Research*, *137*, 362–374. https://doi.org/10.1016/j.watres.2017.12.056

Liu, K., Wu, T. N., Wang, X. H., Song, Z. Y., Zong, C. X., Wei, N. A., & Li, D. J. (2019). Consistent Transport of Terrestrial Microplastics to the Ocean through Atmosphere. *Environmental Science and Technology*, *53*(18), 10612–10619. https://doi.org/10.1021/acs.est.9b03427

Lv, X., Dong, Q., Zuo, Z., Liu, Y., Huang, X. & Wu, W. M. (2019). Microplastics in a municipal wastewater treatment plant: Fate, dynamic distribution, removal efficiencies, and control strategies. *Journal of Cleaner Production*, *225*, 579–586. https://doi.org/10.1016/j.jclepro.2019.03.321

Madhav, S., Ahamad, A., Singh, P. & Mishra, P. K. (2018). A review of textile industry: Wet processing, environmental impacts, and effluent treatment methods. *Environmental Quality Management*, *27*(3), 31–41. https://doi.org/10.1002/tqem.21538

Magnusson, K. & Norén, F. (2014). *Screening of Microplastic Particles in and Down-Stream a Wastewater Treatment Plant*. IVL Swedish Environmental Research Institute: Vol. C 55. https://doi.org/naturvardsverket-2226

McFall-Johnsen, M. (2020). *These facts show how unsustainable the fashion industry is*. World Economic Forum. https://www.weforum.org/agenda/2020/01/fashion-industry-carbon-unsustainable-environment-pollution/

MERMAIDS Consortium. (2017). Microfiber release from clothes after washing: Hard facts, figures and promising solutions. *Position Paper, May*, 1–9. https://www.plasticsoupfoundation.org/wp-content/uploads/2017/08/Position-Paper.Microfiber-release-from-clothes-after-washing.PSF_.pdf

Mintenig, S. M., Int-Veen, I., Löder, M. G. J., Primpke, S. & Gerdts, G. (2017). Identification of microplastic in effluents of waste water treatment plants using focal plane array-based micro-Fourier-transform infrared imaging. *Water Research*, *108*(1), 365–372. https://doi.org/10.1016/j.watres.2016.11.015

Murphy, F., Ewins, C., Carbonnier, F. & Quinn, B. (2016). Wastewater treatment works (WwTW) as a source of microplastics in the aquatic environment. *Environmental Science and Technology*, *50*(11), 5800–5808. https://doi.org/10.1021/acs.est.5b05416

OECD. (2023). *Water - Wastewater treatment - OECD Data*. Data. https://data.oecd.org/water/wastewater-treatment.htm

Palacios-Marín, A. V., Jabbar, A. & Tausif, M. (2022). Fragmented fiber pollution from common textile materials and structures during laundry. *Textile Research Journal*, *92*(13–14), 2265–2275. https://doi.org/10.1177/00405175221090971

Prata, J. C., da Costa, J. P., Duarte, A. C. & Rocha-Santos, T. (2019). Methods for sampling and detection of microplastics in water and sediment: A critical review. *TrAC – Trends in Analytical Chemistry*, *110*, 150–159. https://doi.org/10.1016/j.trac.2018.10.029

Rather, L. J., Jameel, S., Dar, O. A., Ganie, S. A., Bhat, K. A. & Mohammad, F. (2019). Advances in the sustainable technologies for water conservation in textile industries. In *Water in Textiles and Fashion* (pp. 175–194). Woodhead Publishing. https://doi.org/10.1016/b978-0-08-102633-5.00010-5

Rathinamoorthy, R. & Raja Balasaraswathi, S. (2021). *Domestic Laundry and Microfiber Shedding of Synthetic Textiles* (pp. 127–155). https://doi.org/10.1007/978-981-16-0297-9_5

Roos, S., Arturin, O. L. & Hanning, A.-C. (2017). *Microplastics shedding from polyester fabrics*. www.mistrafuturefashion.com

Rudenko, O. (2018). *Apparel and fashion overproduction report with infographic*. Share Cloth. https://sharecloth.com/blog/reports/apparel-overproduction

Ryan, P. G., Suaria, G., Perold, V., Pierucci, A., Bornman, T. G., & Aliani, S. (2020). Sampling microfibres at the sea surface: The effects of mesh size, sample volume and water depth. *Environmental Pollution*, *258*, 113413. https://doi.org/10.1016/j.envpol.2019.113413

Schmiedgruber, M., Hufenus, R. & Mitrano, D. M. (2019). Mechanistic understanding of microplastic fiber fate and sampling strategies: Synthesis and utility of metal doped polyester fibers. *Water Research*, *155*, 423–430. https://doi.org/10.1016/j.watres.2019.02.044

Selonen, S., Dolar, A., Jemec Kokalj, A., Skalar, T., Parramon Dolcet, L., Hurley, R. & van Gestel, C. A. M. (2020). Exploring the impacts of plastics in soil – The effects of polyester textile fibers on soil invertebrates. *Science of the Total Environment*, *700*, 134451. https://doi.org/10.1016/j.scitotenv.2019.134451

Shafiq, A., Johnson, F., Klassen, R. D. & Awaysheh, A. (2016). The impact of supply risk on sustainability monitoring practices and performance. *Academy of Management Proceedings*, *2016*(1), 17571. https://doi.org/10.5465/AMBPP.2016.17

Sun, J., Dai, X., Wang, Q., van Loosdrecht, M. C. M. & Ni, B.-J. J. (2019). Microplastics in wastewater treatment plants: Detection, occurrence and removal. *Water Research*, *152*, 21–37. https://doi.org/10.1016/j.watres.2018.12.050

Swerea IVF. (2018). *Investigation of the occurrence of microplastics from the waste water at five different textile production facilities in Sweden*.

Talvitie, J., Heinonen, M., Pääkkönen, J.-P. P., Vahtera, E., Mikola, A., Setälä, O., Vahala, R., Paakkonen, J. P., Vahtera, E., Mikola, A., Setala, O. & Vahala, R. (2015). Do wastewater treatment plants act as a potential point source of microplastics? Preliminary study in the coastal Gulf of Finland, Baltic Sea. *Water Science and Technology*, *72*(9), 1495–1504. https://doi.org/10.2166/wst.2015.360

Talvitie, J., Mikola, A., Koistinen, A. & Setälä, O. (2017a). Solutions to microplastic pollution – Removal of microplastics from wastewater effluent with advanced wastewater treatment technologies. *Water Research*, *123*, 401–407. https://doi.org/10.1016/j.watres.2017.07.005

Talvitie, J., Mikola, A., Setälä, O., Heinonen, M. & Koistinen, A. (2017b). How well is microlitter purified from wastewater? – A detailed study on the stepwise removal of microlitter in a tertiary level wastewater treatment plant. *Water Research*, *109*, 164–172. https://doi.org/10.1016/j.watres.2016.11.046

Textile Exchange. (2022). *Preferred fiber & materials market report* (Issue October).

Tiffin, L., Hazlehurst, A., Sumner, M., & Taylor, M. (2021). Reliable quantification of microplastic release from the domestic laundry of textile fabrics. *Journal of the Textile Institute*, *113*(0), 558–566. https://doi.org/10.1080/00405000.2021.189230

The Microfibre Consortium. (2019, Feb). *TMC method*. https://www.microfibreconsortium.com

The Microfibre Consortium. (2021). *The Microfibre 2030 Commitment*. https://www.microfibreconsortium.com/2030

The Microfibre Consortium. (2022, May). *Preliminary guidelines: Control of microfibres in wastewater*.

Thompson, R. C., Olson, Y., Mitchell, R. P., Davis, A., Rowland, S. J., John, A. W. G., McGonigle, D. & Russell, A. E. (2004). Lost at sea: Where is all the plastic? *Science*, *304*(5672), 838. https://doi.org/10.1126/science.1094559

UNEP. (2018). *Mapping of global plastics value chain and plastics losses to the environment (with a particular focus on marine environment)*, 1–99. https://gefmarineplastics.org/files/2018 Mapping of global plastics value chain and hotspots - final version r181023.pdf

Vassilenko, E., Watkins, M., Chastain, S., Mertens, J., Posacka, A. M., Patankar, S., & Ross, P. S. (2021). Domestic laundry and microfiber pollution: Exploring fiber shedding from consumer apparel textiles. *PLoS ONE*, *16*(7 July), e0250346. https://doi.org/10.1371/journal.pone.0250346

Wang, F., Wang, B., Duan, L., Zhang, Y., Zhou, Y., Sui, Q., Xu, D., Qu, H. & Yu, G. (2020). Occurrence and distribution of microplastics in domestic, industrial, agricultural and aquacultural wastewater sources: A case study in Changzhou, China. *Water Research*, *182*, 115956. https://doi.org/10.1016/j.watres.2020.115956

Westphalen, H. & Abdelrasoul, A. (2018, March 21). Challenges and treatment of microplastics in water. *Water Challenges of an Urbanizing World*. https://doi.org/10.5772/intechopen.71494

Woodall, L. C., Sanchez-Vidal, A., Canals, M., Paterson, G. L. J., Coppock, R., Sleight, V., Calafat, A., Rogers, A. D., Narayanaswamy, B. E. & Thompson, R. C. (2014). The deep sea is a major sink for microplastic debris. *Royal Society Open Science*, *1*(4), 140317. https://doi.org/10.1098/rsos.140317

Wright, S. L., Thompson, R. C. & Galloway, T. S. (2013). The physical impacts of microplastics on marine organisms: A review. *Environmental Pollution*, *178*, 483–492. https://doi.org/10.1016/j.envpol.2013.02.031

Wright, S. L., Ulke, J., Font, A., Chan, K. L. A., & Kelly, F. J. (2020). Atmospheric microplastic deposition in an urban environment and an evaluation of transport. *Environment International*, *136*, 105411. https://doi.org/10.1016/j.envint.2019.105411

Xu, Z., Bai, X. & Ye, Z. (2021). Removal and generation of microplastics in wastewater treatment plants: A review. *Journal of Cleaner Production, 291*, 125982. https://doi.org/10.1016/j.jclepro.2021.125982

Xu, X., Hou, Q., Xue, Y., Jian, Y. & Wang, L. P. (2018). Pollution characteristics and fate of microfibers in the wastewater from textile dyeing wastewater treatment plant. *Water Science and Technology, 78*(10), 2046–2054. https://doi.org/10.2166/wst.2018.476

Xu, X., Jian, Y., Xue, Y., Hou, Q. & Wang, L. P. (2019). Microplastics in the wastewater treatment plants (WWTPs): Occurrence and removal. *Chemosphere*, 235(June 2018), 1089–1096. https://doi.org/10.1016/j.chemosphere.2019.06.197

Yang, L., Li, K., Cui, S., Kang, Y., An, L. & Lei, K. (2019). Removal of microplastics in municipal sewage from China's largest water reclamation plant. *Water Research, 155*, 175–181. https://doi.org/10.1016/j.watres.2019.02.046

Yuan, F., Zhao, H., Sun, H., Sun, Y., Zhao, J. & Xia, T. (2022). Investigation of microplastics in sludge from five wastewater treatment plants in Nanjing, China. *Journal of Environmental Management, 301*(September 2021), 113793. https://doi.org/10.1016/j.jenvman.2021.113793

ZDHC Foundation. (2016). *Textile industry wastewater discharge quality standard: Literature review, 1*. https://www.roadmaptozero.com/fileadmin/pdf/WastewaterQualityGuidelineLitReview.pdf

Zhang, Q. H., Yang, W. N., Ngo, H. H., Guo, W. S., Jin, P. K., Dzakpasu, M., Yang, S. J., Wang, Q., Wang, X. C., & Ao, D. (2016). Current status of urban wastewater treatment plants in China. *Environment International, 92–93*(July), 11–22. https://doi.org/10.1016/j.envint.2016.03.024

Zhao, J., Ran, W., Teng, J., Liu, Y., Liu, H., Yin, X., Cao, R. & Wang, Q. (2018). Microplastic pollution in sediments from the Bohai Sea and the Yellow Sea, China. *Science of the Total Environment, 640–641*, 637–645. https://doi.org/10.1016/j.scitotenv.2018.05.346

Zhao, S., Zhu, L., Wang, T. & Li, D. (2014). Suspended microplastics in the surface water of the Yangtze Estuary System, China: First observations on occurrence, distribution. *Marine Pollution Bulletin, 86*(1–2), 562–568. https://doi.org/10.1016/j.marpolbul.2014.06.032

Zhou, H., Zhou, L. & Ma, K. (2020). Microfiber from textile dyeing and printing wastewater of a typical industrial park in China: Occurrence, removal and release. *Science of the Total Environment, 739*, 140329. https://doi.org/10.1016/j.scitotenv.2020.140329

Zhu, X., Li, Q., Wang, L., Wang, W., Liu, S., Wang, C., Xu, Z., Liu, L. & Qian, X. (2021). Current advances of polyurethane/graphene composites and its prospects in synthetic leather: A review. *European Polymer Journal, 161*, 110837. https://doi.org/10.1016/J.EURPOLYMJ.2021.110837

Zhu, X., Nguyen, B., You, J. B., Karakolis, E., Sinton, D. & Rochman, C. (2019). Identification of microfibers in the environment using multiple lines of evidence. *Environmental Science and Technology, 53*(20), 11877–11887. https://doi.org/10.1021/acs.est.9b05262

Ziajahromi, S., Neale, P. A., Rintoul, L. & Leusch, F. D. L. L. (2017). Wastewater treatment plants as a pathway for microplastics: Development of a new approach to sample wastewater-based microplastics. *Water Research, 112*, 93–99. https://doi.org/10.1016/j.watres.2017.01.042

Zubris, K. A. V. & Richards, B. K. (2005). Synthetic fibers as an indicator of land application of sludge. *Environmental Pollution, 138*(2), 201–211. https://doi.org/10.1016/j.envpol.2005.04.013

6 Microfiber Contribution from Drying Clothes Is Critical in Estimating Total Microfiber Emissions from Textiles

Kirsten J. Kapp
Central Wyoming College

Rachael Z. Miller
Rozalia Project for a Clean Ocean

6.1 INTRODUCTION

Microfibers are the most prevalent form of microplastics detected in environmental samples. Production of fiber for use in textiles continues to increase, reaching 107 million metric tons in 2018, of which 62% were synthetic (polyester, acrylic, nylon, etc.) (Textile Exchange, 2019). Microfiber emission from washing laundry is a major source of microfiber pollution, and emissions range from 116,000 fibers/kg (Napper and Thompson, 2016) to 3,500,000 fibers/kg (De Falco et al., 2018). Numerous studies show that various conditions during wash cycles influence the amount of fiber shed (such as fabric type, soap used, water temperature, wash cycles, etc.). As a result, various mitigation efforts have been made, from installing washing machine filters to employing devices (e.g., Cora Ball and Guppy Bag) into the washing machine to reduce the amounts of microfibers shed and/or released. However, with most research focusing on the washing of clothes and not on drying methods, the amounts of microfibers released into the environment from clothing may be hugely underestimated. In fact, a study by Lant et al. (2022) suggests that the amounts of microfibers released into the environment by tumble drying in vented dryers are comparable to the amount released by washing, while a study by Kärkkäinen and Sillanpää (2021) suggests that a majority of tested textiles release more fibers when drying compared to washing, which Pirc et al. (2016) quantify as a 3.5-fold increase when compared to washing alone.

While the use and ownership of clothes dryers vary around the world, they are especially common in the US and Canada, and trends are increasing in other countries. For example, in the US, 80.3% (94.9 million households) use their dryer on

DOI: 10.1201/9781003331995-7

average 439 times per year (McCowan et al., 2015), while in Canada, approximately 81% of households have dryers (Statistica, 2020). Dryer use is considerably lower in other countries. For example, 55% of Australian, 42% of German, and 38% of French households have dryers (Kapp and Miller, 2020), but it is expected that the global market will grow at a CAGR of 5.4% and reach $14.77 billion by 2026 (Research and Markets, 2022).

With this increased market growth and the use of clothes dryers, it is essential to consider the microfiber release caused by clothes drying while estimating the contribution of microfiber pollution from laundering textiles. Although microfibers are the most prevalent form of microplastics detected in environmental samples, little is known about the effects of microfiber exposure on aquatic and terrestrial organisms once in the environment. Walkinshaw et al. (2023) observed a significant reduction in growth rate in mussels exposed to more environmentally relevant concentrations (80 fibers/L). Mussels exposed to similar concentrations of cotton fibers also demonstrated a reduction in growth rate, although not significant. Due to their higher surface area to volume ratio, microfibers have the potential to adsorb chemicals such as persistent organic pollutants (POPs) and metals on their surface (Mahbub and Shams, 2022), microfibers may pose a threat not only to the environment but also to human health. Furthermore, in addition to microfibers, substances such as plasticizers, pesticides, and potentially toxic trace elements (such as Lb, Br, and Sb) have been detected in dryer lint, highlighting another exposure pathway for harmful substances to enter the environment (Turner, 2019). When considering strategies, policies, and innovations designed to prevent and mitigate microfiber pollution (such as those implemented for washing machines), understanding the contributions that clothes dryers make to microfiber pollution and how this can be mitigated is critical for both environmental and human health.

6.2 DRYERS ARE A SOURCE OF MICROFIBER POLLUTION

Recent studies have identified clothes dryers as a potentially significant source of microfibers by establishing direct links between vented electric dryers and microfiber emissions into the internal (Lant et al., 2022; O'Brien et al., 2020; and Tao et al., 2022) and external terrestrial environments (Kapp and Miller, 2020), emphasizing that lint traps alone do not effectively stop fibers from entering the environment.

In the study by O'Brien et al. (2020), a 665 ± 6.73 g, blue, 100% polyester blanket was washed and dried consecutively. The dryer (6.5 kg Electrolux sensor dryer) was installed in an isolated room, and the ambient air and lint trap were sampled. An increase in the number of blue fibers in the ambient air was observed with consecutive dry cycles, indicating that some fibers escaped the lint trap filtration system and were directly released into the surrounding environment.

Kapp and Miller (2020) dried hot pink 100% polyester blankets (438 ± 17 g) in two vented dryer models installed at different locations and sampled for fiber deposition in the terrestrial environment at increasing distances (1.52–9.14 m) from the dryer exhaust vent by collecting surface snow samples. The total number of fibers observed in all 14 samples per drying event at each site ranged from 404 ± 192 (Site 1) to $1,169 \pm 606$ (Site 2), with an average of 5.43 ± 8.24 fibers in samples collected

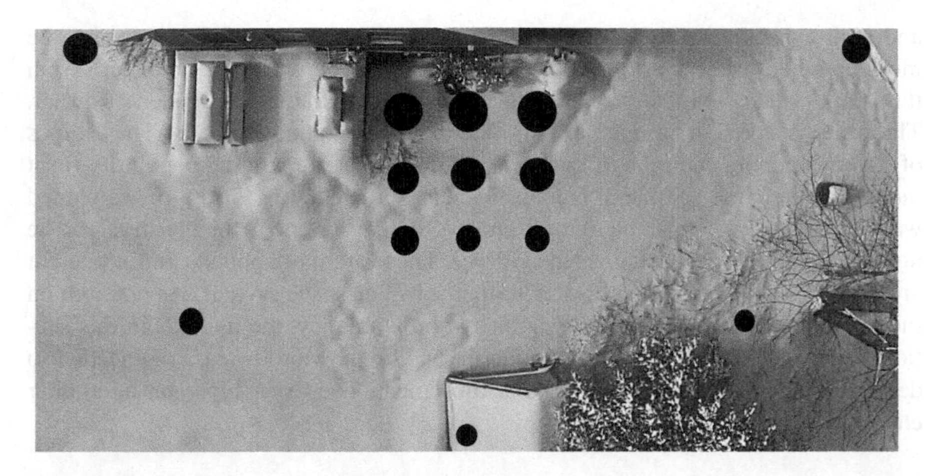

FIGURE 6.1 Microfiber abundance on snow surface samples at 5, 10, 15, and 30 feet from the dryer exhaust vent. (*Source:* modified from "Electric clothes dryers: An underestimated source of microfiber pollution" by K. Kapp and R. Miller, 2020, *PLoS One*, 15(10), 3. CC-BY.)

9.14 m from the vent. Results from this study establish a direct link between vented dryers and microfiber emissions in the terrestrial environment. While the greatest number of fibers were observed closest to the exhaust vent, fibers were also found in each of the five locations 30 feet (9.14 m) from the vent (see Figure 6.1). More research is needed to determine the fate of these fibers once they exit the dryer vent and enter the environment. For example, wind may play an important role in the distance that fibers travel.

While these initial studies show that dryers are a potential source of microfiber pollution, less is known about the specific mechanisms that cause microfiber shedding and/or emissions. Shedding during the drying process may be caused by loose fibers on the fabric as a result of the manufacturing process, fibers broken during the wash cycle and/or those broken during the dry cycle (Kärkkäinen and Sillanpää, 2021). Also, the mechanical stress and rotational forces during tumble drying may have an additive effect on fiber emissions beyond washing as broken fibers attached to the fabric after washing are released (Mahbub and Shams, 2022). In order to develop solutions that reduce additional shedding of microfibers during the clothes drying process, it is necessary to understand whether more fibers are shed during the drying process than during washing and which factors influence the shed rate.

6.3 INFLUENCE OF DRYER DESIGN ON MICROFIBER RELEASE DURING DRYING

Results from Kapp and Miller (2020) suggest that dryer design may influence microfiber emissions, which is therefore necessary to consider during mitigation efforts. Clothes drying methods vary from region to region around the world. Vented tumble dryers release warm, moist air through ducting systems directed to an external wall and emit exhaust directly into the environment. These types of dryers tend to be more common in the US. Condenser and heat pump dryers extract moisture from the clothes, then cool and condense the air, which is drained as condensed water. The water is stored in an internal tank that must be emptied regularly. These types of dryers are more common in European countries but are gaining popularity (heat pump dryers) in the US due to increased energy efficiency. A third type, combined washer-dryers, can either be vented and work like a vented tumble dryer or be non-vented and work like a condenser dryer. These are more popular in Europe but are gaining popularity in the US due to their obvious space savings. As research on the emissions of microfibers from clothes is in its infancy, the findings in this section focus exclusively on vented tumble dryer design. More research is needed to determine the contribution of the other aforementioned dryer types on microfiber emissions.

6.3.1 Venting/Duct System

In the Kapp and Miller (2020) study, two different vented tumble dryers were used, and the weight (of lint collected measured in mg) and number of fibers (counted on the snow surface) were used to indicate microfiber emissions (see Figure 6.1). These two dryers differed in the type and length of ducting used. For example, more pink fibers were emitted and collected on surface snow at the site with shorter and newer duct tubing (1.2 versus 6 m). One possible explanation for this is that as more fibers build up over time along the length of the ducting, resistance increases, and fibers may get trapped in the ducting before leaving the vent. To date, there are no other studies addressing how tubing length and age affect microfiber emissions. More research on the effects of ducting design (shape, length, age, etc.) on microfiber emissions is warranted as this may provide insight into how to reduce overall microfiber emissions from vented tumble dryers.

6.3.2 Lint Filter Design: Surface Area and Pore Size

Both the surface area and pore size of the lint filters installed in vented tumble dryers may influence microfiber emissions. The lint traps installed in the dryers used by Kapp and Miller (2020) varied in surface area (851 versus 282 cm^2), yet both had a 1 mm pore size. The lint trap with the larger surface area consistently collected more lint than the smaller lint trap (68 ± 47 versus 27 ± 22(SD) mg), suggesting that larger lint traps are more effective at removing microfibers before they enter the ducting.

Lant et al. (2022) addressed the effectiveness of lint trap pore size on microfiber emissions by comparing lint traps with pore sizes of 0.2 and 0.04 mm^2. Results showed that the lint trap with a smaller pore size (0.04 mm^2) reduced microfiber emissions from entering dryer ducting and vent exhaust by 34.8%. Results from this study can inform dryer design and significantly reduce fiber emissions by influencing dryer manufacturers to reduce lint trap pore size. Both lint traps installed in the dryers used in the study by Kapp and Miller (2020) had pore sizes much larger than those used by Lant et al. (2022) at 1 mm^2, which is common for most dryers sold in the US market (Lant et al., 2022), yet much variability in lint trap pore size exists. For example, the dryer used in Pirc et al. (2016) had a lint trap pore size of 0.18 mm^2, whereas that used in Kärkkäinen and Sillanpää (2021) was 0.06 mm^2. Based on these data, dryer emissions could be drastically reduced if the dryer appliance industry, particularly in the US, adopted lint traps with smaller pore sizes.

6.4 INFLUENCE OF OTHER PARAMETERS ON MICROFIBER SHEDDING DURING DRYING

The number and weight of fibers released during wash cycles under various conditions have been studied many times (Browne et al., 2011; De Falco et al., 2020; Hernandez et al., 2017; McIlwraith et al., 2019; Napper and Thompson, 2016; and Pirc et al., 2016) and, as a result, upstream mitigation efforts (such as variations in wash conditions, in-drum devices, and after-market filters) are underway to reduce inputs into the environment. However, little is known about how various factors, such as dry temperature, dry time, and fabric care products (e.g., dryer sheets or dryer balls), affect the shed rate of clothing during drying. While both washing machines and clothes dryers use a circular drum, the conditions and processes encountered by clothing during washing and drying are quite different. More research is warranted that will help to establish factors that can be adjusted to reduce overall emissions and help guide recommendations used to mitigate emissions from clothes dryers. Some of these factors are described below, including those addressed by Lant et al. (2022) in the most comprehensive study to date.

6.4.1 Additives: Liquid Fabric Conditioners and Dryer Sheets

In their study, Lant et al. (2022) suggest that fabric care conditioner products may reduce microfiber emissions in dryer exhaust. When applying Ultra Downy® Fabric Conditioner and Lenor® Spring Awakening during washing, an increase in fibers collected in the lint trap was observed, suggesting that fewer fibers are emitted in the vent exhaust when such products are used. Doubling the recommended doses of these products resulted in a significant reduction in microfibers released in exhaust. Interestingly, when using the anti-wrinkle fabric conditioner Downy® Wrinkle Guard, even more significant reductions in dryer vent emissions were observed at all doses (recommended, 1.5× and 2×).

Lant et al. (2022) also found significant reductions (14.1%–34.9%) in the amount of fiber released from dryer exhaust when adding dryer sheets to the dry cycle. When dryer sheets were used in combination with the liquid anti-wrinkle and conditioning products described above, dryer exhaust emissions were reduced by 44.9%. While more research is needed to investigate the effects these types of products have on microfiber emissions, the results are promising that a combination of factors can lead to successful microfiber emission mitigation strategies. Results from this study can also inform the design of reusable devices much like the in-drum devices (e.g., Cora Ball, Guppy Bag, etc.) currently on the market for washing machines.

6.4.2 Fabric Type

The physical and chemical properties of fabric types may play a significant role in the shed rate during dry cycles, similar to washing machine studies (Yang et al., 2019; Zambrano et al., 2019; and De Falco et al., 2020). For example, the diameter and twisting of the fibers, the weave pattern, rigidity, strength, and length of fibers all can determine the pilling tendency of the fabric, and fabrics with high resistance to abrasion, less hairiness, and higher mechanical strength are less likely to shed fibers during washing (Zambrano et al., 2019). For example, Lant et al. (2022) report that cotton is 20 times more sheddable than polyester. A real-world indication of the high sheddability of cotton during wear can be found in Gwinnett and Miller (2021). This paper investigated contamination from researchers during microplastic field studies and found that 23% of the total microparticles (fragments and fibers) found in their samples were cotton fibers from the team's 100% cotton t-shirts. In each case, the samples were exposed to the air for very short periods of time (under 1 minute).

Studies addressing the role that these characteristics play during tumble drying are limited. Mechanical stress during wash and dry cycles can cause fibers on the surface of fabrics to break and, as a result, form pills when they get tangled together. When comparing fleece (fabrics with raised loop piles such as PES-fnap, PES-fap, and PES-ss) to a fluffy, knitted jumper (PAN-je) and technical sports shirts (PES-ts and PA-ts), the least amount of shedding was observed in the technical sports shirts during the first drying only (Kärkkäinen and Sillanpää, 2021). These results confirm that fabric surfaces with raised, lopped piles are more susceptible to breaking off during the mechanical stress of tumble drying. Kärkkäinen and Sillanpää (2021) also suggest that fabrics with raised fiber ends sticking above the fabric surface are susceptible to breaking off during mechanical stress, contributing more to shedding than the technical sports shirts made with fabric containing longer fibers with a tighter knit.

Although research on shed rate specific to fabric type and clothes drying is limited, results show similar patterns to those of washing machines. However, it is still unknown how much just the dry process contributes to the fabric shed rate or whether the dryers are simply releasing the loose fibers that were broken due to mechanical stress during a preceding wash cycle. Less is known about the cumulative effects of both processes when used in the real-world laundering process (i.e., washing followed by drying). Regardless, improving textile designs that focus on reducing the shed rate during drying as well as washing and wearing is of equal importance.

6.4.3 Consecutive Dry Cycles and Fabric Age

Washing machine studies have shown that the amount of fiber shed from a garment decreases with each consecutive wash cycle, and consecutive dry cycles may show a similar pattern. Hartline et al. (2016) observed a 25% increase in fibers released during washing after mechanically aging and suggest that the mechanical action of drying may enhance the aging process of garments and increase sheddability. Yet the results from these first dry studies are inconclusive. Kapp and Miller (2020) and Pirc et al. (2016) observed an overall decrease in the number of fibers shed from 100% polyester fleece blankets after the third and seventh cycles, respectively. Whereas Kärkkäinen and Sillanpää (2021) observed a plateau in the number of fibers shed after five consecutive dry cycles in some textiles but noted considerably more fibers were shed by double-sided fleece textiles (PES-fnap and PES-fap). This is similar to O'Brien et al. (2020), whose results show no reduction/stability in the amount of fiber measured in the ambient atmospheric air after five consecutive dry cycles. More research is needed to determine whether the number of dry cycles and/or the age of textiles result in less microfiber emissions during drying. Gaining a better understanding of the shed pattern and rates throughout a textile's life cycle is necessary to estimate the total amount of microfiber emitted by dryer exhaust.

6.5 POTENTIAL MITIGATIVE MEASURES

Although the topic of microfiber emissions from clothes dryers is under-addressed, solutions working to reduce the contributions of electric clothes dryers to microfiber pollution are currently being designed. For example, since the identification of clothes washing as a major source of microfiber pollution, designing resilient fabrics and/or coatings applied to fabrics that reduce their sheddability has became a focus for textile scientists. Improvements in the design and production of fabrics made from alternative materials are also increasing. Similar to the design of filtration devices for washing machines, strides are being made to design similar devices in clothes dryers, but such devices have not yet been tested to scale or made available for purchase. Finally, engineers are taking a second look at clothes dryer design and working to not only design more efficient and effective dryers (e.g., condensing dryers) but also to reduce the amount of fiber shed during the drying process.

While research on the contribution of electric clothes dryers to microfiber pollution is in its infancy, results from the studies described above demonstrate that some factors may cause more microfibers to be released into the environment than others. More research addressing/identifying factors such as those mentioned above can lead to successfully implemented solutions that mitigate the effects of clothes drying on microfiber release. Figure 6.2 provides a visual summary of factors that have been shown to influence microfiber sheddability, either by increasing or decreasing their release into the environment, which can help inform future solutions.

6.6 CONCLUSIONS AND FUTURE RESEARCH

In conclusion, few studies have addressed microfiber pollution from drying clothes and those that have focused exclusively on electric clothes dryers. Yet, these few studies agree that electric clothes dryers are potentially contributing microfibers from textiles to the environment, and emissions are comparable to, or could be even higher than, those emitted via effluent from washing machines. While Kapp and Miller (2020) observed that microfibers emitted directly into the environment from

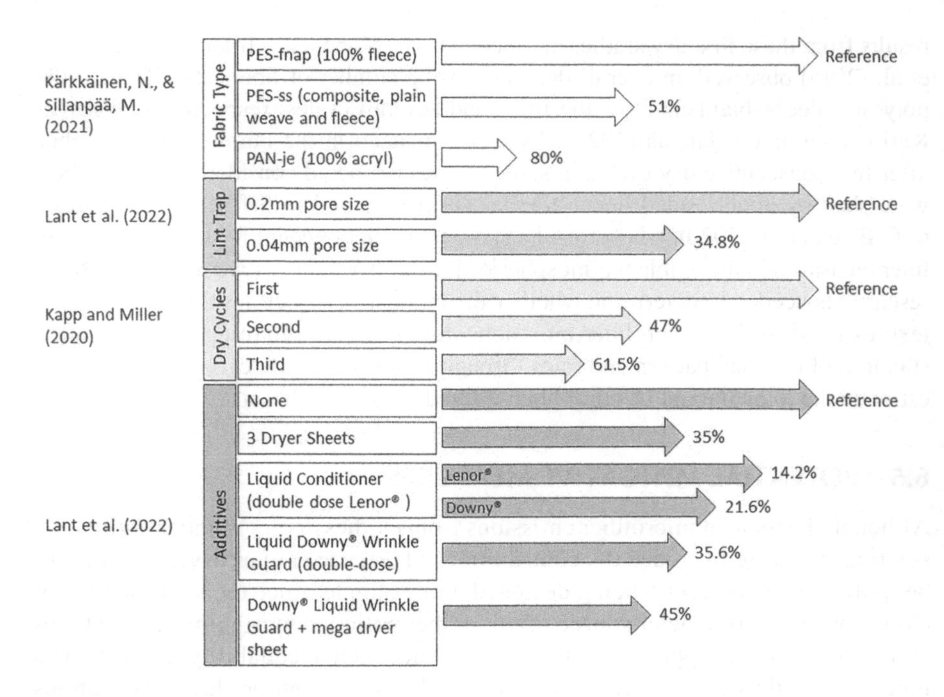

FIGURE 6.2 Factors affecting the sheddability and microfiber emissions of clothes drying using electric clothes dryers.

dryer vent exhaust can travel at least 30 feet (9.14 m) from the vent, more research is needed to determine the fate of microfibers emitted from dryer vents and whether electric clothes dryers are an important source of atmospheric microfiber pollution. Research that addresses the mechanism of transport once fibers are emitted from the dryer vent (such as wind direction, velocity, speed, and precipitation/humidity) and fiber characteristics (such as length, weight, and shape) will make valuable contributions to our overall knowledge of microfiber atmospheric pollution and the extent to which electric clothes dryers are a source.

Results from Kapp and Miller (2020) and Lant et al. (2022) suggest that variations in dryer design, such as lint trap surface area and pore size, influence microfiber emissions. Further work is needed to determine the extent to which factors of dryer design (e.g., dryer power and airflow and temperature/speed settings) and installation (e.g., ducting length/pathways, vent design, and vent height) are most influential in microfiber emissions. Additional research, similar to that of Lant et al. (2022), addressing the use of fabric care products (e.g., dryer sheets, fabric, and softeners) is warranted and can be used to educate individual consumers on ways they can reduce their impact. Results from such studies will be valuable in informing dryer design, consumer behavior, and implementation strategies that can be adopted to reduce microfiber emissions.

This review focuses solely on the microfiber emissions from electric tumble dryers. More research is needed to compare microfiber emissions from alternative drying methods such as line drying, drying closets, condenser/heat pump dryers, combined washer/dryers, and gas dryers. Considering microfiber emissions from all drying methods will provide a more comprehensive understanding of the impacts that washing and drying laundry have on the environment. Such investigations are necessary to drive potential solutions that not only address equipment design and innovation but also legislation, policy, and consumer behavior that successfully reduce and prevent microfiber pollution.

ACKNOWLEDGEMENTS

This publication was made possible, in part, by an Institutional Development Award (IDeA) from the National Institute of General Medical Sciences of the National Institutes of Health under Grant # 2P20GM103432. Its contents are solely the responsibility of the authors and do not necessarily represent the official views of NIH.

REFERENCES

Browne, M. A., Crump, P., Niven, S. J., Teuten, E., Tonkin, A., Galloway, T., & Thompson, R. (2011). Accumulation of microplastic on shorelines worldwide: Sources and sinks. *Environmental Science & Technology*, *45*(21), 9175–9179. https://doi.org/10.1021/es201811s.

De Falco, F., Cocca, M., Avella, M., & Thompson, R. C. (2020). Microfiber release to water, via laundering, and to air, via everyday use: A comparison between polyester clothing with differing textile parameters. *Environmental Science & Technology*, *54*(6), 3288–3296. https://doi.org/10.1021/acs.est.9b06892

De Falco, F., Gullo, M. P., Gentile, G., Di Pace, E., Cocca, M., Gelabert, L., Brouta-Agnésa, M., Rovira, A., Escudero, R., Villalba, R., Mossotti, R., Montarsolo, A., Gavignano, S., Tonin, C., & Avella, M. (2018). Evaluation of microplastic release caused by textile washing processes of synthetic fabrics. *Environmental Pollution*, *236*, 916–925. https://doi.org/10.1016/j.envpol.2017.10.057

Gwinnett, C., & Miller, R. Z. (2021). Are we contaminating our samples? A preliminary study to investigate procedural contamination during field sampling and processing for microplastic and anthropogenic microparticles. *Marine Pollution Bulletin*, *173*, 113095. https://doi.org/10.1016/j.marpolbul.2021.113095

Hartline, N. L., Bruce, N. J., Karba, S. N., Ruff, E. O., Sonar, S. U., & Holden, P. A. (2016). Microfiber masses recovered from conventional machine washing of new or aged garments. *Environmental Science & Technology*, *50*(21), 11532–11538. https://doi.org/10.1021/acs.est.6b03045

Hernandez, E., Nowack, B., & Mitrano, D. M. (2017). Polyester textiles as a source of microplastics from households: A mechanistic study to understand microfiber release during washing. *Environmental Science & Technology*, *51*(12), 7036–7046. https://doi.org/10.1021/acs.est.7b01750

Kapp, K. J., & Miller, R. Z. (2020). Electric clothes dryers: An underestimated source of microfiber pollution. *PLoS One*, *15*(10), e0239165. https://doi.org/10.1371/journal.pone.0239165

Kärkkäinen, N., & Sillanpää, M. (2021). Quantification of different microplastic fibres discharged from textiles in machine wash and tumble drying. *Environmental Science and Pollution Research*, *28*(13), 16253–16263. https://doi.org/10.1007/s11356-020-11988-2

Lant, N. J., Defaye, M. M., Smith, A. J., Kechi-Okafor, C., Dean, J. R., & Sheridan, K. J. (2022). The impact of fabric conditioning products and lint filter pore size on airborne microfiber pollution arising from tumble drying. *PLoS One*, *17*(4), e0265912. https://doi.org/10.1371/journal.pone.0265912

Mahbub, M. S., & Shams, M. (2022). Acrylic fabrics as a source of microplastics from portable washer and dryer: Impact of washing and drying parameters. *Science of the Total Environment, 834*, 155429. https://doi.org/10.1016/j.scitotenv.2022.155429

McCowan, B., Richards, K., & Wacker, M. (2015, April 2). *Residential Electric Clothes Dryer Baseline Study*. North Andover: Energy & Resource Solutions. Retrieved 2020 April 29, from https://neep.org/residential-electric-clothes-dryers-baseline-study

McIlwraith, H. K., Lin, J., Erdle, L. M., Mallos, N., Diamond, M. L., & Rochman, C. M. (2019). Capturing microfibers-marketed technologies reduce microfiber emissions from washing machines. *Marine Pollution Bulletin, 139*, 40–45. https://doi.org/10.1016/j.marpolbul.2018.12.012

Napper, I. E., & Thompson, R. C. (2016). Release of synthetic microplastic plastic fibres from domestic washing machines: Effects of fabric type and washing conditions. *Marine Pollution Bulletin, 112*(1–2), 39–45. https://doi.org/10.1016/j.marpolbul.2016.09.025

O'Brien, S., Okoffo, E. D., O'Brien, J. W., Ribeiro, F., Wang, X., Wright, S. L., Samanipour, S., Rauert, C., Yessenia Alajo Toapanta, T., Albarracin, R., & Thomas, K. V. (2020). Airborne emissions of microplastic fibres from domestic laundry dryers. *Science of the Total Environment, 747*, 141175. https://doi.org/10.1016/j.scitotenv.2020.141175

Pirc, U., Vidmar, M., Mozer, A., & Kržan, A. (2016). Emissions of microplastic fibers from microfiber fleece during domestic washing. *Environmental Science and Pollution Research, 23*(21), 22206–22211. https://doi.org/10.1007/s11356-016-7703-0

Research and Markets. (2022, March). Global electric dryers market report 2022, by type, by type of vent, by distribution channel, by end-user. Retrieved 2022 November 2, from https://www.researchandmarkets.com/reports/5567148

Statistica. (2011, June 30). Percentage of Canadian homes with a clothes dryer from 1998 to 2009 [Internet]. Retrieved 2020 April 27, from https://www.statista.com/statistics/199000/percentage-ofcanadian-homes-with-a-clothes-dryer-since-1998/.

Tao, D., Zhang, K., Xu, S., Lin, H., Liu, Y., Kang, J., … & Leung, K. M. (2022). Microfibers released into the air from a household tumble dryer. *Environmental Science & Technology Letters, 9*(2), 120–126. https://doi.org/10.1021/acs.estlett.1c00911

Textile Exchange. (2019). Preferred fiber and materials market report 2019 [Internet]. [cited 2020 Jun 25]. Available from https://textileexchange.org

Turner, A. (2019). Trace elements in laundry dryer lint: A proxy for household contamination and discharges to waste water. *Science of the Total Environment, 665*, 568–573.

Walkinshaw, C., Tolhurst, T. J., Lindeque, P. K., Thompson, R. C., & Cole, M. (2023). Impact of polyester and cotton microfibers on growth and sublethal biomarkers in juvenile mussels. *Microplastics and Nanoplastics, 3*(1), 1–12.

Yang, L., Qiao, F., Lei, K., Li, H., Kang, Y., Cui, S., & An, L. (2019). Microfiber release from different fabrics during washing. *Environmental Pollution, 249*, 136–143. https://doi.org/10.1016/j.envpol.2019.03.011

Zambrano, M. C., Pawlak, J. J., Daystar, J., Ankeny, M., Cheng, J. J., & Venditti, R. A. (2019). Microfibers generated from the laundering of cotton, rayon and polyester based fabrics and their aquatic biodegradation. *Marine Pollution Bulletin, 142*, 394–407. https://doi.org/10.1016/j.marpolbul.2019.02.062

7 Microfibers Pollution Associated with Disposable PPE Products Driven by the COVID-19 Pandemic and Its Environmental Repercussions

A.D. Forero López
Instituto Argentino de Oceanografía (IADO)

C.E. De-la-Torre
Universidad San Ignacio de Loyola

M.D. Fernández Severini
Instituto Argentino de Oceanografía (IADO)

G.N. Rimondino
Instituto de Investigaciones en Fisicoquímica de Córdoba (INFIQC)

C.V. Spetter
Universidad Nacional del Sur (UNS)

7.1 INTRODUCTION

Microfibers/nanofibers (MFs/NFs) pollution in the environment associated with disposable personal protective equipment (PPE) products driven by the COVID-19 pandemic has been a concern in the academic community due to their contribution to the critical state of plastic pollution in the environment reported before the beginning of the pandemic (Cabrejos-Cardeña et al. 2022). The incorrect disposal of the PPE used by medical establishments and ordinary citizens, as well as the deficiencies in the

DOI: 10.1201/9781003331995-8

Municipal Solid Waste (MSW) management systems in most countries of the world, have brought negative repercussions on the environment (Ardusso et al. 2021). Face masks (e.g., disposal masks, reusable masks, N95 masks, cloth masks) and gloves are commonly employed by the population worldwide (Silva et al. 2021a; Shams et al. 2021), generating hundreds of tons of PPE waste which, in the short and medium term, will cause serious environmental damage. Thus, some investigations have estimated the global demand for PPE during the current pandemic, such as Prata et al. (2020) and Shukla et al. (2022), and the amounts of PPE discharged into the aquatic environment (Oceans Asia 2020). In 2020, the global monthly volume of masks and gloves used was approximately 129 billion and 65 billion, respectively (Prata et al. 2020). Recently, Shukla et al. (2022) estimated that more than 1.5 million face masks are annually generated in 36 countries, China (~0.4 million), India (~0.2 million), and the United States (~0.1 million) are the countries that use face masks the most. Benson et al. (2021) reported that approximately 3.4 billion single-use facemasks/face shields are globally discarded because of the pandemic every day. It is important to mention that these investigations calculated the volume of face mask waste based on the total population of a country or region. Other researchers have calculated the disposal rate of PPE per day/year (Akhbarizadeh et al. 2021) based on their results obtained from monitored PPE waste in coastal or aquatic environments such as Oceans Asia (2020) that estimated that approximately 1.56 billion disposable face masks (DFMs) were discharged into the aquatic environment. However, there is no exact data on PPE production and global and regional waste generated and emitted to the marine environment during the pandemic because of incomplete information in developing countries. Facing this scenario, some studies have focused on monitoring the levels of PPE waste in coastal areas and sandy beaches, and cities from different continents, exhibiting a great abundance of PPE waste (mainly face masks) in these zones (Akarsu et al. 2021; De-la-Torre et al. 2021b; Haddad et al. 2021). According to reports by these authors from various countries, Asia exhibited the highest range of density (total or mean) of face masks between 7×10^{-1} (for Hong Kong) and 9.75×10^{-5} face mask/m^2 (for Iran), followed by Africa, where Ghana had a total density of 3.1×10^{-1} face mask/m^2 and Morocco beaches reported a mean of the density of 1.09×10^{-5} and 1.20×10^{-3} face mask/m^2 (Oliveira et al. 2023). Finally, American countries exhibited the lowest range of density (total or mean) of face masks with 1.70×10^{-3} (for Canada) and 6.23×10^{-4} (for Perú) (De-la-Torre et al. 2022a; Oliveira et al. 2023). According to these results, the high population density, the accessibility to sandy beaches or coastal areas, poor waste management systems, deficiencies in MSW management of each zone monitored, and the low level of recycling culture were the critical factors that influenced the levels of PPE pollution.

The inadequate disposal of PPE waste has generated, in the short term, pressure on the environment because of this new type of litter, which also tends to release large amounts of microfibers (MFs; plastic fibers particles less < than 5 mm in size) and nanofibers (NFs; plastic fibers particles < than 1 um in size) from face masks (Saliu et al. 2021; Wang et al. 2021; Rathinamoorthy & Balasaraswathi 2022). For this reason, some research had been focusing on PPE weathering processes, the release of MFs/NFs, and the leaching of chemical substances from them (e.g., heavy metals, phthalates, and plasticizers), evidencing a new source of MFs/NFs pollution and other pollutants in the environment with potentially hazardous effects on organisms and humans. The present chapter addressed the multiple variables involved in the release of MFs/NFs and plastic micro-fragments from different types of face masks, the estimated magnitude of microplastics/plastic nanoparticles (MPs/PNPs) released, as well as the degradation of PPE waste, and their ecotoxicological implications.

7.2 TYPES, COMPOSITION, AND STRUCTURE OF PPE WASTE

The majority of PPE waste ends up in landfills, in some areas such as beaches or protected areas, or dumped into the sea, either accidentally or intentionally, being face masks and gloves the most common type of PPE litter generated and discharged into the environment. Face masks were commonly used during the pandemic to prevent the spread of the virus and can be divided into three types: (i) disposable, (ii) reusable (e.g., textile), and (iii) respirator masks (e.g., N95, KN95), whose filtration efficiency depends on the type of material, manufacturing method, and mechanical and electrostatic charge of the layers (Konda et al. 2020; Bhattacharjee et al. 2022; Oliveira et al. 2023). Whereas the most common types of gloves used were disposable and they are composed of latex (natural rubber), nitrile, and PP (Jędruchniewicz et al. 2021). The composition of PPE waste in urban environments depends on the availability of the elements or the purchasing power of the population since, in some regions, the use and generation of surgical and textile mask waste have been greater because surgical masks are cheap and disposable, textile ones are reusable whereas respirator masks (N95 or KN95) are expensive and difficult to acquire as it is the case of South American region (Shruti et al. 2020; Ardusso et al. 2021; De-la-Torre et al. 2022a).

In general, surgical face masks consist of three layers; an outer layer of PP nonwoven (water-resistant), a middle one (melt-blown filter of PP, which is the primary filtering layer of the mask), and an inner layer (soft fibers of PP), as well as ear straps of PES or polyamide (Bhattacharjee et al. 2022). However, other types of polymers are also employed in the manufacture of surgical face masks, such as polyethylene (PE), polyurethane (PU), polyacrylonitrile (PAN), and PET, among others (Ardusso et al. 2021; De-la-Torre et al. 2022a) as well as additives such as $CaCO_3$, MgO, or TiO_2 in their polymeric matrix, which are utilized like fillers, extenders, or pigments in the plastic industry (De-la-Torre et al. 2022a). On the other hand, reusable masks or textile masks, can be made of different types of materials such as chiffon, polyester/polyamide, flannel, polyethylene terephthalate (PET), and cotton (Shruti et al. 2020; Ardusso et al. 2021; De-la-Torre et al. 2022a). In addition, some countries developed fabrics treated with antiviral, bactericidal, and fungicidal active ingredients (mainly metal-nanoparticles) for their manufacture. For instance, textile face masks (impregnated with nanoparticles or not) are made of different synthetic or semisynthetic fabrics such as cotton-polyester and polyester (Ardusso et al. 2021; De-la-Torre et al. 2022a). Whereas respiratory masks such as N95 are comprised of four or five layers of plastic: a PP spun-bond outer layer, a second one of cellulose /PES or PET, a third layer of melt-blown filter material, and an inner fourth layer of PP spun-bound, and they are >95% effective in removing 0.3-μm of airborne particles (Konda et al. 2020; Bhattacharjee et al. 2022). On the other hand, surgical masks exhibit between 42% and 80% efficiency (during a period of 4 hours) (Konda et al. 2020; Sankhyan et al. 2021).

However, they have been widely used since they are directly related to the access and purchasing power of the population. Finally, reusable face masks proved to have the lowest filtration efficiency (16%–23%) compared to other types of masks (Sankhyan et al. 2021).

7.3 QUANTITATIVE ESTIMATION OF MICROFIBERS/NANOFIBERS FROM PPE WASTE

PPE waste (particularly face masks) can fragment into small particles or release MFs/NFs from their textile layers into the environment through mechanical (e.g., friction with skin), physical (e.g., tears, cuts, strain), and chemical (hydrolysis) processes or by being exposed to environmental variables such as sunlight, wind, and water, among others (Rathinamoorthy & Balasaraswathi 2022). Degrading and disintegrating the textile fibers in the masks can also begin with the modification and their surface by abiotic processes or by colonization and growth of microorganisms or adhesion of the benthic and algae organisms from aquatic environments (Crisafi et al. 2022; Chomiak et al. 2023). It is well known that textiles are an important source of MPs/PNPs pollution in the environment, fibers being the most common form of MPs originating from textiles. Between 0.2 and 0.5 million tonnes of MPs from textiles have been globally discharged into the oceans annually (Ellen MacArthur Foundation 2017). This is one of the reasons for the growing concern about the impact and negative repercussions of PPE waste in aquatic and terrestrial environments. Facing this scenario, multiple studies have demonstrated that PPE waste, in particular, face masks is a significant source of MFs/NFs and micro-fragments, being PP and PET as the main polymer types released from surgical masks, with estimates from hundreds to millions (per face mask) of MFs/NFs of different sizes, increasing an additional burden on the current MPs/PNPs pollution (Ma et al. 2021; Chen et al. 2021; Liang et al. 2022). In this way, Table 7.1 summarizes various studies that aim to quantify the number of MFs/NFs released by multiple types of face masks under different simulated environmental conditions.

In most of the studies, surgical masks were the most used to investigate the release capacity of MFs/NFs, evidencing that transparent MFs with a size smaller than 10 μm were the most abundant plastic particle released from the middle layer, which is more susceptible to mechanical damage and degradation (Ma et al. 2021; Sun et al. 2021; Wang et al. 2021; Rathinamoorthy & Balasaraswathi 2022). However, MFs with a size range $< 500\,\mu m$ are generally the most abundant microparticle size range reported (see Table 7.1). These particles were evidenced with the use of advanced laser-based techniques such as flow cytometry (Morgana et al. 2021), SEM (Ma et al. 2021), and laser scattering (Wang et al. 2021). For example, Ma et al. (2021) reported the highest levels of MFs released from face masks and N95 masks with a value of 2.23×10^3 MFs per mask and 2.43×10^9 MFs per mask, respectively. Whereas Morgana et al. (2021) stated that the levels of MFs/NFs ranged between $7.6 \pm 4.6 \times 10^8$ items/mask and $3.9 \pm 1.2 \times 10^{10}$ items/mask, depending on the immersion time (1–30 seconds). Moreover, the highest levels of MFs and MFs aggregates were released from the ear straps of masks, which were identified as PET, indicating they also contribute to MFs/NFs pollution (Chen et al. 2021; Ma et al. 2022).

TABLE 7.1
Summarizes Various Studies That Aim to Quantify the Number of MFs/NFs Released by Multiple Types of Face Masks under Different Simulated Environmental Conditions

Face Masks	Conditions of the Experiment	MFs Released	Size Range of MFs/NFs	References
Virgin surgical masks	Stirring at 4,000 rpm for 24 hours	173,000 MFs/per surgical mask per day	1–5 mm 25–500 µm	Saliu et al. (2021)
Virgin disposable face masks	Immersed in an aqueous solution and shaken at 300 rpm for 24 hours.	Outer layer: 4.47 µL/L Middle layer: 1.52 µL/L Inner layer: 0.80 µL/L	Outer layer: 15–50 µm 200–500 µm Middle layer: 10–200 µm Inner layer: 10–200 µm 10–50 µm	Wang et al. (2021)
	Immersed in an aqueous solution with sand and shaken at 300 rpm for 24 hours.	Outer layer: 14.6 µL/L Middle layer: 6.2 µL/L Inner layer: 7.0 µL/L	Outer layer: 20–100 µm 100–500 µm Inner layer: 30–100 µm 100–500 µm	
	Irradiated with UVlight for 36 hours, in an aqueous solution and shaken at 300 rpm for 24 hours.	Outer layer: 70.73 µL/L Middle layer: 60.36 µL/L Inner layer: 11.44 µL/L	Middle layer: 50–200 µm	
	Irradiated with UVlight for 36 hours, in an aqueous solution with sand and shaken at 300 rpm for 24 hours.	Outer layer: 132.52 µL/L Middle layer: 210.23 µL/L Inner layer: 184.94 µL/L	Outer and inner layer: 20–100 µm 100–500 µm Middle layer: 10–200 µm	
Virgin surgical and disposal masks	Shaken at 120 rpm for 24 hours.	183. ± 78.42 pieces per new mask.	100–500 µm	Chen et al. (2021)
N95 face masks	Face masks used for 24 hours and immersed in an aqueous solution and shaken at 120 rpm for 24 hours.	1,246.62 ± 403.50 pieces per used mask	100–500 µm	

(Continued)

TABLE 7.1 (*Continued*)
Summarizes Various Studies That Aim to Quantify the Number of MFs/NFs Released by Multiple Types of Face Masks under Different Simulated Environmental Conditions

Face Masks	Conditions of the Experiment	MFs Released	Size Range of MFs/NFs	References
Virgin surgical face masks Common face mask N95 mask	Immersed in an aqueous solution and shaken at 60 rpm from 0 to 240 hours. Virgin face mask immersed in an aqueous solution with sand and shaken at 60 rpm from 0 to 240 hours	Surgical face mask: 272 ± 12.49 MPs/cm^2 Common mask: 54,400 ± 2,498 MPs N95 mask: 187.9 ± 9.45 MPs/cm^2 Common masks: 68,000 ± 4,808 MPs	0.1–1 mm	Wu et al. (2022)
Virgin surgical masks	Immersed in an aqueous solution and stirred at 120 rpm for 24 hours.	360 items per mask	–	Shen et al. (2021)
	Immersed in an aqueous/alcohol solution and stirred at 120 rpm for 24 hours.	5,400 items per mask	–	
	Immersed in an aqueous/detergent solution and stirred at 120 rpm for 24 hours.	4,400 items per mask	–	
Virgin disposable surgical masks Virgin N95 masks	Shaken vigorously for 3 minutes. This procedure was repeated ten times	2.23×10^3 MPs per mask 2.43×10^9 MPs per mask	<1 µm <1 µm	Ma et al. (2021)

(*Continued*)

TABLE 7.1 (*Continued*)

Summarizes Various Studies That Aim to Quantify the Number of MFs/NFs Released by Multiple Types of Face Masks under Different Simulated Environmental Conditions

Face Masks	Conditions of the Experiment	MFs Released	Size Range of MFs/NFs	References
Surgical masks N95 masks KN95 masks Children masks	Irradiation with UV- light at 25°C for 15 days	Surgical mask 62 MPs per mask N95 mask 24 MPs per mask KN95 mask 180 MPs per mask. Children masks 464 MPs per mask	Mostly 30–500 μm	Ma et al. (2023)
Disposable surgical masks	Immersed in an aqueous solution and exposed to shear damage using a kitchen chopper (from 1 to 120 seconds)	(1 second) 2,600 MPs per mask (120 second) 28,000 MPs per mask	0.1–5 mm	Morgana et al. (2021)

(Continued)

TABLE 7.1 (*Continued*)
Summarizes Various Studies That Aim to Quantify the Number of MFs/NFs Released by Multiple Types of Face Masks under Different Simulated Environmental Conditions

Face Masks	Conditions of the Experiment	MFs Released	Size Range of MFs/NFs	References
Disposable face masks	Face masks on soil and exposed to the environment for 30 days. Face masks (new and naturally weathered) in dry state. Face masks and individual layers of new and naturally weathered masks were immersed in fresh water and seawater for 10 hours in static immersion.	Dry state New face masks 14,031. fibers/mask Weathered face masks 177,601 fibers/mask. Wet state new face masks 2,557.65 fibers/mask Weathered masks 22,525.89 fibers/mask.	New masks: 0.119–2.042 mm Weathered masks: 0.091–2.621 mm	Rathinamoorthy and Balasaraswathi (2022)
Virgin surgical masks Common masks N95 masks	Immersed in an aqueous solution and shaken at 220 rpm for 24 hours	Surgical masks: 1,136 ± 87–2,343 ± 168 MPs/(piece·d) Common masks: 1,034 ± 119–2,547 ± 185 MPs/(piece·d) N95 masks: 801 ± 71–2,667 ± 97 MPs/(piece·d)	Mostly <500 μm	Liang et al. (2022)
Virgin disposable face masks	Immersed in deionized water and shaken at 220 rpm for 24 hours	–	25 μm–2.5 mm	Sullivan et al. (2021)

Likewise, leached MFs/NFs that differed in color and chemical composition from the original face mask were also reported, indicating potential contamination from airborne MFs during their manufacture, manipulation, and use (Chen et al. 2021; Liang et al. 2022). However, the quantity of MFs/NFs released and micro fragments depends on the weathering degree of PPE, the environmental conditions to which they are exposed, the type of face masks, exposure time, and analytical procedures used to quantify, among other factors (Cabrejos-Cardeña et al. 2022; Ma et al. 2023). Some authors have demonstrated not all MFs/NFs released from face masks are, indeed, made out of plastic since non-plastic particles, such as minerals (e.g., quartz, anatase, albite, talc) or β-carotene and microcrystalline cellulose (MCC), representing between 58.0% and 98.8% of the total number of particles, were also registered under lab conditions (Li et al. 2021; Ma et al. 2023). Rathinamoorthy and Balasaraswathi (2022) demonstrated how abrasion during handling, use, and removal led to the release of MFs/NFs from aged or unaged face masks, evidencing a higher amount of MFs/NFs released in dry environments (14,031–177,601 fibers/mask) than in wet ones (2,557–22,525 fibers/mask). Although most investigations have focused on determining the number of MFs/NFs released from different types of masks or their layers, it is of utmost importance to know how aging and the different physico-chemical parameters influence their capacity to release plastic particles. Any type of modification, such as cutting the layers, may directly or indirectly influence the levels of MPs that are released during the experiment (Shen et al. 2021). Most studies have shown that surgical masks are the largest source of MPs; however, in the case of N95, only three studies reported MPs levels but with wide ranges (Ma et al. 2021, 2023; Wu et al. 2022), another article focused on common masks although children's masks evidenced that they are an important MPs source (Ma et al. 2023). Finally, similar to textiles, reusable textile face masks tend to release MFs/NFs during their use (e.g., washing, drying) and final disposal into the environment. Once the PPE waste enters the environment, it goes through weathering processes (UV radiation, mechanical stress, or chemical, physical, and biological degradation) similar to plastic debris (Forero-López et al. 2022).

7.4 DEGRADATION OF PPE WASTE AND ITS SUB-PRODUCTS

PPE undergo degradation processes due to physical, chemical, biological, and biochemical interactions from the time they are used (depending on the type of material) until their disposal in the natural environment. The PPE waste cycle begins (Figure 7.1) when it is randomly disposed of in different ecosystems (urban, aquatic, terrestrial, or mixed) through urban runoff, landfills, or directly due to improper disposal. When face masks and gloves are exposed to weather conditions (e.g., UV light, rain, temperature) and under prolonged exposure to chemical (oxidation), physical (high tidal, mechanical, and sand abrasion), and biological processes in the environment (biodegradation by fungi, bacteria, benthos, plankton, terrestrial invertebrates, etc.), changes are produced in the physical and chemical properties of the material (De-la-Torre et al. 2022c; Pizarro-Ortega et al. 2022; Jiang et al. 2022).

In this way, some researchers have suggested that nearly 15 years would be needed to degrade just half of the face masks in aquatic environments (Chen et al.

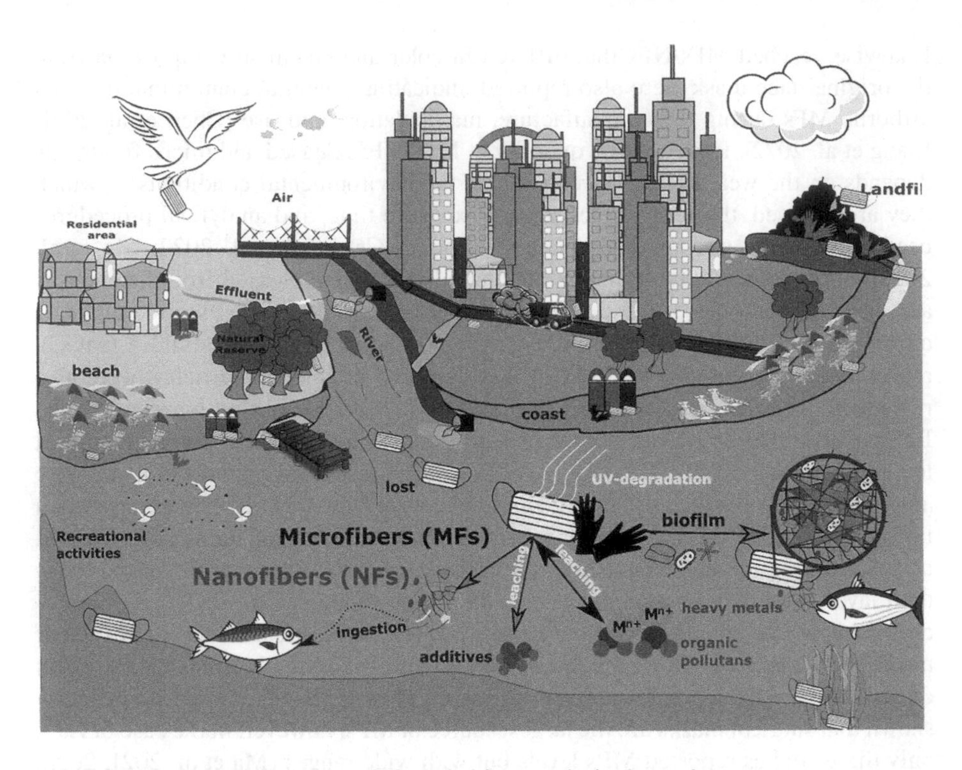

FIGURE 7.1 Fate and behavior of discarded face masks in the environment.

2023). Other authors have evidenced that the aging and degradation of DFMs are faster in a landfill than in water under laboratory conditions (Lyu et al. 2023). Similar to PP plastic waste exposed to the environment, PPE waste comprising the same plastic also tends to suffer a chain scission when in contact with UV light, and subsequent oxidation is produced (Gewert et al. 2018). As a result of this weathering process, multiple changes in their physical and chemical properties and their surface occur, such as changes in the crystallinity regions of used PP face masks (De-La-Torre et al. 2022a). Some authors have reported increased or decreased crystallinity of plastic waste (depending on their nature and the environment exposed), modifying their mechanical properties (e.g., tensile strength) and fracture mechanism (Pizarro-Ortega et al. 2022; Forero-López et al. 2022; Lyu et al. 2023). Other changes are those of the surface chemistry of PP-based face masks, such as the presence of new functional groups (e.g., C=C and C=O) as a product of free-radical chain reactions generated by C–C and C–H bonds broken due to the energy provided by UV radiation (Wang et al. 2021; Pizarro-Ortega et al. 2021). The Modification of the surface morphology of PP-masks (e.g., roughness, cracks, and pits), micro-voids, and cavities on the surface of gloves (De-La-Torre et al. 2022a), as well as changes in the wettability, affect their hydrophobic or hydrophilic nature (Wang et al. 2021; Lyu et al. 2023). During the weathering process, PPE residues tend to release MFs/NFs, plastic micro-fragments, and other chemical substances that can be part of their chemical composition, such as $CaCO_3$, MgO, TiO_2 (De-la-Torre et al. 2022a) or metallic nanoparticles used to confer antiviral properties to the masks (Ardusso et al. 2021; De-la-Torre et al. 2022a). It has been evidenced that other types of contaminants, such as heavy metals (e.g., Pb, Sb, and Zn) (Sullivan et al. 2021; Bussan et al. 2022), phthalate esters and organophosphate esters (OPE) (De-la-Torre et al. 2022b; Wang et al. 2022), can also be leached from PPE waste. All these pollutants, whether released or leached from PPE waste, could further increase the ecotoxicological risk to aquatic or terrestrial organisms and humans. Furthermore, this problem is aggravated by the fact that plastic waste, such as released MFs/NFs, can adsorb and act as carriers of organic pollutants and heavy metals from their disposal site to other regions (Anastopoulos & Pashalidis 2021; Xie et al. 2022). Multiple and complex sorption mechanisms (e.g., electrostatic and π–π interactions, functional groups) between contaminants in aquatic environments and aged plastic waste can occur simultaneously and are controlled by environmental variables such as pH, salinity, and organic matter, among others (Forero López et al. 2022). Similar to this type of plastic waste, face masks can also absorb pollutants, such as heavy metals, on their surface through the interaction of functional groups as products (–COOH; OH) from their weathered surface (Lin et al. 2022). Finally, discarded face masks have the potential to be a vehicle for pathogenic contaminants (as the SARS-CoV-2 virus) and a haven for enriching antibiotic resistance genes (ARGs) in marine microbial (Zhou et al. 2022) as well as a substrate for the growth of an ecological niche, a so-called "plastisphere", due to microbial communities and to the accumulation of contaminants present in bodies of water and living aquatic organisms (De-la-Torre et al. 2022a; Forero López et al. 2022b).

7.5 ENVIRONMENTAL IMPACTS AND ECOTOXICOLOGICAL IMPLICATIONS

As denounced by citizens and researchers from various countries, PPE waste has generated negative repercussions on the environment and living aquatic organisms in the short and medium term. Some of them are the entanglement of biotic organisms in the elastic components of the masks (Silva et al. 2021b; Ammendolia et al. 2022), sometimes causing death by constriction or by ingestion (Neto et al. 2021), as well as pathogenic and plastic contamination in bodies of water (Zhou et al. 2022). At the same time, they can be mistakenly used as nesting material by birds (Neto et al. 2021). Following the cycle of PPE waste previously illustrated, PPE tends to suffer aging processes and will likely undergo fragmentation, and in the case of face masks, it can release MFs/NFs into the environment. However, only a few studies (*n: 6*) addressed this topic. For illustrative purposes, we could mention the research carried out by Mézsaros, Sendra, and Kokalj, who investigated the possible ecotoxicological effects on terrestrial and aquatic organisms and plants by MF as well as the consequence of plastic micro-fragments released from PP surgical masks (Meszáros et al. 2022; Sendra et al. 2022; Kokalj et al. 2022). Although no evidence of lethal effects on the survival of organisms and plants was reported, significant reductions in reproduction, growth and alterations in available energy were observed in *Folsomia candida* and larvae *Tenebrio molitor* (Kokalj et al. 2022). For example, Kwak and An (2021) stated that NFs (~0.1 μm) from melt-blown face mask filters caused adverse effects on soil invertebrates, such as inhibited reproduction and stunted growth in springtail, and decreased intracellular esterase activity in earthworm coelomocytes. Likewise, the dependence between the size range of MFs of surgical masks and the inhibition in shoot seedling root length and seed elongation was also reported (Meszáaros et al. 2022). It should be noted that some parameters, such as the type of polymer and its surface charge, the presence of additives, and the size range of MFs, could play a key role in the results of aggravating ecotoxicological effects (Forero-López et al. 2022; Cabrejos-Cardeña et al. 2022). In particular, the size distribution of plastic fibers may be another important barrier in ecotoxicological studies since in the case of NFs it could be potentially more hazardous than in MFs because they can easily penetrate biological barriers such as cells/tissues, and their high surface area/volume ratio tends to be more reactive than that of larger MFs (Forero-López et al. 2022). Likewise, a wide range of chemical additives used in the polymer matrix of the masks may modify the toxicological profile of MFs/NFs on organisms (Cabrejos-Cardeña et al. 2022). Another concern with PPE material is that face masks are in direct contact with the respiratory system, and therefore MFs/NFs could be directly inhaled or caught in nasal mucus, thus becoming a public health issue (Han & He 2020; Ma et al. 2021). Some studies have provided evidence on the risk of human exposure to microplastics/plastic nanoparticles (MPs/PNPs) (including MFs/NFs) since they could generate pulmonary, kidney, and cardiovascular toxicity, neurotoxicity, hepatotoxicity, and reproductive toxicity through oxidative stress, disrupted metabolism, and microbiota dysbiosis (Ma et al. 2021; Li et al. 2022). However, there is still insufficient data to attribute the occurrence of respiratory and cardiovascular diseases to chronic intake of MPs/NPs (inhaled or ingested). Nevertheless, as the demand for plastic products grows and the amount of waste released to the environment continuously increases, humans are more exposed to these contaminants through multiple pathways that emerge over time.

7.6 INSTRUMENTAL METHODS USED FOR THE DETECTION AND QUANTIFICATION OF MICROFIBERS/NANOFIBERS FROM PPE WASTE

As mentioned above, the growing concern about the impact and repercussions of PPE waste has been evident. For this reason, researchers have employed a set of characterization techniques to study the influence of the weathering degree of face masks and their capacity to release MPs/PNPs and other pollutants in different aquatic and terrestrial environments to provide a basis to understand the scale of the problem of PPE litter and its contribution to MFs/PFs pollution. Scanning Electron Microscope (SEM) coupled with an energy-dispersive X-ray analyzer (EDX), X-ray diffraction (XRD), Fourier-transformed infrared spectroscopy (FTIR), Atomic Force Microscopy (AFM), and Scanning Laser Confocal Microscopy coupled to Raman spectroscopy, among others techniques are used for determining the physicochemical characteristic of the recovered plastic debris and the identification of plastic MFs released from the different type of face masks and gloves (Sullivan et al. 2021; Pizarro-Ortega et al. 2022; Ma et al. 2023). Various methods for simulating weathering process conditions under laboratory conditions, such as irradiation with UV-light, mechanical force with sand abrasion (Wang et al. 2021; Wu et al. 2022) and/or simulating wave influence (e.g., stirring/shaking) (Wang et al. 2021; Chen et al. 2021), and lab steps to obtain MPs in solution (e.g., filtration) have been developed, followed by the use of chemical-analytical techniques to quantify and chemically characterize the MFs/NFs. High levels of MFs/NFs released during the experiments when face mask samples (new or aged) were subjected to mechanical forces such as stirring/shaking (Ma et al. 2021; Saliu et al. 2021) and sand to simulate the effects of waves and mechanic abrasion (Wu et al. 2022). Moreover, depending on the size range and quantity of MFs/NFs released from face masks, other techniques, such as *in-situ* observational technologies and advanced microscopy techniques are also used to facilitate detection and quantification. For example, some researchers have used low-cost techniques such as a stereomicroscope combined with software to visually inspect and quantify MFs/NFs leached from DFMs samples (Sendra et al. 2022; Sun et al. 2021). Other advanced microscopy techniques, such as SEM, to quantify the amount of NFs released from surgical face masks (Ma et al. 2021), and laser *in-situ* scattering have been utilized for rapidly determining in situ particle size distribution and concentration of plastic MFs in dispersed samples with a known volume (Wang et al. 2021). However, some researchers reported that some non-plastic particles released from face masks exhibited similar morphology to MFs, and when analyzed by Raman, these particles exhibited a different chemical composition than that of the plastic which makes up face masks, therefore generating an overestimation of the number of MFs released (Ma et al. 2023). Therefore, without a complete chemical identification of the MFs/NFs released from the face masks, we could indicate an overestimation of the real MFs/NFs values since the masks have adhering impurities on their surface from their manufacture to waste. These impurities may be confusing with plastic microfibers because they have a similar shape to MFs, as it was demonstrated by Ma et al. (2023).

Especially, SEM, AFM, and confocal microscopy help to observe the microstructure, morphology, topography, and texture surface of the micromaterial samples (Ye et al. 2022). In particular, SEM can not only be used to characterize the surface morphology of MFs/NFs but also the combination of this technique with EDX to determine the elemental composition of MFs, including additives into their polymeric matrix, performing a semi-quantitative analysis of the detected elements (Forero-López et al. 2022). However, plastic samples must be plated with gold or another conductive metal to be observed by SEM since they are non-conductive (Ye et al. 2022). In general, these techniques were employed to study the degradation surface of MFs/NFs released from aged masks.

Spectral analytical methods such as FTIR or the micro Fourier Transform Interferometer (μFTIR) and Raman provide information on functional groups and structural information of the MFs/NFs samples (Phan et al. 2022). These spectroscopy techniques are complementary, and their spectra are similar. In the case of face mask waste, μFTIR is widely used for its chemical identification, as well as to study the surface change of the chemical structure of plastics that are generated by weathering processes over a period of time (Akarsu et al. 2021; Ma et al. 2023). In order to determine the degree of surface modification of aged face masks, some authors have determined the carbonyl index (CI) through μFTIR, which is evidenced by changes in the carbonyl band (C=O) in the range from 1,850 to 650 cm^{-1} of the spectra (Akarsu et al. 2021; Phan et al. 2022; Ma et al. 2023). However, this spectroscopy technique should be used together with other characterization surface techniques to corroborate the surface modification of the material, such as X-ray photoelectron spectroscopy (XPS), which is also employed to characterize the chemical composition or chemical changes on the surface of plastic waste (Pizarro-Ortega et al. 2022). Another spectroscopy technique used to characterize MFs/NFs from face masks is Raman, which has a high spatial resolution (less than 1 μm) and is commonly utilized to analyze additives such as pigments and fillers into matrix polymers with a high specificity of the fingerprint spectrum (Phan et al. 2022). However, it is less popular than μFTIR in the identification of plastic particles because of the fluorescence phenomenon caused by inorganic and organic substances and colored additives (Phan et al. 2022; Ye et al. 2022). Some investigations have employed XRD, SEM/EDX, and FTIR to demonstrate crystallographic changes, physical degradation, and surface chemical modification of PPE waste found on the coast and cities worldwide (Arkasu et al. 2021; De-la-Torre et al. 2022a; Pizarro-Ortega et al. 2022). Likewise, the presence of MFs/NFs from airborne and suspended particles, such as quartz, illite, and sand, adhered to the masks has also been reported. It is important to mention that the changes in the chemistry of the surface of any material subjected to some natural or induced modification must be analyzed with at least three characterization techniques to obtain detailed information and avoid false positive interpretations of the spectra of results.

TABLE 7.2
Chemical Analysis Methods Are Used to Quantify, Identify, and Characterize Disposable Face Masks and Their Sub-products as Microfibers/Nanofibers

Method	Type	Advantages	Disadvantages	Reference
		Detection and Quantification		
Stereomicroscope or microscope	Optical microscope	Low-cost technique	Difficult to count large volumes of MFs released from masks	Sendra et al. (2022), Sun et al. (2021), Liang et al. (2022), Saliu et al. (2021), Rathinamoorthy and Balasaraswathi (2022)
SEM	Advanced microscopy	Quantification of NFs and MFs	High operating cost Low sensitivity and sample preprocess	Ma et al. (2021), Shen et al. (2021)

(*Continued*)

TABLE 7.2 (*Continued*)
Chemical Analysis Methods Are Used to Quantify, Identify, and Characterize Disposable Face Masks and Their Sub-products as Microfibers/Nanofibers

Method	Type	Advantages	Disadvantages	Reference
Laser *in-situ* scattering	Laser	Determination of the particle size distribution of MFs/NFs Short measurement times Wide and dynamic size range and high level of reproducibility	Expensive Overestimation of the number of MFs by non-shape analysis	Wang et al. (2021)
Identification and Characterization				
FTIR and/or μFTIR	Spectral analysis	High sensitivity Versatility High throughput screening Accurate results No damage to the sample	Easily disturbed by water Sample needs to be preprocessed or cleaned Low horizontal resolution Limitation in the analysis of small particles ≈100 μm	Ma et al. (2021), Saliu et al. (2021), Wang et al. (2021), Shen et al. (2021), Wu et al. (2022), Rathinamoorthy and Balasaraswathi, (2022), Chen et al. (2023), De-la-Torre et al (2022a, c)
Raman	Spectral analysis	No damage to the sample Analysis of particle size 1 μm Good spatial resolution High specificity of fingerprint spectrum High precision More sensitive to additives	Fluorescence phenomenon Long processing time	Liang et al. (2022)
Scanning laser confocal microscopy coupled to Raman spectroscopy	Spectral analysis	Mapping areas of a sample Depth profiling Analysis of unknown compounds in a small area Spectral and spatial resolution Analysis of small particles	High operating cost	Ma et al. (2023)

(*Continued*)

TABLE 7.2 (*Continued*)
Chemical Analysis Methods Are Used to Quantify, Identify, and Characterize Disposable Face Masks and Their Sub-products as Microfibers/Nanofibers

Method	Type	Advantages	Disadvantages	Reference
SEM/EDX	Microscopy and element analysis	High-resolution images of sample surfaces Chemical analysis of samples Resolution	Long processing time Low sensitivity Sample preprocess Vacuum environment	Ma et al. (2021), Saliu et al. (2021), Sullivan et al. (2021), Wang et al. (2021), Shen et al. (2021), Wu et al. (2022), De la Torre et al. (2022a), Akarsu et al. (2021), Chen et al. (2023), De-la-Torre et al. (2022c)
AFM	Microscopy advanced	Mapping and roughness profiles in three dimensions Vertical sub-nanometric resolution and a lateral resolution Study of surface and mechanical properties such as adhesion, hardness, conductivity, electrical, magnetic, among others.	Relatively scan delay time Image quality Long time to measure large surfaces	Ma et al. (2021), Wang et al. (2021)
Confocal laser scanning microscopy	Microscopy advanced	Obtaining images with 3D information over time Magnification can be adjusted electronically	Scan delay time Lower resolution than camera detection Laser may damage living cells	Wu et al. (2022), Crisafi et al. (2022)

7.7 CONCLUSIONS AND FURTHER RESEARCH

The negative repercussion of the COVID-19 pandemic on the environment is evident because of the huge quantity and poor disposal of PPE waste as well as for the effect of MFs/NFs and other contaminants released, which aggravated the already existing microplastic pollution before the onset of the pandemic. The total amount of PPE waste globally generated is unknown due to the lack of data in many countries. However, it is believed that hundreds of tons of PPE disposal in particular face masks, has been generated. Moreover, scientists have researched how face masks contribute to MFs/NFs pollution in aquatic and terrestrial environments and how PPE weathering influences the amount of release of MFs/NFs and plastic micro-fragments under laboratory conditions. Nevertheless, the ecotoxicological effects of a mix between MFs/NFs and additives employed in PPE manufacture, particularly in face masks, have not been studied in detail as well as their negative impact and bioaccumulation in organisms and humans. Faced with this scenario, many researchers have started to develop environmentally friendly materials such as biopolymers and biodegradable composites (Pandit et al. 2021) for the manufacture of PPE, particularly to be used in face masks. Likewise, new technologies such as catalytic carbonization or using this waste as an additive to mix asphalt have been implemented (Wang et al. 2022). Also, the development of methods and techniques for the sterilization and reusability like hydrogen peroxide vapor (VH_2O_2) and chemicals liquid disinfectant (Ilyas et al. 2020) or the degradation of this waste such as catalytic carbonization (Yu et al. 2021) have been considered. However, all these new technologies are still in their first step. Finally, society is distant from fixing this plastic and microfiber pollution problem generated by the pandemic since it would require a significant operation of different sectors and integration of environmental awareness into people's everyday lives.

REFERENCES

Akarsu, C., Madenli, Ö., & Deveci, E. Ü. 2021. Characterization of littered face masks in the southeastern part of Turkey. *Environmental Science and Pollution Research*, 28(34), 47517–47527.

Akhbarizadeh, R., Dobaradaran, S., Nabipour, I., Tangestani, M., Abedi, D., Javanfekr, F., Faezeh, J., & Zendehboodi, A. 2021. Abandoned Covid-19 personal protective equipment along the Bushehr shores, the Persian Gulf: An emerging source of secondary microplastics in coastlines. *Marine Pollution Bulletin*, 168, 112386–112394.

Ammendolia, J., Saturno, J., Bond, A. L., O'hanlon, N.J., Masden, E. A., Neil, A. J., & Jacobs, S. 2022. Tracking the impacts of COVID-19 pandemic-related debris on wildlife using digital platforms. *Science of Total Environment*, 848, 157614–157624.

Anastopoulos, I., & Pashalidis, I. 2021. Single-use surgical face masks, as a potential source of microplastics: Do they act as pollutant carriers? *Journal of Molecular Liquids*, 326, 115247–1152451.

Ardusso, M., Forero-López, A. D., Buzzi, N. S., Spetter, C. V., & Fernández-Severini, M. D. 2021. COVID-19 pandemic repercussions on plastic and antiviral polymeric textile causing pollution on beaches and coasts of South America. *Science of Total Environment*, 763, 144365–144377.

Benson, N. U., Bassey, D. E., & Palanisami, T. 2021. COVID pollution: Impact of COVID-19 pandemic on global plastic waste footprint. *Heliyon*, 7(2), e06343.

Bhattacharjee, S., Bahl, P., Chughtai, A. A., Heslop, D., & MacIntyre, C. R. 2022. Face masks and respirators: Towards sustainable materials and technologies to overcome the shortcomings and challenges. *Nano Select*, 3(10), 1355–1381.

Bussan, D. D., Snaychuk, L., Bartzas, G., & Douvris, C. 2022. Quantification of trace elements in surgical and KN95 face masks widely used during the SARS-COVID-19 pandemic. *Science of the Total Environment*, 814, 151924–151931.

Cabrejos-Cardeña, U., De-la-Torre, G. E., Dobaradaran, S., & Rangabhashiyam, S. 2022. An ecotoxicological perspective of microplastics released by face masks. *Journal of Hazardous Materials*, 443, 130273–130277.

Chen, X., Chen, X., Liu, Q., Zhao, Q., Xiong, X., & Wu, C. 2021. Used disposable face masks are significant sources of microplastics to the environment. *Environmental Pollution*, 285, 117485–117491.

Chen, C., Yu, G., Wang, B., Li, F., Liu, H., & Zhang, W. 2023. Lifetime prediction of non-woven face masks in ocean and contributions to microplastics and dissolved organic carbon. *Journal of Hazardous Materials*, 441, 129816–129826.

Chomiak, K. M., Eddingsaas, N. C., & Tyler, A. C. 2023. Direct and indirect impacts of disposable face masks and gloves on freshwater benthic fauna and sediment biogeochemistry. *ACS ES&T Water*, 3(1), 51–59.

Crisafi, F., Smedile, F., Yakimov, M. M., Aulenta, F., Fazi, S., La Cono, V., Martinelli, A., Di Lisio, V., & Denaro, R. 2022. Bacterial biofilms on medical masks disposed in the marine environment: A hotspot of biological and functional diversity. *Science of the Total Environment*, 837, 155731–155742.

De-la-Torre, G. E., Dioses-Salinas, D. C., Dobaradaran, S., Spitz, J., Keshtkar, M., Akhbarizadeh, R., Abedi, D., & Tavakolian A. 2022c. Physical and chemical degradation of littered personal protective equipment (PPE) under simulated environmental conditions. *Marine Pollution Bulletin*, 178, 113587–113597.

De-la-Torre, G. E., Dioses-Salinas, D. C., Dobaradaran, S., Spitz, J., Nabipour, I., Keshtkar, M., Akhbarizadeh, R., Tangestani, M., Abedi, D., & Javanfekr, F. 2022b. Release of phthalate esters (PAEs) and microplastics (MPs) from face masks and gloves during the COVID-19 pandemic. *Environmental Research*, 215, 114337–1144347.

De-la-Torre, G. E., Dioses-Salinas, D. C., Pizarro-Ortega, C. I., Fernández Severini, M. D., ForeroLópez, A. D., … & Santillán, L. 2022a. Binational survey of personal protective equipment (PPE) pollution driven by the COVID-19 pandemic in coastal environments: Abundance, distribution, and analytical characterization. *Journal of Hazardous Materials*, 426, 128070–128087.

Ellen MacArthur Foundation. 2017. A new textiles economy: Redesigning fashion's future. https://www.ellenmacarthurfoundation.org/publications.

Forero López, A. D., Fabiani, M., Lassalle, V. L., Spetter, C. V., & Fernandez-Severini, M. D. 2022a. Critical review of the characteristics, interactions, and toxicity of micro/nanomaterials pollutants in aquatic environments. *Marine Pollution Bulletin*, 174, 173276–173296.

Forero-López, A. D., Brugnoni, L. I., Abasto, B., Rimondino, G. N., … & Biancalana, F. 2022b. Plastisphere on microplastics: In situ assays in an estuarine environment. *Journal of Hazardous Materials*, 440, 129737–129753.

Gewert, B., Plassmann, M., Sandblom, O., & MacLeod, M. 2018. Identification of chain scission products released to water by plastic exposed to ultraviolet light. *Environmental Science & Technology Letters*, 5(5), 272–276.

Haddad, M. B., De-la-Torre, G. E., Abelouah, M. R., Hajji, S., & Alla, A. A. 2021. Personal protective equipment (PPE) pollution associated with the COVID-19 pandemic along the coastline of Agadir, Morocco. *Science of the Total Environment*, 798, 149282–149291.

Han, J., & He, S. 2021. Need for assessing the inhalation of micro(nano)plastic debris shed from masks, respirators, and home-made face coverings during the COVID-19 pandemic. *Environmental Pollution*, 268, 115728–115732.

Ilyas, S., Srivastava, R. R., & Kim, H. 2020. Disinfection technology and strategies for COVID-19 hospital and bio-medical waste management. *Science of the Total Environment*, 749, 141652–141653.

Jędruchniewicz, K., Ok, Y. S., & Oleszczuk, P. (2021). COVID-19 discarded disposable gloves as a source and a vector of pollutants in the environment. *Journal of Hazardous Materials*, 417, 125938–125961.

Jiang, H., Luo, D., Wang, L., Zhang, Y., Wang, H., & Wang, C. 2022. A review of disposable facemasks during the COVID-19 pandemic: A focus on microplastics release. *Chemosphere*, 312, 137178–137190.

Kokalj, A. J., Dolar, A., Drobne, D., Škrlep, L., Škapin, A. S., Marolt, G., Nagode, A., & Van Gestel, C. A. 2022. Effects of microplastics from disposable medical masks on terrestrial invertebrates. *Journal of Hazardous Materials,* 438, 129440–129449.

Konda, A., Prakash, A., Moss, G. A., Schmoldt, M., Grant, G. D., & Guha, S. 2020. Response to letters to the editor on aerosol filtration efficiency of common fabrics used in respiratory cloth masks: Revised and expanded results. *ACS Nano*, 14(9), 10764–10770.

Kwak, J. I., & An, Y. J. 2021. Post COVID-19 pandemic: Biofragmentation and soil ecotoxicological effects of microplastics derived from face masks. *Journal of Hazardous Materials*, 416, 126169–126177.

Li, L., Zhao, X., Li, Z., & Song, K. 2021. COVID-19: Performance study of microplastic inhalation risk posed by wearing masks. *Journal of Hazardous Materials*, 411, 124955–124964.

Li, M., Hou, Z., Meng, R., Hao, S., & Wang, B. 2022. Unraveling the potential human health risks from used disposable face mask-derived micro/nanoplastics during the COVID-19 pandemic scenario: A critical review. *Environment International*, 170, 107644–107669.

Liang, H., Ji, Y., Ge, W., Wu, J., Song, N., Yin, Z., & Chai, C. 2022. Release kinetics of microplastics from disposable face masks into the aqueous environment. *Science of Total Environment*, 816, 151650–151657.

Lin, L., Yuan, B., Zhang, B., Li, H., Liao, R., Hong, H., Lu, H., Liu, J., & Yan, C. (2022). Uncovering the disposable face masks as vectors of metal ions (Pb (II), Cd (II), Sr (II)) during the COVID-19 pandemic. *Chemical Engineering Journal*, 439, 135613–135624.

Lyu, L., Wang, Z., Bagchi, M., Ye, Z., Soliman, A., Bagchi, A., Markoglou, N., Yin, J., An, C., Yang, X., Bi, H., & Cai, M. 2023. An investigation into the aging of disposable face masks in landfill leachate. *Journal of Hazardous Materials*, 446, 130671–130685.

Ma, J., Chen, F., Xu, H., Jiang, H., Liu, J., & Pan, K. 2021. Face masks as a source of nanoplastics and microplastics in the environment: Quantification, characterization, and potential for bioaccumulation. *Environmental Pollution*, 288, 117748–117754.

Ma, M., Xu, D., Zhao, J., & Gao, B. 2023. Disposable face masks release micro particles to the aqueous environment after simulating sunlight aging: Microplastics or non-microplastics? *Journal of Hazardous Materials*, 443, 130146–130157

Meszáros, E., Bodor, A., Szierer, Á., Kovács, E., Perei, K., Tölgyesi, C., Bátori, Z., & Feigl, G. 2022. Indirect effects of COVID-19 on the environment: How plastic contamination from disposable surgical masks affect early development of plants. *Journal of Hazardous Materials*, 436, 129255–129265.

Morgana, S., Casentini, B., & Amalfitano, S. 2021. Uncovering the release of micro/nanoplastics from disposable face masks at times of COVID-19. *Journal of Hazardous Materials*, 419, 126507–126515.

Neto, H. G., Bantel, C. G., Browning, J., Della Fina, N., Ballabio, T. A., de Santana, S. T., de Karam e Britto, M., & Barbosa, C. B. 2021. Mortality of a juvenile Magellanic penguin (Spheniscusmagellanicus, Spheniscidae) associated with the ingestion of a PFF-2 protective mask during the Covid-19 pandemic. *Marine Pollution Bulletin*, 166, 112232–112237.

Oceans Asia, 2020, Bondaroff, P., Teale, & Cooke, S. 2020, December. Masks on the beach: The impact of COVID-19 on marine plastic pollution. OceansAsia. https://occansasia.org/covid-19-facemasks/

Oliveira, A. M., Patrício-Silva, A. L., Soares, A. M. V. M., Barceló, D., Armando, D., & Rocha-Santos, T. 2023. Current knowledge on the presence, biodegradation, and toxicity of discarded face masks in the environment. *Journal of Environmental Chemical Engineering*, 109308–109325.

Pandit, P., Maity, S., Singha, K., Uzun, M., Shekh, M., & Ahmed, S. 2021. Potential biodegradable face mask to counter environmental impact of Covid-19. *Cleaner Engineering and Technology*, 4, 100218–100229.

Phan, S., Padilla-Gamiño, J. L., & Luscombe, C. K. 2022. The effect of weathering environments on microplastic chemical identification with Raman and IR spectroscopy: Part I. polyethylene and polypropylene. *Polymer Testing*, 116, 107752–107763.

Pizarro-Ortega, C. I., Dioses-Salinas, D. C., Fernández Severini, M. D., Forero-López, A. D., Rimondino, G. N., Benson, N. U., Dobaradaran S., & De-la-Torre, G. E. 2022. Degradation of plastics associated with the COVID-19 pandemic. *Marine Pollution Bulletin*, 176, 113474–113485.

Prata, J. C., Silva-Patricia, A. L., Walker, T. R., Duarte, A. C., & Rocha-Santos, T. 2020. COVID-19 pandemic repercussions on the use and management of plastics. *Environmental Science Technology*, 54(13), 7760–7765

Rathinamoorthy, R., & Balasaraswathi, S. R. 2022. Mitigation of microfibers release from disposable masks – An analysis of structural properties. *Environmental Research*, 214, 114106–114124.

Saliu, F., Veronelli, M., Raguso, C., Barana, D., Galli, P., & Lasagni, M. 2021. The release process of microfibers: From surgical face masks into the marine environment. *Environmental Advances*, 4, 100042–100048.

Sankhyan, S., Heinselman, K. N., Ciesielski, P. N., Barnes, T., Himmel, M. E., Teed, H., Patel, S., & Vance, M. E. 2021. Filtration performance of layering masks and face coverings and the reusability of cotton masks after repeated washing and drying. *Aerosol and Air Quality Research*, 21(11), 210117–210130.

Sendra, M., Pereiro, P., Yeste, M. P., Novoa, B., & Figueras, A. 2022. Surgical face masks as a source of emergent pollutants in aquatic systems: Analysis of their degradation product effects in Danio renio through RNA-Seq. *Journal of Hazardous Materials*, 428, 128186–128212.

Shams, M., Alam, I., & Mahbub, M. S. 2021. Plastic pollution during COVID-19: Plastic waste directives and its long-term impact on the environment. *Environmental Science*, 5, 100119–100121.

Shen, M., Zeng, Z., Song, B., Yi, H., Hu, T., Zhang, Y., ... & Xiao, R. 2021. Neglected microplastics pollution in global COVID-19: Disposable surgical masks. *Science of the Total Environment*, 790, 148130–148140.

Shruti, V. C., Pérez-Guevara, F., Elizalde-Martínez, I., & Kutralam-Muniasamy, G. 2020. Reusable masks for COVID-19: A missing piece of the microplastic problem during the global health crisis. *Marine Pollution Bulletin*, 161, 111777–111782.

Shukla, S., Khan, R., Saxena, A., & Sekar, S. 2022. Microplastics from face masks: A potential hazard post Covid-19 pandemic. *Chemosphere*, 302, 134805–134814.

Silva, A. L. P., Prata, J. C., Mouneyrac, C., Barcelò, D., Duarte, A. C., & Rocha-Santos, T. 2021b. Risks of covid-19 face masks to wildlife: Present and future research needs. *Science of Total Environment*, 792, 148505–148514.

Silva, A. L. P., Prata, J. C., Walker, T. R., Duarte, A. C., Ouyang, W., Barcelò, D., & Rocha-Santos, T. 2021a. Increased plastic pollution due to COVID-19 pandemic: Challenges and recommendations. *Chemical Engineering Journal*, 405, 126683–126692.

Sullivan, G. L., Delgado-Gallardo, J., Watson, T. M., & Sarp, S. 2021. An investigation into the leaching of micro and nanoparticles and chemical pollutants from disposable face masks linked to the COVID-19 pandemic. *Water Research*, 196, 117033–1177045.

Sun, J., Yang, S., Zhou, G. J., Zhang, K., Lu, Y., Jin, Q., Paul, K. S. L., Kenneth, M. Y. L., & He, Y. 2021. Release of microplastics from discarded surgical masks and their adverse impacts on the marine copepod Tigriopus japonicus. *Environmental Science & Technology Letters*, 8(12), 1065–1070.

Wang, Z., An, C., Chen, X., Lee, K., Zhang, B., & Feng, Q. 2021. Disposable masks release microplastics to the aqueous environment with exacerbation by natural weathering. *Journal of Hazardous Materials*, 417, 126036–126046.

Wang, X., Okoffo, E. D., Banks, A. P., Li, Y., Thomas, K. V., Rauert, C., Aylward, L. L., & Mueller, J. F. 2022. Phthalate esters in face masks and associated inhalation exposure risk. *Journal of Hazardous Materials*, 423, 127001–127010.

Wu, P., Li, J., Lu, X., Tang, Y., & Cai, Z. 2022. Release of tens of thousands of microfibers from discarded face masks under simulated environmental conditions. *Science of the Total Environment*, 806, 150458–150466.

Xie, H., Han, W., Xie, Q., Xu, T., Zhu, M., & Chen, J. 2022. Face mask – A potential source of phthalate exposure for human. *Journal of Hazardous Materials*, 422, 126848–126854.

Ye, Y., Yu, K., & Zhao, Y. 2022. The development and application of advanced analytical methods in microplastics contamination detection: A critical review. *Science of the Total Environment*, 818, 151851–151867.

Yu, R., Wen, X., Liu, J., Wang, Y., Chen, X., Wenelska, K., …& Tang, T. 2021. A green and high-yield route to recycle waste masks into CNTs/Ni hybrids via catalytic carbonization and their application for superior microwave absorption. *Applied Catalysis B: Environmental*, 298, 120544–120553.

Zhou Dan, Y. S., Yang, K., Yang, L. Y., Yang, X. R., Huang, F. Y., Neilson, R., Su, J. Q, & Zhu, Y. G. 2022. Discarded masks as hotspots of antibiotic resistance genes during COVID-19 pandemic. *Journal of Hazardous Materials*, 425, 127774–127784.

Section 2

Sources of Microfibre Pollution and Analytical Tools

8 Quantifying Microfibre Release from Textiles during Domestic Laundering

Challenges and Progress

Alice Hazlehurst, Mark Sumner, and Mark Taylor
University of Leeds

8.1 INTRODUCTION

The quantity of microfibre released from the domestic laundering of textiles has been investigated by numerous studies in recent years. Methods used involve exposing a fabric or garment to simulated domestic laundering and collecting any released microfibres for quantification and analysis. Methodological approaches between studies are varied, with differences in laundering devices, laundering conditions such as temperature and duration, fabric sampling techniques, test liquor filtration and fibre collection methods, and fibre counting. All of which makes meaningful comparison between studies challenging, and reported microfibre release estimates vary dramatically from a few thousands to several millions of fibres released from a single wash. Adoption of a standardised testing method would greatly aid understanding of microfibre release from domestic laundering, providing more reliable quantification estimates and opportunity to compare fabrics under the same conditions of test. This chapter reviews existing approaches and also describes progress in standardisation of a testing method for quantifying microfibre release from domestic laundering.

8.2 METHODS FOR QUANTIFICATION OF MICROFIBRE RELEASE

Table 8.1 shows the diversity of methods, washing conditions and fabric types that have been employed to quantify the release of microfibres from the domestic laundering of textile products. However, this diverse mix of methods can be categorised into broad methodological approaches used for replicating domestic laundering (domestic machine washing or laboratory/simulated washing) and for measuring the release of microfibres (number of fibres or mass of fibres). These categories are discussed below with reference to the studies in Table 8.1.

DOI: 10.1201/9781003331995-10

TABLE 8.1
Comparison of Published Methods for Quantifying Microfibre Release

Author	Test Method	Temp. (°C)	Time (mins)	Agitation (No. Ball Bearings)	Liquor Ratio	Reported Microfibre Release Quantification
Browne **et al.** (2011)	Washing machines – front-loading	40	NS	N/A	NS	>1,900 particles
Dubaish and Liebezeit (2013)	NS	NS	NS	NS	NS	220–260 mg/garment
Karlsson (2015)	Washing machines – front-loading	40	NS	N/A	NS	209,000 fibres/L
Napper and Thompson (2016)	Washing machines – front-loading	30, 40	75	N/A	NS	Poly-cotton: 137,951 fibres/6 kg wash Polyester: 496,030 fibres/6 kg wash Acrylic: 728,789 fibres/6 kg wash
Pirc **et al.** (2016)	Washing machines – front-loading	30	15	N/A	NS	11,300 particles/500 g of fabrics or 6 mg/500 g fabric
Hartline **et al.** (2016)	Washing machines – front- and top-loading	30, 40	30, 48	N/A	NS	Average 1,174 mg/wash
Hernandez **et al.** (2017)	Laboratory method – washtec	25, 40, 60, 80	14, 60, 120, 240, 480	10	7:1 l/m²	Average approx. 0.025 mg/g
Sillanpää and Sainio (2017)	Washing machines – front-loading	40	75	N/A	NS	Polyester: 223,000 particles or 340 mg Cotton: 973,000 or 809 mg
Carney Almroth **et al.** (2018)	Laboratory method – gyrowash	60	30	25	13:1 L/m²	Fleece: 110,000 fibres/garment Knit: 900 fibres/garment
De Falco **et al.** (2018)	Laboratory method – linitest	40, 60	45, 75, 90	0, 10, 20	NS	Polyester: >6,000,000 fibres/5 kg wash
Jönsson **et al.** (2018)	Laboratory method – gyrowash	40	60	25	5:1 L/m²	N/A focus on method development and validation

(Continued)

TABLE 8.1 (*Continued*)
Comparison of Published Methods for Quantifying Microfibre Release

Author	Test Method	Temp. (°C)	Time (mins)	Agitation (No. Ball Bearings)	Liquor Ratio	Reported Microfibre Release Quantification
Belzagui **et al.** (2019)	Washing machines – front-loading	Ambient	15	N/A	NS	Polyester-elastane: 175 fibres/g/wash or 30,000 fibres/m² Acrylic-polyamide: 560 fibres/g/wash or 465,000 fibres/m²
De Falco **et al.** (2019)	Washing machines – front-loading	40	107	N/A	NS	Polyester: 640,000–,100,000 fibres/garment Poly-cotton-modal: 1,500,000 fibres/garment
Haap **et al.** (2019)	Laboratory method – labomat	40	30	50	6:1 L/m²	N/A focus on method development and validation
Kelly **et al.** (2019)	Washing machines – front-loading and laboratory – tergotometer method	30, 15	60, 15	NS	120:1 L/m² and 240:1 L/m²	Standard wash (with detergent): 663,523 fibres/kg Delicate wash (with detergent): 1,474,793 fibres/kg
Yang **et al.** (2019)	Washing machines – front- and top-loading	30, 40, 60	15	N/A	17:1 L/m²	Acetate: up to 74,816 fibres/m² Polyester: up to 72,130 fibres/m²
Zambrano **et al.** (2019)	Washing machines – top-loading and laboratory method – launderOmeter	25, 44	16	25	15:1 L/m²	Cellulose-based fabrics: 0.2–4 mg/g Polyester: 0.1–1 mg/g
Cai **et al.** (2020)	Laboratory method – gyrowash	40	45	0, 10, 20	37.5:1 L/m² and 60:1 L/m²	210–72,000 fibres/g

(*Continued*)

TABLE 8.1 (*Continued*)
Comparison of Published Methods for Quantifying Microfibre Release

Author	Test Method	Temp. (°C)	Time (mins)	Agitation (No. Ball Bearings)	Liquor Ratio	Reported Microfibre Release Quantification
Cesa **et al.** (2020)	Washing machines – top-loading	24	20	N/A	NS	49.8–307.8 mg/wash 18,400–69,600 fibres/wash
Cotton **et al.** (2020)	Washing machines – front-loading	25, 40	30, 85	N/A	NS	NS
Dalla Fontana **et al.** (2020)	Washing machines – front-loading	40, 30	90, 43	N/A	NS	Delicate/silk cycle: 33.86 mg/kg Cotton cycle: 40.19 mg/kg
De Falco **et al.** (2020)	Washing machines – front-loading	40	107	None	NS	Poly-cotton: 3,898 fibres/g Polyester: 709–1,747 fibres/g
Frost **et al.** (2020)	Laboratory method – launderOmeter	20	16	25	NS	NS
Galvão **et al.** (2020)	Washing machines – front-loading	20, 30, 40, 60	NS (automatic)	N/A	NS	Average total MF per wash: 297,400 MF/L (83% from cotton fibres) Estimate for 6 kg wash: 18,000,000 MF/6 kg synthetic fibres 1,842–6,259 mf/g
Kärkkäinen and Sillanpää (2020)	Washing machines – front-loading	40	75	N/A	NS	First wash: 100,000–6,300,000 fibres/k Fifth wash: 19,000–190,000 fibres/kg
Lant **et al.** (2020)	Washing machines – front- and top-loading	40, 15	85, 30	N/A	NS	Approx. 357 mg/3.13 kg wash 96% released fibres = natural (cotton, wool, viscose), 4% = synthetic (acrylic, nylon, polyester)

(*Continued*)

TABLE 8.1 (*Continued*)
Comparison of Published Methods for Quantifying Microfibre Release

Author	Test Method	Temp. (°C)	Time (mins)	Agitation (No. Ball Bearings)	Liquor Ratio	Reported Microfibre Release Quantification
Praveena **et al.** (2020)	Washing machines – majority top-loading (household study)	Room temperature	41–60	N/A	NS	0.0069–0.183 g/m³ Average: 0.068 g/m³
Berruezo **et al.** (2021)	Laboratory method – linitest	40	30	10	50:1 L/m²	NS
Celik (2021)	Washing machine – front-loading	40	53	N/A	NS	3.28–3.91 mg/L 13.1–15.66 mg/kg
Choi **et al.** (2021a)	Washing machine – front-loading	40, 20	80	N/A	NS	Hard twist filament: 51.6 ppm Non-twist filament: 88.7 ppm Spun yarn: 107.7 ppm
Choi **et al.** (2021b)	Washing machine – front-loading	40, 20	80	NA	NS	NS
Dalla Fontana **et al.** (2021)	Washing machine – front-loading	40	90	N/A	NS	38.6–67.9 mg/kg
Özkan and Gündoğdu (2021)	Laboratory method – gyrowash	40	45	10	37.5:1 L/m²	Recycled polyester: 368,094 fibres/kg Virgin polyester: 167,436 fibres/kg
Periyasamy (2021)	Washing machine – front-loading	30, 45, 60	60, 75, 90	N/A	NS	2,305,395–4,874,323 fibres/kg
Tiffin **et al.** (2022)	Laboratory method – gyrowash	40	45	50	15:1 L/m²	N/A focus on method development and validation
Vassilenko **et al.** (2021)	Washing machine – top-loading	41	18	N/A	136.4:1 L/kg	8,809–>6,877,000 fibres 9.6–1,240 mg/kg
Volgare **et al.** (2021)	Washing machine – front-loading	40	107	N/A	NS	0.15 kg wash load: 401 ± 17 mg/kg 0.88 kg wash load: 187 ± 21 mg/kg 1.64 kg wash load: 104 ± 10 mg/kg 2.50 kg wash load: 76 ± 5 mg/kg

(*Continued*)

TABLE 8.1 (*Continued*)
Comparison of Published Methods for Quantifying Microfibre Release

Author	Test Method	Temp. (°C)	Time (mins)	Agitation (No. Ball Bearings)	Liquor Ratio	Reported Microfibre Release Quantification
Zambrano **et al.** (2021)	Laboratory method – launderOmeter	25	16	25	NS	9,000–14,000 particles/g Softener/durable press-treated cotton: 1.30–1.63 mg/g Untreated cotton: 0.73 mg/g
Dreillard **et al.** (2022)	Washing machine – front-loading	30	NS	N/A	23.3:1 L/kg	220,000–2,282,000 fibres/kg
Jabbar **et al.** (2022)	Laboratory method – gyrowash	40	45	0	33:1 L/m² 11:1 L/m²	NS
Kim **et al.** (2022)	Washing machine – top-loading	25	52	N/A	21:1 L/m²	NS
Kumar and Gopalakrishnan (2022)	Washing machine – top-loading	NS	30, 60, 90	N/A	NS	2.24–4.56 g/kg
Mahbub and Shams (2022)	Washing machine – top-loading	20, 40	30, 45, 60	N/A	NS	3.64–133.22 mg/kg
Palacios-Marín **et al.** (2022)	Laboratory method – gyrowash	40	60	50	11:1 L/m²	Flat PET (polyethylene terephthalate): 1,811 fibres/g Textured PET: 440 fibres/g Staple PET: 2,101 fibres/g Cotton: 4,111 fibres/g

(*Continued*)

TABLE 8.1 (*Continued*)
Comparison of Published Methods for Quantifying Microfibre Release

Author	Test Method	Temp. (°C)	Time (mins)	Agitation (No. Ball Bearings)	Liquor Ratio	Reported Microfibre Release Quantification
Raja Balasaraswathi and Rathinamoorthy (2022)	Laboratory method – launderOmeter	30	45	10	6.67:1 L/m²	Open edge: 72.37 fibres/cm² or 0.0239 mg/cm² Finished edge: 18.07 fibres/cm² or 0.0034 mg/cm²
Rathinamoorthy and Raja Balasaraswathi (2022a)	Laboratory method – launderOmeter	30	45	5, 10, 15	2.22:1, 4.44:1, 6.67:1, 8.89:1 L/m²	Machine wash: 18.06 fibres/cm² Gentle hand wash: 17.33 fibres/cm² Intense hand wash: 23.7 fibres/cm²
Rathinamoorthy and Raja Balasaraswathi (2022b)	Laboratory method – launderOmeter	30, 60, 90	30, 45, 60	5, 10, 15	15:1 L/m² 30:1 L/m² 45:1 L/m²	Average 8.255–21.79 fibres/cm² or 8.09–37.18 mg/m²
Sudheshna **et al.** (2022)	Washing machine – front- and top-loading (household study)	NS	NS	N/A	NS	Average of four households: 252,277 fibres/L Sum for total load of 26.2 kg: 33,110,407 fibres

Source: Hazlehurst, A., Tiffin, L., Sumner, M., Taylor, M., 2023. Quantification of microfibre release **from** textiles during domestic laundering. ***Environ. Sci. Pollut. Res.*** 30, 43932–43949.

NS, not stated; N/A, not applicable.

8.2.1 Washing Machine Methods

While the use of domestic washing machines to assess microfibre release is perhaps an obvious choice for providing real-life quantification estimates, there are a number of limitations associated with this approach. The way different domestic washing machines perform and impact fabrics during washing varies greatly between different types, makes and models. Therefore, the quantity of microfibre released also varies depending on the washing machine used.

The type of washing machine (front or top loading) can have an influence on the quantity of microfibres generated during the laundering process. A study comparing front- and top-loading washing machines found for the same fabric and wash conditions: the top-loading machine released 47 mg of microfibres compared with 7.3 mg of microfibres for the front-loading machine per garment per wash (Hartline et al., 2016). It was suggested this was due to the central agitator of the top-loading machine providing greater agitation of the fabrics during laundering compared to the front-loading drum, and the increased agitation led to the production of more microfibres (Hartline et al., 2016).

In domestic washing machines, the level of agitation for a specific washing cycle is dependent on the drum size and configuration, the rotation speed and the wash liquor ratio (ratio of water volume to mass of fabric). When washing loads are large relative to the drum capacity, there is less movement of the fabric and therefore less agitation (Mac Namara et al., 2012; Yun and Park, 2015). Conversely, when washing loads are small relative to the drum capacity, agitation is increased. More movement and agitation of the fabric in the wash liquor tends to increase microfibre release. The wash liquor ratio also influences the movement and friction between fabric in the wash, and so this also influences microfibre release. Similarly, rotational speeds directly impact the level of agitation and microfibre generation. As stated, rotational speeds used in the studies in Table 8.1 ranged from 600 (Browne et al., 2011; Pirc et al., 2016) to 1,400 rpm (revolutions per minute) (Kumar and Gopalakrishnan, 2022; Napper and Thompson, 2016).

In domestic washing machines, the consistency and reproducibility of rotational speeds, water intake/outflow and cycle duration are often poor. This is particularly evident when fuzzy logic is employed in the machine control unit. Fuzzy logic is used to alter the washing conditions dependent on the washing load and other variables (Kosko and Isaka, 1993). As these variables have a major influence on the quantity of microfibre released, test methods based on using domestic washing machines where fuzzy logic control units are used are likely to produce inconsistent results.

There are further, practical limitations to using domestic washing machines for testing microfibre release. Based on UK average load size, replicating a typical domestic washing load would require around 5 kg of test fabric in each experiment (Webber et al., 2016). Washing only single garments or smaller masses of fabric samples for testing could provide misleading results which are not indicative of real washing loads as changing the wash liquor to wash load ratio alters the agitation levels and therefore microfibre release. Additionally, a significant volume of water is needed to complete a single wash cycle, and this has to be collected for analysis; the volumes of water collected in studies from Table 8.1 ranged from 12 L (Mahbub and Shams, 2022) up to 136 L (Hartline et al., 2016). Such large volumes of test water present a practical challenge for analysis, and most of the studies reviewed opted to sample smaller aliquots of liquid for subsequent filtration and analysis. This assumes a homogenous distribution of fibres in the test liquor. A further practical challenge is the large and complex nature of washing machines themselves. The internal pipework of the washing machine may retain microfibres and other contaminants. Most of the studies reviewed included a cleaning cycle between washes for the purpose of removing any residual microfibres between experiments. However, there was little discussion to demonstrate all microfibres released from the drum were captured in the test liquor and that no microfibres were retained in the machine.

8.2.2 Simulated Laundering Device Methods

The action of domestic washing machines can be highly variable due to variations between makes and models, fuzzy logic, etc., making domestic washing machines inherently inconsistent and therefore unreliable when developing a test method for evaluating microfibre release. The inconsistent nature of washing machines makes them difficult to replicate in laboratory environments. Hence, a growing number of studies have opted to use simulated laundering devices that have greater control of experimental variables to approximate the action of domestic washing (Berruezo et al., 2021; Cai et al., 2020; Carney Almroth et al., 2018; De Falco et al., 2018; Frost et al., 2020; Haap et al., 2019; Hernandez et al., 2017; Jabbar et al., 2022; Jönsson et al., 2018; Özkan and Gündoğdu, 2021; Palacios-Marín et al., 2022; Raja Balasaraswathi and Rathinamoorthy, 2022; Rathinamoorthy and Raja Balasaraswathi, 2022a, b; Tiffin et al., 2022; Zambrano et al., 2019, 2021). Simulated laundering devices such as those described by ISO 105-CO6 (British Standards Institution, 2010) and AATCC TM61 (American Association of Textile Chemists and Colorists, 2013) have been used extensively for many years for assessing textile colour fastness to laundering. These devices use stainless steel canisters containing test samples, wash liquor and ball bearings (to replicate in-wash abrasion), which are rotated through a heated water bath within the device. The primary advantage of using these simulated laundering devices is they allow more accurate and reliable control of test variables such as the liquor ratio, wash temperature, rotational speed, agitation and wash duration than typical domestic washing machines. Furthermore, as the wash liquor and test sample are contained within a closed canister, the entirety of the test liquor can also be filtered and analysed, without any issues of contamination or losses within the machine. An obvious limitation is the small canister size necessitates small test sample sizes, meaning that full-scale garments cannot be tested by simulated laundering device methods.

Simulated laundering device methods can not exactly replicate domestic laundering; however, research has suggested there is a reasonable correlation between results obtained from a simulated laundering device and those of a domestic washing machine in terms of the relative amounts of microfibre released for different fabric (Zambrano et al., 2019). A reasonable correlation was also found between domestic washing machine results and results obtained from a laboratory-scale method using a tergotometer, a device that more closely mimics a top-loading washing machine (Kelly et al., 2019). This device is commonly used in the development of laundering care products, but it is not commonly used in textile testing laboratories. A more comprehensive correlation study would certainly be beneficial for the studies of microfibre release and provide greater clarity on the degree of replication between simulated and domestic laundry methods.

8.2.3 Collection of Released Microfibres

Irrespective of the laundering method used, a sample of test liquor is created, which contains the microfibres released from the fabric. In general, most methods require extraction of the microfibres from the test liquor to facilitate further analysis. Almost all of the studies included in Table 8.1 used vacuum-assisted filtration in order to extract and collect microfibres from the test liquor; however, the filters varied in both type and pore size between studies.

Cellulosic filters were used in a number of the studies reviewed (Cotton et al., 2020; Kelly et al., 2019) or polyamide (Belzagui et al., 2019; De Falco et al., 2018, 2019; Hartline et al., 2016; Napper and Thompson, 2016), with typical pore size of around 20 μm. As cellulose is a common fibre type in clothing, the use of these filter bases could cause issues for possible cross-contamination issues during the analysis of results. Other studies used glass fibre, PVDF (polyvinylidene fluoride) or PTFE (polytetrafluoroethylene) filters with typical pore sizes ranging from 0.1 to 5 μm (Carney Almroth et al., 2018; De Falco et al., 2018; Jönsson et al., 2018; Karlsson, 2015; Sillanpää and Sainio, 2017; Yang et al., 2019; Zambrano et al., 2019). The hydrophobic nature of these filter types is an important consideration, ensuring moisture regain does not influence measurements, especially where microfibre release is quantified by mass. A smaller pore size will ensure the capture of smaller fibres; however, only Tiffin et al. (2022) reported any validation of filter retention efficiency to demonstrate that all microfibres released during testing were effectively captured by the filters.

Several research groups also used cascade filtration to indicate the distribution of size/length of the microfibres released (Celi̇k, 2021; De Falco et al., 2019, 2020; Hartline et al., 2016; Kim et al., 2022; Mahbub and Shams, 2022; Volgare et al., 2021). De Falco et al. suggested that most microfibres collected in their study were in the range of 60–400 μm (De Falco et al., 2019, 2020).

8.2.4 Quantification of Collected Microfibres

The quantification of microfibres collected from the test liquors has been done either by fibre count (Belzagui et al., 2019; Browne et al., 2011; Carney Almroth et al., 2018; De Falco et al., 2018; Hernandez et al., 2017; Jönsson et al., 2018; Kärkkäinen and Sillanpää, 2020; Periyasamy, 2021; Raja Balasaraswathi and Rathinamoorthy, 2022; Rathinamoorthy and Raja Balasaraswathi, 2022a, b; Sillanpää and Sainio, 2017; Yang et al., 2019) or by fibre mass (Berruezo et al., 2021; Celi̇k, 2021; Cesa et al., 2020; Choi et al., 2021a, b; Cotton et al., 2020; Dalla Fontana et al., 2020, 2021; De

Falco et al., 2019, 2020; Dreillard et al., 2022; Hartline et al., 2016; Kim et al., 2022; Mahbub and Shams, 2022; Napper and Thompson, 2016; Palacios-Marín et al., 2022; Pirc et al., 2016; Praveena et al., 2020; Volgare et al., 2021; Zambrano et al., 2019).

Counting microfibres is a time-consuming approach, and for this reason, most studies tended to only count the fibres present within selected areas of the filter and considered this as representative of the whole filter. This approach assumes a homogenous distribution of fibres across the whole filter area. Only Hernandez et al. counted all the microfibres collected on their filters (Hernandez et al., 2017). Automated and manual counting also have the potential for errors, as fibres can often be intertwined and overlapping in a three-dimensional mass on the filters making it difficult to identify individual microfibres.

Quantification of microfibre release by mass can offer a more accurate measure of the amount of material released during the washing process. However, this approach is open to errors related to moisture regain of the fibres and the filters. This option has the drawback of not providing any indication of the variability of individual fibres' shape and length, which is important when considering the potential harm of the microfibres if released into the environment.

8.3 INFLUENCE OF METHODOLOGY ON QUANTIFICATION ESTIMATES

Due to differences in methodologies used to quantify microfibre release, direct comparison of the findings between different studies is challenging (Acharya et al., 2021; Cai et al., 2020; Galvão et al., 2020; Gaylarde et al., 2021; Vassilenko et al., 2021). In the studies summarised in Table 8.1, estimates of microfibre release range from a few thousand fibres per wash (Browne et al., 2011) to several million fibres per wash (De Falco et al., 2018, 2019; Dreillard et al., 2022; Kärkkäinen and Sillanpää, 2020; Kelly et al., 2019; Periyasamy, 2021; Vassilenko et al., 2021).

To provide a comparison of the different methodologies, the results in Table 8.1 have been converted into standardised values, depending on whether the study reported release in terms of count of fibres or mass (mg) per kilogram of fabric test sample. The standardised values have been used to calculate an estimate for annual microfibre release from domestic laundering in the UK. The estimates are based on the following assumptions. There are approximately 27.8 million households in the UK (Sanders, 2019), and the average UK household completes 260 wash loads each year. This equates to 7.2 billion wash loads in the UK per year. Using an average wash load of 5 kg of textiles (Webber et al., 2016), the estimated microfibre release from the individual methodologies used in the studies in Table 8.1 are shown in Figure 8.1 (count of microfibres) and Figure 8.2 (mass of microfibres). It is important to note that the standardisation of the results has also assumed the results from each study are representative of a single wash load. However, the methods used for several of the studies did not control for all the laundering variables that can affect microfibre release, and therefore, comparison of the standardised results can only be used to provide an indication of the scale of variation that could be found between methodologies.

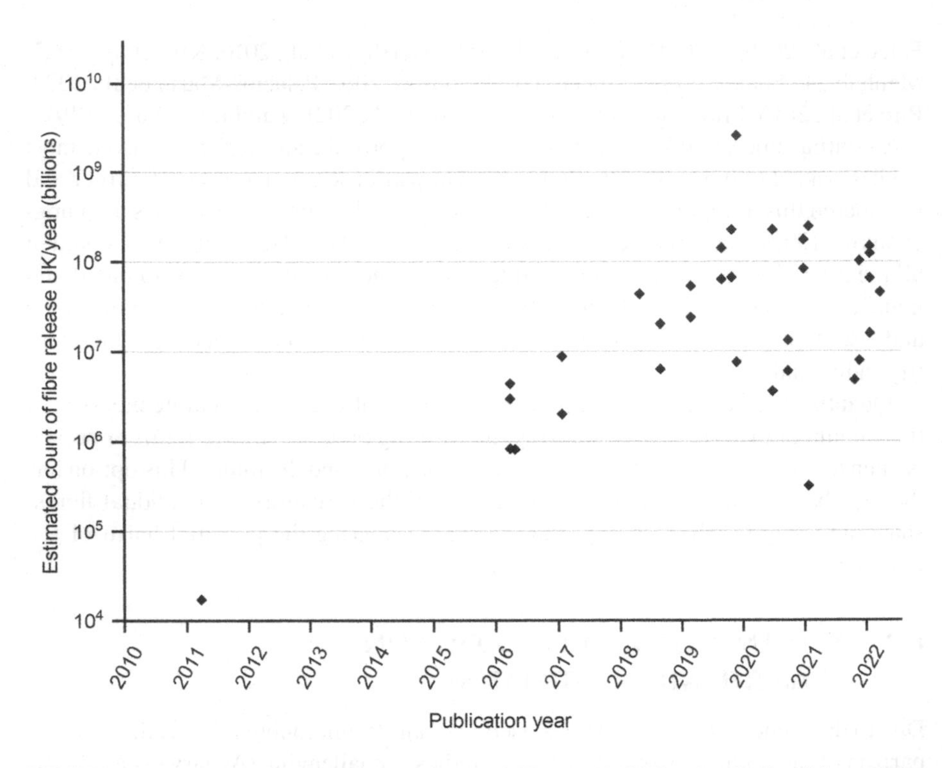

FIGURE 8.1 Estimated count of fibre release UK annual (billions) calculated from published data (logarithmic scale). (*Source:* Hazlehurst, A., Tiffin, L., Sumner, M., Taylor, M., 2023. Quantification of microfibre release from textiles during domestic laundering. *Environ. Sci. Pollut. Res.* 30, 43932–43949.)

The standardised results in Figure 8.1 show the estimated count of microfibres released from the annual domestic laundering in the UK ranges from 17,167 billion [originally reported as >1,900 particles/4 kg wash load (Browne et al., 2011)] to 2,602,080 trillion fibres [originally reported as 72,000 fibres/g (Cai et al., 2020)]. The range in estimated microfibre release is 2,602,063 trillion fibres, and the mean across the studies is 133,718,092 billion fibres.

The estimated mass of microfibre release ranges from 132 tonnes [originally reported as 3.64 mg/kg (Mahbub and Shams, 2022)] to 164,798 tonnes [originally reported as 4.56 g/kg (Kumar and Gopalakrishnan, 2022)] for annual UK domestic laundering (Figure 8.2). The mean across the different methods is 23,562 tonnes, with a range of 164,667 tonnes. To provide some context to these mass values, the UK is reported to dispose of approximately 350,000 tonnes of textile products in landfill annually (Welden, 2019).

It is clear from the studies reviewed in Table 8.1 that the estimates for UK microfibre release vary very significantly, irrespective of whether the count or mass of fibres is measured. However, the methodology used to assess microfibre release is not the only variable that affects the quantification of release for domestic laundry.

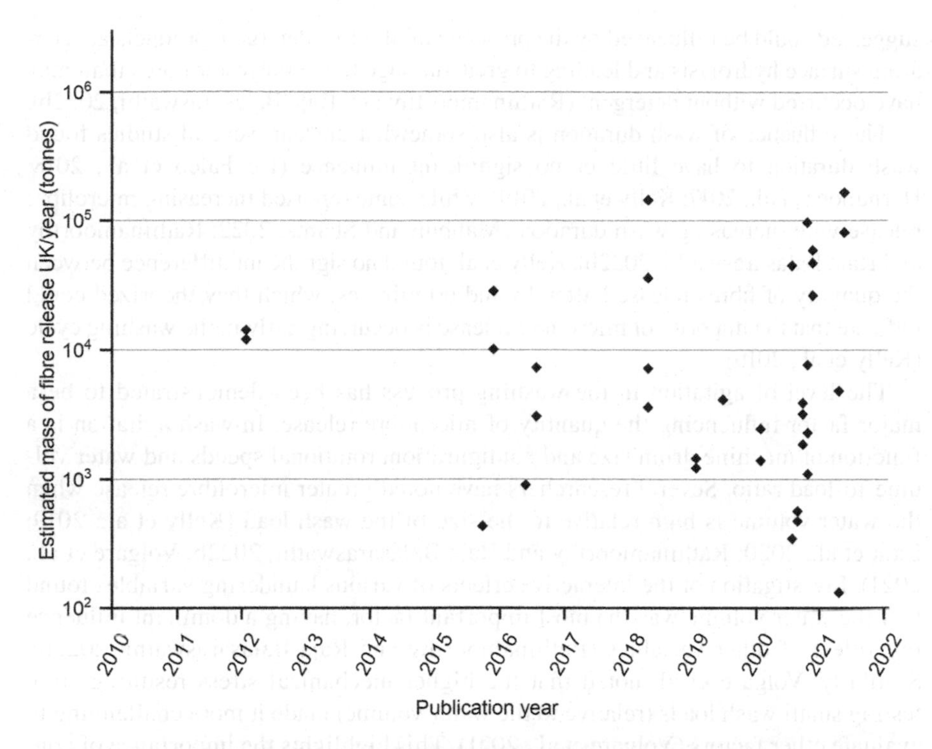

FIGURE 8.2 Estimated mass of fibre release UK annual (billions) calculated from published data (logarithmic scale). (*Source:* Hazlehurst, A., Tiffin, L., Sumner, M., Taylor, M., 2023. Quantification of microfibre release from textiles during domestic laundering. *Environ. Sci. Pollut. Res.* 30, 43932–43949.)

8.4 INFLUENCE OF LAUNDERING VARIABLES

There are several important laundering variables that have been found to influence the quantification of microfibre release in domestic laundering, irrespective of the method used for testing. These include wash temperature, wash duration, liquor-to-textile ratio, in-wash agitation and the presence of detergents or other laundering care products.

Several studies reported increased microfibre release with increasing wash temperature (Choi et al., 2021b; Cotton et al., 2020; Mahbub and Shams, 2022; Napper and Thompson, 2016; Periyasamy, 2021; Rathinamoorthy and Raja Balasaraswathi, 2022b; Yang et al., 2019; Zambrano et al., 2019), while some found that wash temperature had no significant effect on microfibre release (De Falco et al., 2018; Hernandez et al., 2017; Kelly et al., 2019). The divergence in results could be explained by different textile materials, with different thermal properties, being used for testing. Additionally, it could be that the effect of temperature is not dominant in interaction with other in-wash variables. Rathinamoorthy and Raja Balasaraswathi found a significant increase in microfibre release for polyester when washed at 90°C which they suggested could be influenced by the presence of alkaline detergent products accelerating surface hydrolysis and leading to great damage to the polyester fibres than may have occurred without detergent (Rathinamoorthy and Raja Balasaraswathi, 2022b).

The influence of wash duration is also somewhat unclear; several studies found wash duration to have little or no significant influence (De Falco et al., 2018; Hernandez et al., 2017; Kelly et al., 2019), while some reported increasing microfibre release with increasing wash duration (Mahbub and Shams, 2022; Rathinamoorthy and Raja Balasaraswathi, 2022b). Kelly et al. found no significant difference between the quantity of fibres released after 15 and 60 minutes, which they theorized could indicate that the majority of microfibre release is occurring early in the washing cycle (Kelly et al., 2019).

The level of agitation in the washing process has been demonstrated to be a major factor influencing the quantity of microfibre release. In-wash agitation is a function of machine drum size and configuration, rotational speeds and water volume-to-load ratio. Several researchers have noted greater microfibre release when the water volume is high relative to the size of the wash load (Kelly et al., 2019; Lant et al., 2020; Rathinamoorthy and Raja Balasaraswathi, 2022b; Volgare et al., 2021). Investigation of the interactive effects of various laundering variables found that the water volume was the most important factor, having a dominant influence regardless of other variables (Rathinamoorthy and Raja Balasaraswathi, 2022b). Similarly, Volgare et al. noted that the higher mechanical stress resulting from testing small wash loads (relative to the water volume) made it more challenging to evaluate other factors (Volgare et al., 2021). This highlights the importance of considering the wash liquor-to-fabric ratio for testing; studies which only test a single garment in a full-scale washing machine are very likely to overestimate microfibre release for that product in real use.

Hartline et al. also noted the influence of agitation when comparing top-loading and front-loading washing machines; they found greater microfibre release from a top-loading machine which they suggested was due to higher abrasive action from the machine's central agitator compared to the rotating drum of a front-loading machine (Hartline et al., 2016). In simulated laundering devices, a positive correlation between microfibre release and increasing agitation was also noted by Rathinamoorthy and Raja Balasaraswathi who increased agitation by increasing the number of ball bearings used in the test (5, 10, 15) (Rathinamoorthy and Raja Balasaraswathi, 2022a).

Investigating the influence of detergents on microfibre release has been attempted by many researchers, but several found that detergent, particularly powder detergent, can be a significant contaminant on the filters which interfered with the analysis to quantify microfibres released (De Falco et al., 2018; Hernandez et al., 2017; Jönsson et al., 2018; Zambrano et al., 2019). However, where errors due to contamination could be excluded, some studies reported no significant impact of detergent (Kelly et al., 2019; Lant et al., 2020; Napper and Thompson, 2016; Pirc et al., 2016), while others reported significant increase in microfibre release when detergent was used (Carney Almroth et al., 2018; De Falco et al., 2018; Hernandez et al., 2017; Kumar and Gopalakrishnan, 2022; Mahbub and Shams, 2022; Periyasamy, 2021; Yang et al., 2019; Zambrano et al., 2019). Powder detergents appear to have a greater influence than liquid detergents (De Falco et al., 2018; Hernandez et al., 2017; Periyasamy, 2021), which might be explained by small insoluble particles within the powder formulation increasing friction, leading to greater fibre damage and thus microfibre fragmentation (De Falco et al., 2018).

It has been repeatedly reported that the quantity of microfibres released is highest for the first wash of a fabric or garment, and the quantity of microfibres released reduces with subsequent wash cycles, typically levelling out at the third wash (Belzagui et al., 2019; Cai et al., 2020; Carney Almroth et al., 2018; Cesa et al., 2020; Choi et al., 2021b; Dreillard et al., 2022; Jabbar et al., 2022; Kelly et al., 2019; Kim et al., 2022; Kumar and Gopalakrishnan, 2022; Lant et al., 2020; Mahbub and Shams, 2022; Özkan and Gündoğdu, 2021; Palacios-Marín et al., 2022; Pirc et al., 2016; Rathinamoorthy and Raja Balasaraswathi, 2022a; Sillanpää and Sainio, 2017; Vassilenko et al., 2021; Zambrano et al., 2019). This suggests that most of the microfibres present in the fabric are being released in the first few wash cycles, and fewer microfibres are being formed as wash cycles progress. This could be explained by a large quantity of microfibres being formed during the manufacturing process, which remain loosely held in the fabric structure and become detached only when the first laundering process occurs. Some researchers have even suggested that additional washing processes during the manufacturing stage could be implemented to reduce microfibre release in domestic laundering (Cai et al., 2020; Carney Almroth et al., 2018; Cesa et al., 2020). However, while this approach alleviates some of the consumers' responsibility, it still requires responsible management of the microfibre issue within the manufacturing processes. Additionally, in the studies reviewed that completed repeat washing, the samples were not exposed to any additional physical ageing or soiling between wash cycles, as would happen in real-life scenarios. In everyday use and wear of textile items, the fabrics will be subjected to mechanical stress, such as rubbing and flexing, which could cause fragmentation and mobilisation of fibres held within the fabric structure. These fragments might then be released in subsequent laundering.

8.5 INFLUENCE OF FABRIC VARIABLES

Even in cases where the methodology and wash variables are consistent, microfibre release can vary dramatically for different fabrics dependent on their characteristics and properties. Few studies have been dedicated to investigating the influence of different fabric variables, and findings have been largely contradictory; however, some common findings have been reported in recent works.

Several researchers have reported greater microfibre shedding from cellulose-based fabrics compared to synthetic fabrics (Celik, 2021; Cesa et al., 2020; De Falco et al., 2019; Palacios-Marín et al., 2022; Sillanpää and Sainio, 2017; Zambrano et al., 2019). Zambrano et al. noted that this could be related to cellulosic fibres (cotton in particular) producing fabrics with higher hairiness, meaning that more protruding fibres would be available at the fabric surface and vulnerable to release (Zambrano et al., 2019).

The influence of yarn characteristics has also been noted. A number of studies have found greater microfibre release from fabrics constructed from staple spun yarns compared to filament yarns, which can be expected as the fibres in staple spun yarns are shorter and have greater mobility, making them more vulnerable to release (Belzagui et al., 2019; Carney Almroth et al., 2018; Choi et al., 2021a; De Falco et al., 2018; Özkan and Gündoğdu, 2021; Yang et al., 2019). Choi et al. also found greater microfibre release from non-twist filament yarns compared to highly twisted filament yarns, which they theorised to be due to lower inter-fibre friction in yarns without twist leading to greater fibre mobility and potential for fibre damage (Choi et al., 2021a). Jabbar et al. reported greater microfibre release from woven fabrics constructed from conventional ring-spun yarns compared with modified ring-spun yarns [compact, SIRO and SIRO-compact (SIRO spinning system combines spinning and doubling into one process and produces yarns which are typically less hairy than conventional ring-spun yarns)] (Jabbar et al., 2022). They noted modified ring spinning systems produce yarns which have higher compactness than conventional ring spinning, and this would increase cohesive forces between the fibres in the yarn making the release of microfibres less likely (Jabbar et al., 2022).

Several researchers have noted the influence of the tightness or density of the fabric structure, with tighter, more compact structures shown to release fewer microfibres compared to looser, more open structures (Berruezo et al., 2021; Choi et al., 2021b; De Falco et al., 2019; Kim et al., 2022; Raja Balasaraswathi and Rathinamoorthy, 2022; Yang et al., 2019). Considering woven structures, Choi et al. and Berruezo et al. both reported lower microfibre release from plain weave fabrics compared to twill weave fabrics, while, conversely, Kim et al. found twill weave structures to release fewer fibres than plain weave (Berruezo et al., 2021; Choi et al., 2021b; Kim et al., 2022). However, Berruezo et al. noted that the influence of the type of weave structure (plain, twill, satin) was of little significance and that the density of the fabric i.e. compactness of the constituent yarns and the number of interlacing points within the fabric structure were a much better indicator of microfibre release (Berruezo et al., 2021). Raja Balasaraswathi and Rathinamoorthy reported similar findings for knitted fabrics, noting that microfibre release was not dependent on the knit structure as other factors dominated and that increasing the stitch density reduced the quantity of microfibre release (Raja Balasaraswathi and Rathinamoorthy, 2022). Although it may be expected that increasing the density and therefore increasing the number of fibres per unit area in the fabric would increase the quantity of microfibres released, the opposite has been observed. This has been theorised as being due to increased fibre compactness when the fabric structure is denser, leading to reduced freedom and mobility of the fibres making them less likely to be released from the fabric during laundering (Berruezo et al., 2021; Choi et al., 2021b).

8.6 STANDARDISED QUANTIFICATION OF MICROFIBRE RELEASE FROM DOMESTIC LAUNDERING

To address the impact of varied methodologies and laundering parameters on the quantification of microfibre release as highlighted in the results from different studies, a common, standardised test method for microfibre release needs to be adopted. A standard method allows for meaningful comparison across studies as well as a more reliable assessment of individual fabric types under test. The AATCC published their standard method *AATCC TM212–2021 Test Method of Fiber Fragment Release*

FIGURE 8.3 Visual summary of standardised procedure for testing microfibre release from domestic laundering. (*Source:* Adapted from Hazlehurst et al., 2023.)

during Home Laundering in 2021 (American Association of Textile Chemists and Colorists, 2021) and an international standard, *BS EN ISO 4484-1 Textiles and textile products. Microplastics from textiles sources. Part 1. Determination of material loss from fabric during laundering* was published in 2023 (International Organization for Standardization, 2022).

Both standards describe a similar methodological approach, based on the test method described by Tiffin et al. which is summarised below in brief (Tiffin et al., 2022). A visual summary of the testing procedure is also provided in Figure 8.3.

Fabric specimens are prepared by cutting to size and hemming the perimeter to prevent unintentional release of fibre from the cut edge. The mass of the specimens is recorded after oven-drying to negate the influence of moisture retention. Specimens are washed using a simulated laundering device as described in AATCC TM61 (American Association of Textile Chemists and Colorists, 2013). This equipment is typically used for testing colour fastness to laundering and, as such, is already commonplace in most textile testing facilities. Each specimen is placed in a separate canister along with distilled water and stainless steel ball bearings (50 ball bearings) which simulate the abrasive action of domestic washing machines. The canisters are fixed to the rotating shaft of the machine and rotated through the heated water bath (40°C) at a speed of 40 rpm for the duration of the test (45 minutes). The test liquor from each canister is then collected in a separate container (such as a glass beaker), and the specimen and stainless-steel ball bearings are retained using a sieve or similar. The test liquor is filtered through a glass-fibre filter paper using a vacuum-assisted single-stage process to collect any microfibres that were released from the fabric specimen during laundering. The oven-dry mass of the filter is recorded before and after filtration, and the microfibre release is expressed in mg/kg, relative to the initial mass of the fabric specimen before testing.

The reproducibility of this method was validated by an inter-laboratory study according to ASTM E691-18 (ASTM International, 2018). Ten laboratories located in Europe, the USA, and Asia participated in testing three different polyester fabrics using the method described. Results for inter- and intra-laboratory reproducibility demonstrated that the method was repeatable and reproducible, making it suitable for comparison of microfibre release results obtained from different laboratories and tested by different operators (Tiffin et al., 2022). An additional validation study was also conducted as a part of the ISO standardisation process (International Organization for Standardization, 2022).

These standardised methods offer greater reliability and reproducibility than many of the methods previously described, due to the use of standard textile testing equipment and robust validation procedures. This will allow for a more useful comparison of results for different fabrics tested by different groups and laboratories. However, standardised methods are not necessarily intended to be wholly reflective of real-use conditions, and factors such as detergent use, load-size, presence of zips and other garment trims, and level of soiling will likely also influence microfibre release in real laundering conditions.

8.7 CONCLUSION

Domestic laundry has been identified as significant source of microfibres, and the quantification of microfibre released when laundering garments and other textile products has been attempted by many different studies and methodologies for a wide range of wash variables and fabric types. However, it is very difficult to compare the results from these studies, as methodologies, laundering parameters and fabric variables all influence the quantity of microfibre released.

The adoption of a standardised testing method, such as that described by Tiffin et al., offers the opportunity for reliable quantification and a greater understanding of the factors which influence microfibre release from textiles during laundering. This in turn will hopefully facilitate the development of new technologies to minimise microfibre losses from fabrics into the environment.

However, understanding the impact of microfibres on the environment and potential implications for human health will be dependent not only on reliable methods to estimate scale of microfibre pollution from domestic laundering; the quantification of microfibre release from all potential pathways, including non-textile routes, also needs to be estimated using reliable test methods. Additionally, the potential impacts of microfibres will not only be affected by their abundance, but also by their characteristics such as size, shape and material composition. Further work is required to understand how the characteristics of microfibres released are influenced by laundering and fabric variables. Additional research is also required to understand the fate of microfibres beyond the point of release during domestic laundering.

REFERENCES

Acharya, S., Rumi, S.S., Hu, Y., Abidi, N., 2021. Microfibers from synthetic textiles as a major source of microplastics in the environment: A review. *Text. Res. J.* 91, 2136–2156. https://doi.org/10.1177/0040517521991244

American Association of Textile Chemists and Colorists, 2013. *Colour Fastness to Laundering: Accelerated (No. TM61-2013).* AATCC, North Carolina.

American Association of Textile Chemists and Colorists, 2021. *Test Method for Fiber Fragment Release during Home Laundering (No. TM212-2021).* AATCC, North Carolina.

ASTM International, 2018. *Standard Practice for Conducting an Interlaboratory Study to Determine the Precision Method (No. E691-18).* ASTM International.

Belzagui, F., Crespi, M., Álvarez, A., Gutiérrez-Bouzán, C., Vilaseca, M., 2019. Microplastics' emissions: Microfibers' detachment from textile garments. *Environ. Pollut.* 248, 1028–1035. https://doi.org/10.1016/j.envpol.2019.02.059

Berruezo, M., Bonet-Aracil, M., Montava, I., Bou-Belda, E., Díaz-García, P., Gisbert-Payá, J., 2021. Preliminary study of weave pattern influence on microplastics from fabric laundering. *Text. Res. J.* 91, 1037–1045. https://doi.org/10.1177/0040517520965708

British Standards Institution, 2010. *Textiles – Tests for Colour Fastness. Part C06: Colour Fastness to Domestic and Commerical Laundering (No. BS EN ISO 105-C06:2010).* BSI, London.

Browne, M.A., Crump, P., Niven, S.J., Teuten, E., Tonkin, A., Galloway, T., Thompson, R., 2011. Accumulation of microplastic on shorelines woldwide: Sources and sinks. *Environ. Sci. Technol.* 45, 9175–9179. https://doi.org/10.1021/es201811s

Cai, Y., Yang, T., Mitrano, D.M., Heuberger, M., Hufenus, R., Nowack, B., 2020. Systematic study of microplastic fiber release from 12 different polyester textiles during washing. *Environ. Sci. Technol.* 54, 4847–4855. https://doi.org/10.1021/acs.est.9b07395

Carney Almroth, B.M., Åström, L., Roslund, S., Petersson, H., Johansson, M., Persson, N.-K., 2018. Quantifying shedding of synthetic fibers from textiles; a source of microplastics released into the environment. *Environ. Sci. Pollut. Res.* 25, 1191–1199. https://doi.org/10.1007/s11356-017-0528-7

Celik, S., 2021. Microplastic release from domestic washing. *Eur. J. Sci. Technol.* https://doi.org/10.31590/ejosat.933322

Cesa, F.S., Turra, A., Checon, H.H., Leonardi, B., Baruque-Ramos, J., 2020. Laundering and textile parameters influence fibers release in household washings. *Environ. Pollut.* 257, 113553. https://doi.org/10.1016/j.envpol.2019.113553

Choi, S., Kwon, M., Park, M.-J., Kim, J., 2021a. Analysis of microplastics released from plain woven classified by yarn types during washing and drying. *Polymers* 13, 2988. https://doi.org/10.3390/polym13172988

Choi, S., Kwon, M., Park, M.-J., Kim, J., 2021b. Characterization of microplastics released based on polyester fabric construction during washing and drying. *Polymers* 13, 4277. https://doi.org/10.3390/polym13244277

Cotton, L., Hayward, A.S., Lant, N.J., Blackburn, R.S., 2020. Improved garment longevity and reduced microfibre release are important sustainability benefits of laundering in colder and quicker washing machine cycles. *Dyes Pigments* 177, 108120. https://doi.org/10.1016/j.dyepig.2019.108120

Dalla Fontana, G., Mossotti, R., Montarsolo, A., 2020. Assessment of microplastics release from polyester fabrics: The impact of different washing conditions. *Environ. Pollut.* 264, 113960. https://doi.org/10.1016/j.envpol.2020.113960

Dalla Fontana, G., Mossotti, R., Montarsolo, A., 2021. Influence of sewing on microplastic release from textiles during washing. *Water. Air. Soil Pollut.* 232, 50. https://doi.org/10.1007/s11270-021-04995-7

De Falco, F., Cocca, M., Avella, M., Thompson, R.C., 2020. Microfiber release to water, via laundering, and to air, via everyday use: A comparison between polyester clothing with differing textile parameters. *Environ. Sci. Technol.* 54, 3288–3296. https://doi.org/10.1021/acs.est.9b06892

De Falco, F., Di Pace, E., Cocca, M., Avella, M., 2019. The contribution of washing processes of synthetic clothes to microplastic pollution. *Sci. Rep.* 9, 6633. https://doi.org/10.1038/s41598-019-43023-x

De Falco, F., Gullo, M.P., Gentile, G., Di Pace, E., Cocca, M., Gelabert, L., Brouta-Agnésa, M., Rovira, A., Escudero, R., Villalba, R., Mossotti, R., Montarsolo, A., Gavignano, S., Tonin, C., Avella, M., 2018. Evaluation of microplastic release caused by textile washing processes of synthetic fabrics. *Environ. Pollut.* 236, 916–925. https://doi.org/10.1016/j.envpol.2017.10.057

Dreillard, M., Barros, C.D.F., Rouchon, V., Emonnot, C., Lefebvre, V., Moreaud, M., Guillaume, D., Rimbault, F., Pagerey, F., 2022. Quantification and morphological characterization of microfibers emitted from textile washing. *Sci. Total Environ.* 832, 154973. https://doi.org/10.1016/j.scitotenv.2022.154973

Dubaish, F., Liebezeit, G., 2013. Suspended microplastics and black carbon particles in the Jade system, Southern North Sea. *Water. Air. Soil Pollut.* 224, 1352. https://doi.org/10.1007/s11270-012-1352-9

Frost, H., Zambrano, M.C., Leonas, K., Pawlak, J.J., Venditti, R.A., 2020. Do recycled cotton or polyester fibers influence the shedding propensity of fabrics during laundering? *AATCC J. Res.* 7, 32–41. https://doi.org/10.14504/ajr.7.S1.4

Galvão, A., Aleixo, M., De Pablo, H., Lopes, C., Raimundo, J., 2020. Microplastics in wastewater: microfiber emissions from common household laundry. *Environ. Sci. Pollut. Res.* 27, 26643–26649. https://doi.org/10.1007/s11356-020-08765-6

Gaylarde, C., Baptista-Neto, J.A., da Fonseca, E.M., 2021. Plastic microfibre pollution: How important is clothes' laundering? *Heliyon* 7, e07105. https://doi.org/10.1016/j.heliyon.2021.e07105

Haap, J., Classen, E., Beringer, J., Mecheels, S., Gutmann, J.S., 2019. Microplastic fibers released by textile laundry: A new analytical approach for the determination of fibers in effluents. *Water* 11, 2088. https://doi.org/10.3390/w11102088

Hartline, N.L., Bruce, N.J., Karba, S.N., Ruff, E.O., Sonar, S.U., Holden, P.A., 2016. Microfiber masses recovered from conventional machine washing of new or aged garments. *Environ. Sci. Technol.* 50, 11532–11538. https://doi.org/10.1021/acs.est.6b03045

Hazlehurst, A., Tiffin, L., Sumner, M., Taylor, M., 2023. Quantification of microfibre release from textiles during domestic laundering. *Environ. Sci. Pollut. Res.* 30, 43932–43949

Hernandez, E., Nowack, B., Mitrano, D.M., 2017. Polyester textiles as a source of microplastics from households: A mechanistic study to understand microfiber release during washing. *Environ. Sci. Technol.* 51, 7036–7046. https://doi.org/10.1021/acs.est.7b01750

International Organization for Standardization, 2022. *Textiles and Textile Products – Microplastics from Textile Sources – Part 1: Determination of Material Loss from Fabrics during Washing (No. ISO/FDIS 4484–1)*. ISO, Geneva.

Jabbar, A., Palacios-Marín, A.V., Ghanbarzadeh, A., Yang, D., Tausif, M., 2022. Impact of conventional and modified ring-spun yarn structures on the generation and release of fragmented fibers (microfibers) during abrasive wear and laundering. *Text. Res. J.* 004051752211277. https://doi.org/10.1177/00405175221127709

Jönsson, C., Levenstam Arturin, O., Hanning, A.-C., Landin, R., Holmström, E., Roos, S., 2018. Microplastics shedding from textiles-developing analytical method for measurement of shed material representing release during domestic washing. *Sustainability* 10, 2457. https://doi.org/10.3390/su10072457

Kärkkäinen, N., Sillanpää, M., 2020. Quantification of different microplastic fibres discharged from textiles in machine wash and tumble drying. *Environ. Sci. Pollut. Res.* 28, 16253–16263. https://doi.org/10.1007/s11356-020-11988-2

Karlsson, T.M., 2015. *Can Microlitter in Sediment and Biota Be Quantified.* University of Gothenburg.

Kelly, M.R., Lant, N.J., Kurr, M., Burgess, J.G., 2019. Importance of water-volume on the release of microplastic fibers from laundry. *Environ. Sci. Technol.* 53, 11735–11744. https://doi.org/10.1021/acs.est.9b03022

Kim, S., Cho, Y., Park, C.H., 2022. Effect of cotton fabric properties on fiber release and marine biodegradation. *Text. Res. J.* 004051752110687. https://doi.org/10.1177/00405175211068781

Kosko, B., Isaka, S., 1993. Fuzzy logic: The binary logic of modern computers often falls short when describing the vagueness of the real world. Fuzzy logic offers more graceful alternatives. *Sci. Am.* 269, 76–81.

Kumar, A., Gopalakrishnan, M., 2022. Effect of laundry parameters on micro fiber loss during washing and its correlation with carbon footprint. *J. Nat. Fibers* 19, 14744–14754. https://doi.org/10.1080/15440478.2022.2068729

Lant, N.J., Hayward, A.S., Peththawadu, M.M.D., Sheridan, K.J., Dean, J.R., 2020. Microfiber release from real soiled consumer laundry and the impact of fabric care products and washing conditions. *PLoS One* 15, e0233332. https://doi.org/10.1371/journal.pone.0233332

Mac Namara, C., Gabriele, A., Amador, C., Bakalis, S., 2012. Dynamics of textile motion in a front-loading domestic washing machine. *Chem. Eng. Sci.* 75, 14–27. https://doi.org/10.1016/j.ces.2012.03.009

Mahbub, S., Shams, M., 2022. Acrylic fabrics as a source of microplastics from portable washer and dryer: Impact of washing and drying parameters. *Sci. Total Environ.* 8.

Napper, I.E., Thompson, R.C., 2016. Release of synthetic microplastic plastic fibres from domestic washing machines: Effects of fabric type and washing conditions. *Mar. Pollut. Bull.* 112, 39–45. https://doi.org/10.1016/j.marpolbul.2016.09.025

Özkan, İ., Gündoğdu, S., 2021. Investigation on the microfiber release under controlled washings from the knitted fabrics produced by recycled and virgin polyester yarns. *J. Text. Inst.* 112, 264–272. https://doi.org/10.1080/00405000.2020.1741760

Palacios-Marín, A.V., Jabbar, A., Tausif, M., 2022. Fragmented fiber pollution from common textile materials and structures during laundry. *Text. Res. J.* 004051752210909. https://doi.org/10.1177/00405175221090971

Periyasamy, A.P., 2021. Evaluation of microfiber release from jeans: the impact of different washing conditions. *Environ. Sci. Pollut. Res.* https://doi.org/10.1007/s11356-021-14761-1

Pirc, U., Vidmar, M., Mozer, A., Kržan, A., 2016. Emissions of microplastic fibers from microfiber fleece during domestic washing. *Environ. Sci. Pollut. Res.* 23, 22206–22211. https://doi.org/10.1007/s11356-016-7703-0

Praveena, S.M., Syahira Asmawi, M., Chyi, J.L.Y., 2020. Microplastic emissions from household washing machines: Preliminary findings from Greater Kuala Lumpur (Malaysia). *Environ. Sci. Pollut. Res.* 28, 18518–18522. https://doi.org/10.1007/s11356-020-10795-z

Raja Balasaraswathi, S., Rathinamoorthy, R., 2022. Effect of fabric properties on microfiber shedding from synthetic textiles. *J. Text. Inst.* 113, 789–809. https://doi.org/10.1080/00405000.2021.1906038

Rathinamoorthy, R., Raja Balasaraswathi, S., 2022a. Investigations on the impact of handwash and laundry softener on microfiber shedding from polyester textiles. *J. Text. Inst.* 113, 1428–1437. https://doi.org/10.1080/00405000.2021.1929709

Rathinamoorthy, R., Raja Balasaraswathi, S., 2022b. Investigations on the interactive effect of laundry parameters on microfiber release from polyester knitted fabric. *Fibers Polym.* 23, 2052–2061. https://doi.org/10.1007/s12221-022-4929-y

Sanders, S., 2019. *Statistical Bulletin: Families and Households in the UK: 2019*. Office for National Statistics, Newport.

Sillanpää, M., Sainio, P., 2017. Release of polyester and cotton fibers from textiles in machine washings. *Environ. Sci. Pollut. Res.* 24, 19313–19321. https://doi.org/10.1007/s11356-017-9621-1

Sudheshna, A.A., Srivastava, M., Prakash, C., 2022. Characterization of microfibers emission from textile washing from a domestic environment. *Sci. Total Environ.* 852, 158511. https://doi.org/10.1016/j.scitotenv.2022.158511

Tiffin, L., Hazlehurst, A., Sumner, M., Taylor, M., 2022. Reliable quantification of microplastic release from the domestic laundry of textile fabrics. *J. Text. Inst.* 113, 558–566. https://doi.org/10.1080/00405000.2021.1892305

Vassilenko, E., Watkins, M., Chastain, S., Mertens, J., Posacka, A.M., Patankar, S., Ross, P.S., 2021. Domestic laundry and microfiber pollution: Exploring fiber shedding from consumer apparel textiles. *PLoS One* 16, e0250346. https://doi.org/10.1371/journal.pone.0250346

Volgare, M., De Falco, F., Avolio, R., Castaldo, R., Errico, M.E., Gentile, G., Ambrogi, V., Cocca, M., 2021. Washing load influences the microplastic release from polyester fabrics by affecting wettability and mechanical stress. *Sci. Rep.* 11, 19479. https://doi.org/10.1038/s41598-021-98836-6

Webber, D., Payne, C.S., Weedon, O., 2016. *Household Satellite Accounts: 2005 to 2014: Chapter 6: Home Produced 'Clothing' and 'Laundry' Services*. Office for National Statistics, Newport.

Welden, N., 2019. *Textile Derived Microfibre Release: Investigating the Current Evidence Base*. WRAP: Resource Futures, Bristol.

Yang, L., Qiao, F., Lei, K., Li, H., Kang, Y., Cui, S., An, L., 2019. Microfiber release from different fabrics during washing. *Environ. Pollut.* 249, 136–143. https://doi.org/10.1016/j.envpol.2019.03.011

Yun, C., Park, C.H., 2015. The effect of fabric movement on washing performance in a front-loading washer II: Under various physical washing conditions. *Text. Res. J.* 85, 251–261. https://doi.org/10.1177/0040517514545260

Zambrano, M.C., Pawlak, J.J., Daystar, J., Ankeny, M., Cheng, J.J., Venditti, R.A., 2019. Microfibers generated from the laundering of cotton, rayon and polyester based fabrics and their aquatic biodegradation. *Mar. Pollut. Bull.* 142, 394–407. https://doi.org/10.1016/j.marpolbul.2019.02.062

Zambrano, M.C., Pawlak, J.J., Daystar, J., Ankeny, M., Venditti, R.A., 2021. Impact of dyes and finishes on the microfibers released on the laundering of cotton knitted fabrics. *Environ. Pollut.* 272, 115998. https://doi.org/10.1016/j.envpol.2020.115998

9 Forensic Textile and Fibre Examinations for the Purposes of Improved Recovery, Analysis and Interpretation of Microplastic Pollution

Claire Gwinnett
Staffordshire University

9.1 INTRODUCTION

Microplastic (MP) pollution is a key concern of the 21st century. Research to date has shown that this pollutant is present in all environmental compartments (Klingelhofer et al., 2020). Fibres have been found to be a predominant form of MP in the environment (Xu et al., 2021; Wang et al., 2020; Fagiano et al., 2023). For example, fibres have been reported as the most numerous form of this pollution in a study of Belgian marine sediments, accounting for about 59% of the MP particles found (Claessens et al., 2011); the predominant suspended atmospheric MP in Shanghai, comprising 67% of the samples recovered (Liu et al., 2019); and the principal MP type in surface water of the Manas River Basin, China (88% of recovered samples) (Wang et al., 2020).

Since 2004, when the first MP studies were published, there has been a significant increase in studies investigating methods for the retrieval, extraction/separation, quantification and analysis of MPs. This increase in studies has led to many review-based papers that endeavour to compare approaches and suggest global best practices (Hidalgo-Ruz et al., 2012; Luo et al., 2022; Christian and Köper, 2023). However, there is still an acknowledged lack of standardisation between methods used for the sampling, recovery and preparation of MPs from different media, and the analysis, reporting and interpretation of MPs (Zantis et al., 2021; Ding et al., 2022; Cui et al., 2022; Pérez-Guevara et al., 2022). This, in part, is due to different laboratory set-ups/availability of equipment, different aims of the research and expertise of the researchers. This lack of standardisation means that studies are difficult to compare and global datasets problematic to produce.

DOI: 10.1201/9781003331995-11

As MP knowledge has developed over the years, we have seen a greater range of disciplines contributing to these studies, for example, knowledge from the field of sedimentology has been applied to MP pollution (Waldschläger et al., 2022). A multidisciplinary approach that combines the knowledge of textile scientists, polymer chemists, statisticians, ecologists, marine and freshwater biologists, toxicologists and hydrologists has obvious benefits in trying to understand the many factors involved in MP pollution. One discipline that has investigated the many sources of textile fibres along with their recovery, analysis and interpretation is forensic science.

Forensic science may be defined as any science that is used for the purposes of the justice system (Jackson and Jackson, 2017). The sub-disciplines of forensic science are broad and range from marks, such as those from fingerprints and shoes, to trace evidence such as hair, glass, paint and fibres. Of these, textile fibres (both man-made and natural) are one of the more traditional forms of evidence and, as such, is a well-established, research informed and data-driven discipline that has been utilised in courts of law for many decades. This discipline is called forensic fibre examinations and will be referred to as this throughout this chapter.

The interest in man-made fibres by forensic scientists is, of course, not pollution, but depending on the circumstances of a case, such fibres, along with their natural counterparts, can provide evidence of:

- criminal activity,
- a given criminal's identity,
- the sequence of events that occurred during a crime, and
- the time frame within which a crime was committed.

Fibres are deemed evidentially valuable as they are readily transferred from surface to surface, and individual to individual; therefore, they can be very useful in providing intelligence information about who was present at a crime scene. Due to their evidential value, fibres evidence has been used for over 60 years in criminal investigations. During this time, there has been significant research conducted in all aspects of its recovery, analysis and interpretation. Principles such as the mechanisms for transfer of fibres from textiles to surfaces and environments have been studied since 1975 (Pounds and Smalldon, 1975a) and provide important information to those investigating the release of fibres from garments through wear and washing.

There are significant overlaps between forensic fibre analysis and MP pollution studies, including the types of questions being posed to samples.

In forensic fibre examinations, it is common to ask the following of fibres found at crime scenes or on a victim/suspect:

- How many fibres were found?
- What are they made from (e.g. fibre type)/what are their characteristics?
- Where are they from (their source)?
- How did they get there (their transfer mechanism)?
- How long have they been there (their persistence)?

When considering MP studies, similar questions are being asked of these particulates, but from the context of finding them in different environments, for example:

- What is the abundance of these fibres in the marine environment (or in fact any other environment, such as terrestrial, freshwater, urban and atmospheric)?
- What is the polymer type?
- What/where did they originate from, e.g. primary, secondary MP?
- What are their transport mechanisms?
- How long have they been present in the environment?

In forensic fibre examinations, answers to these questions have been sought to aid case investigations and to better understand the evidential value of fibres found in a crime. Research has been dedicated to understanding the optimum methods for fibre recovery, analysis and interpretation, with large datasets being generated, underlying principles for transfer and persistence being developed and interpretation models designed. Considering the significant work put into improving forensic fibre examinations, MP research, in comparison, is at a very early stage. Due to this, forensic fibre knowledge has much to offer MP studies. Figure 9.1 outlines some of the key areas where forensic fibre examination can help inform MP studies; these include methods to prevent contamination during sampling and analysis, applying a range of analytical techniques to better characterise fibres found in the environment to understand their sources; the development of automated systems for the rapid detection and analysis of MPs; knowledge of the transfer of fibres from textiles, including their persistence in the environment; and sophisticated interpretation methods that utilise large datasets.

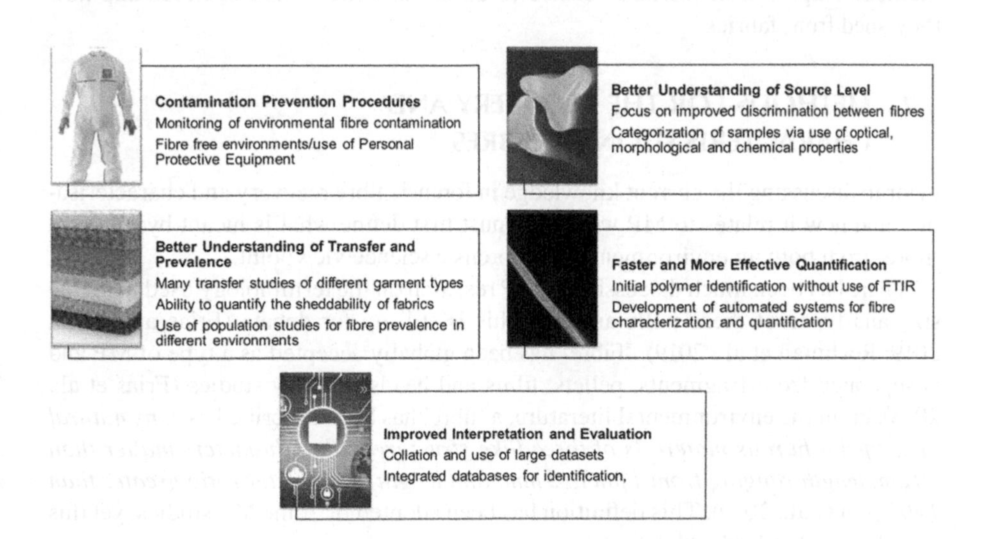

FIGURE 9.1 Areas in which forensic fibre examinations can aid MP studies.

In addition to the above, there are stringent requirements applied to the recovery, analysis and interpretation of fibres by the courts of law; these requirements for rigorous, high-quality data have led to protocols which can stand up to the highest levels of scrutiny.

Despite the clear overlap in forensic fibre examinations and MP studies, there has yet to be a summary of how information from this discipline may be applied to MP work. Although a small number of forensic science studies have been referred to in MP research, these are limited and do not represent the breadth of information that forensic fibre examinations can provide. Consequently, there is a need for a comprehensive review of forensic fibre work in the context of how this may inform MP studies. This chapter takes an interdisciplinary look into the existing literature in forensic fibre examinations, some of which have a long history and underpins the work done today in criminal investigations. In addition, this chapter will summarise the learning from forensic fibre examinations and how this applies to MP research, specifically microfibres found in our environment. This chapter identifies areas of forensic fibre examinations that can aid in MP work and which will facilitate future developments in MP recovery, analysis and interpretation; detail is provided to allow MP analysts to start using these techniques. This chapter will allow current research studies to utilise existing forensic solutions to improve processes and reporting. Although fibres will be focussed upon, certain knowledge areas apply more generally to all forms of MP, for example, contamination prevention procedures.

This chapter is divided into two main sections: (i) methods for the recovery and characterisation of fibres and (ii) interpretation of fibres. The first addresses key methods used in forensic fibres work including microscopical approaches, chemical analysis and contamination prevention procedures. The second section describes methods employed in forensic science to understand the source of fibres and how they shed from fabrics.

9.2 METHODS FOR THE RECOVERY AND CHARACTERISATION OF FIBRES

Prior to discussing the current knowledge in forensic fibre recovery and characterisation and how it relates to MP work, we must first define what is meant by the term 'fibre' from both an environmental and forensic science viewpoint.

There has been much discussion in MP research on the terminology used to define size and types of these pollutants and this is still up for debate (Frias and Nash, 2019; Rochman et al., 2019). 'Fibre' has been globally accepted as a type of MP and is separated from fragments, pellets, films and beads in many studies (Frias et al., 2018). From the environmental literature, a 'fibre' has been described as "*any natural or artificial fibrous materials of threadlike structure with a diameter smaller than 50 μm, length ranging from 1 μm to 5 mm, and length to diameter ratio greater than 100*" (Liu et al., 2019). This definition has been adopted by some MP studies, yet this is still not a standardised definition.

In forensic science, a fibre has been generally defined as "*a substance which has a high length to diameter ratio and is flexible in nature*" (Taylor, 1990; Robertson and Grieve, 1999). Depending on the type of fibre, characteristics of the fibre can differ but generally fibres are acknowledged as having the following qualities:

- structures that are much stronger longitudinally than laterally,
- are flexible and impart extensibility to the fibres,
- natural fibres develop naturally in a fibrous form and will differ according to the variety of plant or breed of animal and external conditions, and
- fibre-forming polymers have streamlined molecular chains and a high molecular weight (Robertson and Grieve, 1999).

This broader description encompasses more characteristics other than solely the dimensions of the particulate; arguably, this is a more comprehensive definition of a 'fibre'.

For the purposes of this chapter, the description originating from forensic science will be used.

9.2.1 Recovery Methods for Fibres

Regardless of whether a scientist is sampling for pollution monitoring or crime investigations, a quick, inexpensive and effective method must be identified to recover particulates of interest from the media being sampled. This method should be discriminatory enough to extract particulates of interest yet remove extraneous material which will inevitably cause problems later in the analysis process.

Typically, in MP studies, media such as water, air, sediment and soil are sampled using appropriate capture devices, for example, manta trawls, steel buckets (Ryan et al., 2020) and pumps for water samples (Du et al., 2022). These devices may facilitate grab samples, also known as a bulk sample, (i.e. where a given volume of media is taken and then filtered within a laboratory setting) or a volume reduction method (i.e. where large amounts of media are filtered during the sampling process, for example through a net, pump or filter). Regardless of which approach it uses, the particulates of interest invariably end up on a substrate ready for searching and analysis, for example, a filter paper.

This process of sampling large areas and recovering particulates of interest onto a substrate to allow for searching is also common to forensic fibre examinations. In addition, it is also required that samples must be quickly and securely taken at crime scenes and then must be transported back to laboratories without losing the integrity of the sample. In forensic science, many methods have been devised for this purpose, including sweeping, combing and vacuuming; however, for most applications, the method of choice is tape lifting (Pounds, 1975; Schotman and van der Weerd, 2015; Robertson et al., 2018; Jones et al., 2019).

Tape lifting is the simple process of adhering a clear, adhesive backed polymer to a surface of interest, which adheres any particulates to its 'sticky' side which then can be secured onto a clear backing. This backing is typically a clear acetate sheet. This method has many benefits for fibre analysis:

- It allows easy searching under both low and high-powered microscopes by containing all particulates of interest in a clear environment that is in the same focal plane.
- It secures the sample in a manner which prevents loss and/or contamination.
- It allows samples to be easily transported from the field to the laboratory or between laboratories which facilitates collaborative working (Gwinnett et al., 2021).

In typical MP studies, filter papers, in which the sample substrate has been filtered on to, must be searched as soon as possible to prevent loss or contamination and then particulates are hand-picked from the filter for further analysis. In comparison, samples encased in a tape lift may be left indefinitely before searching without any sample movement or loss and, depending on the tape used, may be analysed *in-situ* without the need for hand-picking.

Current searching methods used in MP studies, only allow the use of low-powered microscopes (e.g. stereomicroscopes). This is because the filter paper (normally placed in a petri dish or equivalent) cannot be focussed on under a high-powered microscope, as samples must be on a microscope slide or equivalent. Visual inspection at low magnifications, as seen with low-powered microscopes, is likely to mean that many smaller MP particles are missed. The potential for under reporting MP abundance due to using methods that are not able to detect smaller particles has been raised. Pérez-Guevara et al. (2022) identified the variation in the detection limits of visual inspections across studies and noted that this may be between 500 (Hidalgo-Ruz et al., 2012) and 5 μm (Abbasi and Turner, 2021). The Pérez-Guevara et al. (2022) review highlighted the constraints in current methods of extraction, characterisation and identification which lead to data underreporting. An example of this includes a study investigating the MP burden in freshwater fish, which estimated that >95% of particles are likely to be smaller than 40 μm and, thereby, beyond the detection range of most other MP surveys conducted so far (Roch et al., 2019). The combined use of tape lifts and searching under high-powered polarised light microscopes (PLM) is one solution to this problem. Further discussion of the use of polarised light microscopy in the searching process is discussed in Section 9.2.3.1.

The use of tape lifting to recover MPs from filter papers has been investigated in 2021 (Gwinnett et al., 2021). This study developed a workflow that uses a novel lifting tape developed for the forensic industry called Easylift® to recover MPs from filter papers and secure them on microscope slides. Easylift® tape (Figure 9.2) was utilised as it has been developed to allow *in-situ* analysis because it is compatible with a wide range of non-destructive analytical techniques including polarised light microscopy, confocal Raman spectroscopy, fluorescence microscopy, microspectrophotometry (MSP) and hyperspectral microscopy unlike standard tape and acetate backings. Using simulated experiments with known amount of microfibres, it was demonstrated that this tape lifting system had a mean recovery rate of 96.4% ($s_{n-1}=3.5$ percentage points, $n=12$). In addition, by using Easylift®, fibres can be fully characterised, using multiple sequential techniques, without the need for

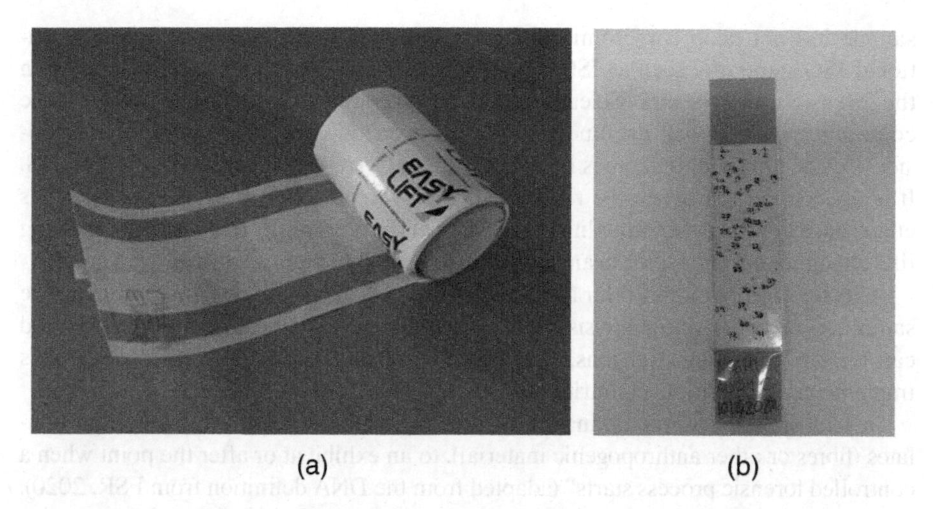

FIGURE 9.2 (a) Easylift® tape on the roll, (b) Easylift® tape mounted on glass slide with identified MPs.

dissection of the particulates which removes the opportunity for contamination and loss of sample; this is an added benefit for both the forensic and environmental sectors. Further discussion of the benefits of using multiple non-destructive techniques for fibre characterisation is discussed in Section 9.2.3.

Other studies than those sampling water, air or other materials, which would traditionally require filtration steps for the recovery of MPs, may also benefit from the use of tape lifting. Studies which test surfaces, such as skin and hair (Abbasi and Turner, 2021) for the presence of MPs, can directly recover particulates via application of Easylift® tape. This removes the need to wash the surface and subsequently filtrate the wash water for MPs, which reduces the opportunity for contamination from the equipment and processing steps; this also speeds up the recovery stages, removes the need for a laboratory environment and makes the study more cost-effective.

The use of tape for securing MP samples has now started to be utilised in MP studies (Abbasi and Turner, 2021). Other forensic lifters have also been tried, for example, gel lifters in a study testing trail surfaces in Australia (Forster et al., 2022); this shows promise for the adoption of forensic recovery approaches in MP work.

9.2.2 Methods for Reducing and Monitoring Procedural Contamination

From the recovery of samples from a crime scene and their analysis in a laboratory to the presentation of the results in court, the quality of procedures must demonstrably meet the high standards of the criminal justice system (CJS). The quality assurance procedures in forensic science are broad and cover everything from ensuring personnel involved are appropriately trained and qualified, to rigorous requirements for the recovery, documentation, storage, analysis and interpretation of the evidence. Evidence must have a 'chain of custody' associated with it, which documents when, where and who has come in contact with it; this is to ensure that the integrity of the sample has not been compromised. In addition, validation of methods to international ISO standards such as ISO 17020 and ISO 17025 is a typical requirement in the forensic science sector (Science and Technology Select Committee, 2019). These comprehensive and well-documented QA/QC procedures are appropriate when considering the potential outcomes involved if these QA/QC procedures fail in the CJS. It is understood that when the risks are higher, the protocols must be stricter. This ethos leads to robust, high-quality procedures and study results that are fit for making life-changing decisions, for example, whether to convict an individual for a crime.

This may seem 'overkill' for MP studies, yet contamination of samples during the sampling, recovery and analysis process is highly likely and the results generated can lead to important decisions, such as policy changes seen with microbead bans implemented in multiple countries.

In forensic science, contamination may be defined as "the introduction of particulates (fibres or other anthropogenic material), to an exhibit at or after the point when a controlled forensic process starts" (adapted from the DNA definition from FSR, 2020). This may be rephrased for MP studies as the introduction of particulates (fibres or other anthropogenic material) to a sample at or after the point when the experiment starts.

The above definition includes from the point of taking the sample, whether that is water, air, soil, sediment, biota or otherwise, all the way through to where any particulates of interest have been fully analysed and are no longer required. Potential routes for contamination include:

- from personnel to the sample,
- from personnel to gloves to the sample,
- from contaminated consumables (for example, sampling devices, glassware, filter papers, personal protective equipment [PPE]/clothing and packaging materials) to the sample,
- from the air to the sample, and
- from sample to sample (FSR, 2020).

This contamination may be introduced from many sources, but predominantly, it will be fibrous in form and likely to have originated from textiles.

The propensity for textiles to shed is well understood in forensic fibre examinations (as is discussed further in Section 9.3.4) and is acknowledged as a major source of contamination which needs to be minimised and monitored. In forensic science, anti-contamination methods either fall into one of the two categories:

1. reduction of the risk of contamination and
2. detection of any contamination present.

There are multiple methods that are commonly used to address the first of these, including:

- the wearing of PPE that is made of known materials and which are low shed, for example, scene of crime suits for sampling at scenes and cotton scrubs for use in the laboratory,

- the use of particulate-free environments for the processing of samples. For example, clean rooms (also known as search rooms) which have strict cleaning and access protocols that also control air flow, and
- the removal of contaminants from consumables and experimental surfaces, via cleaning before and after use.

The factors which affect contamination during forensic fibre examinations have been explored by Roux et al. (2001). This study investigated the effectiveness of a range of cleaning techniques and varying degrees of examiner hygiene by assessing fibre movement from the exhibit being examined. Interestingly, this study also provided information about which fibre types were more easily shed and which were airborne for longer.

Detecting contamination in forensic science involves:

- taking comprehensive and representative control samples of all possible sources of contamination,
- taking control samples of reagents, solutions and solids and processing those for the presence of any contaminants; this is also known as taking blank controls, and
- monitoring for the presence of airborne contaminants, also known as ambient contamination.

The need for ensuring sufficient control samples are taken was demonstrated in a forensic fibre examination conducted by Coyle (2020), where an overlooked potential source of contamination was identified only upon re-examination of the case. The lessons learned in procedural contamination in these forensic fibre case investigations are relevant to MP studies too.

The core of this contaminant detection is the comparison of controls and particulates recovered from blank samples to those that were found in the environmental sample of interest and then removing those which are deemed contaminants from the end results/MP count. For this comparison to be effective and the source of contaminants to be identified correctly, full characterisation of the particulates is required. If only limited characteristics are observed, for example, colour and polymer type, then all particulates matching those two characteristics would be removed from the end results; this will lead to an underestimation of MP abundance as there will be many different sources of microparticulate with the same two observed characteristics. This is further discussed in Rivers et al. (2019) study. This approach leads to items of interest (actual pollution particulates) being discarded along with those that are undesirable (those deemed as procedural contamination). This is known as 'throwing the baby out with the bath water' and is something commonly seen in MP studies.

In early MP studies, minimal anti-contamination procedures were used, and those that did employ anti-contamination procedures, they tended to focus their efforts on laboratory procedures and not field sampling (Gwinnett and Miller, 2021). In early studies, recovery methods were predominantly volume reduction-based methods which used net trawls to filter large volumes of water. As a typical net mesh size ranges between 300 and 500 μm (Ryan et al., 2020), any particulates below this size are not collected, and if particulates were present due to contamination, they should not be counted.

But as techniques for MP sampling and analysis are becoming more sophisticated and methods which allow for the smallest of size fractions to be recovered are being called for and used (such as the Buteler et al., 2023 study), the need for better contamination prevention procedures is required. This evolution towards the need for tighter anti-contamination procedures has also been seen in forensic DNA analysis, where methods progressed to allow single cells to be analysed rather than just visible body fluid stains. This increase in sensitivity means that the likelihood of contamination is much higher, and, therefore, strict anti-contamination steps must be integral at every stage of the process. This is now true for MP studies, which can learn a lot from the work done in forensic science in this area.

The potential for post-sampling contamination in MP studies was acknowledged in 2012 (Hidalgo-Ruz et al., 2012), but it was not until a study in 2015 that forensic contamination procedures were discussed in the context of MP studies (Woodall et al., 2015). This first study presented simple procedures which resulted in low levels of microfibres in the ambient environment by using forensic science approaches. The first test of whether forensic science approaches can be effectively employed in MP studies was in 2021 (Gwinnett and Miller, 2021). This study assessed forensic science anti-contamination procedures during the sampling and processing of water and air samples onboard a research vessel on an MP expedition along the Hudson River, USA. Their method assessed the effectiveness of these procedures by quantifying procedural contamination whilst using these strict protocols during sampling on the deck and processing samples in the galley and then systematically removing those protocols from the deck and galley and identifying any changes in contamination abundance. The difference between using these strict protocols and not using them, was dramatic. When protocols were used, procedural contamination accounted for 33.8% of the total microfibres and MPs found in samples ($n=81$), compared to 70.7% when they were not used ($n=8$). This demonstrates how MP abundance may be hugely overestimated if anti-contamination procedures are not used.

The opportunities for forensic science to improve anti-contamination procedures in MP studies have been highlighted in this 2021 study, yet work is still required to better understand how environmental laboratories, which are not set-up for the strict procedures required in forensic science, can be adapted to facilitate them.

9.2.3 Methods for the Characterisation of Microfibres

The characterisation of MP pollutants serves several purposes. As exemplified by the work of Bergmann et al. (2019), it can aid the understanding of their possible sources and pathways through the environment. It can help understand factors that may contribute to certain ecotoxicological effects or why certain MPs are more likely to be ingested; for example, colour has been identified as a characteristic that may affect the latter (Wright et al., 2013). Multiple studies have highlighted the difficulties in analysing fibres, including Zhu et al. (2019) who noted the small size and the presence of different dyes in the fibres cause difficulties in identification via conventional methods like total attenuated resonance–Fourier transform infrared spectroscopy (FTIR) and Raman spectroscopy. Ryan et al. (2020) noted some key limitations seen in the fibre analysis conducted as part of MP studies, including the difficulty in identifying their composition which has led to some studies simply assuming that all microfibres are MPs.

That notwithstanding, there is a steady increase in the range of types of characteristics being quantified and observed in MP studies, for example, those seen in Zhu et al. (2019). By increasing the range of characteristics analysed, this can allow MPs which share the same polymer type, but which have different morphological, optical and chemical properties, to be differentiated. This more granular characterisation of MPs allows improved contamination detection (as noted in Section 9.2.2) and provides evidence that could inform the inference of source; this is further discussed in Section 9.3.2. Although there are clear benefits, full characterisation of microfibres in environmental studies is still rare.

Currently, forensic examination of fibres is mainly a comparative analysis: a sample of unknown origin is examined to determine whether it is similar (or can be associated) to a sample from a known source (Causin et al., 2005). Alternatively, a fibre analyst may be requested to identify the source or possible end use of the fibre which has been collected as evidence. The outcome of a fibre investigation will be one of the three options:

- association between the unknown fibre and one of a known source,
- elimination, where the unknown fibres are differentiated from the known source, and
- inconclusive, where no conclusion can be made about association or elimination. This may be due to many reasons.

This ability to associate or differentiate between fibres is only possible when multiple 'layers' of information is gathered. It is not acceptable to conclude based on only a few characteristics; therefore, multiple sequential analysis techniques are used. The analysis of multiple characteristics of fibres in MP studies is rare; typically, many only report polymer type, colour, and size. Table 9.1 provides examples of the characteristics of microfibres analysed in a selection of recent environmental studies, which classified fibres from other forms of MP, including those detecting these pollutants in different environments/sample types, ingested by biota or other exposure type studies.

In comparison, in forensic fibre investigations, multiple techniques are used in a sequential manner (Figure 9.3). These analytical techniques have been described in useful handbooks for forensic fibre examiners (Hussain et al., 2020; Jaffe and Menczel, 2020).

Overall, the aims of forensic fibres examinations are:

1. to fully characterise the samples in terms of their morphological, optical and chemical properties,
2. to conduct the analysis by starting with quick and inexpensive techniques and moving towards more complex methods which may ultimately be destructive to the sample, and
3. to ensure that the integrity of the sample is kept throughout the examination so that the history of the sample is known and no morphological, optical or chemical changes occur.

TABLE 9.1
Examples of Environmental Studies Analysing Microfibres and the Characteristics and Techniques used. Please Note; Non-fibre Microplastics may have had different Characteristics Analysed in these Studies

Study	Fibre Characteristics Observed	Techniques Used
Modica et al. (2020)	None (count only)	Stereomicroscope with camera attachment
Suaria et al (2020)	Length and diameter (measured to the nearest 1 μm from digital images)[a] Length:diameter ratio calculated Polymer type (classified as either synthetic, animal or cellulosic fibres (both natural and man-made)[a]	FTIR microscope and operated in attenuated total reflection (ATR) mode
Wang et al. (2020)	Colour Size (one dimension yet unknown as to whether this is length or width) Surface morphology Composition of surface Polymer type[a]	Fluorescent inverted microscope with imaging software Scanning electron microscope with energy disperse spectroscopy (SEM–EDS) μ-FTIR microscopy
Abbasi and Turner (2021)	Colour Size (length or primary diameter) Polymer type	Binocular microscopy, polarised light microscopymicro-Raman spectroscopy
Jiang et al. (2023)	Colour Size (length, 'longest dimension' noted) Polymer type	Micro-FTIR in reflection-point scanning mode with imaging software and measurement tool
Forster et al. (2023)	Colour Size Polymer type[a] Surface morphology[a]	Stereomicroscope Sieved into five groups of 150–250, 250–500, 500–1,000, 1,000–2,000 and 2,000–5,000 μm ATR-FTIR NANO SEM-430 field emission scanning electron microscope (SEM-EDS)

[a] Indicates where only a sub-sample of the particulates were analysed.

The latter aim is important when considering using a method which inherently changes the sample in a manner which prevents specific characteristics being observed. Where possible, these methods are avoided in forensic fibre examinations. Currently in MP studies, the use of Nile red dye as a method for visualising MPs is common (Shruti et al., 2022). The use of this dye artificially makes the MP fluorescent, which masks any original fluorescent properties the sample may have and

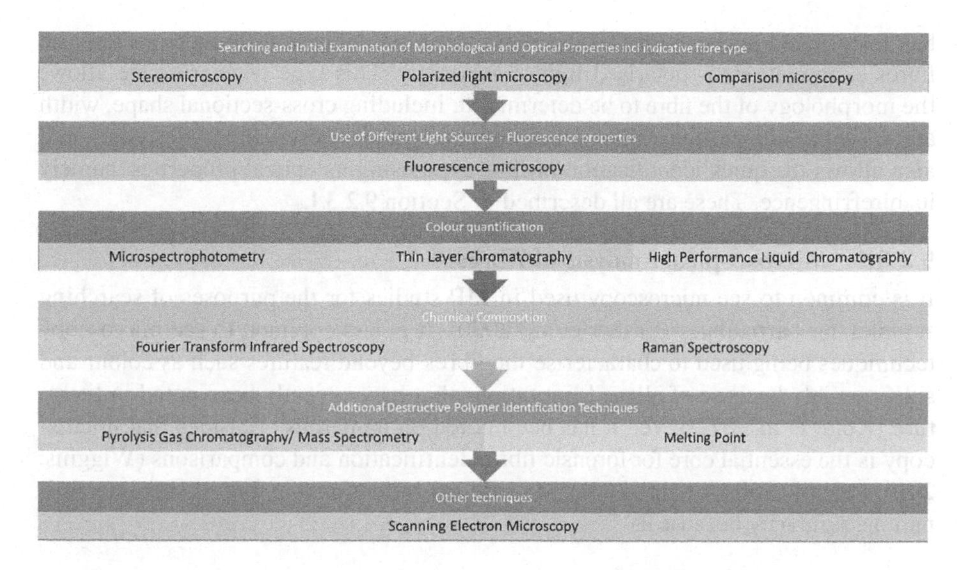

FIGURE 9.3 Sequential analysis techniques used in forensic fibre examinations.

therefore does not allow that characteristic to be used when differentiating between samples. In addition, it has been shown that Nile red also dyes cotton and linen fibres which can confuse analysis for those reliant on this feature to detect synthetic fibres only (Galvão et al., 2023). These limitations are such that it would prevent its use in forensic fibre examinations; and instead, other methods to improve detection of particulates have been developed, such as viewing samples under high magnifications and cross-polar conditions (discussed further in Section 9.2.3.1).

The techniques chosen in forensic fibre investigations to fully characterise samples depend on availability of equipment, training and within country preferences. A study of forensic fibre analysis in Europe and the USA conducted in 2001 looked at the analytical tests performed in 135 laboratories (Wiggins, 2001). It found that fibres from casework were examined using a multitude of techniques, but due to synthetic fibres' optical properties, microscopy was the most common method of analysis, with 98% of laboratories using brightfield microscopy and 93% using polarised light microscopy routinely. The study found that other techniques utilised routinely included: FTIR microspectroscopy, with 52% of laboratories regularly using this method; microspectrophotometry in the visible region (44% of laboratories); solubility testing (35% of laboratories); and thin layer chromatography (TLC) of dyes (15%). Techniques that were identified to only be used occasionally included: pyrolysis--gas chromatography (Pyr-GC); scanning electron microscope–energy dispersive X-ray (SEM-EDX); and melting point (Wiggins, 2001). Based on the authors collaborative work with forensic labs around the world, these preferences for certain techniques still hold true today.

In comparison, a review of techniques used in 90 MP studies of fish and mussels showed that polymer analysis was mainly carried out with FTIR spectroscopy (58% for both mussels and fish) and Raman spectroscopy (14% for mussels, 8% for fish) (Dellisanti et al., 2023). One method which is underutilised in MP studies yet which has been noted above as one of the most common analysis techniques in forensic fibres examinations is polarised light microscopy. This type of microscope allows the morphology of the fibre to be determined, including cross-sectional shape, width and presence of inclusions, all of which are not regularly observed in MP studies, and also allows the quick identification of fibre type using its optical properties, namely its birefringence. These are all described in Section 9.2.3.1.

9.2.3.1 Microscopical Analysis of Fibres

It is common to see microscopy used in MP studies for the purposes of searching samples for particulates (Lusher et al., 2020). It is less common to see microscopy techniques being used to characterise the fibres beyond features such as colour and size, possibly because of alleged limitations that have recently been noted in literature (Kotar et al., 2022). Yet, it has been noted on numerous occasions that microscopy is the essential core for forensic fibre identification and comparisons (Wiggins, 2001; Palenik, 2004; Chabli, 2001). Microscopy is dominant in forensic fibres examinations primarily because it:

- is non-destructive (ENFSI, 2001),
- allows visualisation of even very small fibres,
- is a relatively quick means of sample analysis (Faber et al., 1999) (important for timeliness of analysis, case throughput and if repeat measurements are needed),
- is inexpensive after the initial outlay (very few consumables),
- can analyse the morphology and optical properties of fibres in addition to providing indicative fibre type, including natural types and polymer identification, and
- can distinguish between fibres of the same type but from different sources, for example, different manufacturers.

The popularity of microscopy techniques in forensic fibre analysis was highlighted early on in Wiggins review of fibre examination (Wiggins, 2001). Microscopy is particularly good for the analysis of natural fibres and forms the predominant technique for the identification of these particulates in examination of forensic fibres. Natural fibres can be distinguished from each other by observing the fibres' microscopic morphological characteristics, for example, length of ultimates, presence of nodes and thickness of lumen (Robertson and Grieve, 1999). Subtle differences in the morphology of natural fibres, which can be seen using microscopy techniques, enable analysts to differentiate between natural fibre types, such as different bast fibres (Summerscales and Gwinnett, 2017).

More recently, the usefulness of high-powered microscopy has been demonstrated when observing photodegradation on natural fibres, an observation which is also useful to MP studies investigating degradation/ageing of particulates due to exposure to the environment (DeBattista et al., 2019).

Synthetic fibres cannot be identified on the basis of their morphology in the same way as natural fibres, but synthetic fibres do lend themselves nicely to microscopy methods as they contain many characteristics which can distinguish them from each other. Some of these characteristics are width; birefringence; refractive index (RI); sign of elongation; pleochroism; transverse cross-sectional shape and the presence of inclusions, such as delusterant or flame retardants. These characteristics are regularly the first to be measured/observed as they can be very discriminating and allow the genus (type) of fibre to be identified quickly (Johri, 1979). Some of these morphological characteristics are discussed in Sections of 9.2.3.1.1–9.2.3.1.4.

The polarised light microscope is the most useful and versatile of all the microscope types available to fibre analysts (Palenik, 2004). This microscope allows the same observations as a compound microscope but also permits observations and measurements using plane polarised light and between crossed polars. The polarised light microscope (PLM) allows qualitative information such as sign of elongation to be gained and quantitative information such as birefringence. This information can lead to quick preliminary identification of the generic type of man-made fibres. The determination of birefringence for polymer identification is further discussed in Sections of 9.2.3.1.1–9.2.3.1.4.

Although the PLM is the most useful for fibre characterisation, other microscopes and attachments are also utilised in forensic fibre examination that would also provide further information in MP studies. Some of these microscopes are outlined in Table 9.2.

9.2.3.1.1 Birefringence Analysis for Initial Polymer Identification

Polymer analysis in MP studies heavily relies on techniques such as FTIR, Raman Spectroscopy (Song et al., 2021) and pyrolysis gas chromatography (Pyr-GC) (Picó and Barceló, 2020). Each of these techniques are expensive to purchase, create spectra or chromatographs which are difficult to interpret without a comprehensive database and require awkward sample preparation. It is also problematic to obtain good quality results, which allow identification, from samples which are:

- small in size,
- bio fouled, and
- are not able to be flattened and therefore refract the light (for FTIR and Raman spectroscopy).

The time-consuming nature of these techniques for MP identification has led to only sub-samples of the overall particulates, being found, actually being analysed (Ryan et al., 2020; Sang et al., 2021; Niu et al., 2023).

In comparison, the use of birefringence for initial polymer identification is a quick and effective technique that can be performed on even problematic samples that normally would not be able to be identified via the above methods. For those who conduct birefringence measurements often, a fibre may only take 2–3 minutes to be identified compared to anywhere up to a 1 hour for analysis via FTIR, Raman or Pyr-GC (including sample preparation time). In addition, the birefringence value is a quantitative value that allows direct comparison between unknown and known samples and has shown to be able to be used to differentiate between fibres of the same polymer type but from different manufacturers (Johri, 1979).

TABLE 9.2
Examples of Microscope Types Used in Forensic Fibre Analysis

Microscope Type	Uses	Comments
Stereomicroscope	Primarily used to search, recover and manipulate individual fibres from tapings. Also used for the examination of textile constructions. It is not suitable for accurate identification of fibre type.	Usually low power, for example, a typical range of magnification is between ×0.5 and ×10. Stereomicroscopes should be equipped for observation with both transmitted and reflected light.
Comparison microscope	Provides a side-by-side microscopic comparison between multiple fibres in a single field of view.	A discriminating method for determining if two or more fibres are consistent with originating from the same source.
Fluorescence microscope	Used to search for and observe fluorescence in fibres originating from some dyes, optical brighteners or contaminants, e.g. from washing powders.	The microscope is set up in incident light with a selection of filters that cover the excitation range of ultraviolet through violet, blue and green.
Hot-stage microscope	Melting point determination of fibres. May also be used for RI determination.	Technically an accessory which fits upon the stage of a high-powered light microscope. The hot stage should reach to above 300°C and should allow the user to increase the temperature by 4°C/min or less.
Scanning electron microscope (SEM)	As an imaging tool, it provides high-resolution, three-dimensional images at very high magnifications. SEM can also yield additional physical and analytical information.	SEM utilises high-energy electrons to scan the surface of the fibre. Images of surface topography are either derived from backscattered electrons or secondary electrons.
Interference microscopy	To determine the refractive indices of fibres and combined with a Standort diagram can determine sign of elongation.	This utilizes an interferometer to split polarised light. An interferogram is produced of the recombined light beams to enable the refractive indices of the fibre to be determined. By plotting the refractive indices on a Standort diagram, the sign of elongation can be determined (Heuse and Adolf, 1982).

Nearly all synthetic fibres, other than glass fibres and vinyon HH (Johri, 1979), are anisotropic, meaning they have two refractive indices. The two refractive indices are orientated at 90° to each other, one parallel to the long axis of the fibre(n_e) and one at a right angle to this axis(n_o) (ENFSI, 2001). The two refractive indices are due to the birefringent material splitting light into two components: an ordinary ray and an extraordinary ray which vibrate at 90° to the other and are both plane-polarised. It is the properties of these two rays, specifically their refractive indices, that can aid fibre identification. Birefringence can be determined by accurately measuring n_e and n_o and then calculating the absolute difference between these two refractive indices (De Wael, 2021). This method has been utilised by Grieve for the determination of birefringence of colourless polyester fibres (Grieve, 1983). Birefringence can alternatively be calculated from the retardation (*R*) at a point of the fibre of known thickness (*T*) utilising equation (9.1) (Smith, 1956):

$$\text{Birefringence} = R/(t \times 1{,}000) \tag{9.1}$$

Where retardation (*R*) is in nanometres and thickness (*t*) is in micrometres.

The retardation, also known as optical path difference (OPD) can be determined by using a compensator, Michel-Levy chart and a polarised light microscope (De Wael, 2021).

Compensators consist of thin sections of birefringent minerals such as quartz, gypsum, mica and calcite. Compensators have controlled thicknesses and optical orientation to provide known values of retardation and direction of high and low refractive indices (Delly, 2003). Each compensator type may be utilised for specific fibre types depending on its range of retardation, for example, the Berek compensator is utilised regularly for fibres with retardation values of between 0 and 20λ, where λ is wavelength (Delly, 2003; Sieminski, 1975).

A common method for analysing the retardation value is using a polarised light microscope set up under crossed polars, with the fibre orientated at 45° to the cross-hair, a quartz wedge and a Michel Levy chart (ENFSI, 2001; Smith, 1956; Robinson and Davidson, 2023). The retardation can be estimated by observing the interference colour and order where the thickness of the fibre has been measured and compensation black has been achieved. An additional benefit of using PLM to visualise interference colours in this manner is the significant improvements in contrast of the fibres to the background, especially for colourless particles. This means that searching for fibres under crossed polars (for example, those encased on an Easylift tape ® on a microscope slide as described in Section 9.2.1) is made easier. Figure 9.4 shows the increase in contrast of fibres under crossed polars compared to plane polarised light.

The quartz wedge method only provides the fibre analyst with an approximate birefringence value and may not be appropriate to use for fibres with high OPDs as many quartz wedges only cover four orders of interference colour (ENFSI, 2001). Alternatively, more accurate methods include the use of other types of compensators, including Sénarmont (Delly, 2003; Bloss, 1981; Hartshorne and Stuart, 1971), Ehringhaus or tilting compensator, e.g. Berek (Sieminski, 1975).

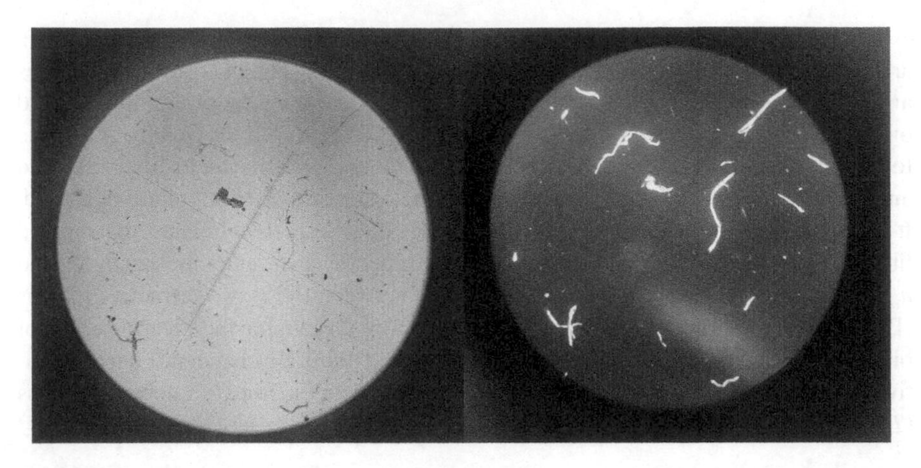

FIGURE 9.4 Left shows anthropogenic fibres under plane polarised light; right shows the same fibres under crossed polars, both at ×200 magnification within Easylift® tape.

Polymer type can be determined from the birefringence values and use of tables of birefringence values and schemes for fibre identification (Stoeffler, 1996). These tables allow indicative polymer types to be given to a fibre, many of which can be made confidently by an experienced examiner. But like all techniques, it has its limitations, these include:

- some fibres do not lend themselves to identification by optical methods, including heavily dyed fibres as the dye can obscure the interference colours,
- difficulty in accurately identifying compensation black due to dispersion of white light (Sieminski, 1975), and
- higher retardation values, for example, greater than 2,200 nm may be more inaccurate.

To overcome the latter limitation, a wedge at the end of the fibre may be cut and the order of interference colour counted (Sieminski, 1975).

The use of birefringence in MP studies is still in its infancy, with only a few studies having used it to characterise microfibres at this point (Sierra et al., 2020; Taylor et al., 2016; Woodall et al., 2015). With an ever-increasing need for more data to better understand MP prevalence, a quicker more cost-effective method to identify fibre type is required. In addition, the ability to train inexperienced personnel to analyse birefringence of fibres has been proven to be possible (Gwinnett, 2009). This bodes well for this technique to be introduced into citizen science programmes globally and to create large MP datasets.

9.2.3.1.2 Width Measurements

In MP studies, it is common to see the size of an MP particle being reported, but this is normally the largest dimension, i.e. it's length rather than its width. Length is of course useful for understanding the size of the polluting particle but provides no information regarding the manufacturing process and the intended use of fibre, with the exception of flock fibres, which are cut to standard lengths (Jones and Coyle, 2010). The width of man-made fibres, if measured accurately, can be very discriminating and allows us to differentiate them from similar fibres whilst providing a better understanding of the potential source. Fibres differ in width due to several reasons; these include manufacturing method; alteration of the fibre for a given effect; and accidental damage either during manufacture or after the manufacturing process. Accidental damage to a textile fibre, although unique in many ways, can inhibit the fibre analyst's ability to compare the sample to control samples or reference samples. Fibre widths produced by the manufacturing process can be varied due to the fibres intended end use. For example, carpet fibres are primarily nylon-type fibres and will be wide in measurement to create a plush consistency (Robertson and Grieve, 1999); whereas, microfibres, as described by their name, are narrow in width which allows for a softer feel (Wiggins and Clayson, 1997).

The width of fibres with round cross-sectional shapes (see Sections of 9.2.3.1.1–9.2.3.1.4 for descriptions of cross-sectional shapes) can be easily measured using a scale in the microscope eyepiece, or via a Filar eyepiece attachment (Raines, 1932). This eyepiece scale must be calibrated using an appropriate stage micrometer. Measurements of round fibres are taken from one edge of the fibre to the other edge of the fibre whilst viewing the fibre longitudinally. Measurements of the width of fibres that are other shapes, such as trilobal, require more consideration. Using a protocol for non-cylindrical fibre measurements as stated by Gaudette (1988), the widths of trilobal, bilobal and irregular fibres can be measured more accurately. If a fibre exhibits multiple dimensions, for example, a bilobal fibre, the range of these dimensions can also be determined.

Inaccuracies in placing fibres in size bins using sieves has been described by Kotar et al. (2022), outlining the potential for improving measurements of fibres using width. Measuring width of fibres is beneficial for environmental studies for other reasons, including:

- improved understanding of the number of potential sources of microfibre pollution, and
- improved characterisation of fibres used in eco-toxicological studies. This is because it is understood that size and surface area may contribute to the uptake of adsorbed chemicals and heavy metals.

9.2.3.1.3 Cross-Sectional Shape

Although 'form' or 'shape' of MP is commonly denoted in MP studies, for example, fragment, film, bead and fibre (Frias and Nash, 2019), it is rare to see further characterisation of the shape of fibres. Man-made fibres have a wide range of cross-sectional shapes that allow further discrimination between samples and can also be used to infer source. The cross-sectional shape of a fibre is described as the shape when observing the fibre in a transverse position (Palenik, 1990). Originally, man-made fibres, in their liquid polymer form, were extruded through circular spinneret holes creating fibres round in cross-section (Taylor, 1990). Non- circular cross-sections have since been developed by either deliberate or accidental changes in the spinning conditions or altering the shape of the spinneret holes (orifices) (Grayson, 1984). For example, wet-spun and dry-spun acrylic fibres have different cross-sectional shapes due to the two different spinning processes, and the polygonal shape of some polyester fibres occurs because of post-extrusion at a high temperature (Palenik, 1990). In man-made fibre manufacture, primarily in melt spun fibres such as nylon, polyester and polypropylene, there are many different engineered cross-sectional shapes. These engineered cross-sectional shapes are determined by the shape and arrangement of spinneret holes. Hollow fibres can be complex in shape and are produced by arranging the spinneret holes so that the liquid polymer stream coalesces below the spinneret (Grayson, 1984). The most common cross-sectional shapes of man-made fibres include round, irregular, flat-ribbon, delta, trilobal, irregular-lobed and bilobal (ENFSI, 2001).

Different cross-sectional shapes have been identified to have useful properties, for example, irregular-shaped fibres have an increased surface area and therefore absorb dyes more readily. The examination and identification of a fibre's cross-sectional shape can be very useful to a fibre analyst and can provide information about the manufacturer of the fibre, the spinning process used to produce the fibre, the physical processing used and the possible end use of the fibre, all of which are valuable for identifying potential sources of the pollution. An example of this are fibres with a triskelion (triple spiral exhibiting rotational symmetry) cross-sectional shape which are most likely to have been produced by Monsanto (Palenik and Fitzimons, 1990a).

The chance of being able to identify the manufacturer of a fibre accurately, and therefore sometimes its end use depends on the choice of method for cross-section determination and the quality of sample preparation. An initial examination of the fibre can be completed whilst the fibre is mounted normally in an appropriate medium [for example, Entellan mounting medium (Wiggins and Drummond, 2007)] and viewing the fibre longitudinally. By viewing the fibre in a longitudinal position, the apparent cross-sectional shape of the fibre can often be determined by a method called optical sectioning; this is where the fibre analyst slowly focuses through the fibre, allowing any twists or lobes to be identified. For irregular-shaped fibres, it is possible to identify longitudinal striations by utilising optical sectioning (Grieve and Deck, 2002). It is important to view the shape along the whole length of the fibre, and on some occasions, it may be helpful to do this under crossed polars. Utilising crossed polars allows the symmetry of the interference colours to be observed, for example, non-symmetrical interference colours usually show a non-circular fibre.

Although the approximate cross-section can be identified by viewing the fibre longitudinally, it is more accurate to make a transverse cross-sectional cut and view the actual cross-section mounted in a section under a microscope (Palenik and Fitzimons, 1990a, b). There are a variety of different methods for preparing a fibre for cross-sectional determination depending upon the number of fibres and the experience of the analyst. The classic method for creating sections uses a microtome. Two other methods utilised by fibre analysts to create fibre sections are the use of a polyethylene film 'sandwich' technique developed by Palenik and Fitzsimons (1990a, b) or the Joliff Plate method developed by E.C. Joliff (Palenik and Fitzimons, 1990b).

To identify the manufacturer of prepared fibre sections, modification ratios of trilobal fibres can be utilised (Palenik and Fitzimons 1990a, b). The modification ratio originated from research conducted by Holland of DuPont to produce a nylon fibre that concealed dirt (Holland, 1960). Other manufacturers produced similar fibres with different modification ratios so as not to violate DuPont's patent. Due to this, specific manufacturers can be identified, for example, Celanese's Type 81 fibre has a modification ratio of 2.1 (Palenik and Fitzimons, 1990a).

By preparing a cross-sectional cut, not only can the shape of the fibre be more accurately identified but other information can be gained, such as the depth of dye penetration, the distribution of delusterant and the presence of spherulites and gas voids (ENFSI, 2001; Palenik and Fitzimons, 1990a). In addition, for MP studies interested in analysing surface area, for example, Wang et al. (2020), noting cross-sectional shape allows more accurate measurements.

Knowing that man-made fibres found in the environment may have different cross-sectional shapes is also important in preventing misidentifications. In the classification diagram, created by Prata et al. (2020) to aid in the identification of natural vs man-made fibres, there is an assumption that a rough irregular surface indicates a natural fibre. This assumption is not entirely true as the previously described irregular-shaped man-made fibres also exhibit a rough surface. In addition, this classification diagram has also assumed that ribbon and rod-shaped fibres are exclusively of natural origin. By observing the cross-sectional shape of fibres using the techniques described here and obtaining a reference set of differently shaped fibres, MP analysts will be better equipped in man-made fibre identification.

Currently, the only MP studies that included cross-sectional shape of microfibres found in the environment are those conducted by this author (Cunningham et al., 2022; Gwinnett and Miller, 2021; Taylor et al., 2016; Woodall et al., 2015). These observations are quick to complete and have benefits to MP studies including:

- better understanding of surface area and depth of dye penetration (if cross-sectional cuts undertaken) for test samples in eco-toxicological studies. This is particularly important when trying to determine what are the factors which are increasing harm from MP ingestion. Studies that focus on leachates of dyes and other additives would particularly find this observation useful, and
- improving opportunities for identifying the source of microfibre pollution for the purposes of more targeted mitigation activities.

9.2.3.1.4 Inclusions and Additives in Fibres

MP studies have more recently become interested in not only the physical polymer of the particulate but also the presence of additional chemicals which may leach out into the environment or cause-specific negative toxicological effects (Kim et al., 2022). These additional chemicals include additives, also known as inclusions, which may be added to fibres for three main reasons: to create a certain aesthetic appearance, to develop a fibre for a specialised use or to compete with cheap textile imports, for example, from Asia or Eastern Europe (Nehse, 2007).

Inclusions within fibres are either introduced deliberately during manufacture or produced accidentally during the manufacturing process; for example, stress marks within synthetic fibres are accidentally produced when the liquid polymer is being extruded. The addition of inclusions or the use of finishings in textile fibre manufacture enables man-made textiles to be improved and/or to be utilised for a specific purpose, for example, the use of delusterant enables synthetic fibres to obtain a more natural appearance. The use of finishings in fibre production creates particular aesthetic appearances or properties. The different finishing processes are utilised to enhance a wide variety of textile properties including crease retention, abrasion resistance, water repellence and flame retardance. Different finishings and inclusions are also employed to create aesthetically pleasing textiles, for example, the utilisation of mechanical processes such as compressive shrinkage and schreinering; the latter produces a silk-like finish.

Delusterant is a common type of inclusion, and its presence or absence is regularly noted in forensic fibre analysis (Grieve, 2000). Delusterant is used to create a matt, non-shiny finish to man-made fibres. The most common type of delusterant is titanium dioxide, but other delusterant types are also employed, for example, paraffin wax delusterant used in nylon fibres made by Monsanto. Noting the presence, size, shape, distribution and abundance of delusterant in a fibre can be evidentially useful. Although delusterant particles cannot identify a particular fibre type, it can indicate the possible end use of the fibre and easily identifies the fibre as man-made.

To identify the presence of delusterant or any other inclusion in a fibre, the fibre must firstly be mounted in a medium which has a similar RI to the fibre (see Wiggins and Drummond, 2007). This allows any inclusions to be more easily seen. Not all particles present in fibres are delusterant; carbon black pigment and flame retardants, such as antimony oxide, are also particulates but tend to be larger in size than delusterant. Delusterant particles are sub-micrometer in size and have a higher RI that the surrounding fibre. A further distinction can be made between carbon black particles and titanium dioxide particles by examining the fibre under reflected light against a dark background. In these conditions, titanium dioxide particles will reflect the light and appear as bright spots whereas carbon black particles will not (ENFSI, 2001).

In addition, other inclusions may be noted which provides further discrimination between microfibre pollution sources and can infer the end use of the fibre. Examples include voids, stress marks or 'fish-eyes', anti-static inclusions and nano-coatings. These inclusions are described in Table 9.3.

9.2.3.2 Quantification of Colour in Fibres

The colour of a fibre is one of the most significant characteristics for comparison and discriminatory purposes in forensic science (ENFSI, 2001; Morgan et al., 2007; Schotman et al., 2017), yet in MP studies, colour is generally low priority. Lusher et al. (2017) noted that colour is not considered to be crucial to MP analysis due to colour differentiation being subjective; this is true when only visual descriptions of colour are provided without the use of objective analytical techniques. This subjectivity in colour analysis was tested by Kotar et al. (2022) who found that colour was regularly mismatched between analysts, for example, noting a gold fibre as orange. The almost infinite colours used in textile fibres means that this characteristic is important in understanding the breadth of pollution sources present in the environment. The idea that 'not all blue fibres are the same' is a crucial principal in forensic fibre examination and for the purposes of better understanding the range of sources

TABLE 9.3
Inclusions Seen in Man-Made Fibres

Inclusion Type	Reasons/Causes	Appearance
Voids	Either added deliberately to man-made fibres to create a practical property, to make dirt less visible, or an accidental defect. Voids are formed by spinning the fibre forming polymer with a blowing agent which causes an air pocket effect.	Air gaps which may appear like bubbles along the fibre. These air gaps may be elongated in shape due to the fibre extrusion.
Stress marks or 'fish eyes'	Thought to be formed due to undissolved pigment or polymer in the fibre-forming mixture. Stress marks are not commonly found in all types of synthetic fibre but can be seen more commonly in polyolefin fibres.	During extrusion, any air gaps around these undissolved compounds become elongated and form the appearance of an 'eye', with the undissolved compound in the centre.
Anti-static inclusions	Synthetic fibres can accumulate high charges of static electricity due to their non-conducting properties; this is due to their containing very little moisture in their dry state (Hall, 1975). To counteract this, some fibres may have a surface treatment that absorbs moisture, resulting in a thin conductive film on the fabric, for example, the fibre Resistat® is a polyamide 6 fibre with a carbon black coating. Alternatively, fillers which are used as electrically conductive additives, may be added inside the fibres, for example, carbon nanotubes (CNTs) or graphene or metal powders. Anti-static inclusions are commonly seen personal protective clothing that requires good anti-static properties (Hufenus et al., 2020).	Spray-on films may not be visually identifiable under normal microscopical conditions but may be identified using other analytical techniques. Fillers may be viewed within the fibre under magnification. Bicomponent fibres may have a filler included in one section, e.g. one lobe of a bilobal fibre.
Nano-coatings	Intelligent textiles are currently being developed in multiple industries including the military.They utilise nanotechnology in the form of nano-particles and nano-coatings to create specific effects, such as anti-microbial and water repellence (Nehse, 2007).	Research conducted by Nehse (2007) showed that nano-coatings may be identified using high magnifications and additional use of dark field microscopy but microscopy alone is not sufficient; therefore additional techniques such as FTIR may be needed.

of pollution; it is also important in MP studies. This is exemplified when we consider the prevalence of indigo blue denim fibres from jeans and the need to better understand the breadth of sources of this particular textile pollutant (Athey et al., 2020). The value of recording colour of MPs has not been completely ignored in MP studies; Frias and Nash (2019) noted the worth of MP colour in studies involving aquatic organisms, where it is thought that some species preferentially ingest MPs based on certain colours (Wright et al., 2013). Due to this, a better understanding of the types of colourants used in fibres and the methods that can quantify colour is beneficial for all MP studies.

Colourants can either be a single component or a mixture of dyes. The analysis of dye components and pigments beyond simply stating a subjective colour category has the potential to increase discrimination by comparing the type and formulation of colourant present, relative amounts of the colourant present and spectral fingerprints of individual extracted dyes (Morgan et al., 2007; Schotman et al., 2017). As fibres found at both crime scenes and in the environment are regularly small in length and usually contain between 2 and 200 ng of dye (Macrae and Smalldon, 1979), techniques employed to analyse these dyes must be sensitive. In forensic fibre examinations, colour may be initially observed using a microscope, in the same manner as in environmental studies, but in forensic examinations, any differences in colour along the length of the fibre will also be noted; this has yet to be seen in MP studies. This initial observation is useful for initial screening purposes, but for a dye to be accurately identified and characterised, alternative techniques must be utilised. Raman spectroscopy and infrared spectroscopy are regularly used for dye analysis, but other techniques, such as TLC, are important methods for discriminating dyes and quantitatively assessing colour. Please see Table 9.4 for a description of these techniques.

For many techniques used in dye analysis, the dye must firstly be extracted from the fibre. Dye extraction from fibres can also help to identify the fibre type and classify the dye type.

9.3 INTERPRETATION OF FIBRES: METHODS USED IN FORENSIC FIBRE EXAMINATIONS

Utilising robust, comprehensive methods for the recovery and analysis of MPs is an important stage in providing rigorous results to research questions, but without appropriate methods for interpreting those results, studies will not reach their maximum potential. How data are reported in MP studies is varied and there has been repeated commentary on the need for standardisation in this aspect, as well as in the recovery and analysis methods (Zantis et al., 2021). The reporting of abundance using concentrations or counts is a simple method used in many studies, for example, reporting a mean number of particles per litre or m^{-3} (Kanhai et al., 2018) and something that forensic fibre examinations do also when reporting how many fibres were found on a particular person or surface, for example, in the Australian case described by Bennett et al. (2010). This section does not reflect on methods for reporting, but rather outlines the questions asked of evidence od fibres and methods used by forensic scientists to gain further meaning from these particulates in criminal cases and makes suggestions on how similar approaches may be useful in MP studies.

TABLE 9.4
Techniques Used to Analyse Colour and Dyes in Forensic Fibre Examinations

Technique	Description	Advantages	Disadvantages
Thin layer chromatography (TLC)	TLC is a separation technique involving the use of a thin layer support medium (TLC plate), in which the dye is applied and an eluent which passes across the plate. This causes the dye components to separate and travel a given distance along the plate. The distance travelled is dependent upon the properties of dye components physical and chemical. TLC was introduced as part of the dye classification scheme by Resua (1980) and is used as part of the dye extraction method. TLC complements microspectrophotometry, especially, if the spectroscopy is restricted to the visible range only and therefore may not detect differences in some dye components (ENFSI, 2001).	TLC is a simple, well-established technique with TLC eluent systems developed for most fibre types (including polyester, nylon, acrylic, (Beattie et al., 1981) cellulosic (Home and Dudley, 1981) and wool (Macrae and Smalldon, 1979) and dyes. TLC can be conducted on single fibres, although minimum fibre lengths apply for different fibre types. The process of extracting the dye from the fibre, using a series of solvents, can give some indication of fibre and dye type (Robertson and Grieve, 1999). TLC has shown to allow dye batch variation to be identified, for example, in viscose fibres, where microscopical and analytical techniques have not allowed the fibres to be distinguished (Wiggins and Holness, 2005).	TLC is a destructive technique. TLC requires relatively large amounts of extracted dye to identify all components (Morgan et al., 2007). If the dye cannot be extracted from the fibre, TLC analysis is not possible. Assessing the colour and intensity of the dye bands on the plate is subjective and some bands may appear faded, but this subjectivity can be partially overcome using densitometry (Golding and Kokot, 1990). Reproducibility of TLC results can be poor due to the lack of stability in eluents and temperature.

(*Continued*)

TABLE 9.4 (*Continued*)
Techniques Used to Analyse Colour and Dyes in Forensic Fibre Examinations

Technique	Description	Advantages	Disadvantages
Micro-spectrophotometry (MSP)	MSP is used for the objective evaluation of colour in trace evidence, including single fibres (Suzuki et al., 2001). MSP allows the measurement of the absorption of electromagnetic radiation by fibre colourants in the visible region of the spectrum and, in some instruments, the ultraviolet region. By utilising the ultraviolet region to analyse colour, information can be gained on the chemical structure of dyes and whitening agents. Use of microspectrofluorimetry also allows comparison of any fluorescent brighteners present (Hartshorne and Laing, 1991).	MSP provides a quick, reproducible and highly discriminating technique to compare fibre colour. This is a widely used, non-destructive technique which does not require large sample sizes. MSP allows discrimination between dye types including dyes which cannot be extracted, for example, vat dyes (Suzuki et al., 2001). MSP can also help to distinguish between undyed polyester fibres and undyed nylon and acrylic fibres (Grieve, 1994). It is thought to be a complementary technique to TLC (Grieve et al., 1988).	MSP is not suited for heavily dyed fibres as these produce featureless spectra or natural fibres as they produce too wide a range of spectral results. Some dyes have similar chemical structures resulting in spectra that appear identical, for example,, synthetic indigo dye and bromine-substituted dyes (Adolf, 1986). Application of MSP has mainly been limited to the visible region, with limited research conducted in the ultraviolet region (Hartshorne and Laing, 1991).

(*Continued*)

TABLE 9.4 (*Continued*)
Techniques Used to Analyse Colour and Dyes in Forensic Fibre Examinations

Technique	Description	Advantages	Disadvantages
High performance liquid chromatography (HPLC)	HPLC is used to analyse a dye class, dye components and dye mixtures from single fibres. HPLC utilises a mobile phase (eluent) and a stationary phase (column) to separate and resolve components of fibre dyes. The time in which the dye components are eluted from the column and detected by the UV detector is termed the retention time. The retention time of the dye components enables identification and comparison. Multiwavelength detectors such as photodiode array detectors (PDA) allow detection of yellow, red and blue components in a single sample injection as they enable the complete UV and visible spectrum to be analysed. This decreases analysis time compared with that needed with the original monochromatic detectors (Grieve, 1994). TLC analysis should be carried out before HPLC to determine dye class so as to allow correct chromatographic conditions to be used (Griffin et al., 1994).	HPLC is a relatively fast and sensitive technique. HPLC allows differentiation of dyes with similar retention times and reproducible retention times of dye components. HPLC has shown that it has greater resolution than TLC, for example, when analysing Acid Black 2 dyes, and allows more components of fibre dyes to be analysed. It also allows easier analysis of yellow dyes than TLC (West, 1981) and also provides a more permanent record than TLC plates.	Dyes in the same class can have very different chemical structures. In terms of HPLC, dyes are either acid, basic or neutral, these can be extremely difficult to separate using a single HPLC system. Method development is time consuming and expensive. Although there are well documented HPLC systems for a small number of dyes or single dyes, there are limited systems available for fibre dyes as the number and combination of dyes within a fibre is unknown. Complex mixtures of dyes are difficult to characterise due to different dyes needing different eluents and experimental conditions. The inability to create a standard method due to the variety of dyes means that spectral databases are difficult to create. Some solvents used for extraction can affect HPLC analysis, for example , some solvents can cause degradation or colour change of dye (West, 1981).

9.3.1 Investigative Star and Research Questions

Interpretation of the analytical findings of a fibres' case can be very difficult and may take several forms; this is also true of MP studies which have gathered complex data and seeks to make sense of it.

The final interpretation and assessment of the significance of fibres' evidence in forensic cases depends on the questions needing to be answered and the factors influencing the significance of the evidence (Roux and Robertson, 2004). As noted at the start of this chapter, there are common questions that may be asked of fibres found in criminal casework and in the environment. In forensic fibre examinations, the questions originate from something called the 'Investigative Star'; this can be seen in Figure 9.5.

The investigative star may be used to pose questions of fibres found in the environment too, yet the methods used to establish answers are different. As a science, forensic fibre investigation is more mature that MP studies and, due to this, has experience in applying different approaches to answering questions posed of the samples. Examples of these questions and the resources/methods used to address them can be seen in Table 9.5. The methods used to answer each of these questions are discussed in Sections 9.3.2–9.3.4. Fibre transfer and persistence studies will be discussed together as the factors involved are related.

9.3.2 What Is the Source of the Fibre?

Textile fibres are ubiquitous in today's society, and the sources of fibres found at crime scenes and in the environment are vast.

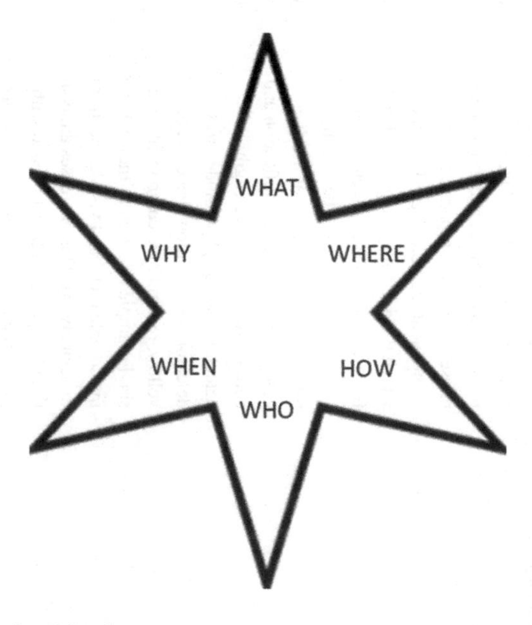

FIGURE 9.5 The investigative star.

TABLE 9.5
Examples of Questions Asked in Fibre Examinations and Resources/Methods Used

Question	Resources and Methods Used
What is the source of the fibre?	Industrial enquiries, reference sets and fibre databases
How common is the fibre in the environment?	Population studies and target fibre studies
How likely is the fibre to shed from a fabric or item?	Fibre transfer studies
How long will a fibre persist in an environment or on a surface?	Fibre persistence studies

In forensic science, the source of fibres is broken down into four main areas:

1. apparel, for example, trousers, shirts, dresses,
2. household textiles, for example, carpets, curtains, bedding,
3. industrial applications, for example, tenting, insulation, filter systems, and
4. other applications, for example, brushes, ropes, twines, tyres, fishing lines (Grayson, 1984).

In MP studies, the possible source of MPs is a common discourse, which is to be expected if MP studies are to move away from solely identifying their presence in the environment to creating effective mitigation activities, yet very few environmental studies attempt to actually identify the specific sources of fibre pollution. Instead, discussions are based around broad causes, activities or environments as potential sources, for example, solid waste and wastewater treatment plant residuals as a source of pollution in landfill leachate (Kabir et al., 2023) or the application of sewage sludge and the wear and tear of tyres as a source of pollution in soils (Surendran et al., 2023). It is rare to see retrospective source identification of fibre samples found in the environment, largely due to the limited resources available for analysts, e.g. fibre databases for comparison and the current limited fibre characteristics examined in environmental studies.

In forensic science, identifying the source of a particular fibre is an important goal; this is generally completed by comparing suspect (unknown) fibres to control (known) fibres from a particular item. In the majority of comparisons, the control samples are taken from particular items of interest in a case, for example, a suspect's garments, but in other instances, these comparisons may be with samples from fibre manufacturers, those contained in reference sets or to data collections created by laboratories.

When comparing fibres to specific control samples, forensic fibre analysts take a minimum of 20–30 fibres from representative areas of the item, ensuring that all fibre types and colours are represented. This may be done by taking the sample from an area believed to be uncontaminated, e.g. a seam of a garment. In addition, any environmental changes to the fibres must also be captured, e.g. colour bleaching by the sun, and therefore, additional control fibres may be taken from the areas exposed to these environmental conditions. To confidently come to the conclusion as to whether unknown fibres may have come from a particular source, the following should be considered:

- analysis methods must be comprehensive enough to fully characterise the fibre and differentiate between those with similar features; these methods are described in Section 9.2.2.
- the commonality of the fibre characteristics must be known to understand how many possible sources of fibres share these features; methods for quantifying this are described in Section 9.3.3.

Identification of a fibre's source via comparison to samples from fibre manufacturers is a much rarer occurrence in forensic fibre examinations as this is very labour intensive and involves mass screening of fibre samples. Due to this, it is reserved for only the most serious of cases. An example of this is the Joan Albert murder investigation in the UK, where fibres found at the scene were compared to numerous samples sent from manufacturers around the world until the source was found by an Italian flock fibre manufacturer (BBC, 2013).

Knowledge of both textile industry and fibre uses, can aid in this type of source identification, for example, some fibre types lend themselves more readily to a certain application more than others. An example of this are nylon fibres, which are the predominant type of fibre used in synthetic carpets; this is due to their excellent wear resistance, low cost, and their ability to be modified in terms of lustre and colour. Some textile fibres are specialist in their applications due to their extreme characteristics, such as high strength and cutting resistance. Kevlar fibres, which are a type of aramid textile fibre, are renowned for these qualities and have been primarily used for protective clothing, such as ballistic protection and items such as tyres for many years (Du Pont de Nemours and Co, 1974). Some applications use multiple types of fibres due to the need for variation in the textile market. The best example of this is in the clothing industry where there is a constant need for new finishes and effects. In environmental studies, this desire for inexpensive, on-trend fabrics is known as 'fast fashion' (Garcia-Ortega et al., 2023). This leads to a huge variety of fibres, with many finishes being present in the environment. These differences can sometimes produce some easily identifiable fibres, for example, the former Imperial Chemical Industries' (ICI) 'Timbrelle' nylon fibre with anti-static finish, but many do not have a particular identifying feature.

Commercial reference sets of fibres such as the Microtrace Forensic Fibre Reference Collection (Microtrace, 2001) are invaluable for training personnel in fibre analysis and provide 200 samples for comparison but are not comprehensive enough to allow source identification of fibres on a day-to-day basis. Instead, a better approach is the development of large collaborative databases which hold information about a vast number of fibre sources. Fibre data collections/databases have been created for many years from fibres that pass through forensic laboratories in connection to normal casework (Home and Dudley, 1980), and there has been a detailed discourse on establishing the most appropriate databases for addressing source-level propositions (Champod et al., 2004).

Several UK forensic laboratories have created their own datasets with the original database being created by the former Forensic Science Service (FSS) from a collection started by Culliford in 1963 (Culliford, 1963). Although this database is no longer in existence after the closure of the FSS, it represented one of the largest databases created; this data collection contained approximately 20,000 fibres (Laing et al., 1987).

Further databases have been produced from casework samples by other organisations, including, the Royal Canadian Mounted Police (RCMP) Forensic Laboratory (Carroll et al, 1988). Data generated from individual research projects can form small collections of fibre information, which are usually started in response to particular casework needs, or research questions, for example, an infrared database created by Tungol et al. in 1990 and the Forensic Automotive Carpet Fiber Identification Database (FACID) (Abendshien et al., 2007). The latter was developed by the Laboratory Division of the Federal Bureau of Investigation (FBI) and holds automotive carpet samples from personally owned vehicles and car manufacturers. The carpet fibres were analysed using microscopy, microspectrophotometry and FTIR techniques and allows questioned fibres to be searched to determine a particular category of vehicle type or to search for vehicle models, year and make to identify possible fibre comparisons (Abendshien et al., 2007).

For both forensic and environmental work, these databases are still not ideal for identifying source-level information as they do not represent the population of fibres due to only containing samples submitted to forensic laboratories. In addition, the data collections are difficult to keep up to date, given that the analysis involved is time consuming. To confidently assign sources to the fibres found, large, easily searched known provenance databases are still required. The development of these can only occur if there is a means of generating a large amount of data on a global scale; this can be achieved by using automated analysis approaches and citizen scientists. In forensic science, citizen-science-type projects are rare but previous research has ascertained that individuals with limited fibre analysis experience can be trained to generate data for fibre databases using traditional analysis methods (Gwinnett, 2009). In addition, work has been done on developing a data structure that can store forensic data obtained by experts from different disciplines and acquired using different instruments (Klaasse et al., 2021) and filtering methods from forensic databases (de Zwart and van der Weerd, 2021). This, along with advances in machine learning for the detection of particulates from images (Wetzer and Lohninger, 2018; Lorenzo-Navarro et al., 2020) and the development of automated microscopy systems (Spectricon, 2023), means that it is only time until these large-scale databases are available for MP work.

9.3.3 How Common Is the Fibre in the Environment?

The proportions of fibre types and or fibre colours and sizes in different environments is a common method of reporting results from MP studies. This information is also sought in forensic fibre examinations. Information about the frequency of fibres enables their rarity to be determined and, therefore, enables analysts to understand the evidential value of the fibres. Basically, the rarer the fibre characteristics, the higher the evidential value in a criminal case as there are limited sources that these fibres could have come from (Roux and Robertson, 2004). Conversely, a mass-produced fibre would hold less evidential value.

In order to identify the frequency of different fibre types, forensic science uses;

- industry enquiries,
- population studies,
- target fibre studies, and
- fibre databases (as described earlier).

Industrial enquiries can involve simple desk-based investigations which may involve gathering publicly available commercial information regarding the quantity of different fibre types being produced per unit time, for example, those reported in textile market reports (Textile Exchange, 2021). This provides high-level figures of textile production quantities which can inform objective opinions about fibre frequencies but does not necessarily provide the granular level information needed to understand how much is used in particular sectors, e.g. garment sales or particular products. This knowledge has attempted to be captured over the years in forensic studies too, such as Grieve in 1992, who reviewed fibre types available from the 1980s to the early 1990s (Grieve, 1992) and Wiggins and Allard in 1987 who investigated the use of fabrics in car seat covers. Car seat fibres have also been surveyed more recently which used more comprehensive analysis techniques than the original 1987 study (Coyle et al., 2012). One of the first large-scale surveys that assessed the frequency of fibre types in clothing worn by the public was conducted by Textile Market Studies (TMS) in 1981 for the FSS (Robertson and Grieve, 1999). The public were interviewed about their clothing, including colour and fibre content. This generated a large amount of information but did not give any information about fibre morphology, optical characteristics or any objective assessment of colour. Since then, data collections of this kind have utilised mail-order catalogues (Biermann and Grieve, 1996a, b; Biermann and Grieve, 1998) and, more recently in 2018, online clothing sales sites (e.g. ASOS) at Staffordshire University, UK.

The industrial enquiries mentioned above are useful in determining the most common fibres used by garment manufacturer, but they do not provide information regarding the prevalence of shed fibres in different populations. For this, forensic science utilises ‘population studies’.

Population studies attempt to quantify the number of fibres of different types (and more importantly, different characteristics) in different environments; these environments are typically different surfaces, given that it is these that are sampled at crime scenes. One of the first of these studies was conducted by Fong and Inami in 1986 where they examined 763 fibres recovered from the surfaces of 40 garments and from their data were able to establish what the most common chance matches were, meaning which fibre types were most commonly found together that had been left on the surface of garments by chance. Since then, population studies have included those investigating fibres found on car seats (Roux and Margot, 1997), undergarments (Grieve and Dunlop, 1992), cinema seats in the Sydney region (Cantrell et al., 2001), bus seats (Was-Gubala, 2000), pub seats (Kelly and Griffin, 1998), outdoor surfaces (Grieve and Biermann, 1997), t-shirts (Marnane et al., 2006; Massonnet et al., 1998) and more recently parapets of high-rise housing in Singapore (Eng and Koh, 2022). Many of these population studies were able to identify the commonest types of fibre and the commonest combinations between fibres, for example, Roux and Margot in 1997 found that the most common fibres were colourless, blue and black cotton but that man-made fibres were relatively rare.

'Target fibre studies' are subtly different from population studies as instead of quantifying any fibres found on different surfaces, it aims to quantify how many of a particular type of fibre can be found (i.e. a target fibre) in different environments. These studies sometimes target very specific fibres, for example, blue wool fibres from Marks and Spencers (UK clothing brand) pullovers or red acrylic fibres from a particular scarf sold in quantities of 5,000–10,000 (Roux and Robertson, 2004). Alternatively, the target is more general, for example, wool and acrylic fibres found in head hair (Cook et al., 1997) and red acrylic and green cotton fibres on cinema seats and car seats (Palmer and Chinherende, 1996).

In forensic fibre examinations, the data and observations from the studies mentioned above are commonly used to create statistical evaluations of fibres evidence. When interpreting fibres evidence for court, it is preferable to give a statistical evaluation beyond just counts as this provides a more objective conclusion to what would otherwise be just an expert's opinion. An example of this is that it is thought that cotton fibres have little evidential value due to being difficult to discriminate from each other and being very common (Grieve et al., 1988). This assessment of cotton fibres is a subjective opinion; although backed up by experience, it does not give a statistical probability of how common the fibre is, i.e. the probability of finding it at a particular crime scene.

This ability to expand on count/concentration data to probabilistically determine the likelihood that a particular fibre may be present in an environment also has potential for MP studies. In forensic fibre examinations, this is achieved by two main methods; statistical probabilities based on fibre frequencies, expressed as statistical odds, or the Bayesian approach which considers the ratio of the likelihood of the results being caused by two different hypotheses, for example, fibre evidence being present if the suspect did or did not commit the crime. In environmental studies, these hypotheses may be formed from likelihoods of given fibres being present in an environment given a particular scenario or activity, for example, likelihood of there being fibres given fishing activity vs likelihood of their being fibres given there being no fishing activity.

The use of these approaches holds challenges, primarily due to the limited large datasets of fibre frequencies and source-level information needed, which was discussed in Section 9.3.2. But the Bayesian approach still holds potential for MP studies, and lessons can be learned from forensic science of its benefits and pitfalls (Champod and Taroni, 1997; Lepot et al., 2023).

9.3.4 How Likely Is the Fibre to Shed from Fabric or Item and How Long Will a Fibre Persist in an Environment or on a Surface?

In criminal investigations, fibres' evidence is particularly useful in identifying how and when individuals were in contact with each other or other objects. This provides something called 'activity level' information which provides investigators evidence of what and when something happened in a crime. Activity level information is gathered by quantifying how many particles transfer under specific conditions and how long those particles are retained via transfer and persistence studies, respectively (Robertson and Lloyd, 1984; Pounds and Smalldon, 1975a; Siegel, 1997; Grieve et al., 1989; Lowrie and Jackson, 1994; Robertson et al., 1982; Sneath et al., 2019; Slot et al., 2017; Suzuki et al., 2009).

Since 1975, there have been 32 published fibre transfer and persistence studies, all of which provide useful information for those interested in microfibre pollution, whether it is to better understand shedding of fibres from textiles or their movement in the environment. Only recently has the mechanisms of how fibres shed from textiles been considered from an MP pollution point of view, yet the factors which affect the transfer of fibres from textiles has been considered for over 35 years in forensic science (Pounds and Smalldon, 1975a).

In the context of forensic science, 'transfer' is defined as the exchange of material between the donor item or person to a recipient surface or item or environment. For fibres, this can include transfer between clothing or objects either directly, known as primary transfer, or indirectly, known as secondary transfer (Hong et al., 2014; Palmer et al., 2017). Whereas 'persistence' refers to the time in which the donor fibres remain on a recipient surface following transfer (Robertson et al., 1982; Ashcroft et al., 1988). Whether a transfer and/or persistence experiment, these studies generally take one or more target textiles which they firstly analyse for their properties, such as fibre type, weave type, elasticity and tightness of weave. Then, these studies either use realistic simulated contacts or controlled contacts between the textiles and another recipient surface(s). Fibres from the original textile are recovered from the recipient surface and counted under magnification or counted in-situ (normally if a fluorescently dyed donor fibre is used). If secondary transfer is also being investigated, the initial recipient surface is contacted again with a second recipient surface and any transferred donor fibres, counted in the same manner as used in the primary transfer experiment. If persistence of the fibres is being investigated, the initial recipient surface is instead subjected to a controlled and monitored activity, for example, normal daily activity (as seen in Palmer and Burch, 2009), or exposed to a particular environment, for example, outdoor conditions (as seen in Krauss and Hildebrand, 1995; Akulova et al., 2002; Palmer and Polwarth, 2011; Prod'Hom et al., 2021), and then remaining donor fibres are counted at different time frames. This process is visualised in Figure 9.6 with examples of factors investigated in previous studies.

This movement of fibres in forensic work is important to understand and quantify if activities are to be reconstructed. In MP studies, this movement of fibres may be thought of as the transport routes from source to sink, which has been previously attempted to be investigated in environmental studies such as Xia et al. (2022).

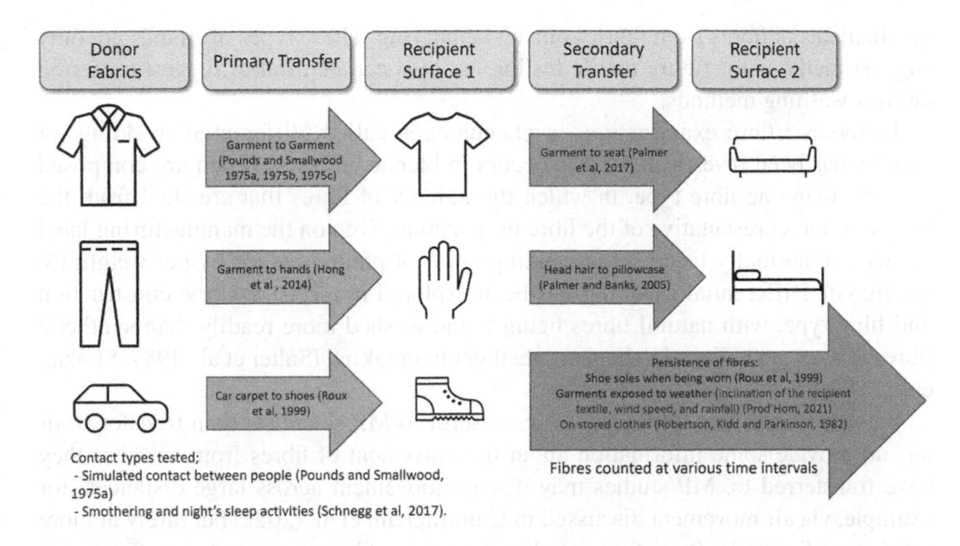

FIGURE 9.6 Common methods for forensic fibre transfer and persistence experiments with examples.

For MP scientists, transfer studies provide information about the mechanisms of fibre fragmentation from textiles under different activities, which may be used to predict common sources of microfibre pollution. Some of these studies, such as those that investigate fabric construction and fibre fragmentation (Pounds and Smalldon, 1975a–c), may be of particular importance for scientists interested in creating sustainable fabrics. These seminal studies provided information on how textile fibres were either pulled out of loose yarns or were directly fragmented when in contact with other fabric surfaces. More recently, statistical approaches have identified that fibre length and number of detached fibres on the textile surface are the most important factors affecting transfer, with fibre diameter and type being secondary to these (Sermier, 2007). Since these studies, methods have been developed to assess the shedability of textiles, including visually observing the surface, taping the surface and using a comparison scale (De Wael et al., 2010) or using a controlled force experiment (Robertson and Grieve, 1999; Coxon et al., 1992). These studies provide MP scientists practical tools to investigate textiles, which currently are not used in experiments. When suggesting simple actions which could reduce fibre pollution, forensic studies which have assessed the effect of fabric softener are useful resources. A study by Bresee and Annis (1991) utilised a transfer abrasion tester to assess the direct effect of fabric softener on fibre transfer, as well as its interactive effects with fibre denier, fabric weave, knit type and fabric thread count. In a similar style to forensic fibre transfer experiments, investigations into the effect of different washing and drying conditions of garments upon fibre shedding have been studied by MP scientists. The effect of handwash and laundry softeners (Rathinamoorthy and Raja Balasaraswathi, 2022) and fabric type and washing conditions (Napper and Thompson, 2016; Sillanpää and Sainio, 2017; De Falco et al., 2019; Zambrano et al., 2019; Cesa et al., 2020) provide invaluable information regarding the transfer mechanisms of fibres from fabrics during laundering. These types of studies not only suggest methods for future textile testing but also can be used to suggest preferred clothes-washing methods.

In forensic fibre examinations, a phenomenon called 'differential shedding' of textiles has been investigated. This occurs in blended fabrics, which are composed of more than one fibre type, in which the number of fibres that are shed from the textile is not representative of the fibre proportions given on the manufacturing label [which are normally listed in descending order of predominance by per weight (% per mass)]. Differential shedding has been explored in terms of fabric construction and fibre type, with natural fibres being found to shed more readily than synthetic fibres as they are believed to be more resilient to breaking (Salter et al., 1987; Skokan et al., 2020).

Persistence studies, although are less useful to MP scientists than transfer studies, do provide some information about the movement of fibres from surfaces they have transferred to. MP studies may discuss movement across large distances, for example, via air movement discussed in Cunningham et al. (2022) but rarely at more local, specific levels. Persistence studies discussing fibre movement away from textiles into the immediate environment around them may be useful when identifying local sources of pollution or activities which move fibres around between surfaces and air. Forensic persistence studies, which have assessed persistence of fibres on fabrics submerged in water (Lepot et al., 2015; Lepot and Driessche, 2015), may be particularly valuable.

9.4 CONCLUSION

Studies in forensic fibre examination extend over 60 years and, during this time, have invested resources into developing and testing improved methods of recovery, analysis and interpretation for the courts of law. The prolific nature of forensic science studies means that there is considerable knowledge that can be utilised in other disciplines, including MP studies. MP studies ask similar research questions to forensic fibre work, including what is the fibre? How many fibres are present? What is the source of the fibre? And how prevalent is it in given environments? In addition, questions regarding the best method(s) to use and how to prevent contamination during analysis are standard in forensic science. The growing number of studies in MPs can also start providing additional information on the presence of fibres in environments for forensic scientists, as seen recently in an Interpol review (Lepot et al., 2023). This knowledge exchange holds much promise for both disciplines in the future.

REFERENCES

Abbasi, S., & Turner, A. (2021). Human exposure to microplastics: A study in Iran. *Journal of Hazardous Materials*, 403.

Abendshien, L. C., Brown, C. J., Williams, D. K., & Shaw, S. (2007). Forensic automotive carpet fiber identification database (FACID) – Preliminary validation and evaluation. In *Proceedings of the Trace Evidence Symposium*, August 2007, Florida.

Adolf, F. P. (1986). Microscope photometry and its application in forensic science – The use for the examination of transparent objects. In *Proceedings of the 6th Meeting of the Scandinavian Forensic Science Laboratories* (pp. 1–10).

Akulova, V., Vasilyauskiene, D., & Talalienė, D. (2002). Further insights into the persistence of transferred fibres on outdoor clothes. *Science and Justice*, 42(3), 165–171.

Ashcroft, C. M., Evans, S., & Tebbett, I. R. (1988). The persistence of fibres in head hair. *Journal of the Forensic Science Society*, 28, 289–293.

Athey, S.N., Adams, J.K., Erdle, L.M., Jantunen, L.M., Helm, P.A., Finkelstein, S.A., & Diamond, M.L (2020) The widespread environmental footprint of indigo denim microfibers from blue jeans. *Environmental Science & Technology Letters*, 7(11), 840–847.

BBC. (2013, August 13). Simon Hall confesses to Joan Albert murder 12 years on. Retrieved from https://www.bbc.co.uk/news/uk-england-suffolk-23611821

Beattie, I. B., Roberts, J. L., & Dudley, R. J. (1981). Thin layer chromatography of dyes extracted from polyester, nylon and polyacrylonitrile fibres. *Forensic Science International*, 17.

Bennett, S., Roux, C. P., & Robertson, J. (2010). The significance of fibre transfer and persistence – A case study. *Australian Journal of Forensic Sciences*, 42(3), 221–228.

Bergmann, M., Mützel, S., Primpke, S., Tekman, M. B., Trachsel, J., & Gerdts, G. (2019). White and wonderful? Microplastics prevail in snow from the Alps to the Arctic. *Science Advances*, 5(8).

Biermann, T. W., & Grieve, M. C. (1996a). A computerised database of mail order garments: A contribution towards estimating the frequency of fibre types found in clothing. Part 1: The system and its operation. *Forensic Science International*, 77, 65–73.

Biermann, T. W., & Grieve, M. C. (1996b). A computerised database of mail order garments: A contribution towards estimating the frequency of fibre types found in clothing. Part 2: The content of the data bank and its statistical evaluation. *Forensic Science International*, 77, 75–91.

Biermann, T. W., & Grieve, M. C. (1998). A computerised database of mail order garments: A contribution towards estimating the frequency of fibre types found in clothing. Part 3: The content of the databank – Is it representative? *Forensic Science International*, 95, 117–131.

Bloss, F. D. (1981). *The Spindle Stage: Principles and Practice*. Cambridge University Press.

Bresee, R.R., & Annis, P.A. (1991). Fiber transfer and the influence of fabric softener. *Journal of Forensic Sciences*, 36, 1699–1713.

Buteler, M., Fasanella, M., Alma, A. M., Silva, L. I., Langenheim, M., & Tomba, J. P. (2023). Lakes with or without urbanization along their coasts had similar level of microplastic contamination, but significant differences were seen between sampling methods. *Science of the Total Environment*, 866.

Cantrell, S., Roux, C., Maynard, P., & Robertson, J. (2001). A textile fibre survey as an aid to the interpretation of fibre evidence in the Sydney region. *Forensic Science International*, 123(1), 48–53.

Carroll, G. R., Lalonde, W. C., Gaudette, B. D., Hawley, S. L., & Hubert, R. S. (1988). A computerized database for forensic textile fibers. *Journal of the Canadian Society of Forensic Science*, 21(1–2), 1–10.

Causin, V., Marega, C., Schiavone, S., & Marigo, A. (2005). Employing glass refractive index measurement (GRIM) in fiber analysis: A simple method for evaluating the crystallinity of acrylics. *Forensic Science International*, 149, 193–200.

Cesa, F.S., Turra, A, Checon, H.H., Leonardi, B, & Baruque-Ramos, J., (2020) Laundering and textile parameters influence fibers release in household washings. *Environmental Pollution,* 257, 113553.

Chabli, S. (2001). Scene of crime evidence: Fibres. In *13th Interpol Forensic Science Symposium*, Lyon France.

Champod, C., Evett, I. W., & Jackson, G. (2004). Establishing the most appropriate databases for addressing source level propositions. *Science & Justice*, 44(3), 153–164.

Champod, C., & Taroni, F. (1997). Bayesian framework for the evaluation of fibre transfer evidence. *Science and Justice*, 37(2), 75–83.

Christian, A. E., & Köper, I. (2023). Microplastics in biosolids: A review of ecological implications and methods for identification, enumeration, and characterization. *Science of the Total Environment*, 864.

Claessens, M., De Meester, S., Van Landuyt, L., De Clerck, K., & Janssen, C. R. (2011). Occurrence and distribution of microplastics in marine sediments along the Belgian coast. *Marine Pollution Bulletin*, 62(10), 2199–2204.

Cook, R., Webb-Salter, M. T., & Marshall, L. (1997). The significance of fibres found in head hair. *Forensic Science International*, 87, 155–160.

Coxon, A., Grieve, M., & Dunlop, J. (1992). A method of assessing the fibre shedding potential of fabrics. *Journal of the Forensic Science Society*, 32(2), 151–158.

Coyle, T. (2020). Practitioner narrative part 2: Fibres stands alone – R-v-Everson a case in double jeopardy. *Forensic Science International: Reports*, 2.

Coyle, T., Jones, J., Shaw, C., & Friedrichs, R. (2012). Fibres used in the construction of car seats – An assessment of evidential value. *Science & Justice*, 52(4), 259–267.

Cui, T., Shi, W., Wang, H., & Lihui, A. N. (2022). Standardizing microplastics used for establishing recovery efficiency when assessing microplastics in environmental samples. *Science of the Total Environment*, 827.

Culliford, B. J. (1963). The multiple entry card index for the identification of synthetic fibres. *Journal of the Forensic Science Society*, 4, 91–97.

Cunningham, M., Rico Seijo, N., Altieri, K., Audh, R., Burger, J., Bornman, T., Fawcett, S., Gwinnett, C., Osborne, A., & Woodall, L. (2022). The transport and fate of microplastic fibres in the Antarctic: The role of multiple global processes. *Frontiers in Marine Science*, 9.

DeBattista, R., Tidy, H., & Clark, M. (2019). Investigating the effect photodegradation has on natural fibres at a microscopic level. *Science & Justice*, 59(5), 498–502.

De Falco, F., Di Pace, E., Cocca, M., & Avella, M. (2019). The contribution of washing processes of synthetic clothes to microplastic pollution. *Scientific Reports*, 9, 6633.

Dellisanti, W., Leung, M. M. L., Lam, K. W. K., Wang, Y., Hu, M., Lo, H. S., & Fang, J. K. H. (2023). A short review on the recent method development for extraction and identification of microplastics in mussels and fish, two major groups of seafood. *Marine Pollution Bulletin*, 186.

Delly, J. G. (2003). Sénarmont compensation: How to accurately measure small relative retardations (0-1). *Modern Microscopy Journal.*

De Wael, K. (2021). *Microscopy in Forensic Fibre Examinations – A Practical Photo Atlas and Training Tool.* Cobalt Blue Coach.

De Wael, K., Lepot, L., Lunstroot, K., & Gason, F. (2010). Evaluation of the shedding potential of textile materials. *Science & Justice*, 50(4), 192–194.

De Zwart, D., & van der Weerd, J. (2021). Extraction of the relevant population from a forensic database. *Science & Justice*, 61(4), 419–425.

Ding, J., Sun, C., Li, J., Shi, H., Xu, X., Ju, P., Jiang, F., & Li, F. (2022). Microplastics in global bivalve mollusks: A call for protocol standardization. *Journal of Hazardous Materials*, 438.

Du, R., Sun, X., Lin, H., & Pan, Z. (2022). Assessment of manta trawling and two newly-developed surface water microplastic monitoring techniques in the open sea. *Science of the Total Environment*, 842.

Du Pont de Nemours and Co (1974). *Properties of Industrial Filament Yarns of Kevlar Aramid Fiber for Tires and Mechanical Rubber Goods.* Bulletin K-1.

Eng, V., & Koh, S. (2022). The population of textile fibres on parapets of high-rise housing in Singapore. *Forensic Science International*, 336.

European Fibres Group, ENFSI. (2001). The manual of best practice for the forensic examination of fibres.

Faber, N. M., Sjerps, M., Leijenhorst, H., & Maljaars, S. E. (1999). Determining the optimal sample size in forensic casework – With application to fibres. *Science and Justice*, 39(2), 113–122.

Fagiano, V., Compa, M., Alomar, C., Rios-Fuster, B., Morató, M., Capó, X., & Deudero, S. (2023). Breaking the paradigm: Marine sediments hold two-fold microplastics than sea surface waters and are dominated by fibers. *Science of the Total Environment*, 858(Part 1).

Fong, W., & Inami, H. (1986). Results of a Study To Determine The Probability Of A Chance Match Occurrences Between Fibres Known To Be From Different Sources. *Journal of Forensic Science*, 31, 65–72.

Forensic Science Regulator. (2020). Guidance the control and avoidance of contamination in scene examination involving DNA evidence recovery FSR-G-206 issue 2.

Forster, N. A., Wilson, S. C., & Tighe, M. K. (2022). Examining sampling protocols for microplastics on recreational trails. *Science of the Total Environment*, 818.

Forster, N. A., Wilson, S. C., & Tighe, M. K. (2023). Trail running events contribute microplastic pollution to conservation and wilderness areas. *Journal of Environmental Management*, 331.

Frias, J. P. G. L., & Nash, R. (2019). Microplastics: Finding a consensus on the definition. *Marine Pollution Bulletin*, 138, 145–147.

Frias, J., Pagter, E., Nash, F., & O'Connor, I. (2018). Standardised protocol for monitoring microplastics in sediments. JPI-Oceans. https://doi.org/10.13140/RG.2.2.36256.89601/1.

Galvão, L. S., Ferreira, R. R., Fernandes, E. M. S., Correia, C. A., Valera, T. S., Rosa, D. D. S., & Wiebeck, H. (2023). Analysis of selective fluorescence for the characterization of microplastic fibers: Use of a Nile Red-based analytical method to compare between natural and synthetic fibers. *Journal of Hazardous Materials*, 443(Part A).

Garcia-Ortega, B., Galan-Cubillo, J., Llorens-Montes, F. J., & de-Miguel-Molina, B. (2023). Sufficient consumption as a missing link toward sustainability: The case of fast fashion. *Journal of Cleaner Production*, 399.

Gaudette, B. D. (1988). The forensic aspects of textile fibre examination. In Saferstein, R. (Ed.), *Forensic Science Handbook* (Vol. 2). Prentice Hall.

Golding, G. M., & Kokot, S. (1990). Comparison of dyes from transferred fibres by scanning densitometry. *Journal of Forensic Science*, 35(6), 1310–1322.

Grayson, J. (1984). *Encyclopaedia of Textile, Fibers and Non Woven Fabrics*. John Wiley & Sons.

Grieve, M. C. (1983). The use of melting point and refractive index determination to compare colourless polyester fibres. *Forensic Science International*, 22, 31–48.

Grieve, M. C. (1992). An index of textile fibres introduced during the last decade. *Journal of the Forensic Science Society*, 32, 35–47.

Grieve.M.C. (1994). Fibers and forensic science-new ideas, developments and techniques. *Forensic Science Review*, 6 (1), 59–80.

Grieve, M. C. (2000). Back to the future – 40 years of fibre examinations in forensic science. *Science and Justice*, 40(2), 93–99.

Grieve, M. C., & Biermann, T. (1997). The population of coloured textile fibres on outdoor surfaces. *Science and Justice*, 37(4), 231–239.

Grieve, M. C., & Deck, S. (2002). Black cellulosic fibres – A "Bete Noire". *Science and Justice*, 42(2), 81–88.

Grieve, M., & Dunlop, J. (1992). A practical aspect of the bayesian interpretation of fibre evidence. *Journal of the Forensic Science Society*, 32, 169–175.

Grieve, M. C., Dunlop, J., & Haddock, P. S. (1988). An assessment of the value of blue, red and black cotton fibres in forensic science investigations. *Journal of Forensic Science*, 33, 1332–1334

Grieve, M.C., Dunlop, J., & Haddock, P.S. (1989). Transfer experiments with acrylic fibres. *Forensic Science International*, 40, 267–277.

Griffin, R. M. E., Speers, S. J., Elliot, C., Todd, N., Sogomo, W., & Kee, T. G. (1994). An improved high performance liquid chromatography system for the analysis of basic dyes in forensic casework. *Journal of Chromatography*, 674, 271–280.

Gwinnett, C. (2009). The use of inexperienced personnel in the production of a synthetic fibres database (Doctoral dissertation, Staffordshire University).

Gwinnett, C., & Miller, R. Z. (2021). Are we contaminating our samples? A preliminary study to investigate procedural contamination during field sampling and processing for microplastic and anthropogenic microparticles. *Marine Pollution Bulletin*, 173(Part B), 112932.

Gwinnett, M. B., Osborne, A. O., & Jackson, A. R. W. (2021). The application of tape lifting for microplastic pollution monitoring. *Environmental Advances*, 5, 100066.

Hall, A. J. (1975). The Standard Handbook of Textiles (8th ed.). Butterworths.

Hartshorne, A. W., & Laing, D. K. (1991). Microspectrofluorimetry of fluorescent dyes and brighteners on single textile fibres: Part 1 – Fluorescence emission spectra. *Forensic Science International*, 51, 203–220

Hartshorne, N. H., & Stuart, A. (1971). Crystals and the polarizing microscope. *Transactions of the American Microscopical Society*, 90(2), 256–257.

Heuse, O., & Adolf, F. P. (1982). Non-destructive identification of textile fibres by interference microscopy. *Journal of the Forensic Science Society*, 22(2), 103–122.

Hidalgo-Ruz, V., Gutow, L., Thompson, R. C., & Thiel, M. (2012). Microplastics in the marine environment: A review of the methods used for identification and quantification. *Environmental Science & Technology*, 46(6), 3060–3075.

Holland, M. C. (1960). Trilobal textile filament (U.S. patent no. 2939201).

Home, J. M., & Dudley, R. J. (1980). A summary of data obtained from a collection of fibres from casework materials. *Journal of the Forensic Science Society*, 20, 253–261.

Home, J. M., & Dudley, R. J. (1981). Thin layer chromatography of dyes extracted from cellulosic fibres. *Forensic Science International*, 17, 71–78.

Hong, S., Han, A., Kim, S., Son, D., & Min, H. (2014). Transfer of fibres on the hands of living subjects and their persistence during hand washing. *Science & Justice*, 54(6), 451–458.

Hufenus, R., Gooneie, A., Sebastian, T., Simonetti, P., Geiger, A., Parida, D., Bender .K., Schäch .G., & Clemens, F. (2020). Antistatic fibers for high-visibility workwear: Challenges of melt-spinning industrial fibers. *Materials*, 13(11), 2645.

Hussain, C., Rawtani, D., Pandey, G., & Tharmavaram, M. (2020). *Handbook of Analytical Techniques for Forensic Samples*. Elsevier.

Jackson, A. R. W., & Jackson, J. M. (2017). *Forensic Science* (4th ed.). Pearson Education Ltd.

Jaffe, M., & Menczel, J. (2020). *Thermal Analysis of Textiles and Fibers*. Woodhead Publishing.

Jiang, S., Wang, J., Wu, F., Xu, S., Liu, J., & Chen, J. (2023). Extensive abundances and characteristics of microplastic pollution in the karst hyporheic zones of urban rivers. *Science of the Total Environment*, 857(Part 3), 144309.

Johri, M. C. (1979). Identification of some synthetic fibres by their birefringence. *Journal of Forensic Science*, 24(3), 692–697.

Jones, J., & Coyle, T. (2010). Automotive flock and its significance in forensic fibre examinations. *Science & Justice*, 50(2), 77–85.

Jones, Z. V., Gwinnett, C., & Jackson, A. R. W. (2019). The effect of tape type, taping method and tape storage temperature on the retrieval rate of fibres from various surfaces: An example of data generation and analysis to facilitate trace evidence recovery validation and optimisation. *Science & Justice*, 59(3), 268–291.

Kabir, M. S., Wang, H., Luster-Teasley, S., Zhang, L., & Zhao, R. (2023). Microplastics in landfill leachate: Sources, detection, occurrence, and removal. *Environmental Science and Ecotechnology*.

Kanhai, L. K., Gårdfeldt, K., Lyashevska, O., Hassellöv, M., Thompson, R. C., & O'Connor, I. (2018). Microplastics in sub-surface waters of the Arctic Central Basin. *Marine Pollution Bulletin*, 130, 8–18.

Kelly, E., & Griffin, R. M. E. (1998). A target fibre study on seats in public houses. *Science and Justice*, 37, 39–44.

Kim, L., Kim, D., Kim, S. A., Lee, T.-Y., An, Y.-J. (2022). Are your shoes safe for the environment? – Toxicity screening of leachates from microplastic fragments of shoe soles using freshwater organisms. *Journal of Hazardous Materials*, 421.

Klaasse, J. R., Alewijnse, L. C., & van der Weerd, J. (2021). TraceBase; A database structure for forensic trace analysis. *Science & Justice*, 61(4), 410–418.

Klingelhofer, D., Braun, M., Quarcoo, D., Brüggmann, D., & Groneberg, D.A (2020). Research landscape of a global environmental challenge: Microplastics. *Water Research*, 170.

Kotar, S., McNeish, R., Murphy-Hagan, C., Renick, V., Lee, C.-F. T., Steele, C., Lusher, A., Moore, C., Minor, E., Schroeder, J., Helm, P., Rickabaugh, K., De Frond, H., Gesulga, K., Lao, W., Munno, K., Thornton Hampton, L.M., Weisberg, S.B., Wong, C.S., Amarpuri, G., Andrews, R.C., Barnett, S.B., Christiansen, S., Cowger, W., Crampond, K., Du, F., Gray, A.B., Hankett, J., Ho, K., Jaeger, J., Lilley, C., Mai, L., Mina, O., Lee, E., Primpke, S., Singh, S., Skovly, J., Slifko, T., Sukumaran, S., van Bavel, B., Van Brocklin, J., Vollnhals, F., Wu, C., & Rochman, C. (2022). Quantitative assessment of visual microscopy as a tool for microplastic research: Recommendations for improving methods and reporting. *Chemosphere*, 308(Part 3)

Krauss, W., & Hildebrand, U. (1995). Fibre persistence on garments under open-air conditions. In *Proceedings of European Fibres Group Meeting* (pp. 32–36).

Laing, D. K., Hartshorne, A. W., Cook, R., & Robinson, G. (1987). A fibre data collection for forensic scientists – Collection and examination methods. *Journal of Forensic Sciences*, 32(2), 364–369.

Lepot, L., & Driessche, T.V. (2015). Fibre persistence on immersed garment-Influence of water flow and stay in running water. *Science and Justice*, 55(6), 431–436.

Lepot, L., Driessche, T.V., Lunstroot, K., Gason, F., & De Wael, K. (2015). Fibre persistence on immersed garment-Influence of knitted recipient fabrics. *Science and Justice*, 55(4), 248–253.

Lepot, L., Vanhouche, M., Vanden Driessche, T., & Lunstroot, K. (2023). Interpol review of fibres and textiles 2019–2022. *Forensic Science International: Synergy*, 6.

Liu, K., Wang, X., Fang, T., Xu, P., Zhu, L., & Li, D. (2019). Source and potential risk assessment of suspended atmospheric microplastics in Shanghai. *Science of the Total Environment*, 675, 462–471. ISSN 0048-9697, https://doi.org/10.1016/j.scitotenv.2019.04.110.

Liu, J., Yang, Y., Ding, J. Zhu, B., & Gao, W (2019) Microfibers: A preliminary discussion on their definition and sources. *Environmental Science and Pollution Research*, 26, 29497–29501.

Lorenzo-Navarro, J., Castrillón Santana, M., Nielsen, E., Zarco, B., Herrera, A., Martinez, I., & Gómez, M. (2020). Deep learning approach for automatic microplastics counting and classification. *Science of the Total Environment*, 765.

Lowrie, C.N., & Jackson, G. (1994). Secondary transfer of fibres. *Forensic Science International*, 64(2–3), 73–82.

Luo, X., Wang, Z., Yang, L., Gao, T., & Zhang, Y. (2022). A review of analytical methods and models used in atmospheric microplastic research. *Science of the Total Environment*, 828.

Lusher, A. L., Bråte, I. L. N., Munno, K., Hurley, R. R., & Welden, N. A. (2020). Is it or isn't it: The importance of visual classification in microplastic characterization. *Applied Spectroscopy*, 74(9).

Lusher, A. L., Welden, N. A., Sobral, P., & Cole, M. (2017). Sampling, isolating and identifying microplastics ingested by fish and invertebrates. *Analytical Methods*, 9.

Macrae, R., & Smalldon, K. W. (1979). The extraction of dyestuffs from single wool fibers. *Journal of Forensic Science*, 24, 109–116.

Marnane, R. N., Elliot, D. A., & Coulson, S. A. (2006). A pilot study to determine the background population of foreign fibre groups on a cotton/polyester t-shirt. *Science and Justice*, 46(4), 215–220.

Massonnet, G., Schiesser, M., & Champod, C. (1998). Population of textile fibres on white T-shirts. In *Proceedings of the 6th European Fibres Group Meeting*, Dundee.

Microtrace. (2001). Forensic fibre reference collection. Retrieved from https://www.microtrace.com/service/forensic-fiber-reference-collection/

Modica, L., Lanuza, P., & García-Castrillo, G. (2020). Surrounded by microplastic, since when? Testing the feasibility of exploring past levels of plastic microfibre pollution using natural history museum collections. *Marine Pollution Bulletin*, 151.

Sermier, F. (2007). Etude Des Mécanismes De Transfert Des Fibres En Sciences Forensiques. (Thèse de doctorat, Université de Lausanne, Lausanne, Suisse).

Morgan, S. L., Vann, B. C., Baguley, B. M., & Stefan, A. R. (2007). Advances in discrimination of dyed textile fibers using capillary electrophoresis/mass spectrometry. In *Proceedings of Trace Evidence Symposium*, Florida.

Napper, I., & Thompson, R.C (2016) Release of synthetic microplastic plastic fibres from domestic washing machines: Effects of fabric type and washing conditions. *Marine Pollution Bulletin*, 112, 39–45.

Nehse, K. (2007). New developments in textile industries – What fiber modifications can we expect to find in the near future. In *Proceedings of the Trace Evidence Symposium*, Florida.

Niu, S., Wang, T., & Xia, Y. (2023). Microplastic pollution in sediments of urban rainwater drainage system. *Science of the Total Environment*, 868.

Palenik, S. (2004). Analytical techniques: Microscopy. In Siegel, J., Saukko, P., & Knupfer, G. (Eds.), *Encyclopaedia of Forensic Sciences* (pp. 161–166). Elsevier.

Palenik, S., & Fitzimons, C. (1990a). Fibre cross-sections: Part 1. *Microscope*, 38, 187–195.

Palenik, S., & Fitzimons, C. (1990b). Fibre cross-sections: Part II. *Microscope*, 38, 313–320.

Palmer, R., & Banks, M. (2005). The secondary transfer of fibres from head hair. *Science & Justice*, 45(3), 123–128.

Palmer, R., & Burch, H. J. (2009). The population, transfer and persistence of fibres on the skin of living subjects. *Science & Justice*, 49(4), 259–264.

Palmer, R., & Chinherende, V. (1996). A target fibre study using cinema and car seats as recipient items. *Journal of the Forensic Science Society*, 41, 802–803.

Palmer, R., & Polwarth, G. (2011). The persistence of fibres on skin in an outdoor deposition crime scene scenario. *Science and Justice*, 51(4), 187–189.

Palmer, R., Sheridan, K., Puckett, J., Richardson, N., & Lo, W. (2017). An investigation into secondary transfer – The transfer of textile fibres to seats. *Forensic Science International*, 278, 334–337.

Pérez-Guevara, F., Roy, P. D., Kutralam-Muniasamy, G., & Shruti, V. C. (2022). Coverage of microplastic data underreporting and progress toward standardization. *Science of the Total Environment*, 829.

Picó, Y., & Barceló, D. (2020). Pyrolysis gas chromatography-mass spectrometry in environmental analysis: Focus on organic matter and microplastics. *TrAC – Trends in Analytical Chemistry*, 130.

Pounds, C. A. (1975). The recovery of fibres from the surface of clothing for forensic examinations. *Journal of the Forensic Science Society*, 15, 127–132.

Pounds, C. A., & Smalldon, K. W. (1975a). The transfer of fibres between clothing materials during simulated contacts and their persistence during wear: Part 1 – Fibres transference. *Journal of the Forensic Science Society*, 15, 17–27.

Pounds, C. A., & Smalldon, K. W. (1975b). The transfer of fibres between clothing materials during simulated contacts and their persistence during wear: Part II – Fibre persistence. *Journal of the Forensic Science Society*, 15(1), 29–37.

Pounds, C. A., & Smalldon, K.W. (1975c). The transfer of fibres between clothing materials during simulated contacts and their persistence during wear: Part III - A preliminary investigation of the mechanisms involved. *Journal of the Forensic Science Society*, 15(3), 197–207.

Prata, J. C., Castro, J. L., da Costa, J. P., Duarte, A. C., Cerqueira, M., & Rocha-Santos, T. (2020). An easy method for processing and identification of natural and synthetic microfibers and microplastics in indoor and outdoor air. *MethodsX*, 7, 100762.

Prod'Hom, A., Werner, D., Lepot, L., & Massonnet, G. (2021). Fibre persistence on static textiles under outdoor conditions. *Forensic Science International*, 318.

Raines, M. A. (1932). A mechanism for reducing vibration in the use of the filar micrometer eyepiece. *Transactions of the American Microscopical Society*, 51(2), 157–158.

Rathinamoorthy, R., & Balasaraswathi, S.R (2022) Investigations on the impact of handwash and laundry softener on microfiber shedding from polyester textiles. *The Journal of the Textile Institute*, 113(7), 1428–1437.

Resua, R. (1980). A semi-micro technique for the extraction and comparison of dyes in textile fibers. *Journal of Forensic Science*, 25, 168–173.

Rivers, M. L., Gwinnett, C., & Woodall, L. C. (2019). Quantification is more than counting: Actions required to accurately quantify and report isolated marine microplastics. *Marine Pollution Bulletin*, 139, 100–104.

Robertson, J., & Grieve, M. (1999). *Forensic Examination of Fibres*. Taylor and Francis.

Robertson, J., Kidd, C. B. M., & Parkinson, H. M. P. (1982). The persistence of textile fibres transferred during simulated contacts. *Journal of the Forensic Science Society*, 22, 353–360.

Robertson, J., & Lloyd, A. (1984). Redistribution of textile fibres following transfer during simulated contacts. *Journal of the Forensic Science Society*, 24, 3–7.

Robertson, J., Roux, C., & Wiggins, K. (Eds.). (2018). *Forensic Examination of Fibres* (3rd ed.). CRC Press.

Robinson, P., & Davidson, M. W. (2023). Michel Levy interference colour chart. Microscopy U. https://www.microscopyu.com/articles/polarized/michel-levy.html

Roch, S., Walter, T., Ittner, L. D., Friedrich, C., & Brinker, A. (2019). A systematic study of the microplastic burden in freshwater fishes of south-western Germany – Are we searching at the right scale? *Science of the Total Environment*, 689, 1001–1011.

Rochman, C.M., Brookson, C., Bikker, J., Djuric, N., Earn, A., Bucci, K., Athey, S., Huntington, A., McIlwraith, H., Munno, K., de Frond, H., Kolomijeca, A., Erdle, L., Grbic, J., Bayoumi, M., Borrelle, S.B., Wu, T., Santoro, S., Werbowski, L.M., Zhu, X., Giles, R.K., Hamilton, B.M., Thaysen, C., Kaura, A., Klasios, N., Ead, L., Kim, J., Sherlock, C., Ho, A., & Hung, C. (2019). Rethinking microplastics as a diverse contaminant suite. *Environmental Toxicology and Chemistry*, 38(4), 703–711.

Roux, C., Huttunen, J., Rampling, K., & Robertson, J. (2001). Factors affecting the potential for fibre contamination in purpose-designed forensic search rooms. *Science & Justice*, 41(3), 135–144.

Roux, C., & Margot, P. (1997). The population of textile fibres on car seats. *Science and Justice*, 37, 25–31.

Roux, C., & Robertson, J. (2004). Fibres – Significance. In *Encyclopaedia of Forensic Sciences* (pp. 829–834).

Ryan, P.G., Giuseppe, S., Perold, V., Pierucci, A., Bornman, T.G., & Aliani, S (2020) Sampling microfibres at the sea surface: The effects of mesh size, sample volume and water depth. *Environmental Pollution*, 258, 113413

Salter, M.T., Cook, R., & Jackson, A.R. (1987). Differential shedding from blended fabrics. *Forensic Science International*, 33(3), 155–164.

Sang, W., Chen, Z., Mei, L., Hao, S., Zhan, C., Zhang, W.B., Li, M., & Liu, J. (2021). The abundance and characteristics of microplastics in rainwater pipelines in Wuhan, China. *Science of the Total Environment*, 755(Pt 2), 142600.

Schnegg, M., Turchany, M., Deviterne, M., Gueissaz, L., Hess, S., & Massonnet, G. (2017). A preliminary investigation of textile fibers in smothering scenarios and alternative legitimate activities. *Forensic Science International*, 279, 165–176.

Schotman, T.G., & Van der Weerd, J. (2015). On the recovery of fibres by tape lifts, tape scanning and manual isolation. *Science & Justice*, 55, 415–421.

Schotman, T. G., Xu, X., Rodewijk, N., & van der Weerd, J. (2017). Application of dye analysis in forensic fibre and textile examination: Case examples. *Forensic Science International*, 278, 338–350.

Science and Technology Select Committee. (2019). Forensic science and the criminal justice system: A blueprint for change (3rd report of session 2017-19- published 1 May 2019- HL paper 333).

Shruti, V.C., Pérez-Guevara, F., Roy, P.D., & Kutralam-Muniasamy, G. (2022). Analyzing microplastics with Nile Red: Emerging trends, challenges, and prospects. *Journal of Hazardous Materials*, 423(Pt B), 127180.

Siegel, J.A. (1997). Evidential value of textile fibres –Transfer and persistence of fibres. *Forensic Science Review*, 9(2), 82–96.

Sieminski, M.A. (1975). A note on the measurement of birefringence in fibers. *Microscope*, 23, 36–35.

Sierra, I., Chialanza, M.R., Faccio, R., Carrizom, D., Fornaro, L., & Pérez-Parada, A. (2020). Identification of microplastics in wastewater samples by means of polarised light optical microscopy. *Environmental Science and Pollution Research*, 27(7), 7409–7419.

Sillanpää, P., & Sainio, P., (2017) Release of polyester and cotton fibers from textiles in machine washings. *Environmental Science Pollution Research International*. 24, 19313–19321.

Skokan, L., Tremblay, A., & Muehlethaler, C. (2020). Differential shedding: A study of the fiber transfer mechanisms of blended cotton and polyester textiles. *Forensic Science International*, 308.

Slot, A., van der Weerd, J., Roos, M., Baiker, M., Stoel, R.D., & Zuidberg, M.C. (2017). Tracers as invisible evidence – The transfer and persistence of flock fibres during a car exchange. *Forensic Science International*, 275, 178–186.

Smith, H. G. (1956). *Minerals and the Microscope*. George Allen and Unwin Publishers Ltd, London.

Sneath, D., Tidy, H., & Wood, B. (2019). The transfer of fibres via weapons from garments. *Forensic Science International*, 301, 278–283.

Song, Y.K., Hong, S.H., Eo, S., & Shim, W.J. (2021). A comparison of spectroscopic analysis methods for microplastics: Manual, semi-automated, and automated Fourier transform infrared and Raman techniques. *Marine Pollution Bulletin*, 173(Pt B), 112874.

Spectricon. (2023). SMMART forensics toolkit. Retrieved from https://spectricon.com/smmart-forensics/

Stoeffler, S. (1996). A flowchart system for the identification of common synthetic fibres by polarized light microscopy. *Journal of Forensic Science*, 41, 297–299.

Suaria, G., Achtypi, A., Perold, V., Lee, J.R., Pierucci, A., Bornman, T.G., Aliani, S., & Ryan, P.G. (2020) Microfibers in oceanic surface waters: A global characterization, *Science Advances*, 6, 23

Summerscales, J., & Gwinnett, C. (2017). Forensic identification of bast fibres. In Ray, D. (Ed.), *Biocomposites for High-Performance Applications* (pp. 125–164), Woodhead Publishing.

Surendran, U., Jayakumar, M., Raja, P., Gopinath, G., & Chellam, P.V. (2023). Microplastics in terrestrial ecosystem: Sources and migration in soil environment. *Chemosphere*, 318(Pt A), 131748.

Suzuki, S., Higashikawa, Y., Sugita, R., & Suzuki, Y. (2009). Guilty by his fibers: Suspect confession versus textile fibers reconstructed simulation. *Forensic Science International*, 189(1–3), e27–e32.

Suzuki, S., Suzuki, Y., Ohta, H., Sugita, R., & Marumo, Y. (2001). Microspectrophotometric discrimination of single fibres dyed by indigo and its derivatives using ultraviolet-visible transmittance spectra. *Science and Justice*, 41(2), 107–111.

Taylor, M.A. (1990). *Technology of Textile Properties* (3rd ed.). Forbes Publications.

Taylor, M., Gwinnett, C., Robinson, L., & Woodall, L. (2016). Plastic microfibre ingestion by deep-sea organisms. *Scientific Reports*, 6(33997), 1–9.

Textile Exchange. (2021). Preferred fiber & materials market report 2021. Retrieved from https://textileexchange.org/app/uploads/2021/08/Textile-Exchange_Preferred-Fiber-and-Materials-Market-Report_2021.pdf

Tungol, M. W., Bartick, E. G., & Montaser, A. (1990). The development of a spectral database for the identification of fibres by infrared microscopy. *Applied Spectroscopy*, 44(4), 543–549.

Waldschläger, K., Brückner, M. Z. M., Carney Almroth, B., Hackney, C. R., Adyel, T. M., Alimi, O. S., Belontz, S.L., Cowger, W., Doyle, D., Gray, A., Kane, I., Kooi, M., Kramer, M., Lechthaler, S., Michie, L., Nordam, T., Pohl, F., Russell, C., Thit, A., Umar, W., Valero, D., Varrani, A., Warrier, A.K., Woodall, L.C., & Wu, N. (2022). Learning from natural sediments to tackle microplastics challenges: A multidisciplinary perspective. *Earth-Science Reviews*, 228.

Wang, G., Lu, J., Tong, Y., Liu, Z., Zhou, H., & Xiayihazi, N. (2020). Occurrence and pollution characteristics of microplastics in surface water of the Manas River Basin, China. *Science of the Total Environment*, 710.

Was-Gubala, J. (2000). A population study of fibres found on bus seats in Cracow. In *Proceedings of the 8th European Fibres Group Meeting* (pp. 82–86), *Cracow*.

West, J. C. (1981). Extraction and analysis of disperse dyes on polyester textiles. *Journal of Chromatography*, 20, 47–54.

Wetzer, E., & Lohninger, H. (2018). Image processing using color space models for forensic fiber detection. *IFAC-PapersOnLine*, 51(2), 445–450.

Wiggins, K. G. (2001). Forensic textile fiber examination across the USA and Europe. *Journal of Forensic Science*, 46, 1303–1308.

Wiggins, K. G., & Allard, J. E. (1987). The evidential value of fabric car seats and car seat covers. *Journal of the Forensic Science Society*, 27, 93–101.

Wiggins, K. G., & Clayson, N. (1997). Microfibers – A forensic perspective. *Journal of Forensic Science Society*, 42, 842–845.

Wiggins, K., & Drummond, P. (2007). Identifying a suitable mounting medium for use in forensic fibre examination. *Science & Justice*, 47(1), 2–8.

Wiggins, K., & Holness, J. A. (2005). A further study of dye batch variation in textile and carpet fibres. *Science and Justice*, 45(2), 93–96.

Woodall, L. C., Gwinnett, C., Packer, M. P., Thompson, R. C., Robinson, L. F., & Paterson, G. L. J. (2015). Using a forensic science approach to minimize environmental contamination and to identify microfibres in marine sediments. *Marine Pollution Bulletin*, 95(1), 40–46.

Wright, S. L., Thompson, R. C., & Galloway, T. S. (2013). The physical impacts of microplastics on marine organisms: A review. *Environmental Pollution*, 178, 483–492.

Xia, F., Liu, H., Zhang, J., & Wang, D. (2022). Migration characteristics of microplastics based on source-sink investigation in a typical urban wetland. *Water Research*, 213.

Xu, Y., Chan, F. K. S., Stanton, T., Johnson, M. F., Kay, P., He, J., Wang, J., Kong, C., Wang, Z., Liu, D., & Xu, Y. (2021). Synthesis of dominant plastic microfibre prevalence and pollution control feasibility in Chinese freshwater environments. *Science of the Total Environment*, 783.

Zambrano, M.C., Pawlak, J. J., Daystar, J., Ankeny, M., Cheng, J.J., & Venditti, R.A. (2019) Microfibers generated from the laundering of cotton, rayon and polyester based fabrics and their aquatic biodegradation. *Marine Pollution Bulletin* 142, 394–407.
Zantis, L. J., Carroll, E. L., Nelms, S. E., & Bosker, T. (2021). Marine mammals and microplastics: A systematic review and call for standardisation. *Environmental Pollution*, 269.
Zhu, X., Nguyen, B., You, J. B., Karakolis, E., Sinton, D., Rochman, C. (2019). Identification of microfibers in the environment using multiple lines of evidence. *Environmental Science & Technology*, 53(20), 11877–11887.

10 Dynamic Image Analysis for Determination of Textile Fibers

Edith Classen and Jasmin Jung
Hohenstein Institut für Textil Innovation GmbH

10.1 INTRODUCTION

One source of the so-called microplastic in the environment is fiber release from textiles during the use phase. Textiles have the potential to release fibers during production, their use phase (e.g., wearing, washing, and drying), and disposal (Almroth et al. 2018, Bahners et al. 1994, Wang 2010, Carr 2017). Shed fibers can enter the environment via direct pathways; however, the presence of synthetic fibers in the environment is mostly attributed to the laundry process (Carr 2017, Browne et al. 2011). Studies have shown that wastewater treatment plants have a limited retention capacity for microplastic particles and fibers (Mintenig 2017, Ziajahromi et al. 2017). There is no effective way to remove fibers from the wastewater of washing processes as the lint filters in washing machines are not designed to totally hold back the entire particle/fiber discharge (Cesa et al. 2017).

The detection of fiber debris in the environment is important to determine the contribution to microplastic pollution. The method of choice of research groups is often the filtration of wastewater, with subsequent manual counting of the fibers recovered on a filter material by light microscopy and scanning electron microscopy (SEM) (Almroth et al. 2018, Hernandez et al. 2017, Sillanpää and Sainio 2017, De Falco et al. 2018). Fiber counting is mostly applied manually by the naked eye, and it is highly time-consuming. Improvements to counting are the counting of fibers on SEM images acquired in the orthogonal direction of the filter and extrapolating to the complete surface (De Falco et al. 2018) and an automated software-based fiber counting of the filter surface using microscopy (Jönsson 2018). However, visual counting methods entail the risk of uncertainty. An overload of fibers on the filter may lead to the overlapping of individual fibers (Hernandez et al. 2017, Jönsson 2018). The number of fibers is potentially underestimated. For this reason, a homogeneous distribution of fibers on the filter surface has to be ensured (Hernandez et al. 2017).

Gravimetric analysis of the filtered material is another approach used (Napper and Thompson 2016, Hartline 2016, Sillanpää and Sainio 2017, Pirc et al. 2016). The first step is a filtration step, followed by gravimetric mass determination. Fibers are separated from the liquid mechanically through a filter, usually added by a vacuum pump, and the mass of the recovered weight is determined. For the analysis, the

DOI: 10.1201/9781003331995-12

selection of the appropriate filter is important. For weight analysis and counting, the selection of the filter pore size impacts the results significantly. Small pore sizes increase the retention capacity of fibers; however, clogging effects are more likely to occur (Pirc et al. 2016). Sometimes, cascade filtering is applied to prevent filter blockage (Hartline 2016, Jönsson 2018). The selection of an appropriate filter material and sample handling in terms of drying, conditioning, and weighting factors can influence quantitative results. Furthermore, material loss and side contaminations are critical points and imply a strong need for repeated negative control tests (Zeng 2018). Though gravimetric evaluations are comparatively straightforward and often used, they do not provide information about the morphological properties and dimensions of fibers. These characteristics are important for causal research in terms of fiber release and risk assessments (Covernton et al. 2019). The lack of visualization leads to the assumption that the entire material recovered on the filter is attributed to fibers only. Textiles are an emitting source of fibers as fiber shedding occurs and serve as a carrier for the adhered foreign fibers and particles of various materials, shapes, and sources. Dust, production residues, and contaminants from the surroundings are examples of fibers and particles adhering to the textile surface and may be released in the washing process (Jönsson 2018, Mellin 2016). Consequently, there is a significant risk of overestimating the results obtained by weight analysis.

Dynamic image analysis (DIA), which is an optical detection system, is a powerful method for the analysis of particles in fluids and has the potential to overcome the drawbacks of the abovementioned methods of quantification. So far, this method has not been used for the analysis of textile fibers, and this was the basis for the evaluation of this method in the first step with dispersed defined fibers and in the second step with wastewater from laundering. This chapter describes how DIA can be used for the characterization of fiber debris from textiles.

10.2 DYNAMIC IMAGE ANALYSIS

DIA is an optical detection system for particle characterization commonly applied in pharmacy, the food industry, and geology (Yu 2008, Zhao and Wang 2016, Karam et al. 2017). The measuring system allows the online detection of particles in a solid state or particles suspended in liquids without specific sample pretreatment. DIA is a rapid and non-destructive technology for the characterization and quantification of fiber debris and for fiber analysis in liquids with high precision and reliability. In this method, the particles, which have to be well dispersed, are streamed continuously through the measuring volume controlled by a high-resolution camera.

Motion blur during image acquisition is minimized using a pulsed laser as a light source with an exposure time of approximately 1 ns. Image analysis algorithms enable statistical analysis regarding the number of fibers and their morphological characteristics, e.g., length, diameter, straightness, and elongation. The free movement leads to the random orientation of the particles. From different perspectives, their actual shape and size distribution can be accurately determined. With a continuous feed of dispersed particles, reliable and representative results are achieved based on a statistically significant number of particles. The overlay of particles can

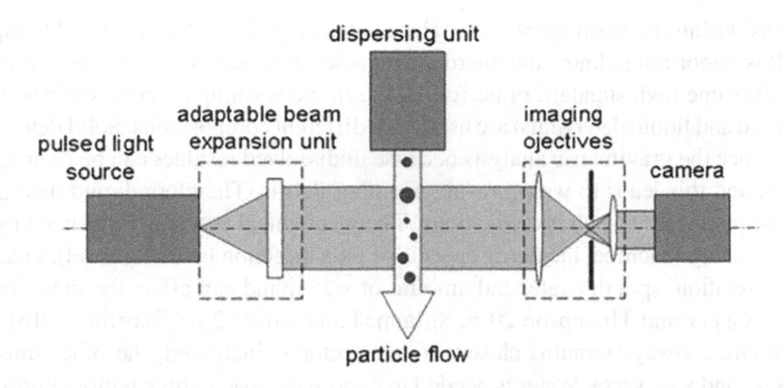

FIGURE 10.1 Dynamic image analysis setup for suspension measurements (Sympatex 2017) and the measurement setup.

be prevented by closely controlling the concentration of the particle flow. Particle size distribution is determined by the characteristic diffraction pattern of a particle collective using laser diffraction. Image analysis captures the physical properties of every single particle, and even the smallest amounts of over- or undersized particles may be detected. Single particles with specific geometric properties such as aggregates, fractures, and foreign particles are traceable. A digital camera with special optics captures the particles within the frame. Physical information about particle properties is transmitted to a computer. For every single particle in the image, size and shape descriptors are determined by evaluation software. Figure 10.1 illustrates the working flow of a DIA system.

10.3 WASHING OF TEXTILES

Textiles from the clothing sector are the most washed items. Typically, these textiles are made out of five main polymer types: polyester (PES), cotton (CO), polyamide, wool, and silk (Kothari 2008). Additionally, the apparel design, including the textile characteristics, e.g., fiber type, yarn construction, and surface treatment, has considerable potential to impact fiber release (Hernandez et al. 2017, Hartline 2016). In most studies, various synthetic textiles made from PES were selected for investigations (Almroth et al. 2018, Hernandez et al. 2017, De Falco et al. 2018, Jönsson 2018). PES/CO blends reflect the most manufactured fiber composition in the clothing industry, and there is a lack of information regarding the shedding behavior of fiber mixtures during laundry (De Silva et al. 2014). In one investigation, the wastewater effluent of PES/CO garments was analyzed; however, only the total amount of fibers was determined, and there was no distinction between synthetic and natural sources (Napper and Thompson 2016).

During washing, textiles are exposed to different strains, which may cause fibers to be released into the washing effluent (Hernandez et al. 2017). The washing process can be described with the Sinner's Circle and the four Sinner's parameters: chemistry, temperature, time, and mechanical power. These Sinner's parameters influence fiber emissions during the washing process (Hernandez et al. 2017, Napper and Thompson 2016). It is important to know the interdependencies of different parameters in order to ensure the same high standard of performance. In the washing process, various detergents (solid and liquid detergents) are used with different compositions. Solid detergents can influence the gravimetric analysis because undissolved residues can be retained on the filter, and this leads to wrong results for fiber debris. Therefore, liquid detergents should be preferred in such investigations. The mechanical power is influenced by the type of washing machine, including aspects of axis direction (front/top load), machine capacity, rotation, spin dry rate, and amount of water, and can affect the mass loss of textiles (Napper and Thompson 2016, Sillanpää and Sainio 2017, Hartline 2016). The Sinner's circle always remains closed: if one factor is increased, the other must be decreased, and vice versa. Water is needed to expose the load to high temperatures and dissolve chemicals and soil, and it plays a significant role in applying mechanical force. Water is the binding factor that not only combines the effects of all other factors but also transports their effects to the load to achieve high washing performance.

Most of the studies available reported fiber release during domestic laundry, but only a minor number of publications investigated industrial washing processes in their research (Hernandez et al. 2017, De Falco et al. 2018). Domestic laundry and industrial laundry differ in washing parameters. The washing conditions of industrial laundry are harsher than those of domestic washing, e.g., the washing machine has a higher wash drum volume; the washing time is shorter; the composition of the detergents differs; and the wash temperature is often higher. At present, the comparability of different studies is hindered by the diversity in analytical technologies and pretreatment procedures, as well as the diversity in selected textiles and applied washing processes (Cesa et al. 2017). In this study, DIA was used for the first time to analyze the microfiber content in wastewater samples from textile laundry, and the methods for the analysis were evaluated.

10.4 METHODS AND MATERIALS

10.4.1 Textile Material

Polyamide 6.6 fibers (*Borchert + Moller GmbH*) with a length of 0.3 mm and a diameter of 15 μm (1.7 dtex fiber fines) were used as reference fibers to validate different methods. For quantitative fiber analysis, aqueous suspensions of the uniform reference material were prepared. 3.15 mg (S1) and 6.17 (S2) mg of fibers were suspended in 500 mL detergent/water solution, respectively (1 g of detergent in 500 mL of water, see Section 10.4.3). The textile material was a workwear fabric made out of 50/50 PES/CO. The woven fabric (S-twill 2/1) was comprised of an intimate mixture (warp: 65/35% PES/CO, weft: 35/65% PES/CO) of staple yarn with a surface weight of 215 g/m^2 with a yarn count warp: 40 Nm, weft: 24 Nm.

10.4.2 Pretreatment and Preparation of the Test Specimens

Before testing, the fabrics were pre-washed to remove non-permanent textile auxiliaries and residues from production. The pre-wash was carried out according to the standard EN ISO 15797:2004 but without additional textile ballast, to avoid cross-contamination of the samples. Test specimens were prepared with dimensions of 170 × 165 mm. Each test specimen was cleaned with a lint roller on the inner side and then folded once in half. The final specimen size was 85 × 165 mm. The corners and edges of the test specimens were sealed using a textile glue (*Skinotex, Weniger*) to prevent fiber loss. For this purpose, the specimens were immersed in the adhesive by 0.5 mm on each side. The prepared specimens were air-dried overnight in a dust-free environment prior to washing.

10.4.3 Washing Procedure of the Test Specimens

The laboratory washing process was carried out using a laboratory washing test for industrial laundry processes based on ISO 105-C12:2004, which is used for colorfastness analysis in industrial washing processes. This washing process is designed to mimic the mechanical stress of 5–10 washing cycles in an industrial washing machine. For the washing process, the laboratory washing machine "Labomat" (*BFA, Werner Mathis AG*) equipped with 8 × 500 mL steel canisters was used under controlled conditions. The turning velocity was set to 40 ± 2 rpm with a reversing time of 30 seconds with a left and right turn. The mechanical stress was simulated by 50 stainless steel balls with a diameter of 6 mm. The washing temperature was set to 40 ± 2°C with a heating rate of 1.5°C/min. The washing duration was 30 minutes in total, including the heating period. In addition, 160 mL of washing liquid containing 2 g/L of liquid color-safe detergent (*Derval Rent, Kreussler, pH = 11*) without an optical brightener applied for industrial purposes was used. Two test specimens were washed in one canister. Each washing experiment was repeated three times. After washing, the samples were removed and rinsed with ultrapure water from both sides using a wash bottle. The steel balls were separated from the washing liquid by a strainer and rinsed with ultrapure water. All washing liquids were collected, and the solution was diluted to a total volume of 300 mL prior to DIA measurements (Haap et al. 2019).

10.4.4 Contamination Prevention

To minimize the risk of sample contamination during analysis, the working space was cleaned thoroughly using lint-free wipes. Furthermore, no CO lab coat was used; instead, a lab coat made of a polyamide monofilament commonly used for surgical gowns was worn by the analysts. This lab coat is designed to have a low-lint surface. Additionally, arm sleeves made of polyethylene foil and gloves were worn as preventative measures. Glassware used in experiments was cleaned with ultrapure water prior to use, oven-dried, and stored in a dust-free environment. Glassware was kept covered to prevent airborne contamination. To ensure particle- and fiber-free water quality, ultrapure water (0.055 µS/cm) from a TKA-GenPure system (Thermo Scientific) with an additional 0.2-µm particle filter was used. Negative controls were applied regularly to control the laboratory and working conditions (Haap et al. 2019).

10.4.5 Pretreatment of Polyester–Cotton Blends

The samples were treated with sulfuric acid based on DIN EN ISO 1833-11 to achieve a chemical separation of PES and CO. Using the sulfuric acid treatment, the CO component was hydrolyzed, while PES was left over. The wastewater from the washing process was filtered through a cellulosic filter (Whatman® grade 40, pore size 8 µm, diameter 90 mm) under vacuum conditions. The filter was rinsed with 100 mL ultrapure water to remove excess detergent and dried for 15 minutes at 50°C in a glass Petri dish. Then, 20 mL sulfuric acid was added and the suspension was incubated for 1 hour at 50°C under shaking in regular time intervals of approximately 20 minutes. After the treatment, the acid solution was neutralized with a 25% ammonia solution in 150 mL of water. Each sample was diluted to a total volume of 300 mL and cooled down to room temperature prior to analysis.

10.4.6 Dynamic Image Analysis

A DIA device (*Qicpic, Sympatec GmbH*) was used for fiber detection with Windox 5.10.0.3 software to acquire and analyze the data. The measuring system was equipped with a beaker containing an outlet tap on the bottom as a reservoir for the dispersion. To prevent sedimentation in the beaker, a magnetic stirrer was installed in the beaker. The liquid was magnetically stirred repeatedly for 5 seconds within intervals of 30 seconds. The reservoir was connected to the flow-through cuvette of the detection system (0.2 mm thickness) by a peristaltic pump (MCP Process with pump head SB 3V, Ismatec). Pulsation during liquid transportation was reduced by a pulsation dampener consisting of a T-piece adapter and a tube that was 30 cm in length. To obtain an air pocket for an appropriate dampening function, the end of the tube was sealed. The volume flow of the pump was set to 172 mL/min, and image acquisition was performed at a frame rate of 85 fps. Figure 10.2 shows the laboratory setup for fiber quantification and characterization using DIA. Before starting the measurement, the liquid was circulated to flush the system from the drainage beaker to the measuring unit.

The measurement was started, and after a lead time of 3 seconds, the tube was laid from the drainage beaker into a collecting vessel. The measurement was stopped when the cuvette was fully drained with liquid. Each sample was analyzed two times. The images were converted into binary images by the integrated software. Eventually, fiber doublings, which occurred due to inhomogeneous volume flow events, were subtracted manually from the counting results.

Figure 10.3 shows a schematic cross-section of the flow-through cuvette. The cuvette was comprised of a circular glass window (3.4 cm in diameter), and the total horizontal axis was 4.1 cm in length. The cuvette width can be divided into 11 equally sized detection frames. Each square was 3.75 × 3.75 mm and corresponded to the dimensions of the detector size. The detection volume was 2.9 µL. The measurement is usually performed at the cuvette center (position 0). However, the setup allowed the cuvette position to be changed in one direction. The colored areas are the possible measuring positions, and the negative position numbers indicate the moving direction of the detection focus starting from the central point position 0. Gauge blocks

FIGURE 10.2 The measurement setup of the dynamic image analysis (DIA). 1 – DIA (measurement unit); 2 – a drainage beaker with a magnetic stirrer; 3 – a pump with pulsation; and 4 – results of the data analysis with the software.

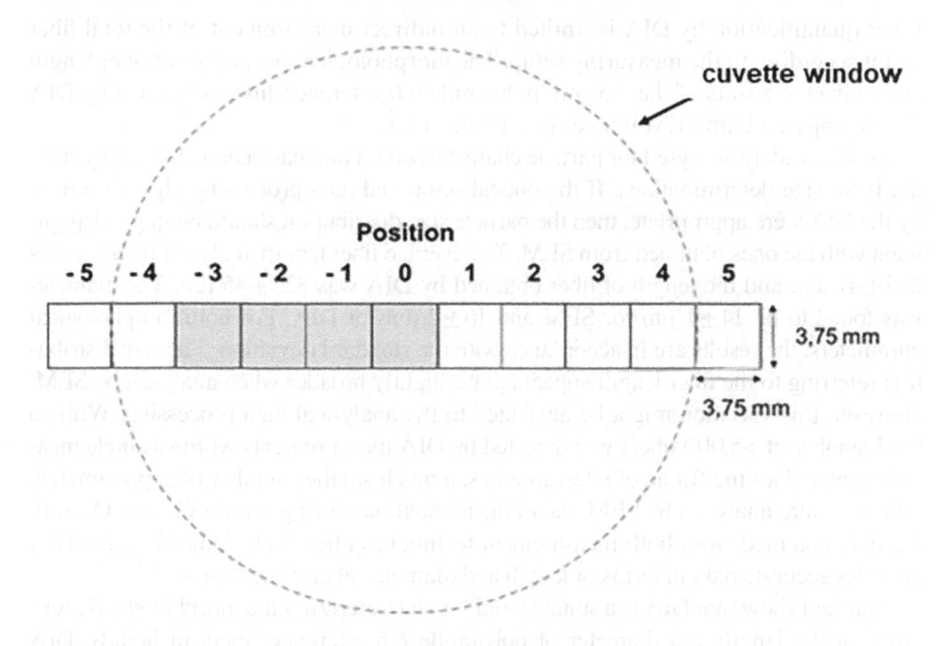

FIGURE 10.3 Cross-section of the dynamic image analysis flow-through cuvette. The horizontal axis is divided into 3.75×3.75 mm frames corresponding to the detection window. Positions 0 to −3 (yellow boxes) represent the experimentally accessible areas, whereas the remaining frames are not usable due to setup limitations.

were used to ensure a defined cuvette movement in 3.75-mm steps. Fiber analysis was performed for cuvette positions 0–3. For the evaluation process, a minimum fiber length of 50 μm and a minimum diameter of 7 μm were defined. Additionally, shape parameters straightness, convexity, and aspect ratio were included in the evaluation process.

10.4.7 Fiber Quantification Using Microscopy Methods

Fiber counting was performed on the filter and in suspension. On the filter, the reference fibers were analyzed using an SEM (JSM-5610 LV, Jeol). The dry fibers were mounted on an adhesive tape and coated with gold under vacuum conditions before imaging to prevent electrostatic charging of the surface (Cressington Sputtercoater 108 auto, 40 mA, <0.1 mbar, 90 s). SEM observations and image acquisition were performed using an accelerating voltage of 1–2.5 kV. Fiber length and diameter were determined by the freeware image processing software ImageJ (release 1.49 v).

Fiber counting in suspensions was additionally performed by visual microscopy using a digital microscope (VHX-S15, Keyence). For this, 50 μL aliquots were withdrawn from the homogenized model wastewater sample, and the number of fibers within this volume was counted.

10.5 RESULTS AND DISCUSSION

Fiber quantification by DIA is limited to an indirect measurement of the total fiber count according to the measuring setup. The morphological properties of fiber length and diameter results of the 0.3-mm polyamide 6.6 reference fibers obtained by DIA were compared with SEM images (see Figure 10.4).

SEM is widely accepted for particle characterization and has been validated by standards for size determinations. If the optical setup and data processing algorithm used by the DIA were appropriate, then the particle size distribution should be in good alignment with the ones obtained from SEM. The average fiber length analyzed by SEM was 297 ± 41 μm, and the length of fiber obtained by DIA was 325 ± 45 μm. The diameter was found to be 14 ± 1 μm for SEM and 16 ± 2 μm for DIA. For both morphological parameters, the results are in accordance with the standard deviation. The size distribution referring to the fiber length appears to be slightly broader when analyzed by SEM. However, this variation might be attributed to the analytical data processing. While a total number of >5,000 fibers was detected by DIA measurements within a single measurement and a time frame of a few minutes, a much smaller number of approximately 70 fibers were analyzed by SEM due to the manual measuring effort of hours. Overall, the data obtained from both measurement techniques align well, indicating that DIA provides accurate data in terms of length and diameter fiber dimensions.

The data show that DIA is a suitable tool for characterizing the morphological properties of the length and diameter of polyamide 6.6 reference fibers in liquids. DIA only allows the analysis of a limited detection volume during a single measurement. Consequently, the data need to be extrapolated to the entire sample volume. To study the effects occurring when micro-sized fibers are probed at different detection positions of the DIA cuvette, a suspension of reference fibers was used. Figure 10.5 presents the number of fibers obtained from DIA in correlation to the cuvette position. The highest number of fibers was measured in the central positions 0 and 1 of the cuvette. The farther the detection focus is shifted away from the center, the lower the number of fibers measured. For positions 0 to –3, three replicates were analyzed. Data points of positions –4 and –5 were calculated. Due to the axial symmetry of the cuvette, a symmetric liquid flow was assumed. The distribution of the fiber amounts correlating

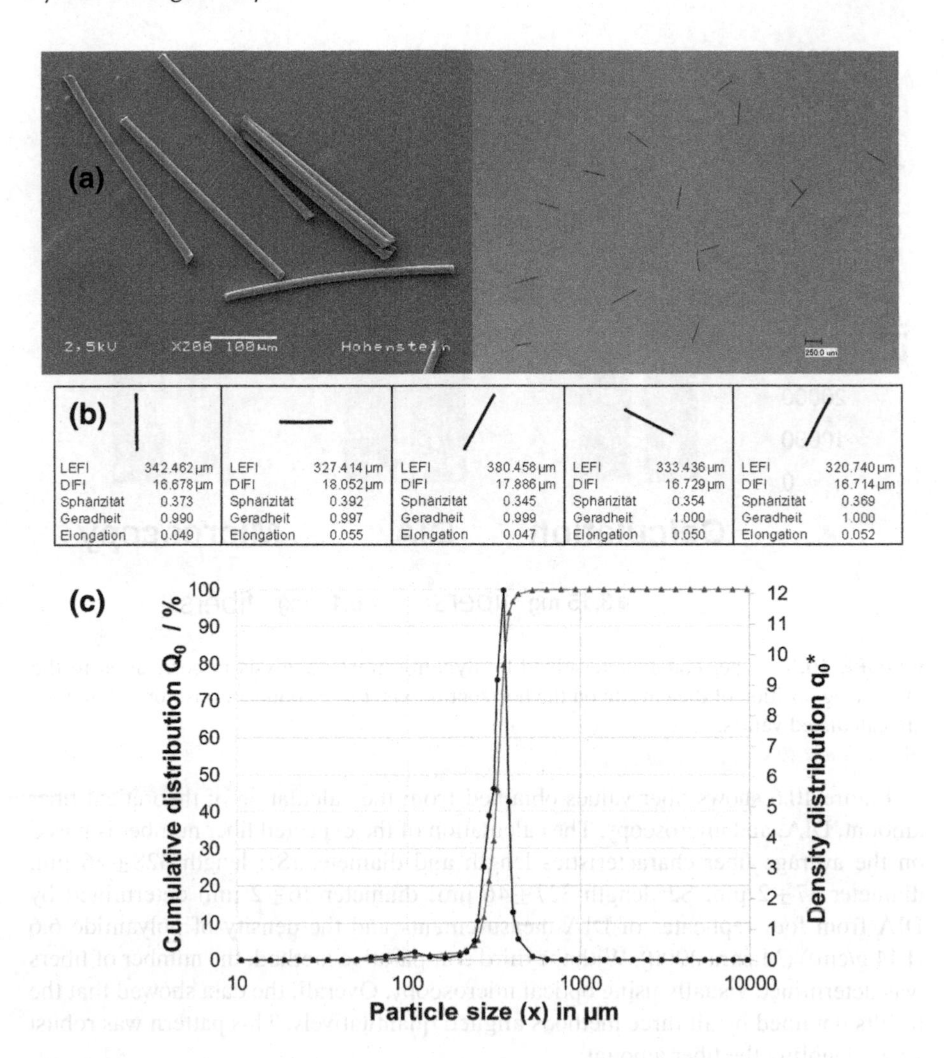

FIGURE 10.4 Pictures from SEM and microscopy of the polyamide 6.6 reference fibers (a), the shape and size measured in DIA (b), and the cumulative and density distributing depending on the particle size (c).

to the measuring position indicates a flow rate distribution within the cuvette matching a parabolic flow profile. Shear stress influences the fluid dynamics causing a lower fluid velocity at the cuvette wall regions and leading to a decrease in fiber amount counted on the outer sides (laminar flow). The available data allow for establishing a correlation between the measuring position and fiber counts. Based on these experimental results, a formula was derived that permits quantifying the total fiber amount in a suspension by one single DIA measurement at the cuvette center. A symmetric fiber distribution was assumed, and a quadratic curve fitting was applied. The validity of the calculation concept was verified by a comparison performed with two fiber suspensions, 3.15 mg (S1) and 6.17 mg (S2) reference fibers, respectively.

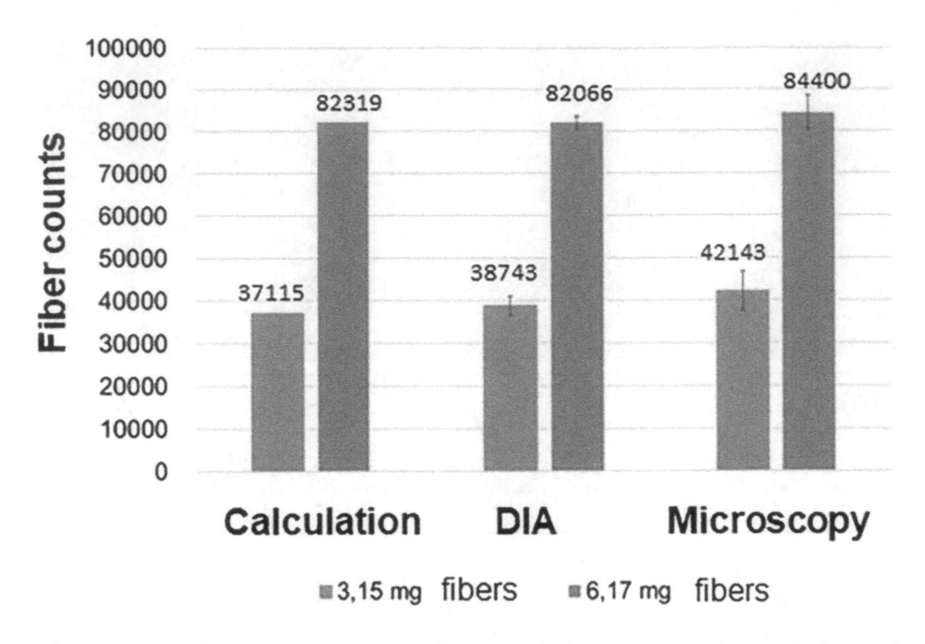

FIGURE 10.5 Fiber counts determined by dynamic image analysis in correlation to the measuring position of the cuvette on the horizontal axis. Fiber counts of position −4 and −5 are calculated values.

Figure 10.6 shows fiber values obtained from the calculation of theoretical fiber amount, DIA, and microscopy. The calculation of the expected fiber number is based on the average fiber characteristics length and diameter (S1: length 328 ± 46 μm, diameter 17 ± 2 μm, S2: length 327 ± 46 μm, diameter 16 ± 2 μm) determined by DIA from four replicates of DIA measurements and the density of polyamide 6.6 (1.14 g/cm³) (Mishra 2000). With the third comparative method, the number of fibers was determined visually using optical microscopy. Overall, the data showed that the results obtained by all three methods aligned quantitatively. This pattern was robust when doubling the fiber amount.

The main differences between DIA and static microscopy were observed during the data acquisition process. While microscopic determination requires a measuring effort of hours, DIA requires only a few minutes to return exactly the same results. The reason for the time-consuming nature of microscopy analysis is the manual counting procedure and the small area that is visually accessible each round (Jönsson 2018). Microscopic analysis is also prone to errors and subjective assessments of the experimenter, e.g., three different results can be obtained if three persons analyze using the same optical method for the same sample (Dekiff et al. 2014). Microscopic measurements require a homogeneous and representative distribution of fibers on the substrate (Hernandez et al. 2017, Jönsson 2018). A high fiber content entails the risk of overlapping and crossing fibers. This increases the chance of underestimating the total number of fibers in the samples (Hernandez et al. 2017). To overcome this issue, adapted sample preparation procedures are required, e.g., the additional application of a cascade filtering procedure prior to microscopy (Jönsson 2018). However, any

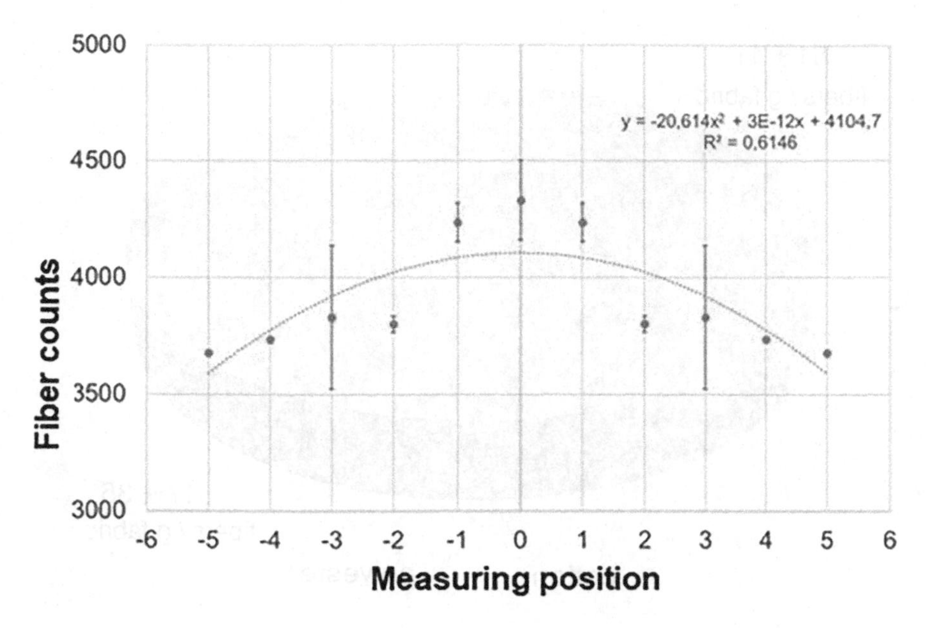

FIGURE 10.6 Results of the quantitative fiber determination of suspended polyamide 6.6 (PA-) fibers (0.3 mm length, 1.7 dtex). Fiber counts were determined for 3.15 and 6.17 mg fiber suspensions by three methods: calculation of the theoretical number of fibers, dynamic image analysis, and optical microscopy counting.

additional step increases the complexity and may influence the uncertainty and the error of the analytical method. In contrast, no specific sample pretreatment is needed for DIA. In the case of high fiber loads, a homogeneous distribution during measurement can be achieved by a simple dilution combined with sufficient stirring of the analyzed wastewater sample. Furthermore, DIA measurement is non-destructive, and the results are not influenced by subjective evaluations of the experimenter.

In a further step, the wastewater from laundering the fabric was investigated regarding the composition of fiber debris. Figure 10.7 shows the fiber counts for CO and PES fibers in wastewater. Expressing the results in percentage, the proportion of released fibers is 14% ± 3% PES and 86% ± 3% CO. In this case, the amount of CO fibers is six times higher compared to the amount of PES fibers in the mixture. Existing studies focusing on the shedding behavior of PES/CO blends did not distinguish between the fiber types (Napper and Thompson 2016, Zambrano et al. 2019). Instead, the total number of fibers was evaluated. However, the observations in this study can be attributed to differences in shedding behavior.

10.6 CONCLUSION

In this study, DIA has been introduced as an alternative method for fiber characterization and quantification in water and wastewater from laundry processes. This method was analytically validated by comparing fiber morphology and quantitative analysis achieved using this method with those of conventional optical microscopy methods.

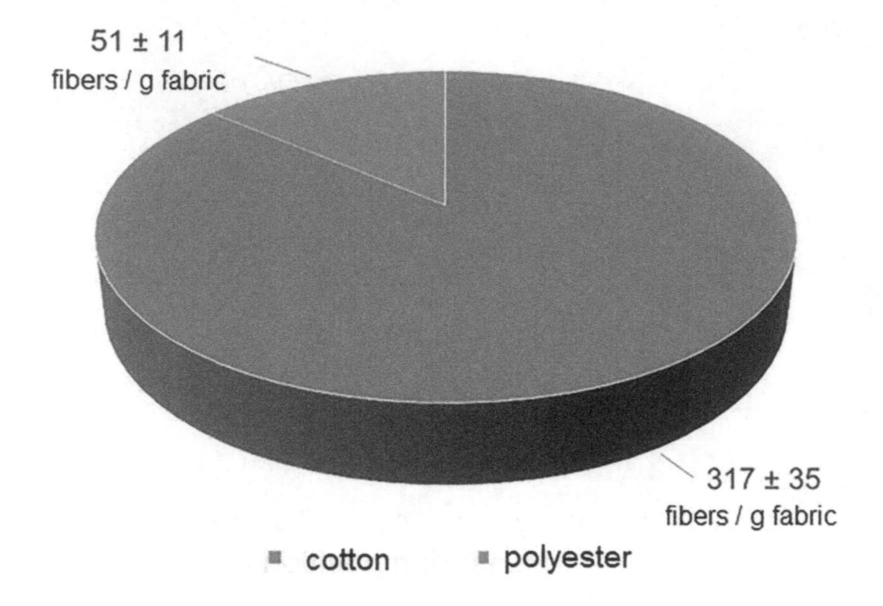

FIGURE 10.7 Fiber counts per gram of fabric of polyester and cotton fibers regarding the total amount of shed fibers. These results refer to a wastewater sample from a washing process of a 50/50 polyester–cotton intimate blended fabric analyzed by dynamic image analysis.

The main advantages of DIA presented compared to commonly applied analytics are as follows:

- the measurement of a larger data set for high numbers of fibers within a measuring time of a few minutes is possible;
- the evaluation of qualitative results and morphological properties is automated with the software;
- this is a non-destructive analysis;
- the sample pretreatment is simple.

The capability of DIA to simultaneously measure fiber morphology characteristics and fiber counts within short measuring times could help achieve valuable insights into a variety of textile applications with regard to fiber release and support new product and process developments. At present, the comparability of different studies is hindered by the diversity in analytical technologies and pretreatment procedures, as well as the diversity in selected textiles and applied washing processes (Cesa et al. 2017). Establishing a harmonized protocol for the analysis of fibers in water is a growing need to generate a comprehensive understanding of fiber shedding during washing and to develop feasible solutions.

ACKNOWLEDGEMENT

The IGF research project no 19219 N of the research association Forschungskuratorium Textil e.V., Berlin/Germany, was funded via the AiF within the framework of the program for funding the industrial cooperative research and development (IGF) by the German Federal Ministry of Economic Affairs and Energy (BMWi) based on a decision of the German Federal Parliament.

REFERENCES

Almroth, B.M.C. et al. (2018) Quantifying shedding of synthetic fibers from textiles; a source of microplastics released into the environment. *Environ. Sci. Pollut. Res.* 25, 1191–1199, doi:10.1007/s11356-017-0528-7.

Bahners, T.; Ehrler, P.; Hengstberger, M. (1994), Erste Untersuchungen zur Erfassung und Charakterisierung textiler Feinstäube. *Melliand Textilberichte* (1), 24–30.

Browne, M.A. et al. (2011) Accumulation of microplastic on shorelines woldwide: Sources and sinks. *Environ. Sci. Technol.* 45, 9175–9179, doi:10.1021/es201811s.

Carr, S.A. (2017) Sources and dispersive modes of micro-fibers in the environment. *Integr. Environ. Assess. Manag.* 13, 466–469, doi:10.1002/ieam.1916.

Cesa, F.S.; Turra, A.; Baruque-Ramos, J. (2017) Synthetic fibers as microplastics in the marine environment: A review from textile perspective with a focus on domestic washings. *Sci. Total Environ.* 598, 1116–1129, doi:10.1016/j.scitotenv.2017.04.172

Covernton, G.A. et al. (2019) Size and shape matter: A preliminary analysis of microplastic sampling technique in seawater studies with implications for ecological risk assessment. *Sci. Total Environ.* 667, 124–132, doi:10.1016/j.scitotenv.2019.02.346.

De Falco, F. et al. (2018) Evaluation of microplastic release caused by textile washing processes of synthetic fabrics. *Environ. Pollut.* 236, 916–925, doi:10.1016/j.envpol.2017.10.057.

Dekiff, J.H. et al. (2014) Occurrence and spatial distribution of microplastics in sediments from Norderney. *Environ. Pollut.* 186, 248–256, doi:10.1016/j.envpol.2013.11.019.

De Silva, R.; Wang, X.; Byrne, N. (2014) Recycling textiles: The use of ionic liquids in the separation of cotton polyester blends. *RSC Adv.* 4, 29094–29098, doi:10.1039/C4RA04306E.

Haap J. et al. (2019) Microplastic fibers released by textile laundry: A new analytical approach for the determination of fibers in effluents. *Water*. 11, 2088

Hartline, N.L. (2016) Microfiber masses recovered from conventional machine washing of new or aged garments. *Environ. Sci. Technol.* 50, 11532–11538, doi:10.1021/acs.est.6b03045.

Hernandez, E.; Nowack, B.; Mitrano, D.M. (2017) Polyester textiles as a source of microplastics from households: A mechanistic study to understand microfiber release during washing. *Environ. Sci. Technol.* 51, 7036–7046, doi:10.1021/acs.est.7b01750.

Jönsson, C. (2018) Microplastics shedding from textiles – Developing analytical method for measurement of shed material representing release during domestic washing. *Sustainability* 10, 2457, doi:10.3390/su10072457.

Karam, M.C. et al. (2017) Effect of whey powder rehydration and dry-denaturation state on acid milk gels characteristics. *J. Food Process. Preserv.* 41, e13200, doi:10.1111/jfpp.13200.

Kothari, V. (2008) Polyester and polyamide fibers-apparel applications. In *Polyesters and Polyamides*; Woodhead: Cambridge, England, pp. 419–440.

Mellin, P. (2016) Nano-sized byproducts from metal 3D printing, composite manufacturing and fabric production. *J. Clean. Prod.* 139, 1224–1233, doi:10.1016/j.jclepro.2016.08.141.

Mintenig, S. (2017) Identification of microplastic in effluents of waste water treatment plants using focal plane array-based micro-Fourier-transform infrared imaging. *Water Res.* 108, 365–372, doi:10.1016/j.watres.2016.11.015.

Mishra, S. (2000) *A Text Book of Fibre Science and Technology*; New Age International: New Delhi, India.

Napper, I.E.; Thompson, R.C. (2016) Release of synthetic microplastic plastic fibres from domestic washing machines: Effects of fabric type and washing conditions. *Mar. Pollut. Bull.* 112, 39–45, doi:10.1016/j.marpolbul.2016.09.025.

Pirc, U. et al. (2016) Emissions of microplastic fibers from microfiber fleece during domestic washing. *Environ. Sci. Pollut. Res.* 23, 22206–22211, doi:10.1007/s11356-016-7703-0.

Sillanpää, M.; Sainio, P. (2017) Release of polyester and cotton fibers from textiles in machine washings. *Environ. Sci. Pollut. Res.* 24, 19313–19321, doi:10.1007/s11356-017-9621-1.

Sympatex. (2017) sympatex.com.

Wang, Y. (2010) Fiber and textile waste utilization. *Waste Biomass Valoriz.* 1, 135–143, doi:10.1007/s12649-009-9005-y.

Yu, W. (2008) Hancock, B.C. Evaluation of dynamic image analysis for characterizing pharmaceutical excipient particles. *Int. J. Pharm.* 361, 150–157, doi:10.1016/j.ijpharm.2008.05.025.

Zambrano, M.C. et al. (2019) Microfibers generated from the laundering of cotton, rayon and polyester based fabrics and their aquatic biodegradation. *Mar. Pollut. Bull.* 142, 394–407, doi:10.1016/j.marpolbul.2019.02.062.

Zeng, E.Y. (2018) *Microplastic Contamination in Aquatic Environments: An Emerging Matter of Environmental Urgency*; Elsevier: Amsterdam, The Netherlands.

Zhao, B.; Wang, J. (2016) 3D quantitative shape analysis on form, roundness, and compactness with μCT. *Powder Technol.* 291, 262–275, doi:10.1016/j.powtec.2015.12.029.

Ziajahromi, S. et al. (2017) Wastewater treatment plants as a pathway for microplastics: Development of a new approach to sample wastewater-based microplastics. *Water Res.* 112, 93–99, doi:10.1016/j.watres.2017.01.042.

Section 3

Impact of Microfibre Pollution

11 The Occurrence of Natural and Synthetic Fibers in the Marine Environment

Giuseppe Suaria
CNR-ISMAR, Istituto di Scienze Marine,
Consiglio Nazionale delle Ricerche

11.1 INTRODUCTION

According to the general definition proposed by Liu et al. (2019), microfibers are "*any natural or artificial fibrous materials of threadlike structure with a diameter smaller than 50 μm, length ranging from 1 μm to 5 mm, and length to diameter ratio greater than 100*". These fibers originate from textile materials, whose shedding is mainly due to washing and wear and tear of fabrics during their regular use. The shedding of textile fibers depends on several factors such as textile characteristics (age, finishing, material, coatings, etc.), washing conditions and UV exposure (Cesa et al. 2017; Carney Almroth et al. 2018; De Falco et al. 2020; Sørensen et al. 2021). In recent years, increasing production and widespread consumption of textile materials have led to the accumulation of substantial amounts of microfibers in the marine environment (Arias et al. 2022). Global fiber production, both synthetic and natural, has almost doubled in the past 20 years, reaching 116 million tons in 2022, and is expected to reach 147 million tons in 2030, if business as usual continues (Textile Exchange 2023). Synthetic polymers dominated the textile market since the mid-1990s when they overtook cotton as the dominant fiber type. Largely driven by the production of polyester, synthetic fibers now account for almost two-thirds of global fiber production and 14.5% of plastic production by mass (Geyer et al. 2017). Clothing and garments are the primary applications for both natural and synthetic fibers, followed by home and office furnishings, transportation, and a wide range of other industrial uses like filtration, personal care, and construction. Large numbers of fibers enter the environment mainly through aerial deposition and wastewater effluent, or through the application of contaminated sludge on agricultural soils (Carr 2017). As a result, textile microfibers currently account for 80–90% of all microplastic counts and are the most common form of anthropogenic particle discovered by microplastic pollution surveys around the world (Gago et al. 2018).

DOI: 10.1201/9781003331995-14

It is now clear that microfibers are a widespread contaminant in the marine environment. Substantial concentrations of this emerging pollutant have been detected in most marine ecosystems surveyed to date, including surface and subsurface waters (Suaria et al. 2020; Pedrotti et al. 2021), sea ice (Tirelli et al. 2022), deep-sea sediments (Sanchez-Vidal et al. 2018) and coastal environments worldwide (Gago et al. 2018). Several studies also reported the presence of microfibers in wet and dry atmospheric deposition (Dris et al. 2017; Finnegan et al. 2022), as well as in the digestive tracts of many marine and terrestrial organisms, from commercial fish species to seabirds and deep-sea invertebrates (Liu et al. 2022). When in suspension in the water column, microfibers are often more abundant close to the coast than in offshore environments (Desforges et al. 2014). Maximum densities of microfibers in beach, coastal and deep-sea sediments are often lower than those reported for traditional microplastic fragments (Fagiano et al. 2023), but microfibers are more widespread and homogeneously distributed in the natural environment than most of the other microplastic's typologies (Liebezeit and Dubaish 2012). The size of the microfibers found in the marine environment can vary widely, but most authors report diameters comprised between 5 and 30 μm (most commonly around 10–20 μm) and microfiber lengths varying between 50 μm and 5 mm, which is currently considered the upper size limit of the microplastics definition (Frias & Nash 2010). Although there have been observations of adverse health effects in marine invertebrates caused by the ingestion of microfibers (e.g., Siddiqui et al. 2023), there is currently no evidence of harm to wild organisms exposed to environmentally relevant concentrations of fibers (Bucci et al. 2020). Nonetheless, our comprehension of the impacts of microfibers on natural populations remains incomplete. Moreover, the use of a wide range of chemicals during natural and synthetic textile production, such as dyes, additives, and flame retardants, raises concerns about the role of fibers as carriers of hazardous substances in the marine environment (Sait et al. 2021). The purpose of this chapter is to present up-to-date information on the presence, properties, fate, toxicity, uptake and persistence of both synthetic and natural fibers in the marine environment. Emphasis is also placed on sampling and analytical techniques to study microfibers in the marine environment, as well as on prospective areas for future research.

11.2 SOURCES OF MICROFIBERS IN THE MARINE ENVIRONMENT

The sources of microfibers in the marine environment have never been fully investigated in detail, although this is critical if effective control and remediation measures are to be developed. Indeed, as recently underlined by Carr (2017), it is very likely that laundry discharges have been deemed to be the primary source of fibers found in aquatic environments mainly because of our poor understanding and limited knowledge of atmospheric fibers deposition, coupled with a much greater familiarity with laundry fibers. The assumption that Wastewater Treatment Plants (WWTPs) are the main source of fibers to the marine environment has remained basically unchallenged for several years and this misleading notion has been quoted multiple times in the scientific literature dealing with marine microplastic pollution. Several studies quantified the number of microfibers released during machine washing of a variety of textiles (e.g., De Falco et al. 2018; Dreillard et al. 2022). Recent studies however, clearly demonstrated that wastewater treatment plants, do efficiently remove large amounts of microfibers from effluent waters, with efficiencies which are often greater than 80–90% (Ngo et al. 2019; Sun et al. 2019; Iyare et al. 2020; Liu et al. 2021), basically discrediting the long-lived assumption that most laundry-discharged fibers were simply passing through WWTPs and getting released into the aquatic environment. At the same time, striking evidence started to appear in the scientific literature, showing that textile fibers are extremely abundant in atmospheric fallout samples, both in urban and rural areas (e.g., Dris et al. 2016; Finnegan et al. 2022). These findings, coupled with the detection of relatively high concentrations of textile microfibers in remote environments (i.e., Antarctica, Arctic, Mountain Glaciers, etc.), in laboratory or field contamination blanks (Torre et al. 2016; Prata et al. 2020b; Belontz and Corcoran 2021), as well as in both indoor and outdoor environments (Dris et al. 2017), further supported the conclusion that WWTPs were not the most important source of microfibers to the environment and that other significant sources and dispersal routes should exist (Carr 2017). This has been recently and unambiguously demonstrated by De Falco et al. (2020), who provided evidence that the release of textile fibers into the air just by wearing a polyester garment, is one order of magnitude greater than the release of fibers into the water which occurs when the same garment is washed (i.e., 2.98×10^8 fibers to water vs 1.03×10^9 fibers to air). More recently, Napper et al. (2023) measured and compared the quantities of microfibers entering the marine environment via both wastewater discharge and atmospheric deposition in the same study site, and clearly concluded that atmospheric deposition was the dominant dispersal pathway, releasing textile fibers at a rate which was several orders of magnitude greater than via treated wastewater effluent. So, from the evidence available thus far, we can safely deduce that atmospheric deposition is the major source of microfiber contamination in the marine environment, although direct or inferred estimates of the global relevance of this process are currently missing. Within this context, it is becoming increasingly clear that microfibers are continuously and diffusely released by a wide variety of garments and textile products (i.e., carpets, clothes, flags, furniture, etc.) during daily use and normal wear and tear (Yang et al. 2023), and not only during washing machine cycles. After being generated, these fibers – depending on their size and density – will arrive at the marine environment either by becoming directly airborne after shedding and then getting further dispersed by atmospheric currents before settling on the sea surface (sometimes at hundreds if not thousands of miles further away from their sources), or they will arrive at sea through the terrestrial water cycle, either by land runoff of those fibers previously deposited on urban and terrestrial environments, or through direct input of fibers in rivers and effluents, especially in those areas where untreated effluents are directly discharged at sea or in river basins. Unfortunately, we are currently unable to provide precise estimates of the relative importance and contribution of all these pathways to the global microfibers cycle. However, it is hoped that future research will shed light and provide better global estimates on the relative contribution of most of these often-neglected sources of textile fibers to the world's oceans.

11.3 TRANSPORT AND DEGRADATION OF MICROFIBERS IN THE MARINE ENVIRONMENT

Regardless of how textile microfibers arrive in the marine environment, understanding their distribution patterns and their residence time in each marine compartment is critical if a deeper understanding of their impacts on marine ecosystems wants to be achieved. Once in the air, aero-transported fibers will be generally dispersed over large distances through atmospheric currents or during wind or storm events, often covering hundreds if not thousands of kilometers (Evangeliou et al. 2020; Liu et al. 2020; Allen et al. 2022). After a certain time, some fibers will start settling and will be deposited on the sea-surface microlayer, where indeed substantial concentrations have been already measured by multiple authors (e.g., Birkenhead et al. 2020). On the sea surface, fibers can be transported by surface currents, following the main water mass movements and accumulating along density fronts and/or mesoscale or sub-mesoscale structures (Suaria et al. 2021), like any other floating material. After surface-tension breaks, some fibers will inevitably start sinking into the water column. However, field studies showed that in general microfibers concentrations are considerably higher in surface waters if compared to sub-surface waters (e.g., Ryan et al. 2020; Suaria et al. 2020), so there might be either a constant aerial deposition over the sea surface, or there must be some retention mechanism which is efficiently preventing, to a certain extent, the sinking of these fibers in the water column. The dynamics and vertical velocities associated with the sinking of microfibers in viscous fluids are currently unclear and extremely complicated to model (see e.g., Lindström and Uesaka 2007; Bagaev et al. 2017; du Roure et al. 2019), however once at sea, being generally denser than seawater (Suaria et al. 2020), an unknown fraction of these fibers will inevitably start sinking to the seafloor. As a matter of fact, textile microfibers have been already detected in the water column by multiple authors, with records coming from all over the world, including the Atlantic Ocean (Reineccius et al. 2020) and the Mediterranean Sea (Rios-Fuster et al. 2022) at various depths. In the water column, sinking fibers can be also advected by deep or intermediate currents (Paluselli et al. 2022), before being deposited over the seafloor which can be considered as their final sink (Woodall et al. 2014). As a matter of fact, substantial concentrations of microfibers have been detected in shallow and deep-sea sediments all over the world, from coastal regions to abyssal plains (e.g., Sanchez-Vidal et al. 2018; Adams et al. 2021). Density increases due to fouling microorganisms and ingestion and egestion by marine biota have been also proposed as additional mechanisms which could potentially increase the proportion of microplastic fragments exported to the deep sea (Halsband 2022). However, it is currently unclear whether this is also the case for microfibers and very few data are currently available about the colonization dynamics of natural and synthetic microfibers by microorganisms in the marine environment. That said, ingestion and entanglement of microfibers by zooplankton are demonstrated (Desforges et al. 2015; Kang et al. 2020), and so there is no reason to believe that ingestion of microfibers by zooplankton or other marine organisms (see Section 11.4), and subsequent incorporation into sinking fecal pellets are not contributing to the sequestration of large amounts of microfibers in marine sediments, exactly as it happens for microplastic fragments (Coppock et al. 2019; Compa et al. 2022).

Once a microfiber is introduced to the marine environment, it is immediately subjected to physical, chemical, and biological agents that contribute to its degradation. The speed and the efficacy of these processes, largely depend on the material properties and the physical and chemical conditions of the receiving environment (Lott et al. 2022). The main agent causing the degradation of textile fibers in the marine environment are mechanical stress, UV radiation, hydrolysis, oxidation or biotic processes, which are often mediated by a wide range of microorganisms. At present, however, very little is known about the actual degradation rates and residence times in the marine environment of most natural and synthetic textile materials currently available on the market. As reported by Lott et al. (2022), despite a wealth of claims based on observations and indirect measurements exist, direct proof for the biodegradation of conventional synthetic polymers is still missing and there is no evidence for biodegradation of un-weathered conventional plastic polymers in the open environment at a substantial rate compared to human timescale. At the same time, the issue of the degradation of natural fibers (i.e., mainly wool and cellulosic fibers), is more controversial. Rayon and cotton yarns are often processed, finished, dyed, and coated with a wide range of chemicals including resins, softeners, coatings, antimicrobial agents and flame retardants, which may considerably slow their remineralization (Li et al. 2010), to the extent that a dyed cotton waistcoat recovered from a deep-ocean shipwreck showed almost no sign of degradation after 133 years of submersion (Chen and Jakes 2001). Within this context, the high abundance of cellulosic fibers often observed in marine environments, both in seawater as well as in sediments and organisms (Remy et al. 2015; Sanchez-Vidal et al. 2018; Savoca et al. 2019; Avio et al. 2020; Suaria et al. 2020), can be explained by the various chemical treatments commonly used during the textile finishing process. Research has shown that these treatments can inhibit or delay the biodegradation of cellulosic fibers (Lykaki et al. 2021). However, the underlying mechanisms are still largely unexplored and the real half-life of natural and synthetic fibers in the environment is currently unknown. To better understand fiber degradation and residence times of these anthropogenic contaminants in the marine environment, future studies focusing on giving accurate weathering and ageing indicators for cellulosic and synthetic fibers will be very helpful.

11.4 INGESTION OF MICROFIBERS BY MARINE ORGANISMS

Studies are increasingly reporting the ingestion of textile fibers by marine fishes and other organisms. Indeed, when delving into the microplastic ingestion literature, it becomes immediately clear to the reader that the vast majority of the so-called "microplastic particles" ingested by marine organisms are actually textile fibers of different natural and synthetic materials (Rebelein et al. 2021), often accounting from 40–50% up to 90–100% of all ingested anthropogenic particles (Santini et al. 2022). It should be also pointed out that, similarly to what is found in seawater and sediment samples, a very large portion of all microfibers extracted from biota is of natural origin, and mainly composed of animal fibers (such as wool) and cellulose-based fibers, which consists of both dyed natural cellulose (i.e., cotton, flax, hemp, etc.) and regenerated (man-made) cellulosic fibers such as rayon, viscose, and cellulose acetate (Savoca et al. 2019; Avio et al. 2020; Santini et al. 2022).

Absolute amounts of microfibers taken up by organisms generally show a high temporal and spatial variability (Gouin 2020). Microfiber abundance in the tissues of marine organisms seems to be correlated with the concentration of microfibers in the surrounding seawater (Sui et al. 2020). In the Mediterranean Sea, microfibers uptake by fish was found to be positively correlated to coastal human population, river inputs, and shipping lanes (Sbrana et al. 2020). In addition, due to their opportunistic feeding strategies, a greater proportion of passive suspension feeders take up microplastic particles and textile fibers compared to more selective feeding taxa, such as crustaceans or fish (Bour et al. 2018). Ingestion of microfibers, however, seems to be widespread across different ecological compartments, trophic levels, and feeding strategies. Microfiber ingestion has been reported to occur globally, by a wide range of marine organisms, from zooplankton (Desforges et al. 2015; Zheng et al. 2020) to many fish species, including benthic, demersal, and pelagic species (Savoca et al. 2019; Capillo et al. 2020; Rodríguez-Romeu et al. 2020; Macieira et al. 2021), deep-sea organisms (Taylor et al. 2016), shellfish and mussels (Santonicola et al. 2021; Volgare et al. 2022), seabirds (Provencher et al. 2018), marine mammals (Perez-Venegas et al. 2018), and several other marine invertebrate species (e.g., Gusmão et al. 2016; Lourenço et al. 2017; Piarulli et al. 2019). Considerable numbers of microfibers, largely made of cellulosic materials, were also found in commercially important Mediterranean fish species including sardines (*Sardina pilchardus*), anchovies (*Engraulis encrasicolus*), and European hakes (*Merluccius Merluccius),* with this raising important concerns about food safety and human consumption of fiber-contaminated species (Compa et al. 2018; Avio et al. 2020). Lastly, multiple reports also exist about the ingestion of microfibers by wild Antarctic (e.g., Le Guen et al. 2020; Bottari et al. 2022; Bergami et al. 2023) or Arctic biota (Morgana et al. 2018), providing further evidence about the global pervasiveness of this emerging contaminant (Alurralde et al. 2022).

It is still unclear if the ingestion of microfibers can have detrimental impacts on the health of marine organisms. Several contradicting evidences are currently present in the scientific literature. Gusmão et al. (2016) for instance, showed that some species of meiofauna collected from sandy beaches are able to egest microfibers with no obvious physical injury, suggesting that although microfibers were rapidly egested with no apparent harm, there was still the potential for trophic transfer along marine food webs. On the other hand, the ingestion of natural and synthetic microfibers altered growth and behavior in the early life stages of estuarine organisms (Siddiqui et al. 2023). Hu et al. (2020) reported structural changes in the gills of adult Japanese medaka after exposure to polyester and polypropylene microfibers. Similarly, a lower energy budget for reproduction and reduced food consumption was found in shore crabs (*Carcinus maenas*) fed with microfibers generated by a polypropylene rope (Watts et al. 2015). Lower body conditions were also observed in fish specimens that had ingested a higher content of textile fibers in the wild (Mizraji et al. 2017). The causes for this effect are unknown, however, the authors speculated that chronic consumption of a microfiber-rich diet can cause deprived nutrient availability via gut blockage, false satiety sensation, or physical injury of the digestive tract (Mizraji et al. 2017). It has been also shown that zooplankton can accumulate fibers in the digestive tract breaking them down into nanofibers, which can cause cytotoxicity through phagocytotic absorption (Mishra et al. 2019). Lastly, it should be noted that microfibers can also adsorb and accumulate organic or inorganic pollutants from the surrounding seawater, potentially transferring them across food webs, together with the dyes, chemical additives and finishing agents used during their industrial production (Rochman et al. 2015; Li et al. 2023). Whether ingested natural and synthetic microfibers (or their chemical additives) can be bioaccumulated or biomagnified along the food chain is still debated in the scientific literature. Bioaccumulation generally implies higher intake than the egestion of microfibers within an organism, while when a predator's microfiber concentration is higher than that of its prey, the term "biomagnification" is generally used. Observations in the field however don't seem to currently support these hypotheses (Gouin 2020). The effects of textile microfibers on the marine ecosystem, including trophic transfer, accumulation, and potentially toxic effects, are still poorly understood. Most microfiber toxicity studies have investigated the effects of synthetic microfibers, while the effects of cellulosic and natural-based fibers remain greatly understudied (Athey and Erdle 2022). More research and dedicated laboratory experiments encompassing a wide range of fiber sizes and materials at environmentally relevant concentrations are needed to better clarify the global importance of these processes.

11.5 SAMPLING TOOLS AND ANALYTICAL METHODS TO STUDY MICROFIBER POLLUTION

Methodological issues historically led to a great underestimation of the total amount of textile fibers in the aquatic environment. Especially in the early days of microplastic research, most surface water samples were taken with neuston or manta nets, which are usually equipped with mesh sizes of 200–333 μm. Thus, items smaller than 200 μm (like most microfibers), were considered to originate from external contamination and were intentionally excluded from the total microplastic counts. Subsequently, higher concentrations of microplastics (particles and fibers) started to be reported by those authors using smaller mesh sizes (5–50 μm) or bulk water sampling methods (Athey and Erdle 2022). Shortly after, more rigorous contamination control protocols and blank procedures started to be implemented and more widely adopted, and it soon became clear that these anthropogenic particles were particularly abundant in the marine environment (Figure 11.1).

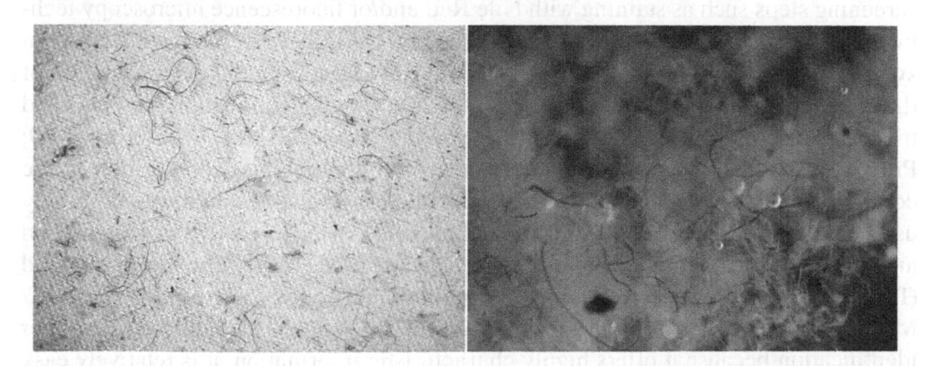

FIGURE 11.1 Two images of textile microfibers collected in marine surface water samples during a large-scale survey of microfiber pollution carried out in offshore oceanic environments. (*Source:* Photo credit: Giuseppe Suaria.)

Right now, dedicated sampling protocols are being developed and scientific publications focusing exclusively on the issue of textile fibers are becoming more common. It is now clear that for the detection of textile microfibers in seawater samples, much lower mesh sizes are needed as, due to their diameters (10–20 μm), most fibers can pass through larger sampling nets. As a matter of fact, increasing concentrations of fibers are detected in water sampled with decreasing mesh filter sizes (Barrows et al. 2018; Whitaker et al. 2019; Ryan et al. 2020). Furthermore, major losses of textile fibers from water samples can also occur when using surface tow nets, as large volumes of sampled water facilitate the flushing of fibers through coarse net meshes (Ryan et al. 2020). Textile fibers can be sampled from the atmosphere or in indoor environments using passive or active air samplers, which can be also employed in the marine environment (e.g., from research vessels). Fibers can be also extracted from organic-rich sediment or biota samples using a variety of digestion and density separation protocols (see Athey and Erdle 2022 for a comprehensive review). Care is needed, however, since many natural-based polymers such as cotton or wool, can be easily degraded by the most commonly used digestion protocols (e.g., KOH or H_2O_2 solutions), thus resulting in an underestimation of cellulosic microfiber abundance (Collard et al. 2015; Rodríguez-Romeu et al. 2020; Prata et al. 2020b). Other problems related to the detection and quantification of textile fibers in the marine environment are the use of different units (e.g., fibers per area, mass, or volume) and a lack of standardization concerning sampling methods, extraction, and analysis of microfibers, as it has been already emphasized and discussed multiple times (Gago et al. 2018). Furthermore, several studies are still reporting total 'microplastic' concentrations without discriminating between microplastic particles and fibers. Further challenges to overcome during sampling are the lack of an appropriate number of replicates and the non-homogeneous distribution of textile fibers in the water column (Ryan et al. 2020). Overall, sampling for textile fibers necessitates a strong study design with careful consideration of sampling mesh sizes, depth, and volume.

Once a sample is concentrated on a filter, most studies examining the occurrence of textile fibers in the marine environment use optical tools for fiber quantification, which is often performed under a traditional dissecting stereomicroscope (Gago et al. 2018). Fiber quantification can be complemented by the use of intermediate screening steps such as staining with Nile Red and/or fluorescence microscopy techniques. The main shortcoming of these methods is the inaccurate distinction between synthetic and natural-based fibers, which can lead to a substantial overestimation of the actual microfibers counts (Suaria et al. 2020). Visual identification protocols and morphometric/morphologic workflows are being developed (e.g., Zhu et al. 2019; Prata et al. 2020a), however, to reliably determine the polymer composition of the collected fibers, chemical characterization is the most reliable and most commonly used method for final fiber identification. This often involves costly instrumentation and elaborated validation techniques such as Raman or Fourier Transform Infrared (FTIR) spectroscopy, or pyrolysis gas chromatography–mass spectrometry (Pyr-GC/MS). FTIR is widely used, and it is a particularly suitable method for microfiber identification because it offers highly characteristic information, it is relatively easy to use, and most of all, is a non-destructive technique (Comnea-Stancu et al. 2017; Peets et al. 2017; Käppler et al. 2018; Cai et al. 2019). These methods are, however, costly and time-consuming and as such, they are often used only for the verification of a randomly selected representative sub-sample. Additional challenges of fiber characterization relevant to all analytical methods are precise counting and length measurements, especially of twisted, entangled, and overlaying fibers (Primpke et al. 2019). Furthermore, airborne fibers are a steady source of potential contamination during sampling or analysis in the laboratory (Woodall et al. 2015). Textile fibers of lab coats and clothes underneath can accidentally get introduced into samples and make the adoption of contamination control procedures strictly necessary when analyzing microfibers (Torre et al. 2016). When possible, it is best to carefully implement procedural blanks, negative control samples, and positive (fiber-spiked) control samples throughout the monitoring procedure. All steps besides environmental sampling should be conducted in clean conditions, such as clean air facilities and ideally making use of a forensic approach (Gwinnett and Miller 2021). Basically, when dealing with microfibers, it is imperative to adopt robust quality assurance/quality control (QA/QC) protocols to reduce, assess, and characterize the extent of external sample contamination (Prata et al. 2021).

11.6 OUTLOOK AND FINAL REMARKS

Microfibers are an "*unavoidable byproduct of contemporary lifestyles*" (Carr 2017). A large number of textile fibers have been found in seawater, sediments, and organisms from almost every marine habitat studied around the world. The environmental abundance of these fibers has been reported to be more than five times that of traditional secondary plastic fragments. Given the projected increase in textile production (Textile Exchange 2023), there is little doubt that these numbers will continue to increase for the foreseeable future. Our best hope for reducing the environmental loads of this emerging contaminant may lie in improving textile manufacturing technologies, favoring materials, finishings and textile characteristics which improve yarn durability and reduce microfiber shedding during use (Cesa et al. 2020; Kim et al. 2022). Another possibility is improving the biodegradability of polymers currently used in the manufacturing of fabrics and textiles in order to reduce their residence time in the marine environment (Lott et al. 2022). There is still a lack of consistency in the sampling and extraction techniques used to quantify microfibers in the marine environment due to the rapid development of microplastic research. It is frequently impossible to compare the microplastic concentrations reported by various studies due to the wide variation in the analytical protocols used. The majority of these discrepancies are explained by (i) variations in the lower and upper size limits used, (ii) the sensitivity of the applied extraction technique, and (iii) variations in the sampling technique leading to a wide range of reporting units. To ensure that reliable monitoring programs are implemented, there is an urgent need for standardization of microfiber analytical protocols, which is of vital importance for identifying hotspot areas and organisms or ecosystems at particular risk. It is challenging to control the presence of microfibers in our daily lives, and cross-contamination affects every step of the sampling procedure. Rigorous QA/QC procedures ensuring minimal contamination levels are critical to properly assess the risk that microfibers pose to the marine environment.

The widespread distribution and accumulation of microfibers in the marine environment raise concerns regarding the interaction and potential effects of microfibers on marine biota (Kwak et al. 2022). Both suspension and deposit feeders may unintentionally or deliberately consume microfibers as a result of their abundance in plankton, sediments, and pelagic organisms. When microfibers are ingested, the leaching of adsorbed contaminants and additives could represent a source of toxic substances that affect the organisms and enter the food chain that ultimately feeds humans. Measures to reduce the input of microfibers into the marine environment are urgently needed. The use of natural fibers is being advocated as a strategy to reduce inputs and risks of microfibers for the marine ecosystem (Henry et al. 2019). However, animal and cellulosic fibers are still greatly underrepresented in the environmental pollution literature, despite consistently representing the majority of the fibers usually retrieved from natural environments (Athey and Erdle 2022). Microfiber prevalence, fate, and effects research is still in its infancy, and it frequently favors synthetic polymers. The degradation of natural fibers in comparison to synthetic polymers in a variety of different environmental conditions clearly requires more research. Studies have also shown that it is paramount to develop reliable methods to quickly and reliably distinguish natural fibers from synthetic ones. Future research should also more thoroughly examine how natural and synthetic fibers affect marine biota. Since there is little doubt that the ingestion of textile fibers is now a widespread global phenomenon. For the purpose of evaluating their potential effects on the marine environment and on ecosystems all over the world, it is essential to comprehend the ecological effects and biodegradation rates of natural and synthetic fibers in a variety of environmental conditions.

REFERENCES

Adams, J. K., Dean, B. Y., Athey, S. N., Jantunen, L. M., Bernstein, S., Stern, G., Diamond, M. L., & Finkelstein, S. A. (2021). Anthropogenic particles (including microfibers and microplastics) in marine sediments of the Canadian Arctic. *Science of the Total Environment*, *784*, 147155. https://doi.org/10.1016/j.scitotenv.2021.147155

Allen, D., Allen, S., Abbasi, S., Baker, A., Bergmann, M., Brahney, J., Butler, T., Duce, R. A., Eckhardt, S., Evangeliou, N., Jickells, T., Kanakidou, M., Kershaw, P., Laj, P., Levermore, J., Li, D., Liss, P., Liu, K., Mahowald, N., … Wright, S. (2022). Microplastics and nanoplastics in the marine-atmosphere environment. *Nature Reviews Earth & Environment*, *3*(6), 393–405. https://doi.org/10.1038/s43017-022-00292-x

Alurralde, G., Isla, E., Fuentes, V., Olariaga, A., Maggioni, T., Rimondino, G., & Tatián, M. (2022). Anthropogenic microfibres flux in an Antarctic coastal ecosystem: The tip of an iceberg? *Marine Pollution Bulletin*, *175*, 113388. https://doi.org/10.1016/j.marpolbul.2022.113388

Arias, A. H., Alfonso, M. B., Girones, L., Piccolo, M. C., & Marcovecchio, J. E. (2022). Synthetic microfibers and tyre wear particles pollution in aquatic systems: Relevance and mitigation strategies. *Environmental Pollution*, *295*, 118607. https://doi.org/10.1016/j.envpol.2021.118607

Athey, S. N., & Erdle, L. M. (2022). Are we underestimating anthropogenic microfiber pollution? A critical review of occurrence, methods, and reporting. *Environmental Toxicology and Chemistry*, *41*(4), 822–837. https://doi.org/10.1002/etc.5173

Avio, C. G., Pittura, L., d'Errico, G., Abel, S., Amorello, S., Marino, G., Gorbi, S., & Regoli, F. (2020). Distribution and characterization of microplastic particles and textile microfibers in Adriatic food webs: General insights for biomonitoring strategies. *Environmental Pollution*, *258*, 113766. https://doi.org/10.1016/j.envpol.2019.113766

Bagaev, A., Mizyuk, A., Khatmullina, L., Isachenko, I., & Chubarenko, I. (2017). Anthropogenic fibres in the Baltic Sea water column: Field data, laboratory and numerical testing of their motion. *Science of the Total Environment*, *599–600*, 560–571. https://doi.org/10.1016/j.scitotenv.2017.04.185

Barrows, A. P. W., Cathey, S. E., & Petersen, C. W. (2018). Marine environment microfiber contamination: Global patterns and the diversity of microparticle origins. *Environmental Pollution*, *237*, 275–284. https://doi.org/10.1016/j.envpol.2018.02.062

Belontz, S. L., & Corcoran, P. L. (2021). Prioritizing suitable quality assurance and control standards to reduce laboratory airborne microfibre contamination in sediment samples. *Environments*, *8*(9), 89. https://doi.org/10.3390/environments8090089

Bergami, E., Ferrari, E., Löder, M. G. J., Birarda, G., Laforsch, C., Vaccari, L., & Corsi, I. (2023). Textile microfibers in wild Antarctic whelk Neobuccinum eatoni (Smith, 1875) from Terra Nova Bay (Ross sea, Antarctica). *Environmental Research*, *216*, 114487. https://doi.org/10.1016/j.envres.2022.114487

Birkenhead, J., Radford, F., Stead, J. L., Cundy, A. B., & Hudson, M. D. (2020). Validation of a method to quantify microfibres present in aquatic surface microlayers. *Scientific Reports*, *10*(1), 17892. https://doi.org/10.1038/s41598-020-74635-3

Bottari, T., Nibali, V. C., Branca, C., Grotti, M., Savoca, S., Romeo, T., Spanò, N., Azzaro, M., Greco, S., D'Angelo, G., & Mancuso, M. (2022). Anthropogenic microparticles in the emerald rockcod Trematomus bernacchii (Nototheniidae) from the Antarctic. *Scientific Reports*, *12*(1), 17214. https://doi.org/10.1038/s41598-022-21670-x

Bour, A., Avio, C. G., Gorbi, S., Regoli, F., & Hylland, K. (2018). Presence of microplastics in benthic and epibenthic organisms: Influence of habitat, feeding mode and trophic level. *Environmental Pollution*, *243*, 1217–1225. https://doi.org/10.1016/j.envpol.2018.09.115

Bucci, K., Tulio, M., & Rochman, C. M. (2020). What is known and unknown about the effects of plastic pollution: A meta-analysis and systematic review. *Ecological Applications*, *30*(2). https://doi.org/10.1002/eap.2044

Cai, H., Du, F., Li, L., Li, B., Li, J., & Shi, H. (2019). A practical approach based on FT-IR spectroscopy for identification of semi-synthetic and natural celluloses in microplastic investigation. *Science of the Total Environment*, *669*, 692–701. https://doi.org/10.1016/j.scitotenv.2019.03.124

Capillo, G., Savoca, S., Panarello, G., Mancuso, M., Branca, C., Romano, V., D'Angelo, G., Bottari, T., & Spanò, N. (2020). Quali-quantitative analysis of plastics and synthetic microfibers found in demersal species from Southern Tyrrhenian Sea (Central mediterranean). *Marine Pollution Bulletin*, *150*, 110596. https://doi.org/10.1016/j.marpolbul.2019.110596

Carney Almroth, B. M., Åström, L., Roslund, S., Petersson, H., Johansson, M., & Persson, N.-K. (2018). Quantifying shedding of synthetic fibers from textiles; a source of microplastics released into the environment. *Environmental Science and Pollution Research*, *25*(2), 1191–1199. https://doi.org/10.1007/s11356-017-0528-7

Carr, S. A. (2017). Sources and dispersive modes of micro-fibers in the environment: Environmental Microfiber Sources. *Integrated Environmental Assessment and Management*, *13*(3), 466–469. https://doi.org/10.1002/ieam.1916

Cesa, F. S., Turra, A., & Baruque-Ramos, J. (2017). Synthetic fibers as microplastics in the marine environment: A review from textile perspective with a focus on domestic washings. *Science of the Total Environment*, *598*, 1116–1129. https://doi.org/10.1016/j.scitotenv.2017.04.172

Cesa, F. S., Turra, A., Checon, H. H., Leonardi, B., & Baruque-Ramos, J. (2020). Laundering and textile parameters influence fibers release in household washings. *Environmental Pollution*, *257*, 113553. https://doi.org/10.1016/j.envpol.2019.113553

Chen, R., & Jakes, K. A. (2001). Cellulolytic biodegradation of cotton fibers from a deep-ocean environment. *Journal of the American Institute for Conservation*, *40*(2), 91–103. https://doi.org/10.1179/019713601806113076

Collard, F., Gilbert, B., Eppe, G., Parmentier, E., & Das, K. (2015). Detection of anthropogenic particles in fish stomachs: An isolation method adapted to identification by Raman spectroscopy. *Archives of Environmental Contamination and Toxicology*, *69*(3), 331–339. https://doi.org/10.1007/s00244-015-0221-0

Comnea-Stancu, I. R., Wieland, K., Ramer, G., Schwaighofer, A., & Lendl, B. (2017). On the identification of rayon/viscose as a major fraction of microplastics in the marine environment: Discrimination between natural and manmade cellulosic fibers using Fourier transform infrared spectroscopy. *Applied Spectroscopy*, *71*(5), 939–950. https://doi.org/10.1177/0003702816660725

Compa, M., Alomar, C., Ventero, A., Iglesias, M., & Deudero, S. (2022). Anthropogenic particles in the zooplankton aggregation layer and ingestion in fish species along the Catalan continental shelf. *Estuarine, Coastal and Shelf Science*, *277*, 108041. https://doi.org/10.1016/j.ecss.2022.108041

Compa, M., Ventero, A., Iglesias, M., & Deudero, S. (2018). Ingestion of microplastics and natural fibres in Sardina pilchardus (Walbaum, 1792) and Engraulis encrasicolus (Linnaeus, 1758) along the Spanish Mediterranean coast. *Marine Pollution Bulletin*, *128*, 89–96. https://doi.org/10.1016/j.marpolbul.2018.01.009

Coppock, R. L., Galloway, T. S., Cole, M., Fileman, E. S., Queirós, A. M., & Lindeque, P. K. (2019). Microplastics alter feeding selectivity and faecal density in the copepod, Calanus helgolandicus. *Science of the Total Environment*, *687*, 780–789. https://doi.org/10.1016/j.scitotenv.2019.06.009

De Falco, F., Cocca, M., Avella, M., & Thompson, R. C. (2020). Microfiber release to water, via laundering, and to air, via everyday use: A comparison between polyester clothing with differing textile parameters. *Environmental Science & Technology*, *54*(6), 3288–3296. https://doi.org/10.1021/acs.est.9b06892

De Falco, F., Gullo, M. P., Gentile, G., Di Pace, E., Cocca, M., Gelabert, L., Brouta-Agnésa, M., Rovira, A., Escudero, R., Villalba, R., Mossotti, R., Montarsolo, A., Gavignano, S., Tonin, C., & Avella, M. (2018). Evaluation of microplastic release caused by textile washing processes of synthetic fabrics. *Environmental Pollution*, *236*, 916–925. https://doi.org/10.1016/j.envpol.2017.10.057

Desforges, J.-P. W., Galbraith, M., Dangerfield, N., & Ross, P. S. (2014). Widespread distribution of microplastics in subsurface seawater in the NE Pacific Ocean. *Marine Pollution Bulletin*, *79*(1–2), 94–99. https://doi.org/10.1016/j.marpolbul.2013.12.035

Desforges, J.-P. W., Galbraith, M., & Ross, P. S. (2015). Ingestion of microplastics by zooplankton in the northeast Pacific Ocean. *Archives of Environmental Contamination and Toxicology*, *69*(3), 320–330. https://doi.org/10.1007/s00244-015-0172-5

Dreillard, M., Barros, C. D. F., Rouchon, V., Emonnot, C., Lefebvre, V., Moreaud, M., Guillaume, D., Rimbault, F., & Pagerey, F. (2022). Quantification and morphological characterization of microfibers emitted from textile washing. *Science of the Total Environment*, *832*, 154973. https://doi.org/10.1016/j.scitotenv.2022.154973

Dris, R., Gasperi, J., Mirande, C., Mandin, C., Guerrouache, M., Langlois, V., & Tassin, B. (2017). A first overview of textile fibers, including microplastics, in indoor and outdoor environments. *Environmental Pollution*, *221*, 453–458. https://doi.org/10.1016/j.envpol.2016.12.013

Dris, R., Gasperi, J., Saad, M., Mirande, C., & Tassin, B. (2016). Synthetic fibers in atmospheric fallout: A source of microplastics in the environment? *Marine Pollution Bulletin*, *104*(1–2), 290–293. https://doi.org/10.1016/j.marpolbul.2016.01.006

du Roure, O., Lindner, A., Nazockdast, E. N., & Shelley, M. J. (2019). Dynamics of flexible fibers in viscous flows and fluids. *Annual Review of Fluid Mechanics*, *51*(1), 539–572. https://doi.org/10.1146/annurev-fluid-122316-045153

Evangeliou, N., Grythe, H., Klimont, Z., Heyes, C., Eckhardt, S., Lopez-Aparicio, S., & Stohl, A. (2020). Atmospheric transport is a major pathway of microplastics to remote regions. *Nature Communications*, *11*(1), 3381. https://doi.org/10.1038/s41467-020-17201-9

Fagiano, V., Compa, M., Alomar, C., Rios-Fuster, B., Morató, M., Capó, X., & Deudero, S. (2023). Breaking the paradigm: Marine sediments hold two-fold microplastics than sea surface waters and are dominated by fibers. *Science of the Total Environment*, *858*, 159722. https://doi.org/10.1016/j.scitotenv.2022.159722

Finnegan, A. M. D., Süsserott, R., Gabbott, S. E., & Gouramanis, C. (2022). Man-made natural and regenerated cellulosic fibres greatly outnumber microplastic fibres in the atmosphere. *Environmental Pollution*, *310*, 119808. https://doi.org/10.1016/j.envpol.2022.119808

Frias, J. P., & Nash, R. (2019). Microplastics: Finding a consensus on the definition. *Marine Pollution Bulletin*, *138*, 145–147.

Gago, J., Carretero, O., Filgueiras, A. V., & Viñas, L. (2018). Synthetic microfibers in the marine environment: A review on their occurrence in seawater and sediments. *Marine Pollution Bulletin*, *127*, 365–376. https://doi.org/10.1016/j.marpolbul.2017.11.070

Geyer, R., Jambeck, J. R., & Law, K. L. (2017). Production, use, and fate of all plastics ever made. *Science Advances*, *3*(7), e1700782. https://doi.org/10.1126/sciadv.1700782

Gouin, T. (2020). Toward an improved understanding of the ingestion and trophic transfer of microplastic particles: Critical review and implications for future research. *Environmental Toxicology and Chemistry*, *39*(6), 1119–1137. https://doi.org/10.1002/etc.4718

Gusmão, F., Domenico, M. D., Amaral, A. C. Z., Martínez, A., Gonzalez, B. C., Worsaae, K., Ivar do Sul, J. A., & Cunha Lana, P. da. (2016). In situ ingestion of microfibres by meiofauna from sandy beaches. *Environmental Pollution*, *216*, 584–590. https://doi.org/10.1016/j.envpol.2016.06.015

Gwinnett, C., & Miller, R. Z. (2021). Are we contaminating our samples? A preliminary study to investigate procedural contamination during field sampling and processing for microplastic and anthropogenic microparticles. *Marine Pollution Bulletin*, *173*, 113095. https://doi.org/10.1016/j.marpolbul.2021.113095

Halsband, C. (2022). Effects of biofouling on the sinking behavior of microplastics in aquatic environments. In T. Rocha-Santos, M. Costa, & C. Mouneyrac (eds.), *Handbook of Microplastics in the Environment* (pp. 1–13). Springer International Publishing. https://doi.org/10.1007/978-3-030-10618-8_12-1

Henry, B., Laitala, K., & Klepp, I. G. (2019). Microfibres from apparel and home textiles: Prospects for including microplastics in environmental sustainability assessment. *Science of the Total Environment*, *652*, 483–494. https://doi.org/10.1016/j.scitotenv.2018.10.166

Hu, L., Chernick, M., Lewis, A. M., Ferguson, P. L., & Hinton, D. E. (2020). Chronic microfiber exposure in adult Japanese medaka (Oryzias latipes). *PLoS One*, *15*(3), e0229962. https://doi.org/10.1371/journal.pone.0229962

Iyare, P. U., Ouki, S. K., & Bond, T. (2020). Microplastics removal in wastewater treatment plants: A critical review. *Environmental Science: Water Research & Technology*, *6*(10), 2664–2675. https://doi.org/10.1039/D0EW00397B

Kang, J.-H., Kwon, O.-Y., Hong, S. H., & Shim, W. J. (2020). Can zooplankton be entangled by microfibers in the marine environment? : Laboratory studies. *Water*, *12*(12), 3302. https://doi.org/10.3390/w12123302

Käppler, A., Fischer, M., Scholz-Böttcher, B. M., Oberbeckmann, S., Labrenz, M., Fischer, D., Eichhorn, K.-J., & Voit, B. (2018). Comparison of μ-ATR-FTIR spectroscopy and py-GCMS as identification tools for microplastic particles and fibers isolated from river sediments. *Analytical and Bioanalytical Chemistry*, *410*(21), 5313–5327. https://doi.org/10.1007/s00216-018-1185-5

Kim, S., Cho, Y., & Park, C. H. (2022). Effect of cotton fabric properties on fiber release and marine biodegradation. *Textile Research Journal*, *92*(11–12), 2121–2137. https://doi.org/10.1177/00405175211068781

Kwak, J. I., Liu, H., Wang, D., Lee, Y. H., Lee, J.-S., & An, Y.-J. (2022). Critical review of environmental impacts of microfibers in different environmental matrices. *Comparative Biochemistry and Physiology Part C: Toxicology & Pharmacology*, *251*, 109196. https://doi.org/10.1016/j.cbpc.2021.109196

Le Guen, C., Suaria, G., Sherley, R. B., Ryan, P. G., Aliani, S., Boehme, L., & Brierley, A. S. (2020). Microplastic study reveals the presence of natural and synthetic fibres in the diet of King Penguins (Aptenodytes patagonicus) foraging from South Georgia. *Environment International*, *134*, 105303. https://doi.org/10.1016/j.envint.2019.105303

Li, L., Frey, M., & Browning, K. J. (2010). Biodegradability study on cotton and polyester fabrics. *Journal of Engineered Fibers and Fabrics*, *5*(4), 155892501000500. https://doi.org/10.1177/155892501000500406

Li, Y., Lu, Q., Xing, Y., Liu, K., Ling, W., Yang, J., Yang, Q., Wu, T., Zhang, J., Pei, Z., Gao, Z., Li, X., Yang, F., Ma, H., Liu, K., & Zhao, D. (2023). Review of research on migration, distribution, biological effects, and analytical methods of microfibers in the environment. *Science of the Total Environment*, *855*, 158922. https://doi.org/10.1016/j.scitotenv.2022.158922

Liebezeit, G., & Dubaish, F. (2012). Microplastics in beaches of the East Frisian Islands spiekeroog and kachelotplate. *Bulletin of Environmental Contamination and Toxicology*, *89*(1), 213–217. https://doi.org/10.1007/s00128-012-0642-7

Lindström, S. B., & Uesaka, T. (2007). Simulation of the motion of flexible fibers in viscous fluid flow. *Physics of Fluids*, *19*(11), 113307. https://doi.org/10.1063/1.2778937

Liu, J., Liu, Q., An, L., Wang, M., Yang, Q., Zhu, B., Ding, J., Ye, C., & Xu, Y. (2022). Microfiber pollution in the earth system. *Reviews of Environmental Contamination and Toxicology*, *260*(1), 13. https://doi.org/10.1007/s44169-022-00015-9

Liu, K., Wang, X., Song, Z., Wei, N., Ye, H., Cong, X., Zhao, L., Li, Y., Qu, L., Zhu, L., Zhang, F., Zong, C., Jiang, C., & Li, D. (2020). Global inventory of atmospheric fibrous microplastics input into the ocean: An implication from the indoor origin. *Journal of Hazardous Materials*, *400*, 123223. https://doi.org/10.1016/j.jhazmat.2020.123223

Liu, J., Yang, Y., Ding, J., Zhu, B., & Gao, W. (2019). Microfibers: A preliminary discussion on their definition and sources. *Environmental Science and Pollution Research*, *26*(28), 29497–29501. https://doi.org/10.1007/s11356-019-06265-w

Liu, W., Zhang, J., Liu, H., Guo, X., Zhang, X., Yao, X., Cao, Z., & Zhang, T. (2021). A review of the removal of microplastics in global wastewater treatment plants: Characteristics and mechanisms. *Environment International*, *146*, 106277. https://doi.org/10.1016/j.envint.2020.106277

Lott, C., Eich, A., & Weber, M. (2022). Degradation of fibrous microplastics in the marine environment. In *Polluting Textiles: The Problem with Microfibres* (1st ed., pp. 185–213). Routledge. https://www.taylorfrancis.com/chapters/edit/10.4324/9781003165385-10/degradation-fibrous-microplastics-marine-environment-christian-lott-andreas-eich-miriam-weber

Lourenço, P. M., Serra-Gonçalves, C., Ferreira, J. L., Catry, T., & Granadeiro, J. P. (2017). Plastic and other microfibers in sediments, macroinvertebrates and shorebirds from three intertidal wetlands of southern Europe and West Africa. *Environmental Pollution*, *231*, 123–133. https://doi.org/10.1016/j.envpol.2017.07.103

Lykaki, M., Zhang, Y.-Q., Markiewicz, M., Brandt, S., Kolbe, S., Schrick, J., Rabe, M., & Stolte, S. (2021). The influence of textile finishing agents on the biodegradability of shed fibres. *Green Chemistry*, *23*(14), 5212–5221. https://doi.org/10.1039/D1GC00883H

Macieira, R. M., Oliveira, L. A. S., Cardozo-Ferreira, G. C., Pimentel, C. R., Andrades, R., Gasparini, J. L., Sarti, F., Chelazzi, D., Cincinelli, A., Gomes, L. C., & Giarrizzo, T. (2021). Microplastic and artificial cellulose microfibers ingestion by reef fishes in the Guarapari Islands, southwestern Atlantic. *Marine Pollution Bulletin*, *167*, 112371. https://doi.org/10.1016/j.marpolbul.2021.112371

Mishra, S., Rath, C. Charan, & Das, A. P. (2019). Marine microfiber pollution: A review on present status and future challenges. *Marine Pollution Bulletin*, *140*, 188–197. https://doi.org/10.1016/j.marpolbul.2019.01.039

Mizraji, R., Ahrendt, C., Perez-Venegas, D., Vargas, J., Pulgar, J., Aldana, M., Patricio Ojeda, F., Duarte, C., & Galbán-Malagón, C. (2017). Is the feeding type related with the content of microplastics in intertidal fish gut? *Marine Pollution Bulletin*, *116*(1–2), 498–500. https://doi.org/10.1016/j.marpolbul.2017.01.008

Morgana, S., Ghigliotti, L., Estévez-Calvar, N., Stifanese, R., Wieckzorek, A., Doyle, T., Christiansen, J. S., Faimali, M., & Garaventa, F. (2018). Microplastics in the Arctic: A case study with sub-surface water and fish samples off Northeast Greenland. *Environmental Pollution*, *242*, 1078–1086. https://doi.org/10.1016/j.envpol.2018.08.001

Napper, I. E., Parker-Jurd, F. N. F., Wright, S. L., & Thompson, R. C. (2023). Examining the release of synthetic microfibres to the environment via two major pathways: Atmospheric deposition and treated wastewater effluent. *Science of the Total Environment*, *857*, 159317. https://doi.org/10.1016/j.scitotenv.2022.159317

Ngo, P. L., Pramanik, B. K., Shah, K., & Roychand, R. (2019). Pathway, classification and removal efficiency of microplastics in wastewater treatment plants. *Environmental Pollution*, *255*, 113326. https://doi.org/10.1016/j.envpol.2019.113326

Paluselli, A., Suaria, G., Musso, M., Bassotto D., Vitale, G., Borghini, M., & Aliani, S. (2022, November 14–18). Vertical distribution and transport of textile microfibers (MFs) in the Mediterranean water column. In *MICRO 2022 International Conference on Plastic Pollution – From Micro to Nano*. Online Atlas Edition.

Pedrotti, M. L., Petit, S., Eyheraguibel, B., Kerros, M. E., Elineau, A., Ghiglione, J. F., Loret, J. F., Rostan, A., & Gorsky, G. (2021). Pollution by anthropogenic microfibers in North-West Mediterranean Sea and efficiency of microfiber removal by a wastewater treatment plant. *Science of the Total Environment*, *758*, 144195. https://doi.org/10.1016/j.scitotenv.2020.144195

Peets, P., Leito, I., Pelt, J., & Vahur, S. (2017). Identification and classification of textile fibres using ATR-FT-IR spectroscopy with chemometric methods. *Spectrochimica Acta Part A: Molecular and Biomolecular Spectroscopy*, *173*, 175–181. https://doi.org/10.1016/j.saa.2016.09.007

Perez-Venegas, D. J., Seguel, M., Pavés, H., Pulgar, J., Urbina, M., Ahrendt, C., & Galbán-Malagón, C. (2018). First detection of plastic microfibers in a wild population of South American fur seals (Arctocephalus australis) in the Chilean Northern Patagonia. *Marine Pollution Bulletin*, *136*, 50–54. https://doi.org/10.1016/j.marpolbul.2018.08.065

Piarulli, S., Scapinello, S., Comandini, P., Magnusson, K., Granberg, M., Wong, J. X. W., Sciutto, G., Prati, S., Mazzeo, R., Booth, A. M., & Airoldi, L. (2019). Microplastic in wild populations of the omnivorous crab Carcinus aestuarii: A review and a regional-scale test of extraction methods, including microfibres. *Environmental Pollution*, *251*, 117–127. https://doi.org/10.1016/j.envpol.2019.04.092

Prata, J. C., Castro, J. L., da Costa, J. P., Duarte, A. C., Cerqueira, M., & Rocha-Santos, T. (2020a). An easy method for processing and identification of natural and synthetic microfibers and microplastics in indoor and outdoor air. *MethodsX*, *7*, 100762. https://doi.org/10.1016/j.mex.2019.11.032

Prata, J. C., Castro, J. L., da Costa, J. P., Duarte, A. C., Rocha-Santos, T., & Cerqueira, M. (2020b). The importance of contamination control in airborne fibers and microplastic sampling: Experiences from indoor and outdoor air sampling in Aveiro, Portugal. *Marine Pollution Bulletin*, *159*, 111522. https://doi.org/10.1016/j.marpolbul.2020.111522

Prata, J. C., Reis, V., da Costa, J. P., Mouneyrac, C., Duarte, A. C., & Rocha-Santos, T. (2021). Contamination issues as a challenge in quality control and quality assurance in microplastics analytics. *Journal of Hazardous Materials*, *403*, 123660. https://doi.org/10.1016/j.jhazmat.2020.123660

Primpke, S., Dias, P., & Gerdts, G. (2019). Automated identification and quantification of microfibres and microplastics. *Analytical Methods*, *11*(16), 2138–2147. https://doi.org/10.1039/C9AY00126C

Provencher, J. F., Vermaire, J. C., Avery-Gomm, S., Braune, B. M., & Mallory, M. L. (2018). Garbage in guano? Microplastic debris found in faecal precursors of seabirds known to ingest plastics. *Science of the Total Environment*, *644*, 1477–1484. https://doi.org/10.1016/j.scitotenv.2018.07.101

Rebelein, A., Int-Veen, I., Kammann, U., & Scharsack, J. P. (2021). Microplastic fibers – Underestimated threat to aquatic organisms? *Science of the Total Environment*, *777*, 146045. https://doi.org/10.1016/j.scitotenv.2021.146045

Reineccius, J., Appelt, J.-S., Hinrichs, T., Kaiser, D., Stern, J., Prien, R. D., & Waniek, J. J. (2020). Abundance and characteristics of microfibers detected in sediment trap material from the deep subtropical North Atlantic Ocean. *Science of the Total Environment*, *738*, 140354. https://doi.org/10.1016/j.scitotenv.2020.140354

Remy, F., Collard, F., Gilbert, B., Compère, P., Eppe, G., & Lepoint, G. (2015). When microplastic is not plastic: The ingestion of artificial cellulose fibers by macrofauna living in seagrass macrophytodetritus. *Environmental Science & Technology*, *49*(18), 11158–11166. https://doi.org/10.1021/acs.est.5b02005

Rios-Fuster, B., Compa, M., Alomar, C., Fagiano, V., Ventero, A., Iglesias, M., & Deudero, S. (2022). Ubiquitous vertical distribution of microfibers within the upper epipelagic layer of the western Mediterranean Sea. *Estuarine, Coastal and Shelf Science*, *266*, 107741. https://doi.org/10.1016/j.ecss.2022.107741

Rochman, C. M., Tahir, A., Williams, S. L., Baxa, D. V., Lam, R., Miller, J. T., Teh, F.-C., Werorilangi, S., & Teh, S. J. (2015). Anthropogenic debris in seafood: Plastic debris and fibers from textiles in fish and bivalves sold for human consumption. *Scientific Reports*, *5*(1), 14340. https://doi.org/10.1038/srep14340

Rodríguez-Romeu, O., Constenla, M., Carrassón, M., Campoy-Quiles, M., & Soler-Membrives, A. (2020). Are anthropogenic fibres a real problem for red mullets (Mullus barbatus) from the NW Mediterranean? *Science of the Total Environment*, *733*, 139336. https://doi.org/10.1016/j.scitotenv.2020.139336

Ryan, P. G., Suaria, G., Perold, V., Pierucci, A., Bornman, T. G., & Aliani, S. (2020). Sampling microfibres at the sea surface: The effects of mesh size, sample volume and water depth. *Environmental Pollution*, *258*, 113413. https://doi.org/10.1016/j.envpol.2019.113413

Sait, S. T. L., Sørensen, L., Kubowicz, S., Vike-Jonas, K., Gonzalez, S. V., Asimakopoulos, A. G., & Booth, A. M. (2021). Microplastic fibres from synthetic textiles: Environmental degradation and additive chemical content. *Environmental Pollution*, *268*, 115745. https://doi.org/10.1016/j.envpol.2020.115745

Sanchez-Vidal, A., Thompson, R. C., Canals, M., & de Haan, W. P. (2018). The imprint of microfibres in southern European deep seas. *PLoS One*, *13*(11), e0207033. https://doi.org/10.1371/journal.pone.0207033

Santini, S., De Beni, E., Martellini, T., Sarti, C., Randazzo, D., Ciraolo, R., Scopetani, C., & Cincinelli, A. (2022). Occurrence of natural and synthetic micro-fibers in the Mediterranean sea: A review. *Toxics*, *10*(7), 391. https://doi.org/10.3390/toxics10070391

Santonicola, S., Volgare, M., Di Pace, E., Cocca, M., Mercogliano, R., & Colavita, G. (2021). Occurrence of potential plastic microfibers in mussels and anchovies sold for human consumption: Preliminary results. *Italian Journal of Food Safety*, *10*(4). https://doi.org/10.4081/ijfs.2021.9962

Savoca, S., Capillo, G., Mancuso, M., Faggio, C., Panarello, G., Crupi, R., Bonsignore, M., D'Urso, L., Compagnini, G., Neri, F., Fazio, E., Romeo, T., Bottari, T., & Spanò, N. (2019). Detection of artificial cellulose microfibers in Boops boops from the northern coasts of Sicily (Central mediterranean). *Science of the Total Environment*, *691*, 455–465. https://doi.org/10.1016/j.scitotenv.2019.07.148

Sbrana, A., Valente, T., Scacco, U., Bianchi, J., Silvestri, C., Palazzo, L., de Lucia, G. A., Valerani, C., Ardizzone, G., & Matiddi, M. (2020). Spatial variability and influence of biological parameters on microplastic ingestion by Boops boops (L.) along the Italian coasts (Western Mediterranean sea). *Environmental Pollution, 263*, 114429. https://doi.org/10.1016/j.envpol.2020.114429

Siddiqui, S., Hutton, S. J., Dickens, J. M., Pedersen, E. I., Harper, S. L., & Brander, S. M. (2023). Natural and synthetic microfibers alter growth and behavior in early life stages of estuarine organisms. *Frontiers in Marine Science*, *9*, 991650. https://doi.org/10.3389/fmars.2022.991650

Sørensen, L., Groven, A. S., Hovsbakken, I. A., Del Puerto, O., Krause, D. F., Sarno, A., & Booth, A. M. (2021). UV degradation of natural and synthetic microfibers causes fragmentation and release of polymer degradation products and chemical additives. *Science of the Total Environment*, *755*, 143170. https://doi.org/10.1016/j.scitotenv.2020.143170

Suaria, G., Achtypi, A., Perold, V., Lee, J. R., Pierucci, A., Bornman, T. G., Aliani, S., & Ryan, P. G. (2020). Microfibers in oceanic surface waters: A global characterization. *Science Advances*, *6*(23), eaay8493. https://doi.org/10.1126/sciadv.aay8493

Suaria, G., Berta, M., Griffa, A., Molcard, A., Özgökmen, T. M., Zambianchi, E., & Aliani, S. (2021). *Dynamics of Transport, Accumulation, and Export of Plastics at Oceanic Fronts*. Springer: Berlin Heidelberg. https://doi.org/10.1007/698_2021_814

Sui, M., Lu, Y., Wang, Q., Hu, L., Huang, X., & Liu, X. (2020). Distribution patterns of microplastics in various tissues of the Zhikong scallop (Chlamys farreri) and in the surrounding culture seawater. *Marine Pollution Bulletin, 160*, 111595. https://doi.org/10.1016/j.marpolbul.2020.111595

Sun, J., Dai, X., Wang, Q., van Loosdrecht, M. C. M., & Ni, B.-J. (2019). Microplastics in wastewater treatment plants: Detection, occurrence and removal. *Water Research, 152*, 21–37. https://doi.org/10.1016/j.watres.2018.12.050

Taylor, M. L., Gwinnett, C., Robinson, L. F., & Woodall, L. C. (2016). Plastic microfibre ingestion by deep-sea organisms. *Scientific Reports*, *6*(1), 33997. https://doi.org/10.1038/srep33997

Textile Exchange (2023). Materials Market Report. December 2023. https://textileexchange.org/app/uploads/2023/11/Materials-Market-Report-2023.pdf

Tirelli, V., Suaria, G., & Lusher, A. L. (2022). Microplastics in polar samples. In T. Rocha-Santos, M. F. Costa, & C. Mouneyrac (eds.), *Handbook of Microplastics in the Environment* (pp. 281–322). Springer International Publishing. https://doi.org/10.1007/978-3-030-39041-9_4

Torre, M., Digka, N., Anastasopoulou, A., Tsangaris, C., & Mytilineou, C. (2016). Anthropogenic microfibres pollution in marine biota. A new and simple methodology to minimize airborne contamination. *Marine Pollution Bulletin, 113*(1–2), 55–61. https://doi.org/10.1016/j.marpolbul.2016.07.050

Volgare, M., Santonicola, S., Cocca, M., Avolio, R., Castaldo, R., Errico, M. E., Gentile, G., Raimo, G., Gasperi, M., & Colavita, G. (2022). A versatile approach to evaluate the occurrence of microfibers in mussels Mytilus galloprovincialis. *Scientific Reports, 12*(1), 21827. https://doi.org/10.1038/s41598-022-25631-2

Watts, A. J. R., Urbina, M. A., Corr, S., Lewis, C., & Galloway, T. S. (2015). Ingestion of plastic microfibers by the crab *carcinus maenas* and its effect on food consumption and energy balance. *Environmental Science & Technology*, *49*(24), 14597–14604. https://doi.org/10.1021/acs.est.5b04026

Whitaker, J. M., Garza, T. N., & Janosik, A. M. (2019). Sampling with Niskin bottles and microfiltration reveals a high prevalence of microfibers. *Limnologica, 78*, 125711. https://doi.org/10.1016/j.limno.2019.125711

Woodall, L. C., Gwinnett, C., Packer, M., Thompson, R. C., Robinson, L. F., & Paterson, G. L. J. (2015). Using a forensic science approach to minimize environmental contamination and to identify microfibres in marine sediments. *Marine Pollution Bulletin*, *95*(1), 40–46. https://doi.org/10.1016/j.marpolbul.2015.04.044

Woodall, L. C., Sanchez-Vidal, A., Canals, M., Paterson, G. L. J., Coppock, R., Sleight, V., Calafat, A., Rogers, A. D., Narayanaswamy, B. E., & Thompson, R. C. (2014). The deep sea is a major sink for microplastic debris. *Royal Society Open Science*, *1*(4), 140317. https://doi.org/10.1098/rsos.140317

Yang, T., Gao, M., & Nowack, B. (2023). Formation of microplastic fibers and fibrils during abrasion of a representative set of 12 polyester textiles. *Science of the Total Environment*, *862*, 160758. https://doi.org/10.1016/j.scitotenv.2022.160758

Zheng, S., Zhao, Y., Liangwei, W., Liang, J., Liu, T., Zhu, M., Li, Q., & Sun, X. (2020). Characteristics of microplastics ingested by zooplankton from the Bohai Sea, China. *Science of the Total Environment*, *713*, 136357. https://doi.org/10.1016/j.scitotenv.2019.136357

Zhu, X., Nguyen, B., You, J. B., Karakolis, E., Sinton, D., & Rochman, C. (2019). Identification of microfibers in the environment using multiple lines of evidence. *Environmental Science & Technology*, *53*(20), 11877–11887. https://doi.org/10.1021/acs.est.9b05262

12 Dynamics of Microfibers Discharged into the Urban and Suburban Environment

Satoshi Nakao
Osaka City Research Center of Environmental Science

Tetsuji Okuda
Ryukoku University

12.1 INTRODUCTION

Microfibers (MFs), a type of fibrous microplastics (MPs), have not received much attention, even after MP research became popular worldwide due to their ubiquity. This may be because Fourier-transform infrared (FTIR) microscopy, which is more expensive than FTIR, is required to identify the components of MFs, and analytical methods for MFs have not been established. However, as the effects of MFs on the environment are being elucidated, their toxicity to living organisms has been identified (Kim et al., 2021). Since many MFs are present in the clothes we wear (Liu et al., 2021), they came to be widely studied. However, compared with MP research, our knowledge on the environmental fate of MFs is still insufficient, and many unknowns remain, including their health effects on humans. Therefore, the study of MFs in the environment is the most advanced research in the field of MPs. In Japan, MF research has not yet been conducted, and further research on MFs environmental dynamics remains to be performed. In this chapter, we introduce the investigation our research group has conducted to understand the dynamics of MFs in the environment in urban and suburban areas in western Japan.

When we talk about MFs, an anthropogenic pollutant, in the same context as MPs, it is often assumed that MFs are a type of MP. However, there is no firm definition of MFs (Athey & Erdle, 2022), and MFs are treated differently among researchers. Some researchers have discussed MFs by differentiating between plastic, semi-plastic, and natural (Barrows et al., 2018; Athey & Erdle, 2022). Our research group has taken the position that MFs are plastics (Nakao et al., 2021); however, we have struggled with the definition for many years. Even if we simply assume that fibrous MPs with a length of less than 5 mm are MFs, the authors cannot determine whether it is appropriate to ignore non-plastic MFs. However, please note that the authors' research

DOI: 10.1201/9781003331995-15

group has defined MFs as "fibrous plastics" since the beginning of our research, and the main research presented in this chapter focuses on synthetic MFs. We are not ignoring non-synthetic MFs such as viscose, which account for 56.9% of the MFs in deep sea sediment and are twice as abundant as synthetic MFs (Henry et al., 2019; Zambrano et al., 2019). Instead, we base our argument on research conducted around 2015, when limited knowledge was available, which indicated that synthetic MFs, similar to MPs, are prone to adsorbing hydrophobic persistent organic pollutants. In any case, MFs are considered anthropogenic pollutants and have received a lot of attention in recent years due to the growing enthusiasm for MP research. However, it is still unclear how MFs are released into the environment, how they behave in the environment, and how they affect the ecosystem, including humans. MFs have also been detected in the air (Dris et al., 2016; Cai et al., 2017) and can enter human bronchi through the air and cause lesions (Gasperi et al., 2018). Therefore, MFs, similar to MPs, require cross-disciplinary studies, such as in the environmental and medical fields, to understand their environmental dynamics, as discussed in this chapter.

Fibrous MPs or MFs, emitted into the environment are mostly degraded polyesters used in synthetic fibers that have become finer. Most polyesters are recognized as polyethylene terephthalate (PET) (Fontana et al., 2020). However, other MFs such as polypropylene (PP), polyurethane (PU), and polyacrylonitrile (PAN) also exist. Their sources are common items such as clothing made of polyester, fishing nets, and rope. In particular, the amount of MFs generated from clothing is considered to be immense. It has been noted that they mostly either enter the sewage system through treated laundry water or are discharged directly into untreated water systems (Acharya et al., 2021). Most MFs that enter the sewage system are expected to be removed by sewage treatment (Nakao et al., 2021), while those that are not removed are discharged into rivers and marine waters (Acharya et al., 2021). The MFs are also believed to exist in the atmosphere and have been detected in the free troposphere (Allen et al., 2019). This suggests that global contamination by MFs is occurring. In addition, MFs have been detected in soil and agricultural land. In the European case, the MFs migration phenomenon from dry soils to the atmosphere and water bodies has been noted due to the direct return of sewage sludge to agricultural land (Athey & Erdle, 2022). Even in countries where sewage sludge is not directly returned to farmland, it is possible that similar contamination is caused by composting sewage sludge and returning it to farmland (El Hayany et al., 2022). The MFs are also known to be present in road dust, and although their origin is unknown, at some sites, more than half of the detected MPs are MFs on a piece count basis (Nakao et al., 2022a); whereas in the ocean, MFs are thought to occur mostly from fishing nets, and it could be speculated that the occurrence of MFs from washed up fishing gear on the coast is high. Thus, MFs, similar to MPs, are ubiquitous and exist in every environment on earth.

Although the mechanism of MF generation may vary, we believe that it is similar to that of MPs. For example, if PET ropes are left on the road or shore, they would be exposed to ultraviolet radiation from the sun, and physical forces such as wind, rain, and waves would break them down to finer fibers (Song et al., 2017). Following these repeated processes, the finer PET fibers become less than 5 mm in length and then become MFs. In addition, most clothing that is commonly worn contains polyester (mostly PET), and it is possible that MFs fall directly into the environment while being worn, as clothing deteriorates after repeated washing (Liu et al., 2021). Resin nets can also be a source of MF contamination. Resin nets are widely used in our livelihoods, agriculture, and fishing because they are inexpensive and durable; however, if they deteriorate, they become a source of MF emission.

12.2 IMPACT ON ECOSYSTEMS

Given the ubiquity of MFs in the environment, it could be assumed that contamination of the organisms that inhabit the environment is also ubiquitous. Notably, MFs have been detected in many organisms, and as with MPs, it is unlikely that any organism, including humans, has not been contaminated with MFs. However, similar to MPs, the biological and ecological impacts of MFs are often unknown, and their impact on ecosystems is not well understood. Furthermore, adverse physiological and reproductive effects of MPs have been noted, and it is known that the impact of MFs on ecosystems is not zero. Research that not only focuses on MF cases of detection from individual organisms but also on their effects on the ecosystem from a macro perspective is awaited. In addition, a study by Qin et al. (2021) states that MPs produced from biodegradable plastics behave similarly to normal MPs; therefore, it is important to study MFs produced when biodegradable fibers are decomposed.

The authors detected many MFs in bivalves inhabiting tidal flats in Osaka Bay in western Japan (Nakao et al., 2020; Figure 12.1). However, at the time, it was unclear whether the detected fibers were plastic MFs or not, as FTIR microscopy, which is an instrument that identifies MFs with plastic, could not be used. The fact that MFs were detected in many bivalves, which could not be identified as plastic or not, suggests that bivalves may easily ingest and accumulate fibrous materials in their bodies

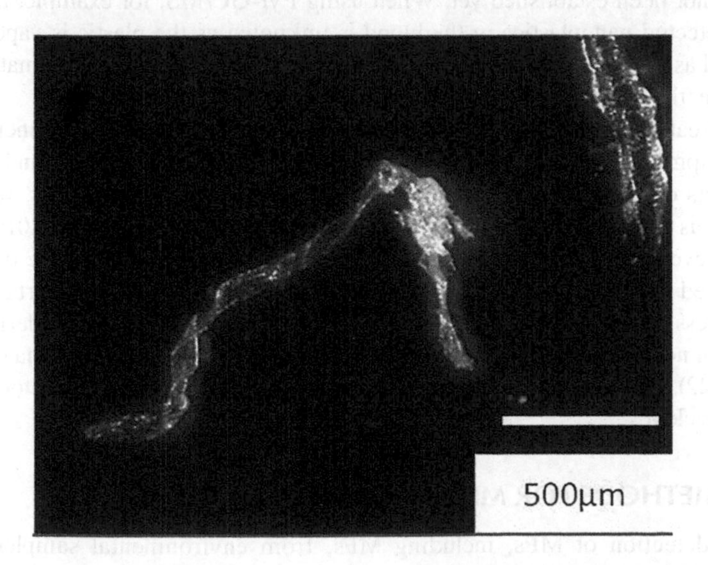

FIGURE 12.1 Fibers detected in bivalves (not known if plastic or not). (*Source:* Nakao et al., 2020.)

during filter-feeding. If MFs absorb toxic chemicals and are selectively ingested by bivalves, the toxic chemicals may be desorbed in the body of the bivalves, and they would potentially become toxic. In addition, bivalves are possibly adversely affected physiologically by an overabundance of MFs, which reduces their numbers. Therefore, the more MPs, including MFs, are present in the environment, the more the abovementioned problems are expected to occur, in addition to concerns about ecosystem disturbance and damage to biodiversity. In fact, findings supporting these events are accumulating. For example, MPs have been detected in marketed pelagic fish (horse mackerel, herring, and merluca) and coastal fish (lingcod, marblefish, and pehlay) (Pozo et al., 2019), and MFs have been confirmed in in vivo feeding experiments with humphead beetles, medaka, and zebrafish (Ma et al., 2021).

12.3 EFFECTS ON THE HUMAN BODY

Unfortunately, the effects of MPs, including MFs, on the human body are still not well understood. However, the World Wide Fund for Nature (WWF) reports that humans ingest 5 g (equivalent to one credit card) of plastic per week (WWF, 2019). The possible routes of ingestion include bottled water and seafood; however, the quantities of MFs within these are unknown. Over time, MPs can become finer nanoplastics, less than 1 μm, which have been detected in human blood (Leslie et al., 2022). Such nanoplastics could flow through blood vessels and affect various organs. Although the shape of nanoplastics has been observed by SEM analysis in some cases, no reports of detected fibrous nanoplastics have been identified. This is because although nanoplastics can be detected by methods such as Raman spectroscopy and pyrolysis gas chromatography mass spectrometry (Pyr-GC/MS), Raman spectroscopy can determine the fiber shape; however, it is expensive, and the method for nanoplastic detection has not been established yet. When using Pyr-GC/MS, for example, the shape of the detected nanoplastics in the blood is unknown as the plastic is vaporized in the blood as the plastic and its respective mass are identified using chromatography. Therefore, the Pyr-GC/MS method is a destructive test.

The greatest concern about the effects of MFs on the human body is their impact on the respiratory system. Asbestos is known as a fibrous particle that can have serious effects on the human respiratory system, and it is feared that MFs, which are also fibrous MPs, may have similar effects as asbestos (Gasperi et al., 2018). When MFs of several tens of micrometers in length enter the alveoli, they are thought to be removed from the body by macrophages; however, when MFs are short (<10 μm), they are less likely to be removed from the body, similar to asbestos. Air-derived MFs have been noted to become lodged in the alveoli and cause mesothelioma (Wieland et al., 2022). However, the effects of MPs and MFs on the human respiratory system remain unclear.

12.4 METHODS FOR MF DETECTION

For the detection of MPs, including MFs, from environmental samples (water, sludge, soil, etc.), it is essential to remove as much non-plastic organic matter as possible. The most commonly used method is the use of a Fenton reaction with 30%

hydrogen peroxide and a small amount of ferric sulfate. Our experience revealed that the above process does not seem to cause further miniaturization of MPs; however, this has been noted in previous studies (Kataoka et al., 2022). It is the author's experience that hydrogen peroxide treatment removes stains from the surface of MPs and increases the intensity of the IR spectrum.

12.4.1 Detection by Fluorescent Staining

This is a simple method to detect MFs with a fluorescence stereomicroscope using Nile Red, one of the fluorescent stains. The MPs, including MFs, are stained with Nile Red, excited blue by a fluorescence stereomicroscope, and observed through an orange absorption filter, which allows for their identification by their red or yellow-green glow (Nakajima & Yamashita, 2020). However, in cases where a type of plastic is difficult to dye with Nile Red, skilled techniques are required. Since the FTIR microscopy described below is more accurate for identification, it should be regarded as a simplified method.

12.4.2 Detection by FTIR Microscopy

The identification of fibrous MPs, such as MFs, is difficult with the attenuated total reflection (ATR) method using conventional FTIR analysis due to their fineness. As the name suggests, FTIR microscopy is a device that displays a microscopic image of an analyte on a monitor for analysis.

The selection of ATR, transmission, and reflection methods of the FTIR microscopy measurement should be determined by the filter that captures the particles in the environmental sample. The reflection method would be suitable for the metal filters used, and the transmission method would be an option if the filter does not interfere with IR transmission. However, the ATR method in FTIR microscopy is more effective for the identification of rubber, which is more susceptible to degradation than plastic, but since the probe is in direct contact with the particles, there are concerns about contamination and particle loss. In addition, since the germanium ATR prism is easily damaged and expensive (approximately $2,000), it would not be appropriate to use it except when analyzing fine rubber particles exclusively. Pyr-GC/MS and Raman spectroscopy are available for the analysis of MPs; however, these are not discussed in detail in this section.

12.5 DYNAMICS OF MFs IN THE URBAN ENVIRONMENT

12.5.1 Characteristics of the Study Area

The study area is Osaka City, the central area of Osaka Prefecture (population approximately 8.8 million), located in western Japan (Figure 12.2). Osaka Prefecture has a long north-south topography (area: 1,899 km^2), with Osaka Bay to the west, plains to the north leading to Kyoto, and mountains to the east and south. The entire area consists of vast flatlands. Osaka City is a commercial city with a population of approximately 2.7 million (area: 223 km^2). Although it is smaller in size than central Tokyo as the

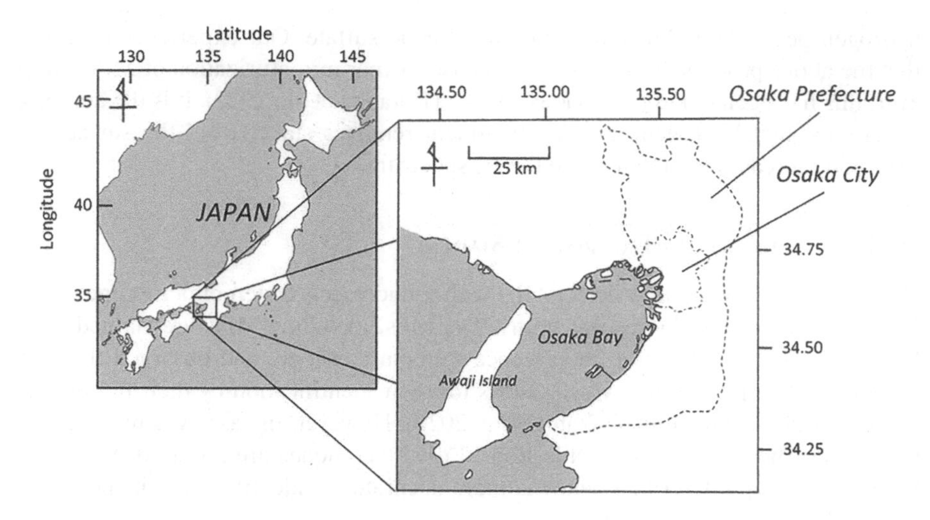

FIGURE 12.2 Location map of Osaka City in Japan.

center of eastern Japan, Osaka City has long flourished as the central city of western Japan. Osaka City has the largest population of any municipality in western Japan, which means that economic activity is very active, and the city has long been involved in and solved pollution problems. Many apparel companies originated in Osaka City due to its history as a thriving textile industry. Here, our findings on the dynamics of MFs in water, air, and urban road dust in Osaka City are discussed.

12.5.2 Aquatic Environment

Although the Osaka City rivers are naturally considered to be contaminated by MFs, the authors have not directly detected MFs in the rivers. However, MFs have been detected in rivers in other regions of Japan (the Seta River in this study (Section 12.6)), and therefore MFs are likely considered to be present in rivers in Osaka City. We detected MFs in river sediments in Osaka City, where MPs in river water would be concentrated (Figure 12.3). Figure 12.3a shows non-plastic MFs, while Figure 12.3b shows PET MFs. Therefore, in the Osaka City rivers, as the sediments in the river are concentrated with plastics of high specific gravity, the rivers in Osaka City are certain to be contaminated with MFs, as in other areas of Japan.

Sewage treatment plants are a possible source of MF contamination in the aquatic environment. We developed a detailed MPs balance sheet for the sewage and sewage sludge treatment processes at a sewage treatment plant (Figure 12.4, Nakao et al., 2021). The MPs were found to circulate in the sewage and sewage sludge treatment processes, and MFs were particularly difficult to remove during the sludge treatment process, contributing significantly to the circulation of MPs in the system. In addition, an attempt was made to improve the MP removal rate by adding a coagulant in the sewage sludge thickening process, but the results confirmed that MFs remained difficult to remove even with the addition of coagulants and circulated within the sewage treatment plant process (Nakao et al., 2022b). Although our study did not

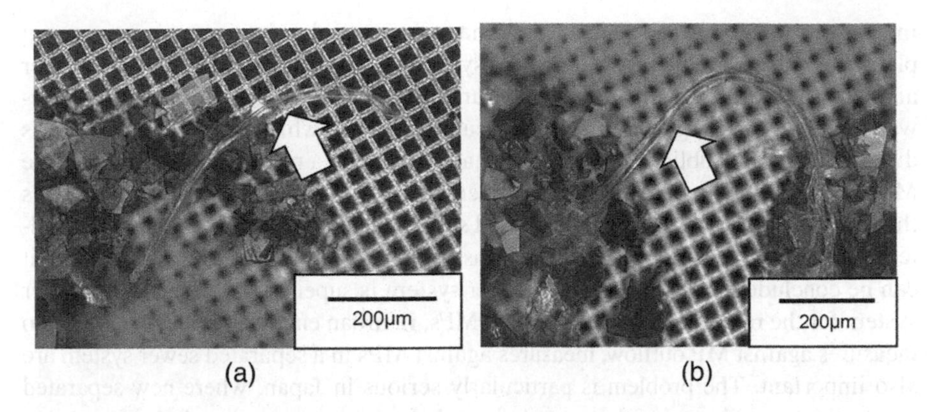

FIGURE 12.3 (a) Cellulose fibers detected in river sediment, (b) PET MFs detected in river sediment.

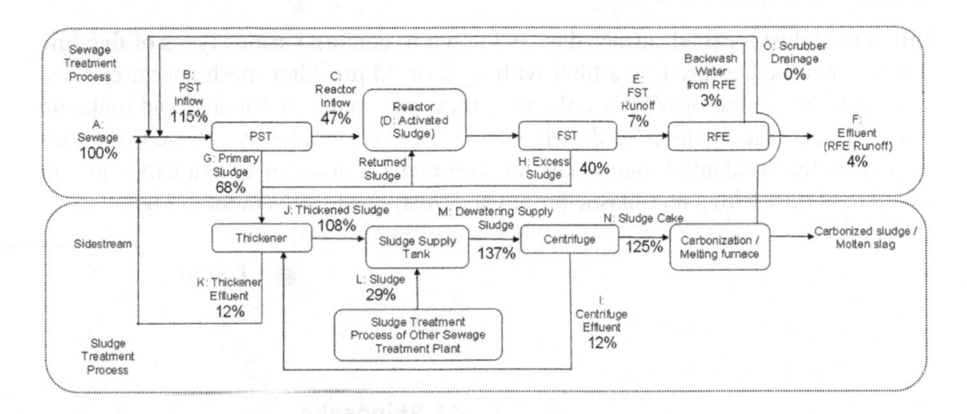

FIGURE 12.4 Behavior of MPs in sewage treatment and sewage sludge treatment processes (case study of a Japanese sewage treatment plant) (MPs load in sewage A is set as 100%). (*Source:* Nakao et al., 2021.)

provide direct evidence of MF discharge from sewage treatment plants, it is possible that MFs are removed in the sand filtration process during the latter stage of the sewage treatment process. Alternatively, MFs are miniaturized again and discharged into public waters as they circulate within the sewage treatment plant process. However, MF emissions have been confirmed in sewage treatment plants in other countries and regions, and their complete removal from conventional sewage treatment systems would be difficult. Membrane-based wastewater treatment, such as membrane bioreactors (MBRs), can remove 99.9% of MPs (Talvitie, 2018), making the introduction of MBRs an effective solution for MF contamination of aquatic environments.

When considering the MF contamination of the aquatic environment in urban environments, we believe that the degree of MF contamination varies depending on the sewage collection system. There are two types of sewage collection systems: combined and separated sewer systems. In the combined sewer system, domestic and industrial wastewater and rainwater drainage flow together into a sewage treatment plant. In contrast, in the separated sewer system, domestic and industrial wastewater and rainwater drainage are collected separately, and domestic and industrial wastewater is then discharged to a sewage treatment plant, while rainwater drainage is discharged into a public water body such as a river. As a result, MPs, including fine MFs, are accumulated in road dust and then discharged into sewage treatment plants through rainfall in the case of a combined sewer system or into rivers and other bodies of water without being treated in the case of a separate sewer system. Therefore, it can be concluded that the combined sewer system is superior to the separated sewer system for the removal of MPs, including MFs, in urban environments. In addition to measures against MP outflow, measures against MPs in a separated sewer system are also important. The problem is particularly serious in Japan, where new separated sewer systems and most of the recently installed sewers are separated sewer systems.

12.5.3 Urban Road Dust

Urban road dust, or road surface dust, is known to contain various types of fine (the size that can be captured by a filter with a 10 or 20 μm filter mesh aperture) MPs. Although the source of MFs is unknown, they are evenly distributed and make up a high percentage of the road dust in Osaka City. We randomly selected five sites in each of the residential, industrial, and commercial areas of Osaka City and collected road dust with a non-plastic broom and dustpan (15 sites in total, Figure 12.5).

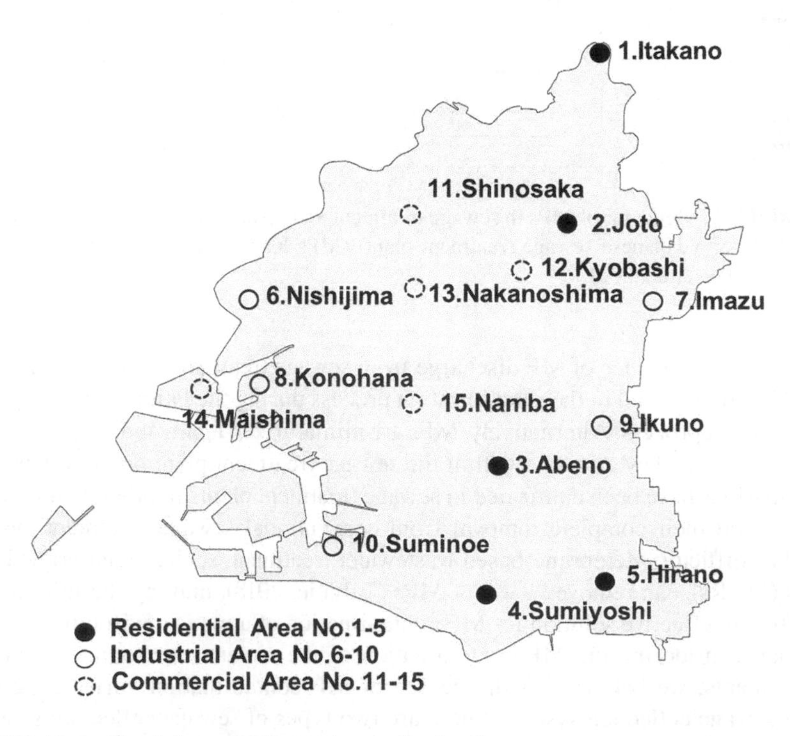

FIGURE 12.5 Road dust sampling points in Osaka City.

The average size of detected MPs (arithmetic mean of the long and short diameters) from the residential area was 71 (21–810) μm (n = 148), 78 (17–1,100) μm (n = 155) for the industrial area, and 77 (12–1,300) μm (n = 201) for the commercial area. A comparison of the detected MP sizes from each region showed no significant differences in both median (Kruskal–Wallis test) and variance (Leven test). Figure 12.6 shows the percentage of MPs in each region by shape. For convenience, the shapes were classified into three types: fragments, MFs (short), and MFs (long). MPs with an aspect ratio (length to diameter) greater than 10 and length less than 500 μm were considered MFs (short), those with a length greater than 500 μm were considered MFs (long), and all others were considered fragments. The proportion of MFs in residential areas was about 30%, and the proportion of MFs between sites did not differ considerably, while the proportion of MFs in industrial and commercial areas varied widely between sites, ranging from about 25% to 65%. This is thought to be the case because residential areas are less affected by people and automobiles, while industrial and commercial areas are more diverse than residential areas in terms of the types of businesses and sources of emissions within the area.

12.5.4 Atmospheric Environment

Three sites in Osaka City in Nishiyodogawa Ward (sampling altitude: approximately 3 m), Joto Ward (16 m), and Tennoji Ward (30 m) were selected (Figure 12.7), and fallout was collected for approximately 1 month. Table 12.1 shows the results of the airborne MP analysis of the dust fallout. The number of airborne MPs per m^2 was higher in months with increased precipitation. This is thought to be due to airborne MPs suspended in the atmosphere falling by rainfall. The Nishiyodogawa area was located near an automobile gas emission measurement station at an altitude of about 3 m and was therefore considered to be readily affected by road dust roll-up. In October, when precipitation was low, no considerable differences were observed among the three sites. For reference, the dust fallout unit area in large cities in other countries was 2–355/m^2·day in the urban and suburbs of Paris, France (airborne MPs size of 50 μm or more) (Dris et al., 2016), and 175–313/m^2·day in Dongguan, Guangdong, China, in 2017 (airborne MPs size of 1.0 μm or more) (Cai et al., 2017), which were lower than those in the current study. The fraction of MFs detected was low, and they were not detected in high concentrations as the fallout dust in the European case (Figure 12.8). In Europe, sewage sludge is often returned directly to farmland, where it dries and MFs diffuse into the atmosphere. In Japan, however, sewage sludge is incinerated or melted in principle; therefore, the atmosphere is not as contaminated with MFs as in Europe. However, there are cases in Japan where sewage sludge is composted and returned to farmland; therefore, it is highly likely that farmland fertilized with sewage sludge compost is contaminated with MFs, although to a lesser extent than in Europe. In fact, our research group detected MFs in sewage sludge compost in Japan (Figure 12.9), and it is possible that Japanese agricultural land is contaminated with MFs. The variety of MPs detected in the fallout dust and the lack of variation in the number of MPs detected from an altitude of 3–30 m suggested that the origin of the MPs in the fallout dust was not from transboundary transport but rather from road dust that was rolled up by vehicle traffic and other events.

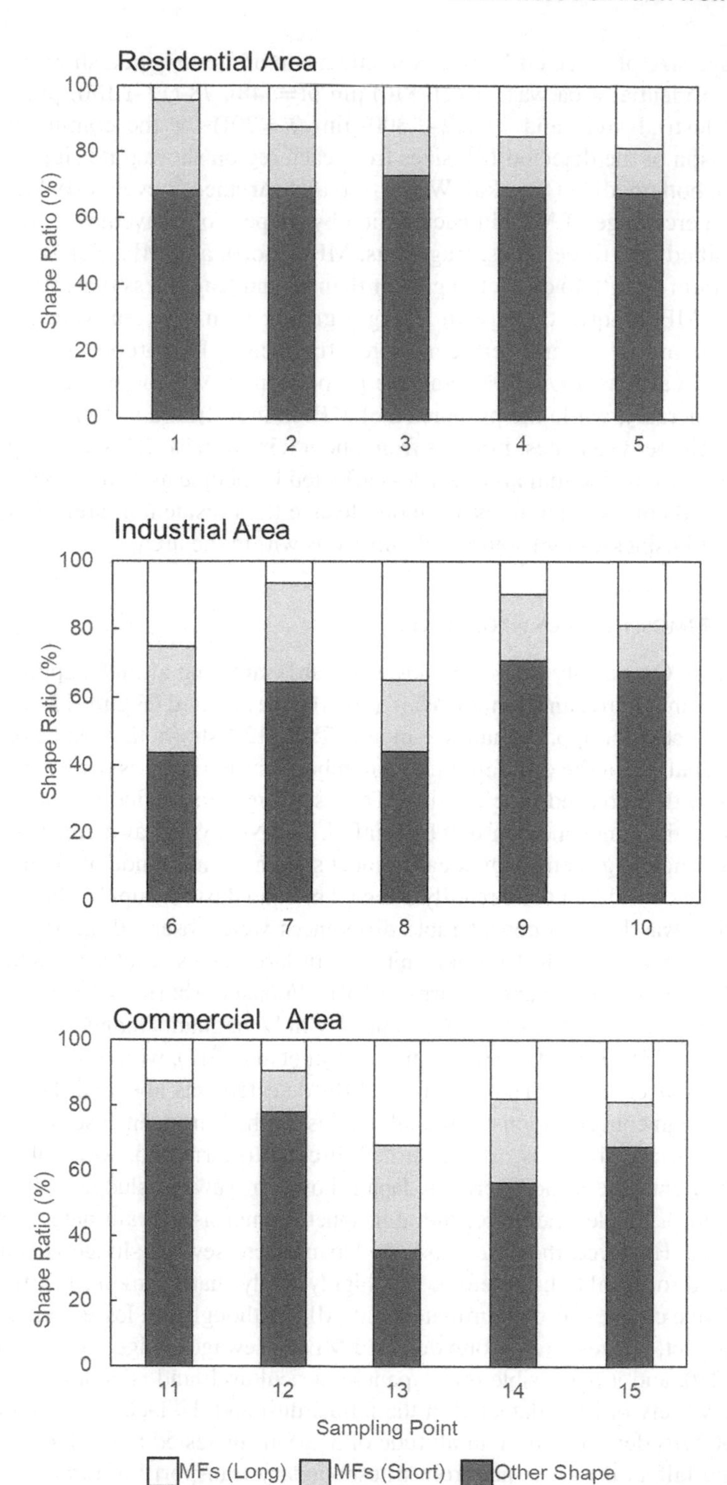

FIGURE 12.6 Shape fraction of MPs detected in road dust in Osaka City.

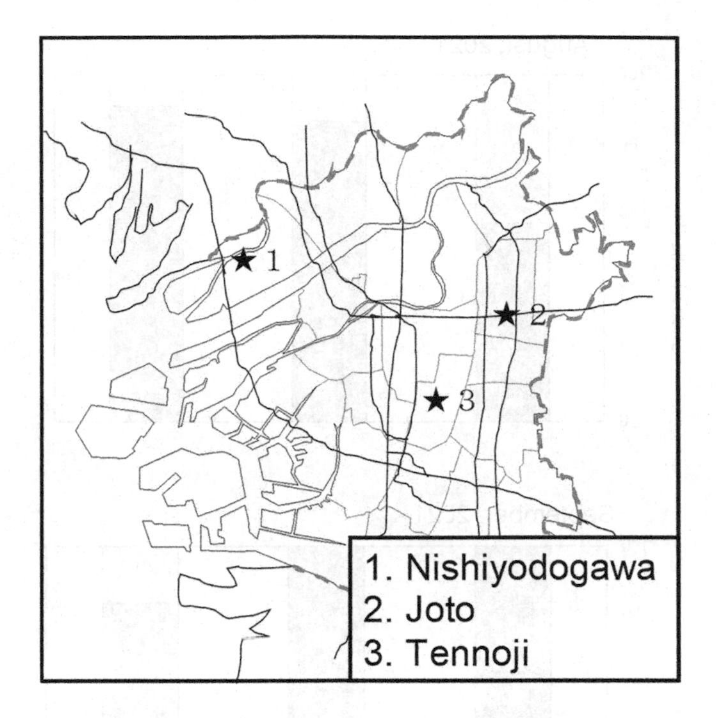

FIGURE 12.7 Fallout collection points in Osaka City.

TABLE 12.1
Daily Area Unit of Microplastics in Fallout

	Airborne MPs (pieces/m²/day)	Precipitation (mm)
	August, 2021	
1. Nishiyodogawa	5,800	314.5
2. Joto	2,600	
3. Tennoji	1,500	
	September, 2021	
1. Nishiyodogawa	3,200	192.0
2. Joto	190	
3. Tennoji	670	
	October, 2021	
1. Nishiyodogawa	390	68.5
2. Joto	260	
3. Tennoji	300	

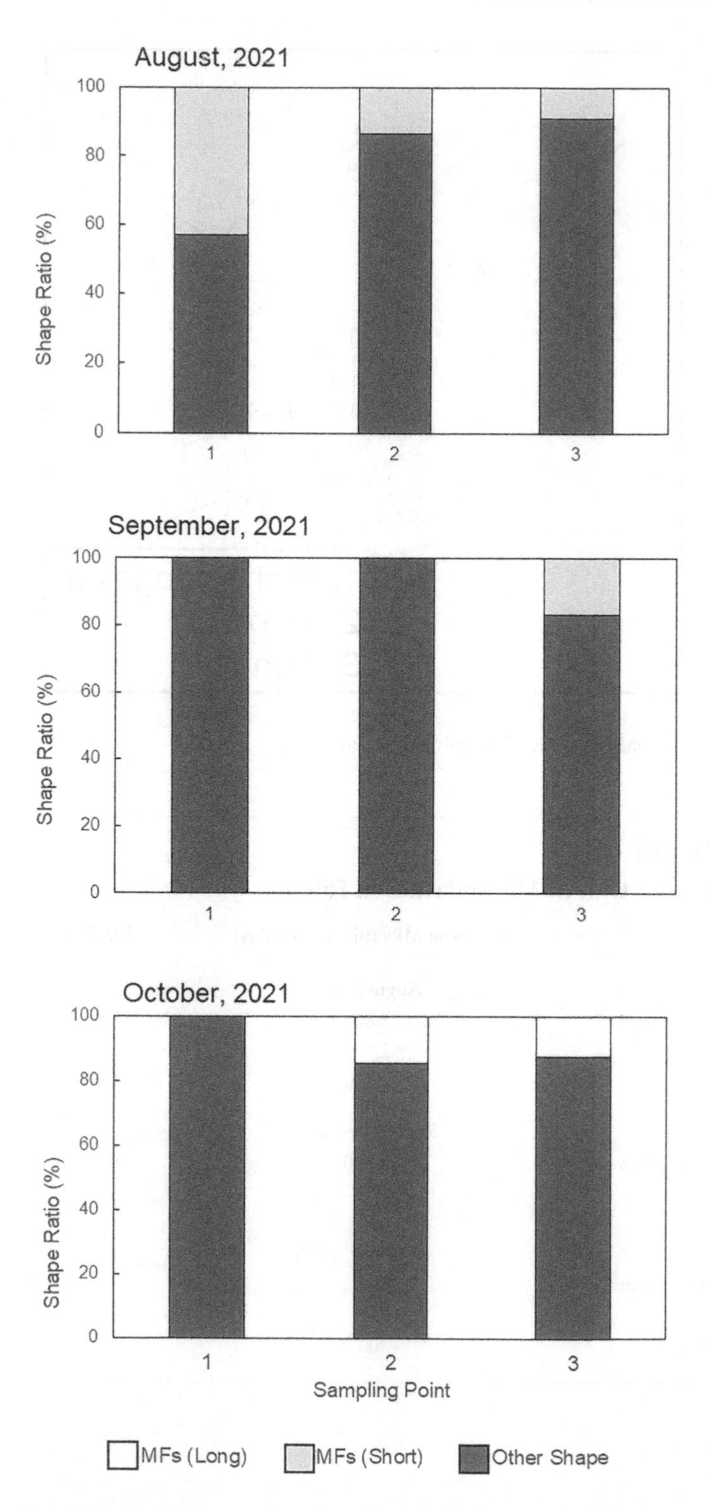

FIGURE 12.8 Shape fraction of MPs detected in the fallout in Osaka City.

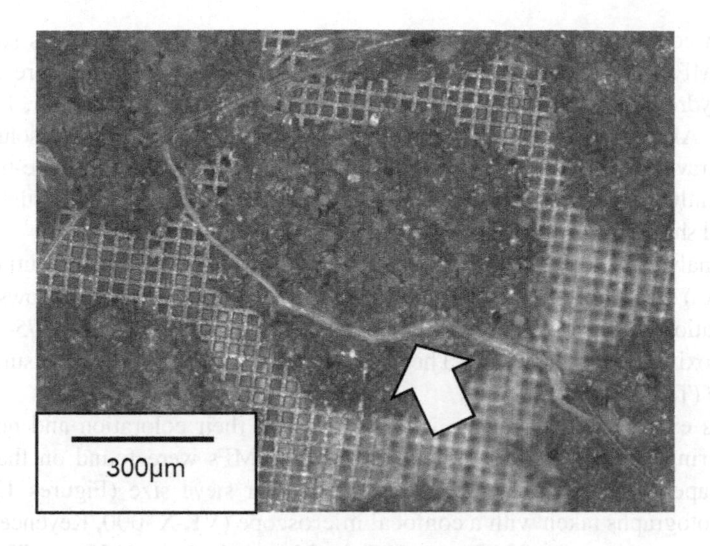

FIGURE 12.9 MFs (PU) detected in sewage sludge compost.

12.6 DYNAMICS OF MFs IN THE SUBURBAN ENVIRONMENT

12.6.1 Characteristics of the Suburban Area Studied

The study area of this section is Lake Biwa, which is located upstream of the large cities studied in Section 12.5. Lake Biwa is the largest lake in Japan, with an area of approximately 670 km^2, its shoreline extends approximately 235 km, and its maximum depth is approximately 104 m. In addition, it is surrounded by several cities with a maximum population of approximately 300,000. The basic water flow is from north to south, and the distance from north to south is longer, 60 km, than that from east to west. Its reservoir volume is approximately 28 billion tons, and approximately 15 million people utilize its water for domestic water and other purposes. The catchment area includes the surrounding mountains of Ibuki, Suzuka, and Hira, where more than 1 million people live. Uniquely, the outflow from Lake Biwa is only the Seta River, except for an artificial canal to Kyoto city. Lake Biwa has the same watershed (Yodo-gawa river system) as the river flowing into Osaka Bay, as described in Section 12.5.

In this section, the results of our surveys conducted in the Seta River from Lake Biwa are presented. In addition, some findings of the MF dynamics of the area, including Lake Biwa, are summarized, including the results of surveys conducted by other researchers.

12.6.2 Aquatic Environment in the Studied Suburban Area

As many previous reports related to MPs, including MFs, targeted around 100 μm, we collected river water MFs using a 75–125 μm aperture size sieve. Here, MF samples were collected by the filtration of approximately 3 tons of river water. This multilayer-filtration method was adopted because it was less likely to cause aggregation of the fibers during collection compared with general studies (plankton-net); therefore, it was suitable to observe MFs. After the filtration (screening), all particles and fibers were collected using a hydrophilic polytetrafluoroethylene filter (25 or 47 mm in diameter, 1 μm pore diameter, ADVANTEC) after oxidation with a 30% hydrogen peroxide solution and specific gravity separation with a 5.3 M sodium iodide (NaI) solution as a pre-treatment. Subsequently, plastic species were identified by FTIR microscopy and visual analysis (color and shape), and the results of the visual analysis are presented here.

The analysis of the Seta River water (approximately 1 km after the outflow from Lake Biwa) was conducted in 2019, and the winter (November) result shows that the concentration of MPs, including fragment MPs, captured by the sieve (75–300 μm) was approximately 40 pieces/m^3. These values were similar to those measured at the same site (Tanaka et al., 2020).

In this experiment, MFs were judged based on their coloration and non-living shape during the microscopic observation. Many MFs were found on the 75 and 125 μm aperture size sieves, compared with other sieve size (Figures 12.10 and 12.11: photographs taken with a confocal microscope (VK-X3000, Keyence, Osaka, Japan)). The percentage of MFs was 60% in 23 samples/m^3 of MPs on 75 and 125 μm sieve. In the survey conducted in the summer (July), approximately 80% of the samples were MFs for all MPs.

The above results are underestimated values because only the MFs color and shape (after oxidation treatment and specific gravity separation) were used for identification. In contrast, some research groups reported doubtful MF counts owing to increased MF contamination during laboratory analysis (Mishima et al., 2022). Therefore, we conducted a control test by filtering approximately 100 L of tap water, following the same method as the river sampling, and no MFs were detected. Therefore, MF contamination was limited in our experiments.

Figure 12.10 shows MFs captured by a 75 μm sieve; however, it is approximately 1,000 μm in length. This 1,000 μm fiber passed through a 125 μm sieve, meaning that the MFs would be able to pass through the mesh with approximately 1/10 the size of their length if their diameter was smaller than the mesh size. Therefore, only MFs with a much longer fiber length than each sampling mesh (filter and net) size would be collected, indicating the difficulty of collecting and measuring MFs.

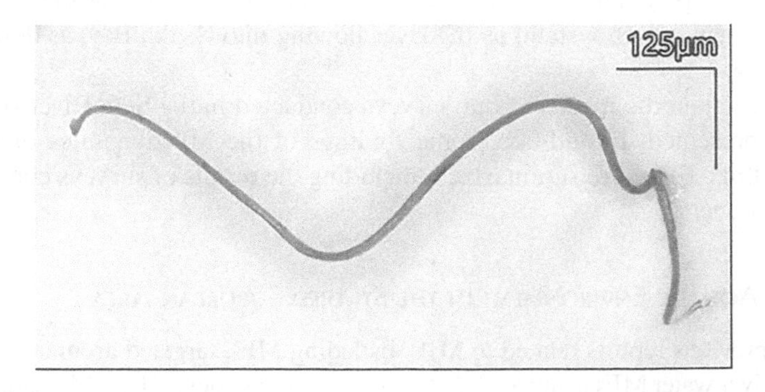

FIGURE 12.10 MFs collected on a 75 μm sieve (Seta River: November 2019). Scale size of 125 μm means that this fiber passed the upper sieve size.

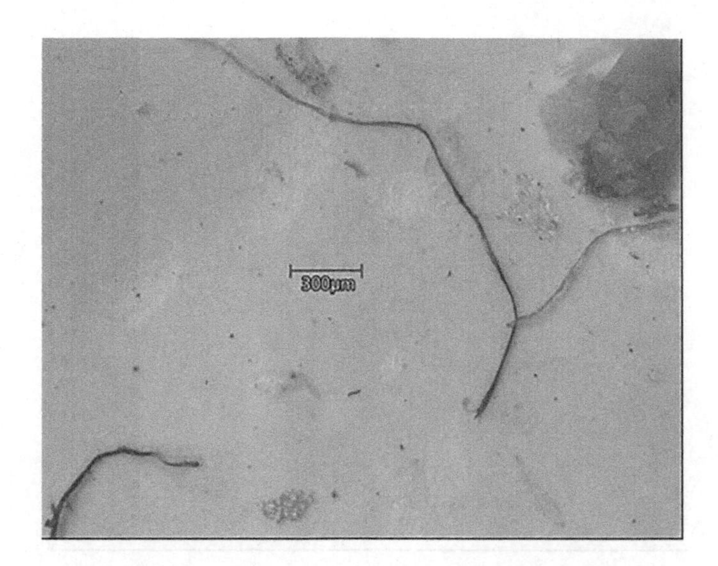

FIGURE 12.11 MFs collected on a 125 μm sieve (Seta River: November 2019). Scale size of 300 μm means that this fiber passed the upper sieve size. Left-bottom fiber is black (judged as not-plastic), center-long fiber is blue, and right-side fiber is transparent (judged as not-plastic).

The main sources of MFs originate from clothing, which disperses in the air and then migrates to the aquatic environment. Another main method is through ground and surface water, dumped plastic products, which accumulate on roads and get crushed and/or decomposed to generate the MFs, can easily be washed out by rain. Another reasonable source is water discharged from sewage treatment plants, as described in Section 12.5. Approximately 2 km upstream of the sampling point in the Seta River (the average water volume of the Seta River is approximately 14 million m^3/day), an inflow of sewage treatment plant water discharge is present (separated sewer system, projected treatment water volume: about 400,000 m^3/day). Therefore, we conducted a similar survey of MFs in lake water (southernmost part of Biwa Lake) further upstream from the discharge point of the treated sewage water (approximately 4 km upstream from the sampling point of the Seta River).

The results showed that 52% of approximately 20 pieces/m^3 of MPs (filtered more than 3 t of surface water in Biwa Lake) were MFs, and there was no significant difference in the concentration or percentage of MFs in the MPs before and after the sewage discharge point. The shape of typical MFs (Figures 12.11 and 12.12) was single fibrous, not floc-like, as reported in previous research (Tanaka et al., 2019), which studied the same treatment plant. Although there is a possibility that the pre-treatment (oxidation) had an effect in our experiment, it is possible that single fibers were detected downstream (Seta River) since the MFs aggregated as floc are likely to precipitate at the lake and/or in the treatment plant. However, since the MF concentration did not vary before and after the sewage discharge point, the main source (route) of MFs to the river seems to be the atmosphere (fallout) and rain, including wash-out from the road. Sewage treatment plants discharge water that contains MFs, especially smaller (shorter) MFs, as reported (Tanaka et al., 2019), which enter the

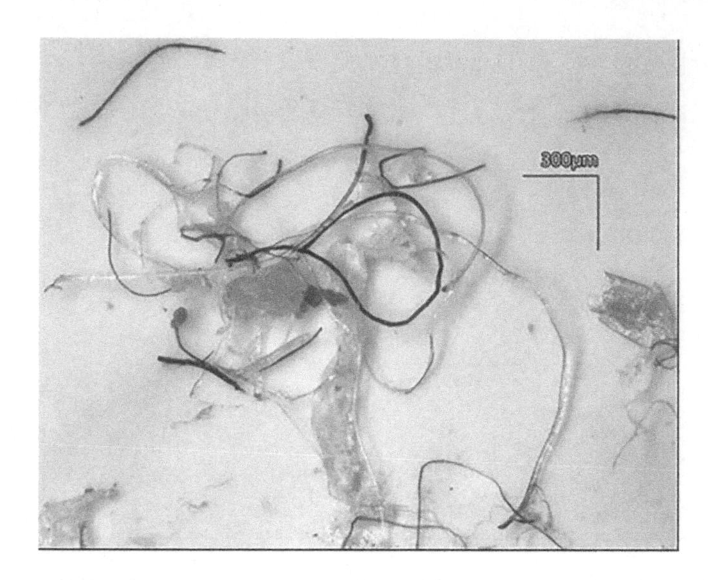

FIGURE 12.12 MFs collected on a 125 μm sieve (South Lake: November 2019). We judged non-plastic for 90% of fibers in this photo by color and shape.

water environment. However, MFs of 75 μm or greater in sewage-treated water do not much contribute to increasing its concentration in the Seta River because the dilution rate of the treated sewage water is large (about 30 times) at the sampling site. Since the lake surface (which directly captures MFs from the air) and the catchment area are also large, we believe that the main sources are fibers dissociated from clothing during human movement, from plastic (synthetic polymer) fibers, and from strings used in gardens, agriculture, and other areas. In addition, compared to other MPs, fibrous plastics might have less settling potential in Lake Biwa, and they can reach and flow out to the Seta River because of their shape.

To think more about the MF sources, especially sewage effluent, we introduce an example of a study of changes in MF concentrations in a river through a city in the suburban area, even if not our results (Yamasaki et al., 2022). The report is a case study of the results of a river in Hiroshima Prefecture, which is located in western Japan, and the authors well know the river (a second-class river flowing through Higashi-Hiroshima City: the Kurose River). Yamazaki et al. judged MFs as plastic based on microscopic observations, similar to our study, but they fractionated MPs into six species: 'pellets', 'fragments', 'rubber', 'foam', 'film', and 'fiber' based on the shapes observed under the microscope. The concentration of MFs collected by a plankton-net with 100 μm mesh did not show a tendency to increase uniformly with the flow (about 0.3 pieces/m^3), although it slightly increased in the urban area during the flow.

In terms of shape, MFs were about 20% in the urban area at the middle point of river flow (annual average flow rate of the river: approx. 140,000 m^3/day), where the highest concentrations of MPs were observed; among nine sampling sites from the source of river and sea (estuary), MF concentration was very high (100%)

in the site at the dam, probably in a slower flow; it was similar to our summer (July) results shown in this section. This might indicate that fibrous plastics have higher mobility compared to other MPs and that MFs are preferentially detected in river water, especially when the river flows slowly, for example, with no rainwater influence. Incidentally, just after (downstream) the point of treated water effluent from the sewage treatment plant of the middle point urban area (separated sewer system, projected treatment water volume: approximately 70,000 m^3/day), the concentration of MPs was reduced to almost half, and MFs were not detected. It is possible that the concentration of MFs in treated sewage effluent is low and might have a dilution effect on MFs at the river point.

Here, Yamazaki et al. also tried MF collection using elastic mesh (stockings) for the same nine samples and reported that more MFs were collected by the elastic mesh than in the plankton net. As the concentrations were lower than plankton nets at each sampling point, indicating fragment MPs passed the elastic mesh than with plankton nets, the results of different collection apparatuses (nets) should not be compared because the collection ratio is different in MFs due to a larger ratio of length to diameter and flexibility of MFs.

12.6.3 Other Emission Sources Such as Atmospheric Origin in Suburban Area

As discussed in the above section, the contribution of atmospheric origin would be large for MPs in an aquatic environment; however, it has been reported that urban airborne MPs have few fibrous forms (Okochi et al., 2022). In contrast, in the results of the observation of fallout MPs around Paris, France (Dris et al., 2016), introduced in Section 12.5, it was reported that the deposition (number of pieces) in the suburban area was about half of that in urban areas (53 ± 38 pieces/m^2/day) and that most of them were fibers. Results from Asaluyeh County on the southern coast of Iran (Abbasi et al., 2019) also reported that the major MPs were MFs (0.3–1.1 pieces /m^3 at 20–100 μm). However, these results do not identify plastics (MFs) by FTIR. It is expected that there are many natural fibers, such as cellulose, in the particles of the fallout, and there are currently no reports that confirm the origin of the fibers. Okochi et al. mentioned that the detection of many fibers in these previous studies was due to the larger particle size in their analysis. Compared with the research results, most MFs size ranged from 5–70 μm in urban areas. Bigger MFs around 100 μm might be few in the urban atmosphere because of trapping (falling down to the ground) by artificial structures.

Another possible emission source pathway in suburban areas is the agricultural area. Fibrous synthetic polymer and textile fabric sheets are often used in the open field in farmlands; thus, they would be damaged by UV rays from the sun, and MFs may be present in the runoff from these areas. Lake Biwa is famous for water recreational activities such as fishing, and MFs from fishing lines and nets are possible sources.

The main pathway is road dust, which is discussed in Section 12.5 and, only concentrations (emissions) may differ from urban areas due to differences in population density and the number of vehicles between urban and suburban areas. Therefore, the behavior of MFs via road dust is the same as discussed in the previous section. The pathways are part of the atmospheric pathway because most MFs on the road would come from the atmosphere.

The concentration of MPs on the road in the middle urban area beside Biwa Lake (in Kusatsu City, Shiga Prefecture, a city located slightly upstream of the above sampling point in the southernmost part of Biwa Lake) has been investigated (Yukioka et al., 2020). The concentration of MPs per road area was reported as 2.0 ± 1.6 pieces/m^2. Although they do not report the morphology ratio of MFs, many plastics were identified to be tire fragments (rubber) and PP and PET from crushed plastic bags and containers. Normally, PET resin is often found as MFs, but it was not detected in the range of 100–1,000 μm. Therefore, in the road environment, there is a possibility that MFs move mainly through the air, but there is a difference from the results of this study in urban areas (Osaka City) and Okochi et al. (2022), so further investigation is necessary.

12.7 CONCLUSIONS AND FUTURE RECOMMENDATIONS

The results of the study presented in this chapter can be summarized as follows. MFs were detected in river sediments in Osaka City, a large city located in western Japan, suggesting that the river water in Osaka City is contaminated with MFs. The source of the MFs might be the sewage treatment plant, but MFs may be retained and circulated in the sewage treatment process. When they leave the circulation, MFs are removed by the sand filtration process at a later stage of the sewage treatment process. Thus, the emissions of MFs from sewage treatment plants with more advanced treatment systems may be lower than previously thought. For the treatment of MFs in sewage treatment plants, membrane-based treatment methods such as MBRs may be effective. In suburban areas, 60% of the MPs detected in the river water were fibrous MFs, suggesting that the MFs in the river water were not derived from treated sewage water. The percentage of MFs to MPs in residential areas remained constant at about 30%, whereas this percentage in industrial and commercial areas ranged from ~25% to 65%. The reason for this may be that residential areas are less affected by people and automobiles, while industrial and commercial areas are more diverse than residential areas in terms of the types of businesses and human activity within the area.

The contribution of MF contamination by atmospheric fallout dust was examined. The concentration of MPs, including MFs, in the fallout dust in the urban area (Osaka City) ranged from 190 to 5,800 pieces/m^2/day and in the suburban area (Otsu) from 0.18 to 1.2 pieces/m^2/day. These results suggest that MFs in the road dust in urban areas may be transported to the air as dust due to the winds caused by driving vehicles and other factors. In suburban areas, sample collection at a large water surface with no human disturbance within a few meters of the periphery, which is less susceptible to road dust, was found to result in extremely low fallout compared with urban areas. These fallout materials accumulate directly in rivers and lakes or as road dust, and then some of them are transferred to the aquatic environment by rainwater. This study shows that even in suburban areas with low fallout, direct loading to lakes and rivers may also be one of the main pathways for the inflow of MPs and MFs in streams and rivers. In other words, MPs and MFs are thought to enter rivers and lakes from the degradation of treated sewage water and outdoor products, improper management of discharged garbage (e.g., spreading of garbage), and degradation and runoff of plastic products used in agricultural land and materials used in civil engineering works. In addition, the contribution of road dust retained on land that is discharged into the aquatic environment during rainfall events may be important. Paradoxically, this may indicate that measures such as removing road dust before rain could substantially reduce the reintroduction of MPs and MFs to atmospheric fallout and their transfer to the aquatic environment.

What measures should be taken to prevent further contamination by MFs? Information on the sources, amounts, and types of MFs emitted into the environment is scarce, and countermeasures are not yet available. Further progress in the study of MFs in various countries, regions, and environments is crucial. In addition, the PET ropes and plastic nets used regularly in our livelihoods must be converted to non-plastic materials.

It will take time to develop alternative materials to plastics and to make these alternatives part of our daily lives; however, we need to be proactive in dealing with the widespread contamination of MFs and their potential to disturb the ecosystem and have adverse effects on human health.

In the future, we will conduct research on (i) understanding the long-term variations in the concentration of MFs in the atmosphere, (ii) estimating the source of MFs in the atmosphere, and (iii) identifying the sites of action of MFs on the respiratory system to further our understanding of the dynamics of MFs in the environment and their effects on human health.

ACKNOWLEDGMENTS

This research was carried out under the grant ‘the research grant of restoration and creation in a coastal environment in the Osaka Bay area’ by the Osaka Bay Regional Offshore Environmental Improvement Center, JSPS KAKENHI, Grant Numbers 17K06628 and 22K12421, the Ryukoku University Science and Technology Fund, and the Joint Research Center for Science and Technology of Ryukoku University. We express our deepest gratitude to these institutions.

DATA AVAILABILITY

The datasets generated and/or analyzed during the current study are available from the corresponding author on reasonable request.

REFERENCES

Abbasi, S., Keshavarzi, B., Moore, F., Turner, A., Kelly, J. F., Dominguez, O. A., & Jaafarzadeh, N. (2019). Distribution and potential health impacts of microplastics and microrubbers in air and street dusts from Asaluyeh County, Iran. *Environmental Pollution, 244*, 153–164. doi:10.1016/j.envpol.2018.10.039.

Acharya, S., Rumi, S. S., Hu, Y., & Abidi, N. (2021). Microfibers from synthetic textiles as a major source of microplastics in the environment: A review. *Textile Research Journal, 91*(17–18), 2136–2156. doi:10.1177/0040517521991244.

Allen, S., Allen, D., Phoenix, V. R., Le Roux, G., Durántez Jiménez, P., Simonneau, A., Binet, S., & Galop, D. (2019). Atmospheric transport and deposition of microplastics in a remote mountain catchment. *Nature Geoscience*, *12*(5), 339–344. doi:10.1038/s41561-019-0335-5

Athey, S. N., & Erdle, L. M. (2022). Are we underestimating anthropogenic microfiber pollution? A critical review of occurrence, methods, and reporting. *Environmental Toxicology and Chemistry*, *41*(4), 822–837. doi:10.1002/etc.5173.

Barrows, A. P. W., Cathey, S. E., & Petersen, C. W. (2018). Marine environment microfiber contamination: Global patterns and the diversity of microparticle origins. *Environmental Pollution*, *237*, 275–284. doi:10.1016/j.envpol.2018.02.062.

Cai, L., Wang, J., Peng, J., Tan, Z., Zhan, Z., Tan, X., & Chen, Q. (2017). Characteristic of microplastics in the atmospheric fallout from Dongguan city, China: Preliminary research and first evidence. *Environmental Science and Pollution Research*, *24*, 24928–24935. doi:10.1007/s11356-017-0116-x.

Dris, R., Gasperi, J., Saad, M., Mirande, C., & Tassin, B. (2016). Synthetic fibers in atmospheric fallout: A source of microplastics in the environment? *Marine Pollution Bulletin*, *104*(1–2), 290–293. doi:10.1016/j.marpolbul.2016.01.006.

El Hayany, B., Rumpel, C., Hafidi, M., & El Fels, L. (2022). Occurrence, analysis of microplastics in sewage sludge and their fate during composting: A literature review. *Journal of Environmental Management*, *317*, 115364. doi:10.1016/j.jenvman.2022.115364.

Fontana, D.G., Mossotti, R., & Montarsolo, A. (2020). Assessment of microplastics release from polyester fabrics: The impact of different washing conditions. *Environmental Pollution*, *264*, 113960. doi:10.1016/j.envpol.2020.113960

Gasperi, J., Wright, S. L., Dris, R., Collard, F., Mandin, C., Guerrouache, M., Langlois, V., Kelly, F. J., & Tassin, B. (2018). Microplastics in air: Are we breathing it in? *Current Opinion in Environmental Science & Health*, *1*, 1–5. doi:10.1016/j.coesh.2017.10.002.

Henry, B., Laitala, K., & Klepp, I. G. (2019). Microfibres from apparel and home textiles: Prospects for including microplastics in environmental sustainability assessment. *Science of the Total Environment*, *652*, 483–494. doi:10.1016/j.scitotenv.2018.10.166

Kataoka, H., Tanaka, S., Yukioka, S., Moriya, A., Wase, N., & Takada, H. (2022). Degradation characteristics of various plastics in pretreatment processes of microplastics analysis by hydrogen peroxide and fenton reaction. *The 56th Annual Conference of Japan Society on Water Environment 2022.online*, 1-I-11-3 (Japanese).

Kim, D., Kim, H., & An, Y. J. (2021). Effects of synthetic and natural microfibers on Daphnia magna – Are they dependent on microfiber type? *Aquatic Toxicology*, *240*, 105968. doi:10.1016/j.aquatox.2021.105968

Leslie, H. A., Van Velzen, M. J., Brandsma, S. H., Vethaak, A. D., Garcia-Vallejo, J. J., & Lamoree, M. H. (2022). Discovery and quantification of plastic particle pollution in human blood. *Environment International*, *163*, 107199. doi:10.1016/j.envint.2022.107199.

Liu, J., Liang, J., Ding, J., Zhang, G., Zeng, X., Yang, Q., Zhu, B., & Gao, W. (2021). Microfiber pollution: An ongoing major environmental issue related to the sustainable development of textile and clothing industry. *Environment, Development and Sustainability*, *23*, 11240–11256. doi:10.1007/s10668-020-01173-3.

Ma, C., Li, L., Chen, Q., Lee, J. S., Gong, J., & Shi, H. (2021). Application of internal persistent fluorescent fibers in tracking microplastics in vivo processes in aquatic organisms. *Journal of Hazardous Materials*, *401*, 123336. doi:10.1016/j.jhazmat.2020.123336.

Mishima, S., Ozawa, K., Nakayama, S., Kikuchi, H., Namba, A., Kataoka, T., & Nihei, Y. (2022). Comparative analysis of plastic pieces in basin, river and coast: Case study in the Hikiji River Basin, Kanagawa Prefecture. *Journal of Japan Society on Water Environment*, *45*(1), 11–19 (Japanese). doi:10.2965/jswe.45.11.

Nakajima, R., & Yamashita, R. (2020). Methods for sampling, processing, identification, and quantification of microplastics in the marine environment. *Oceanography in Japan*, *29*(5), 129–151 (Japanese).

Nakao, S., Akita, K., Ozaki, A., Masumoto, K., & Okuda, T. (2021). Circulation of fibrous microplastic (microfiber) in sewage and sewage sludge treatment processes. *Science of the Total Environment*, *795*, 148873. doi:10.1016/j.scitotenv.2021.148873.

Nakao, S., Akita, K., Ozaki, A., Masumoto, K., & Okuda, T. (2022a). Presence of microplastics in road dust in Osaka City. *The 56th Annual Conference of Japan Society on Water Environment 2022. Online*, 1-I-11-1 (Japanese).

Nakao, S., Akita, K., Ozaki, A., Masumoto, K., & Okuda, T. (2022b). Fate of microplastics flowing into sewage treatment plant (4th report (final report). *The 59th Annual Technical Conference on* Sewerage, N-3-1-2 (Japanese).

Nakao, S., Ozaki, A., Yamazaki, K., Masumoto, K., Nakatani, T., & Sakiyama, T. (2020). Microplastics contamination in tidelands of the Osaka Bay area in western Japan. *Water and Environment Journal*, *34*(3), 474–488. doi:10.1111/wej.12541.

Okochi, H., Yoshida, N., Zhao, H., Fujikawa, M., Tani, Y., Katsumi, N., Miyazaki, A., Takada, H., Itaya, Y., Ogata, H., Niida, Y., Umezawa, N., Kobayashi, H., & Urayama, N. (2022). Research trends and challenges of airborne microplastics. *Journal of Japan Society for Atmospheric Environment*, *57*(3) (Japanese).

Pozo, K., Gomez, V., Torres, M., Vera, L., Nuñez, D., Oyarzún, P., Mendoza, G., Clarke, B., Fossi, M. C., Baini, M., Přibylová, P., & Klánová, J. (2019). Presence and characterization of microplastics in fish of commercial importance from the Biobío region in central Chile. *Marine Pollution Bulletin*, *140*, 315–319. doi:10.1016/j.marpolbul.2019.01.025.

Qin, M., Chen, C., Song, B., Shen, M., Cao, W., Yang, H., Zeng, G., & Gong, J. (2021). A review of biodegradable plastics to biodegradable microplastics: Another ecological threat to soil environments? *Journal of Cleaner Production*, *312*, 127816. doi:10.1016/j.jclepro.2021.127816.

Song, K. Y., Hong, H. S., Jang, M., Han, M. G., Jung, W. S., & Shim, J. W. (2017). Combined effects of UV exposure duration and mechanical abrasion on microplastic fragmentation by polymer type. *Environmental Science and Technology*, *51*(8), 4368–4376. doi:10.1021/acs.est.6b06155.

Talvitie, J. (2018). Wastewater treatment plants as pathways of microlitter to the aquatic environment. *Aalto University Publication Series, Doctoral Dissertations 86/2018*. doi:10.13140/RG.2.2.34370.27848

Tanaka, S., Kakita, M., Yukioka, S., Suzuki, Y., Fujii, S., & Takada, H. (2019). Behavior of microplastics in wastewater treatment processes and estimation of its loading to Lake Biwa. *Journal of Japan Society of Civil Engineers, Series G (Environmental Research) (Web)*, *75*(7), 35–40. doi:10.2208/jscejer.75.7_III_35.

Tanaka, S., Yukioka, S., Bouche, L., Wang, M., Nabetani, Y., Ushijima, Kakita, M., Okamoto, M., Fujii, S., & Takada, H. (2020). Current situation of microplastics contamination in urban water circulation systems. *Journal of Environmental Conservation Engineering*, *49*(6), 296–300. doi:10.5956/jriet.49.6_296.

Wieland, S., Balmes, A., Bender, J., Kitzinger, J., Meyer, F., Ramsperger, A. F., Roeder, F., Tengelmann, C., Wimmer, B. H., Laforsch, C., & Kress, H. (2022). From properties to toxicity: Comparing microplastics to other airborne microparticles. *Journal of Hazardous Materials*, 128151. doi:10.1016/j.jhazmat.2021.128151.

WWF. (2019). *Plastic Ingestion by Humans Could Equate to Eating a Credit Card a Week*. World Wildlife Foundation.

Yamasaki, H., Nakamura, A., Hisanori, K., & Hiratani, A. (2022). Simple river microplastics survey method for environmental education. *Japanese Journal of Environmental Education*, *31*(4), 4_40–4_47 (Japanese). doi:10.5647/jsoee.31.4_40.

Yukioka, S., Tanaka, S., Nabetani, Y., Suzuki, Y., Ushijima, T., Fujii, S., Takada, H., Tran, Q. V., & Singh, S. (2020). Occurrence and characteristics of microplastics in surface road dust in Kusatsu (Japan), Da Nang (Vietnam), and Kathmandu (Nepal). *Environmental Pollution*, *256*, 113447. doi:10.1016/j.envpol.2019.113447.

Zambrano, M. C., Pawlak, J. J., Daystar, J., Ankeny, M., Cheng, J. J., & Venditti, R. A. (2019). Microfibers generated from the laundering of cotton, rayon and polyester based fabrics and their aquatic biodegradation. *Marine Pollution Bulletin*, *142*, 394–407. doi:10.1016/j.marpolbul.2019.02.062

13 Indoor Microplastics and Microfibers

Sources and Impacts on Human Health

Mansoor Ahmad Bhat
Eskişehir Technical University

ABBREVIATIONS

ABS	Acrylonitrile butadiene styrene
ALK	Alkyd
AR	Acrylic
ATR-FTIR	Attenuated total reflectance-Fourier transform infrared spectroscopy
BW	Body weight
CO	Cotton
CP	Cellophane
CV	Rayon
EDI	Estimated daily intake
EPR	Ethylene propylene
EVA	Ethylene-vinyl acetate
FPA	Focal plane array
HDPE	High-density polyethylene
HPLC-MS	High-performance liquid chromatography-mass spectrometry
LC/ESI-MS/MS	Liquid chromatography-electrospray ionization tandem mass spectrometric
LDPE	Low-density polyethylene
LI	Linen
MPs	Microplastics
PA	Polyamide
PA 6	Polyamide 6
PA 66	Polyamide 66
PAN	Polyacrylonitrile
PA resin	Polyamide resin
PC	Polycarbonate
PCL	Polycaprolactone

DOI: 10.1201/9781003331995-16

PE	Polyethylene
PEI	Polyethylenimine
PEO	Polyethylene oxide
PES	Polyester
PET	Polyethylene terephthalate
PM	Particulate matter
PMMA	Polymethacrylate
PP	Polypropylene
PS	Polystyrene
PSU	Polysulfone
PTFE	Polytetrafluroethylene
PUR	Polyurethane
PVAc	Polyvinyl acetate
PVA	Polyvinylalcohol
PVC	Polyvinyl chloride
Pyr-GCMS	Pyrolysis–gas chromatography-mass spectrometry
RB	Rubber
RC	Resin
SBR	Styrene butadiene rubber
SEM-EDS	Scanning electron microscopy and energy-dispersive X-ray spectroscopy
TN	Tencel
TPE	Thermoplastic elastomer
VI	Viscose
WO	Wool

13.1 INTRODUCTION

Microplastic (MP) contamination, particularly microfibers, is an increasing concern for environmental and human health and has been designated one of the most pressing ecological concerns. MPs and microfibers have the potential to harm organisms and impede ecosystem functions when they are released into the environment. Measuring MP and microfiber toxicity in organisms and the environment is challenging and inconclusive. Research on the effects of MPs and microfibers on organisms and the environment does not yield consistent results, owing to differences in polymers (e.g., polyethylene (PE), polystyrene (PS), polyvinyl chloride, polyethylene terephthalate (PET) etc.), shapes (e.g., films, fragments, fibers, etc.), organisms (e.g., annelids, crustaceans, mollusks, etc.), and varying doses and exposure times used in the experiments (Bucci et al., 2020).

MPs were primarily an issue in marine ecosystems on a worldwide scale. But in recent years, it has become clear that MPs and microfibers represent a new class of particle contaminants everywhere in the environment. People in industrialized regions spend much time indoors; thus, there needs to be much consideration given to indoor air and household dust exposure (Bhat et al., 2021; Bhat 2024). Additionally, it should be recognized that an indoor setting does not always refer to a private dwelling. Instead, indoor areas include office spaces, governmental structures, educational institutions, sports arenas, hospitals, etc. For decades, indoor airborne particles and household dust have been the subject of much research. With the exception of asbestos and mineral fibers, tests for airborne particles were primarily done based on mass and number concentration. In the case of household dust, chemical, and biological components have typically received attention (Mølhave et al., 2000). Examining dust and particles for plastics has recently become popular (Bhat et al., 2022a; Bhat 2024). This late realization is surprising because there are many direct and indirect sources of plastics and their additives in the indoor environment (Bhat et al., 2022b; Bhat; 2024; Eraslan et al., 2021). These include personal care products and paints, artificial turf in sports halls, and wear and tear from floors, furniture, and fabrics (Bhat et al., 2021). The use of thermoplastics inside has increased a lot because of 3D printing. Through air exchange, plastics can also be brought in from outside. Car tires' wear and tear are considered massive MP sources. Also, many plastics have additives that can give off volatile and semivolatile organic compounds when used (Salthammer, 2020).

Most MPs and microfibers in indoor air come from synthetic textiles used in clothes, such as acrylic (AR), polyamide (PA), or polyester (PES) (Bhat et al., 2021; Chen et al., 2020). Microfibers rip from clothing when worn, cleaned, and dried. Microfibers emitted by these materials are often longer and more hazardous to people. When MPs and microfibers smaller than 5 μm in diameter are breathed, they are not filtered out by the nose and may become trapped deep inside the lungs, producing a variety of health problems from a simple cough to lung infections like pneumonia (OMEGA, 2017). Particles less than 2.5 μm in diameter can trigger lifelong lung damage. They can potentially enter the bloodstream and create significant health problems such as cardiovascular disease or cancer (Kevin Luo, 2018). Airborne MPs and microfibers may absorb and transport toxic chemicals or matter, such as germs or viruses (Enyoh et al., 2019). Because MPs and microfibers are bio-persistent, they cannot be ejected or broken down once they enter the human body. Individuals in developed countries spend more than 90% of their time in indoor environments (Bhat et al., 2022; Enyoh et al., 2019); the prevalence of MPs and microfibers in the indoor environment and their influence on human health are critical. This chapter reviews the present state of the art on the sources and impacts on the human health of MPs and microfibers in indoor environments.

13.2 MICROPLASTICS AND MICROFIBERS IN THE INDOOR ENVIRONMENT

Nowadays, both inorganic and organic substances are frequently measured in indoor environments. Therefore, it is often difficult to tell which compounds originate from MPs and other sources. Figure 13.1 gives a rough summary of possible MP sources in indoor environments as a basis for the subsequent discussion. In general, there are two types of indoor environments (industrial and non-industrial). The MPs found in these indoor environments can be either primary or secondary. These primary and secondary MPs in indoor environments have different sources, from personal care products and packaging materials to cleaning products. If these MPs are inhaled, they can reach lower (1–5 μm) and upper airways (5–30 μm) and can have harmful impacts on human health (Figure 13.1).

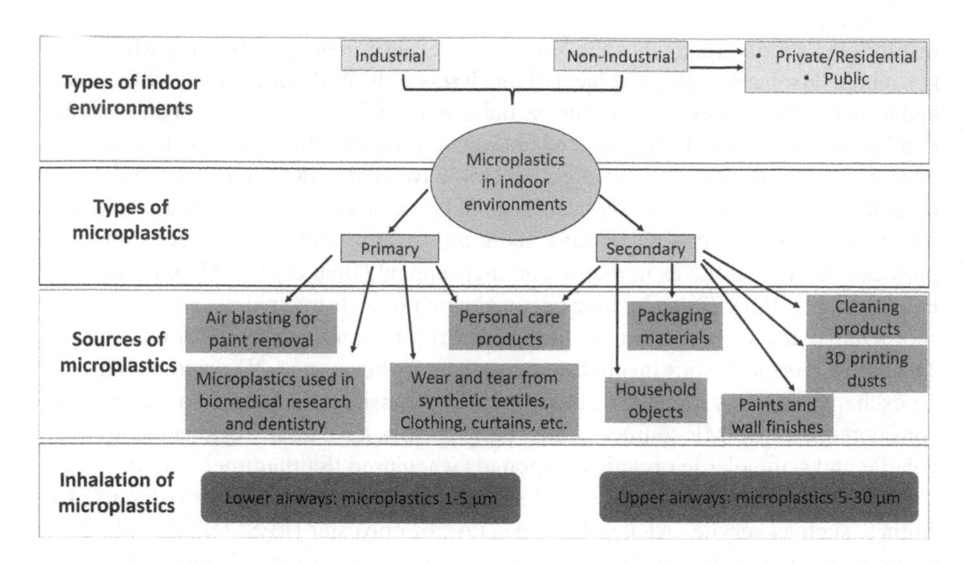

FIGURE 13.1 General view of types and sources of microplastics in indoor environments.

13.2.1 Outdoor Environment

Many overviews are available on the presence of MPs in outdoor air (Amato-Lourenço et al., 2020; Bhat et al., 2022b; Gaston et al., 2020). The most substantial contributors are tire abrasion (primarily styrene butadiene rubber (SBR)), road dust, waste management, sports fields, macroplastic debris, and building work. In general primary and secondary MPs are two of the origins of the MPs found in outdoor habitats. Common sources include personal care products with microbeads (Napper et al., 2015), washing synthetic textiles that release microfibers (Browne et al., 2011), plastic pellets or nurdles used in manufacturing processes (Cole et al., 2011), tire wear because of the presence of synthetic rubber (Kole et al., 2017), road markings and construction materials that deteriorate and abrade (Kreider et al., 2010). Other recently discussed potential sources of MPs include emissions from recycling polymeric materials (Bhat et al., 2022a; Bhat 2024; Thacharodi et al., 2024). In addition, little or no information exists regarding the combustion process as a source of polymeric particle emissions. Recently, MPs have been discovered in the bottom residue of incineration facilities (Shen et al., 2021). MPs were abundant in fly ash, bottom ash, and soil at 23, 171, and 86 particles/kg concentrations, respectively (Shen et al., 2021). While analyzing the atmospheric MPs, Dris et al. (2017) classified fibers based on their size. Klein and Fischer (2019) distinguish between fragments and fibers based on their morphology and categorize them by size. Xie et al. (2022) thoroughly examine MPs regarding chemical analysis, color, shape, and size. Through infiltration, tracking, and penetration, particles can infiltrate the indoor environment. Infiltration is the direct exchange of air via open windows and doors, as well as mechanical ventilation systems without particle filters. Tracking refers to particle entry through footwear and clothing, whereas penetration refers to particle entry through small gaps in the building exterior. The efficacy of the mechanisms varies depending on the circumstances.

13.2.2 Abrasion

Abrasion is the mechanical wear of tissue components induced by rubbing against another surface. This causes the discharge of particles (fragments) and fibers, typically seen in household dust. Today, many products are subject to specific minimum requirements in abrasion resistance (Sinclair, 2015), but only a limited statement can be made about their properties under actual conditions. Furthermore, the standardized test techniques primarily address the material surface rather than fiber release. However, it is reasonable to anticipate that mechanical stress on surfaces contributes to a sizeable part of the MPs seen in household dust. Mainly applicable to flooring and furnishings. The dust particles can be mobilized by resuspension. Thatcher and Layton (1995) discovered the highest resuspension rates for particle sizes between 5 and 25 μm.

13.2.3 Clothing/Laundry

It is well known that washing textiles can significantly increase the number of microfibers in wastewater (De Falco et al., 2019; Henry et al., 2019; Napper & Thompson, 2016). However, very few studies have examined how washing affects air quality. O'Brien et al. (2020) have studied mechanical drying processes and concluded that it is an emission source of MP fibers into the ambient air. De Falco et al. (2020) demonstrated that the immediate release of microfibers from clothes to air due to daily wear is equally essential to discharge wastewater by laundering. In particular, several studies have shown that 1 kg of synthetic fabric textiles releases 23,333–116,666 microfibers per wash cycle (Yang et al., 2019), with a length of mostly less than 5 mm (Pirc et al., 2016). Additionally, it has been confirmed that using pulsator laundry machines rather than platen laundry machines has increased the release of microfibers (Yang et al., 2019). Yang et al. (2019) found that an increase in temperature and the presence of detergent increases the shedding of microfibers from textiles. Recently, Bhat et al. (2022c) collected dust samples from the laundry and dry cleaning environments. During the density separation of collected samples, the upper portion of the samples was relocated for MP investigation. Otherwise, the characterization of micro and nanoplastics would have been hampered by these millimeter-sized MPs occupying the maximum filter surface. They found that the laundry and dry cleaning environments are the sources of MPs. The number of MPs seen was between 10 and 23. Fragment, pellet, film, and foam were the most prevalent MPs. As the bigger size MPs were analyzed, this might be why fibers were not dominant. At the same time, their dominant colors were red, white, transparent, and black. The size of the detected MPs was between 1.2 and 5 mm. Polyvinyl acetate (PVAc), Tencel (TN), Polyamide resin (PA resin), Polyamide-11, Polyurethane (PUR), Ethylene-vinyl acetate (EVA), and PET were the prevalent types of MPs seen in this investigation.

13.2.4 3D Printing

There is growing interest in 3D printing as a source of indoor polymers, monomers, and additives. The fused filament manufacturing method, which is popular in the commercial sector, is used to treat thermoplastic polymers. Acrylonitrile butadiene styrene is the common polymer used in fused filament fabrication, with the product of the main emission being styrene. A more environmentally friendly option is polylactic acid, which releases lactide, the cyclic diester of lactic acid. Other common thermoplastics for fused filament fabrication are Polyamide 6 (PA 6) (caprolactam), high-impact PS (styrene), and PVA (acetic acid) (Salthammer, 2022). Even though many chemical compounds are released during 3D printing, only a small number of them can be attributed directly to the polymers or their additives (Azimi et al., 2016; Davis et al., 2019; Gu et al., 2019). However, this also includes many novel substances not previously recognized to be present in indoor air, such as stabilizers. With 3D printing, particulate emissions always happen. The particles cannot be directly categorized as MPs because their normal size range is 10–200 nm (Salthammer, 2022). In experiments with different polymers, Gu et al. (2019) discovered that the particles evaporate as the temperature rises and vanish between 250°C and 300°C. Around the glass transition temperature of the corresponding thermoplastic, the volatilization process starts. Using thermogravimetric studies, Ding et al. (2019) reached similar conclusions. Carbon nanotubes are added to 3D printer filaments to boost their conductivity, as pointed out by Potter et al. (2021). Researchers Bossa et al. (2021) showed that abrasion releases MPs from 3D printing polymers, including multiwall carbon nanotubes. Overall, the chemical composition of the particles produced by 3D printing is extremely different, and the filament influences it, the chemicals in the printer, and the working conditions. As a result, such particles' chemical fingerprints cannot be ascribed to a single polymer (Tang & Seeger, 2022).

13.2.5 Sport Arenas

Artificial turf and sand/fiber blends are two significant sources of MPs and microfibers. The components of contemporary artificial grass include a bonded substrate (base layer), a synthetic elastic layer (often made of polypropylene (PP)), fibers (PE, PP, PA), and filler granules (SBR, PUR, occasionally used in conjunction with sand) (Salthammer, 2022). Aggregates are often added to equestrian or any arenas with a sand foundation to increase the floor's longevity. Shredded carpets, non-woven materials, garments, and automobile tire residues are examples. They still have the original ingredients in them. Another issue is that severe mechanical stress further crushes them. Recently, biobased fibers and wood chips have been added to sand in horse arenas. One study revealed that a 7,881 m^2 football field's artificial turf contained three distinct granular materials: 51 tons of SBR, 61 tons of EPDM, and 87 tons of TPM (Miljöinstitutet, 2012). Furthermore, artificial grass in Norway included 90% SBR and 10% ethylene propylene diene monomer and thermoplastic elastomer (TPE). TPE is extensively used in indoor sports facilities for artificial turf (Simon Magnusson, 2015). TPE particles will stick to shoe tread and be released into the environment in sporting venues using TPE artificial grass. Significantly, MPs may be found in the filler particles of artificial grass. MPs emitted by artificial grass will also enter the environment when the turf ages and degrades. In one study, artificial turf lost 0.38–0.63 kg/m^3 of rubber annually (Magnusson et al., 2016).

13.2.6 Other Indoor Sources

Numerous interior goods either consciously or unconsciously release MPs and microfibers. The main components of dispersion paints are pigments, a binder: typically PVA or AR, and additives. MPs and microfibers are also discharged during application or cleaning (Bhat et al., 2021, 2022d). In general, spray paints and polymers for spray application are also important sources. MPs (also known as beads) in cosmetics, personal care items (Anagnosti et al., 2021; Lei et al., 2017), detergents, and cleaning goods are hotly debated. The degree of abrasion produced by shoes, many of which have PUR soles, is also noticeable. Despite the fact that the sources of fluorinated MPs, notably PTFE, have significantly diminished, they are nevertheless used in cosmetics, water-repellent garments, and nonstick coatings. Last but not least, it should be noted that food and drink might expose humans to MPs (De Frond et al., 2022; Dessì et al., 2021; Hernandez et al., 2019). More detailed information about the sources of MPs in food and drink has already been highlighted (Ageel et al., 2022; Kwon et al., 2020; Toussaint et al., 2019).

Table 13.1 summarizes publicly available data (indoor house dust, indoor air, and indoor fallout) on the presence of MPs indoors and indicates that the concentrations and rates of deposition extend a broad range making exposure prediction challenging. This opens up many possibilities for interpretation. Until now, units for MP deposition or indoor concentration environments have not been formulated, and studies are using their units for the MPs (Table 13.1), like ng, mg, kg, items, numbers, etc. The usage of different types of units in expressing the concentration of MPs and microfibers in indoor environments shows that methods have not been developed until now. The research on the MPs and microfibers in indoor environments is still early. Minimal studies have been done on the presence of MPs and microfibers in these indoor environments, although humans spend most of their time from homes to workplaces, and their air quality is essential for human health, as these MPs and microfibers are very harmful to human health (Bhat et al., 2022a; Eraslan et al., 2023). More inter-laboratory work should be done on the indoor microfibers to get comparative data for the human health risk assessment. The work on indoor MPs and microfibers has been done in minimal countries. A limited number of indoor environments have been selected to investigate MPs and microfibers. It is believed that indoor environments are rich sources of microfibers due to the usage of textile items (Bhat et al., 2021). Most studies have focused on the standard instruments for analyzing MPs and microfibers, like a light microscope, stereomicroscope, fluorescence microscope, Fourier-transform infrared spectroscopy (FTIR), and Raman. These instruments give us the quality frame of MPs and microfibers while their concentration and presence of additives are essential. Instruments like Liquid chromatography-electrospray ionization tandem mass spectrometric (LC-ESI-MS) and Pyrolysis gas chromatography-mass spectrometry (Pyr-GC/MS) etc. should be used. Studies have been focusing on common MPs and microfibers like PA, PP, PET, Polycarbonate (PC), and PE. A higher number of types of MPs and microfibers should be focused on to get more accurate results about the presence of their sources in different indoor environments and their mitigation measures.

TABLE 13.1
Literature on Indoor Microplastic and Microfiber Studies in House Dust, Indoor Air, and Indoor Fallout

Sample Type	Location	Sampling Site	Instruments Used	Concentration	MPs Identified	Reference
House dust	France	Apartment and office	ATR-FTIR	190–670 fibers/mg	PA, PP, and PE	Dris et al. (2017)
	China	Home	Light microscope, μFTIR, and LC-MS/MS	1,550–120,000 mg/kg PET; <1-≈100 mg/kg PC	PET and PC	Liu et al. (2019)
	China	Dormitory	LC-MS/MS	18–43 mg/kg PA 6; 54–321 mg/kg PA 66	PA6 and PA66	Peng et al. (2020)
	12 countries	Bedroom and living room	HPLC-MS/MS and ESI-MS/MS	38–120,000 mg/kg PET; <0.11–1,700 mg/kg PC	PET and PC	Zhang et al. (2020a)
	Iran	School	Optical microscope, SEM-EDS, and Raman	10–635 MPs/g	PET, PP, PS, and PA	Nematollahi et al. (2021)
	China	Apartment, office, hotel, classroom, and dormitory	μFTIR,	62–3,861 MPs/g	PA, PS, PC, PES, PP, PE, PVC, PET, AR and CP	Zhu et al. (2022)
	Netherlands	House	LC-ESI-MS	<0.31–305 mg/g PET	PET	Tian et al. (2022)
	Iran	Classroom	Binocular microscope, μRaman, and SEM-EDS	81–55,830 MP/g	PET, PP, PS, and PA.	Abbasi et al. (2022)
	Iran	Hospital, mosque, kindergarten, university, and house	Binocular microscope, μRaman, and SEM-EDS	48.6–139 items/mg	PE, PC, PP, PET, and PA	Kashfi et al. (2022)
	Pakistan	House	Stereomicroscope and FTIR	29–636 fibers/m^2	PES, PE, EPR, PTFE, PUR, PET, and PP	Aslam et al. (2022)
Indoor air	France	Apartment and office	ATR-FTIR	1–60 fibers/m^3	PA, PP, and PE	Dris et al. (2017)
	Denmark	Apartment (breathing thermal manikin)	FPA-μFTIR	1.7–16.2 N_{MP}/m^3	PES, PA, PS, PE, and PUR	Vianello et al. (2019)

(*Continued*)

TABLE 13.1 (*Continued*)
Literature on Indoor Microplastic and Microfiber Studies in House Dust, Indoor Air, and Indoor Fallout

Sample Type	Location	Sampling Site	Instruments Used	Concentration	MPs Identified	Reference
	Portugal	House	ATR-FTIR	6 fibers/m^3	PES, PA,CO , WO, LI, VI, and CV	Prata et al. (2020)
	California United States	University and hospital	Fluorescence microscope, μFTIR, and μRaman	Fibers 3.3 ± 2.9 and 12.6 ± 8.0 fragments/m^3	PVC, PE, PS, PC, PA, ABS, RC, PET, and AR	Gaston et al. (2020)
	China	Apartment, office, classroom, hospital, and transit station waiting hall	Fluorescence stereomicroscope and μFTIR	1,583 ± 1,180 m^{-3}	PES, PA, PP, PE, PS, and PVC	Liao et al. (2021)
	China	Living room and office room	Raman	16–93 N/m^3	PE, PES, RC, PVC, CO, PP, PUR, and rubber	Xie et al. (2022)
	Taiwan	Nail salon	ECHO revolve microscope and FTIR	46 ± 55 MPs/m^3	PUR, PVC, PCL, PEI, PVA, AR, and RB	Chen et al. (2022)
	South Korea	House	FTIR	0.49–6.64 MPs/m^3	PP, PES, PS, PTFE, PVC, ALK AR, PA, PU, and PE	Choi et al. (2022)
	Turkey	University (laboratory, corridor, and office)	Optical microscope, μRaman, and SEM-EDS	12.03–18.51 MPs/m^3	PA 6, PTFE, HDPE, PEO, PVC, PP, and PVA	Bhat et al. (2022c)
	Kuwait	Government buildings, residential dwellings, hospitals, and mosque	Fluorescence stereomicroscope, UV stereomicroscope and, μRaman	3.24–27.13 MP m^3	PES and PA	Uddin et al. (2022)
Indoor fallout	France	Apartment and office	ATR-FTIR	1,586–11,130 fibers/m^{2}day	PA, PP, and PE	Dris et al. (2017)

(*Continued*)

TABLE 13.1 (*Continued*)
Literature on Indoor Microplastic and Microfiber Studies in House Dust, Indoor Air, and Indoor Fallout

Sample Type	Location	Sampling Site	Instruments Used	Concentration	MPs Identified	Reference
	China	Dormitory, office, and corridor	Stereomicroscope and μFTIR	Dormitory (9.9×10^3), office (1.8×10^3) and corridor (1.5×10^3) MPs/m^2/day	PES, CV, AR, PR, PP, PS, and PA	Zhang et al. (2020b)
	UK	House	Stereomicroscope and μFTIR analysis	$1,414 \pm 1,022$ MPs/m^{2}day	PET, PA, AR, PP, PAN, PE, and PMMA	Jenner et al. (2021)
	Australia	House	Stereomicroscope, fluorescence stereomicroscope, and μFTIR	22–6,169 fibers/m^{2}day	PE, PES, PET PA, and PVC	Soltani et al. (2021)
	Brazil	University (reception room)	Fluorescence microscope and ATR-FTIR	309 ± 215 MPs/m^{2}day	PP, PES, LDPE, and PS	Amato-Lourenço et al. (2022)
	USA	University (office, hallway, classroom) and house	Stereomicroscope and Raman	$(6.20 \pm 0.57) \times 10^3$–$(1.96 \pm 1.09) \cdot 10^4$ fibers/m^{2}day	PS, PET, PE, PVC, and PP	Yao et al. (2021)
	China	Dining room in an apartment, dining hall on campus, restaurant, office, classroom	Fluorescence stereomicroscope and μFTIR	$(7.6 \pm 3.9) \times 10^5$ MPs/m^{2}day	PET, PE, and PA	Fang et al. (2022)

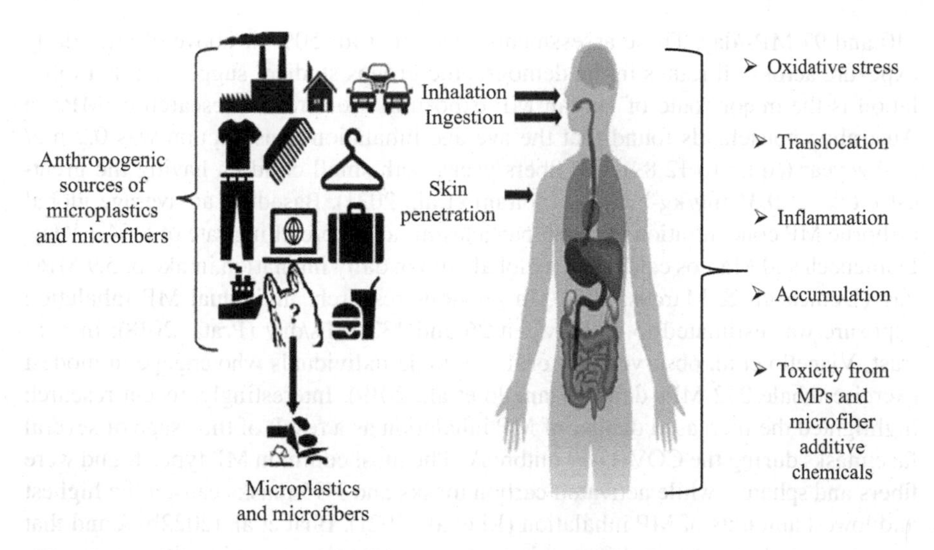

FIGURE 13.2 Pathways of human exposure to MPs and microfibers and their associated risks. (***Source***: Adopted and modified from Bhat et al., 2022a.)

13.3 HUMAN EXPOSURE TO MICROPLASTICS AND MICROFIBERS

Currently, little is known about human MP and microfiber exposure, mainly indoor exposure, compared to studies on outdoor environments. This might be owing to a number of reasons, including a lack of standardized procedures for quantifying MPs and microfibers in diverse matrices important to human exposure (e.g., indoor dust, air, food, water, and personal care products). It is now believed that human exposure to MPs and microfibers occurs through a mix of inhalation (air), ingestion (food, beverages, dust), and skin contact (dust, clothes, cosmetics) (Bhat et al., 2021, 2022a). Recent studies document microfibers in human intestinal tracts, placental tissue, blood, and lung tissue, though effects on humans remain largely unknown (Ibrahim et al., 2021; Leslie et al., 2022; Pauly et al., 1998; Ragusa et al., 2021). Figure 13.2 demonstrates the many routes via which humans are exposed to MPs and microfibers, as well as the dangers associated with such exposure. The widespread impacts of MPs and microfibers on human health are oxidative stress, translocation, inflammation, accumulation, and toxicity from MPs and microfibers additive chemicals. In this chapter, human exposure to MPs and microfibers will be discussed separately under inhalation, ingestion, and dermal contact headings.

13.3.1 INHALATION

Since MPs are consistently found in high quantities in indoor and outdoor air, inhalation has become a significant pathway for human exposure to MPs (Bhat et al., 2023; Li et al., 2021; Zhang et al., 2020b). Cox et al. (2019) identified an average concentration of 9.8 MPs/m^3; the results showed that males and females were exposed to 170 and 132 MPs/day, respectively, while male and female children were exposed to

110 and 97 MPs/day. These assessments accounted for 50% or above of total daily exposure across all routes in the demographic groups studied, suggesting that inhalation is the major route of human MP exposure. More recent research of MPs in Australian households found that the average inhalation consumption was 0.2 mg/kg-bw/year (equal to 12,891 MP fibers/year), with small children having the greatest intake at 0.31 mg/kg-bw/year (Soltani et al., 2021). Based on an average global airborne MP concentration of 0.685 particles/m^3 and a breathing rate of 8.64 m^3/day, Domenech and Marcos calculated a global human daily inhalation intake of 5.9 MPs/day (Domenech & Marcos, 2021). In previous research, individual MP inhalation exposure was estimated to be between 26 and 132 MPs/day (Prata, 2018); in contrast, Vianello et al. observed that ordinary male individuals who engage in modest exercise inhale 272 MPs daily (Vianello et al., 2019). Interestingly, recent research highlighted the increased danger of MP inhalation as a result of the usage of several face masks during the COVID-19 outbreak. The most common MP types found were fibers and spheres, while activated carbon masks and N95 masks caused the highest and lowest amounts of MP inhalation (Li et al., 2021). Bhat et al. (2022b) found that employees and students in different indoor environments on a university campus are exposed to a concentration of indoor airborne 12.03–18.51 MPs/m^3.

It is important to note that the various sampling, processing, and exposure assessment procedures make it challenging to compare the limited research on MP and microfiber inhalation exposure currently available. The units used to report airborne MP concentrations also differ in this way. Studies have reported concentrations as MPs per m^2, mg/kg, MPs/g, items/mg, N/m^3, MPs/m^2/day, and MPs/m^3, respectively (Table 13.1). Developing standardized procedures for measuring MPs and microfibers in the air is critical for obtaining valid and consistent exposure assessment results. Presently, little is known regarding the inhalation exposure to MPs and microfibers generated by indoor settled dust resuspension. Several studies have shown experimental evidence that resuspended home dust contributes to inhalable airborne particulate matter (PM). When previously deposited dust particles get dislodged from surfaces (such as carpets and floors), they are re-entrained into interior air by human actions, including walking, crawling, and vacuuming (Bhat et al., 2021; Hyytiäinen et al., 2018; Lewis et al., 2018; Rasmussen et al., 2018; Wang et al., 2021). Interestingly, toddlers' exposure to pollutants in resuspended dust particles through crawling was observed to be greater than adult exposure through walking (Hyytiäinen et al., 2018). The inhalation intake of several resuspended particulate-phase pollutants in indoor dust has been evaluated in several investigations (e.g., PM_{10}, $PM_{2.5}$, bacteria) (Knibbs et al., 2012; Mitro et al., 2016; Wu et al., 2018). It was found that the source intensity of $PM_{2.5}$ from the resuspension of home dust ranged from 0.03 to 0.5 mg/min. Furthermore, it was shown that the main source of $PM_{2.5}$ and PM_{10} exposure in the personal cloud (i.e., personal breathing zone) of residents of retirement centers was resuspension from walking on carpets (Knibbs et al., 2012; Mitro et al., 2016; Rasmussen et al., 2018; Rodes et al., 2001; Wang et al., 2021; Wu et al., 2018). Although it is hypothesized that MPs and microfibers would act similarly, there is little information on the inhalation exposure of MPs and microfibers from settled indoor dust that is resuspended. This demonstrates a significant information vacuum and suggests that existing reports of human inhalation exposure to MPs and microfibers, especially in toddlers, may be overstated.

13.3.2 Ingestion

13.3.2.1 Diet

Until recently, it was commonly believed that ingestion was the primary method of human exposure to MPs (Bergmann et al., 2015). This was sparked by a few early investigations that discovered MPs in various foods (Cauwenberghe & Janssen, 2014; De Witte et al., 2014). Multiple studies have identified MPs in different foods, including fish, shellfish, table salt, sugar, honey, milk, and beer, at varying amounts (Diaz-Basantes et al., 2020; Hernandez et al., 2019). This will result in the dietary exposure of humans to these MPs. Few studies, nevertheless, have calculated this dietary exposure. According to Cox et al. (2019), American male adults, female adults, male children, and female children consume 142, 113, 126, and 106 MPs/day, respectively, from their diets. Another study found that European adult exposure to MPs by table salt intake was between 37 and 100 MPs/person/year (Karami et al., 2017). Abbasi et al. (2018) measured concentrations of MPs in fish and shellfish samples from Iran could contribute to an average daily adult exposure of 5 particles/day.

Due to the absence of information on MPs and microfibers in important food categories, the current information about dietary exposure to MPs and microfibers is far from comprehensive. (e.g., meat, vegetables, oil, dairy products etc.). More extensive dietary studies are needed to adequately evaluate the risk associated with human food exposure to MPs and microfibers. Another important aspect that has to be researched further is the leakage of MPs and microfibers from plastic packaging into food and drinks, which may enhance exposure. Hot beverage research sparked concerns by noting that boiling one plastic teabag at 95°C releases around 11.6 billion MPs into a single cup of beverage (Hernandez et al., 2019). The methodologies and procedures used in this study that resulted in such a high estimate of MPs were called into doubt (Busse et al., 2020), and the possible release of MPs from plastic teabags has been confirmed. Another concerning research found that frequent sterilization and exposure to hot water might cause MP concentrations as high as 16 million particles per liter in PP newborn feeding bottles. The estimated exposure varied from 14,600 to 455,000 particles per capita per day, indicating that newborn exposure to MPs may be larger than previously thought (Li et al., 2020). Recent research found MPs from PS packing trays in meat products at amounts ranging from 4 to 18.7 MPs/kg of packed meat. These particles were challenging to rinse off and were probably cooked before consumption (Kedzierski et al., 2020). μRaman identification results revealed that thermoplastic sulfone polymers (polyethersulfone and polysulfone (PSU)) were common types of MPs in milk samples, which are highly used membrane materials in dairy processes (Kutralam-Muniasamy et al., 2020). Furthermore, mechanical stress has been found to be a component that may impact human exposure to MPs through bottled beverages. According to Winkler et al. (2019), repeated opening and shutting of plastic bottles increases the number of MPs released into bottled water via deterioration, increasing the likelihood of ingesting MPs.

These studies provide compelling evidence of human exposure to MPs via food in a variety of countries and age groups ranging from breastfeeding babies to adults. There is even evidence of MP exposure during pregnancy via the placenta (Ragusa et al., 2021). However, more extensive, large-scale investigations are needed to understand better the amount, consequences, and risks associated with human dietary exposure to MPs and microfibers.

13.3.2.2 Dust

Much attention has been focused over the past two decades on the role of indoor dust as a significant matrix for human exposure to a wide range of hazardous substances. This has prompted several recent investigations to look into possible human exposure to MPs and microfibers through unintended intake of indoor dust. According to research on PET and PC MPs in indoor and outdoor dust samples from 39 Chinese cities, the estimated daily intake ranges from 6,500 to 89,700 ng/kg-bw/day for various age groups. Children were the most exposed age group, with a mean estimated daily intake (EDI) of 17,300 ng/kg-bw/day, while teens and adults had mean estimated daily intakes of 7,270 and 6,500 ng/kg-bw/day, respectively. A comparison of indoor and outdoor dust exposure by EDI of PET MPs in all age groups found increased exposure (almost twofold) via indoor dust (Liu et al., 2019). Another study of MPs in house dust from 12 different countries verified the prevalence of human exposure to MPs via indoor dust, as well as higher intake and risk in younger age groups. MPs were found in high amounts (38–120,000 μg/g), resulting in median EDI values of 4,000–150,000 ng/kg-bw/day for newborns. Adults were less exposed, with median EDI values ranging from 360 to 12,000 ng/kg-bw/day, owing to a combination of higher dust intake rates and lower body weight in infants (Zhang et al., 2020). A recent investigation on MPs in indoor dust from 32 Australian households found similar results, with estimated mean ingestion rates of 6.1 mg/kg-bw/year (EDI 16,712 ng/kg-bw/day) for children and 0.5 mg/kg-bw/year (EDI 141,370 ng/kg-bw/day) for adults (Soltani et al., 2021).

Despite the small number of studies, it is plausible to conclude that human exposure to MPs and microfibers by ingestion of indoor dust, as well as the potential impact of such exposure, should not be neglected, especially in children. More research is needed to understand the spatial and temporal variability of such exposure, as well as the impact of individuals' time-activity patterns (e.g., the proportion of time spent in homes, offices, outdoors, cars, schools, and other types of microenvironments) on overall daily exposure to MPs and microfibers via dust ingestion in different age groups.

13.3.3 Dermal Contact

Currently, no studies analyze human dermal exposure to MPs and microfibers and their dangers. However, given the prevalence of MPs and microfibers in indoor dust, atmospheric deposition from both indoor and outdoor air, widespread use of microbeads in cosmetics, and ongoing degradation of microfibers from textiles, it is acceptable to consider skin contact as a route of human exposure to MPs (Bhat et al., 2022a; Domenech & Marcos, 2021). MP beads (microbeads with a diameter of < 1 mm) have been widely used in dermal exfoliating and cleaning products, as well as toothpaste and denture fillings (Anagnosti et al., 2021). Few studies have tried to estimate the amount of microbeads each individual uses daily using particular personal care products. According to research on several types of face scrubs sold in the UK, the MP content ranged from 10 to 100 g/L, with daily consumption per person being between 40.5 and 215 mg (Napper et al., 2015) and according to Gouin et al. (2011) using liquid soap results in an average daily intake of 2.4 mg MPs for the US population.

These few studies show that the dermal route cannot be neglected, even if they do not offer a complete picture of human skin exposure to MPs. Although studies reveal that only particles < 100 nm (i.e., nanoplastics) may directly breach the dermal barrier, human skin can nevertheless function as an effective barrier against the entry of big particles (Revel et al., 2018), the transdermal penetration of bigger particles may also occur through open skin wounds, sweat glands, or hair follicles (Schneider et al., 2009). Additionally, cutaneous exposure to MPs has been linked to skin damage brought on by inflammation and oxidative stress (Wright & Kelly, 2017). Therefore, additional study is needed to identify the relevance of this exposure route and the health risks associated with it, as well as human skin exposure to MPs and microfibers through contact with cosmetics, settled dust particles, textile fibers, etc.

13.4 CONCLUSION AND FUTURE RECOMMENDATIONS

Over the past decade, much research has been done to identify and describe MPs and microfibers. This research has shown that MPs and microfibers are everywhere in the environment. Despite the fact that the majority of the research has focused on the aquatic environment, a few recent studies have highlighted concerns regarding the existence of MPs and microfibers in interior spaces at levels harmful to human exposure. One must understand, nevertheless, that indoor contamination with MPs and microfibers is by no means a recent issue; it has lasted since plastics were first utilized in construction materials and everyday home items. Potential issues have been identified, such as artificial grass and sand/fiber blends. On the other hand, emerging sources, such as 3D printing, should not be neglected in terms of their impact on indoor air quality. Using home dust analysis and biomonitoring, trends in reducing or raising exposure to the target compound can be discovered. Whether MPs and microfibers seriously endanger human health has not yet been resolved scientifically. Strategies must be created to lessen MP production indoors and use MP beads. Their usage is technically not required in many household and personal care products.

According to current knowledge, human exposure to MPs and microfibers indoors can happen through skin contact, ingestion, and inhalation. While the presence of MPs and microfibers in indoor dust, food, and beverages suggests that ingestion is possible, the presence of MPs and microfibers in indoor air and atmospheric deposition samples suggests that inhalation is a substantial exposure route. Furthermore, owing to the use of MPs in cosmetics and personal care goods, as well as their release through the breakdown of microfibers from domestic textiles, the dermal exposure route cannot be overlooked. People spend up to 90% of their time inside, which has recently been exacerbated by COVID-19 pandemic lockdown techniques, which makes this alarming. The lack of knowledge on human exposure to MPs and microfibers inside and the danger involved has been made clear by critically examining the current literature in the present study. The demand for regulators and policymakers to take prompt action to alleviate such deleterious impacts is rising due to the growing public concern regarding MP and microfiber pollution and its possible negative implications on human health and the environment. The information basis for making educated decisions to protect public health against MPs and microfiber contamination is presently insufficient. There are considerable research gaps in our knowledge of the origins and amount of human MP and microfiber exposure, as well as the proportional contributions of different exposure routes to MP and microfiber body burdens. Therefore, it is advised that the following issues be given priority in MP and microfiber research.

- There is a critical need for verified, standardized methodologies for assessing MP and microfiber concentrations in indoor air and dust samples. An agreement on the units for reporting these concentrations on a mass basis that are acceptable for exposure assessment models must also be obtained. The harmonization of units reporting airborne concentrations of MPs and microfibers is essential.
- Despite the few studies that reveal MP concentrations in human exposure-relevant media (e.g., air, dust, and food), nothing is known about the quantity of exposure in real-life situations and across age/gender groups. As a result, comprehensive human exposure assessment studies in various countries for different age groups are critical for estimating individual- and population-level exposure to MPs via total diet studies, market basket studies, dermal contact scenarios, and/or time-activity patterns (i.e., time spent in homes, offices, schools, nurseries, cars, outdoor exercise, and so on).
- Evidence on the numerous sources and methods that may contribute substantially to human exposure to MPs and microfibers is limited. More research is required to assess the resuspension of MP particles from settled indoor dust into adults' and children's personal clouds, and hence its contribution to MP inhalation exposure in humans. Personal activity (e.g., walking adult, crawling child), interior features (e.g., carpeted vs. hardwood floors), vacuuming pace, temperature, and ventilation should all be included in the research.
- While a few recent studies have looked at human MP exposure through food (primarily seafood), little is known about the function of food packaging and food-contact materials in human MP consumption. Future studies should investigate the effects of packaging (e.g., plastic boxes, films, covers, plates, and utensils) on MPs and microfiber concentrations in food and subsequent consumption by various age groups and cultures. More research is also needed to understand the role of plastic bottles and caps in MPs and microfiber exposure through drinking, as well as the effect of beverage composition (e.g., probiotic drinks, multivitamin drinks) and temperature (e.g., hot coffee vs. cold juice) on MPs leaching from the bottle and subsequent human exposure through drinking these beverages.

REFERENCES

Abbasi, S., Soltani, N., Keshavarzi, B., Moore, F., Turner, A., & Hassanaghaei, M. (2018). Microplastics in different tissues of fish and prawn from the Musa Estuary, Persian Gulf. *Chemosphere*, *205*, 80–87. https://doi.org/10.1016/j.chemosphere.2018.04.076

Abbasi, S., Turner, A., Sharifi, R., Nematollahi, M. J., Keshavarzifard, M., & Moghtaderi, T. (2022). Microplastics in the school classrooms of Shiraz, Iran. *Building and Environment*, *207*, 108562. https://doi.org/10.1016/j.buildenv.2021.108562

Ageel, H. K., Harrad, S., & Abdallah, M. A.-E. (2022). Occurrence, human exposure, and risk of microplastics in the indoor environment. *Environmental Science: Processes & Impacts, 2006*. https://doi.org/10.1039/d1em00301a

Amato-Lourenço, L. F., dos Santos Galvão, L., de Weger, L. A., Hiemstra, P. S., Vijver, M. G., & Mauad, T. (2020). An emerging class of air pollutants: Potential effects of microplastics to respiratory human health? *Science of the Total Environment, 749*, 141676. https://doi.org/10.1016/j.scitotenv.2020.141676

Amato-Lourenço, L. F., dos Santos Galvão, L., Wiebeck, H., Carvalho-Oliveira, R., & Mauad, T. (2022). Atmospheric microplastic fallout in outdoor and indoor environments in São Paulo megacity. *Science of the Total Environment, 821*, 153450. https://doi.org/10.1016/j.scitotenv.2022.153450

Anagnosti, L., Varvaresou, A., Pavlou, P., Protopapa, E., & Carayanni, V. (2021). Worldwide actions against plastic pollution from microbeads and microplastics in cosmetics focusing on European policies. Has the issue been handled effectively? *Marine Pollution Bulletin, 162*, 111883. https://doi.org/10.1016/j.marpolbul.2020.111883

Aslam, I., Qadir, A., & Ahmad, S. R. (2022). A preliminary assessment of microplastics in indoor dust of a developing country in South Asia. *Environmental Monitoring and Assessment, 194*(5). https://doi.org/10.1007/s10661-022-09928-3

Azimi, P., Zhao, D., Pouzet, C., Crain, Neil, E., & Stephens, B. (2016). Emissions of ultrafine particles and volatile organic compounds from commercially available desktop 3D printers with multiple filaments. *Environmental Science & Technology, 50*(3), 1260–1268.

Bergmann, M., Gutow, L., & Klages, M. (2015). Micro- and nano-plastics and human health. In M. Bergman, L. Gutow, & M. Klages (Eds.), *Marine Anthropogenic Litter* (1st ed., pp. 343–366). Springer, Cham. https://doi.org/10.1007/978-3-319-16510-3

Bhat, M. A., Eraslan, F. N., Awad, A., Malkoç, S., Üzmez, O. O., Döğeroğlu, T., & Gaga, E. O. (2022). Investigation of indoor and outdoor air quality in a university campus during COVID-19 lock down period. *Building and Environment Journal, 219*, 109176. https://doi.org/10.1016/j.buildenv.2022.109176

Bhat, M. A., Eraslan, F. N., Gaga, E. O., & Gedik, K. (2023). Scientometric analysis of microplastics acroos the globe. In M. Vithanage, & M. N. V. Prasad (Eds.), *Microplastics in the Ecosphere: Air, Water, Soil, and Food* (1st ed., pp. 3–14). John Wiley & Sons. https://doi.org/10.1002/9781119879534.ch1

Bhat, M. A., Eraslan, F. N., Gedik, K., & Gaga, E. O. (2021). Impact of textile product emissions: Toxicological considerations in assessing indoor air quality and human health. In J. A. Malik, & S. Marathe (Eds.), *Ecological and Health Effects of Building Materials* (1st ed., pp. 505–541). Springer Nature, Switzerland. https://doi.org/https://doi.org/10.1007/978-3-030-76073-1_27

Bhat, M. A., Gedik, K., & Gaga, E. O. (2022a). Atmospheric micro (nano) plastics: Future growing concerns for human health. *Air Quality, Atmosphere & Health*. https://doi.org/10.1007/s11869-022-01272-2

Bhat, M. A., Gedik, K., & Gaga, E. O. (2022b). Characterization of indoor airborne microplastics in a university campus. In *10th International Symposium on Atmospheric Sciences, October*, 1–2.

Bhat, M. A., Gedik, K., & Gaga, E. O. (2022c). Environmental toxicity of emerging micro and nanoplastics: A lesson learned from nanomaterials. In A. H. Dar, & G. A. Nayik (Eds.), *Nanotechnology Interventions in Food Packaging and Shelf Life* (1st ed., pp. 311–337). Taylor & Francis (CRC Press). https://doi.org/10.1201/9781003207641-18

Bhat, M. A., Gedik, K., & Gaga, E. O. (2022d). Laundry and dry cleaning environments as a source of microplastics. *MICRO 2022, Online Atlas Edition: Plastic Pollution from MACRO to Nano, Online*, 17 November, 1. https://doi.org/https://doi.org/10.5281/zenodo.7217111

Bhat, M. A. (2024). Unveiling the overlooked threat: Macroplastic pollution in indoor markets in an urban city. *Case Studies in Chemical and Environmental* Engineering, *9*, 100558. https://doi.org/10.1016/j.cscee.2023.100558

Bossa, N., Sipe, J. M., Berger, W., Scott, K., Kennedy, A., Thomas, T., Hendren, C. O., & Wiesner, M. R. (2021). Quantifying mechanical abrasion of MWCNT nanocomposites used in 3D printing: Influence of CNT content on abrasion products and rate of microplastic production. *Environmental Science and Technology*, *55*, 10332–10342. https://doi.org/10.1021/acs.est.0c02015

Browne, M. A., Crump, P., Niven, S. J., Teuten, E., Tonkin, A., Galloway, T., & Thompson, R. (2011). Accumulation of microplastic on shorelines woldwide: Sources and sinks. *Environmental Science and Technology*, *45*(21), 9175–9179. https://doi.org/10.1021/es201811s

Bucci, K., Tulio, M., & Rochman, C. M. (2020). What is known and unknown about the effects of plastic pollution: A meta-analysis and systematic review. *Ecological Applications*, *30*, 1–16. https://doi.org/10.1002/eap.2044

Busse, K., Ebner, I., Humpf, H. U., Ivleva, N., Kaeppler, A., Oßmann, B. E., & Schymanski, D. (2020). Comment on "plastic teabags release billions of microparticles and nanoparticles into tea". *Environmental Science and Technology*, *54*, 14134–14135. https://doi.org/10.1021/acs.est.0c03182

Cauwenberghe, L. Van, & Janssen, C. R. (2014). Microplastics in bivalves cultured for human consumption. *Environmental Pollution*, *193*, 65–70. https://doi.org/10.1016/j.envpol.2014.06.010

Chen, E. Y., Lin, K. T., Jung, C. C., Chang, C. L., & Chen, C. Y. (2022). Characteristics and influencing factors of airborne microplastics in nail salons. *Science of the Total Environment*, *806*, 151472. https://doi.org/10.1016/j.scitotenv.2021.151472

Chen, G., Feng, Q., & Wang, J. (2020). Mini-review of microplastics in the atmosphere and their risks to humans. *Science of the Total Environment*, *703*, 135504. https://doi.org/10.1016/j.scitotenv.2019.135504

Choi, H., Lee, I., Kim, H., Park, J., Cho, S., Oh, S., Lee, M., & Kim, H. (2022). Comparison of microplastic characteristics in the indoor and outdoor air of urban areas of South Korea. *Water, Air, and Soil Pollution*, *233*(5), 1–10. https://doi.org/10.1007/s11270-022-05650-5

Cole, M., Lindeque, P., Halsband, C., & Galloway, T. S. (2011). Microplastics as contaminants in the marine environment : A review. *Marine Pollution Bulletin*, *62*, 2588–2597. https://doi.org/10.1016/j.marpolbul.2011.09.025

Cox, K. D., Covernton, G. A., Davies, H. L., Dower, J. F., Juanes, F., & Dudas, S. E. (2019). Human consumption of microplastics. *Environmental Science and Technology*, *53*(12), 7068–7074. https://doi.org/10.1021/acs.est.9b01517

Davis, A. Y., Zhang, Q., Wong, J. P. S., Weber, R. J., & Black, M. S. (2019). Characterization of volatile organic compound emissions from consumer level material extrusion 3D printers. *Building and Environment*, *160*, 106209. https://doi.org/10.1016/j.buildenv.2019.106209

De Falco, F., Cocca, M., Avella, M., & Thompson, R. C. (2020). Microfiber release to water, via laundering, and to air, via everyday use: A comparison between polyester clothing with differing textile parameters. *Environmental Science and Technology*, *54*(6), 3288–3296. https://doi.org/10.1021/acs.est.9b06892

De Falco, F., Di Pace, E., Cocca, M., & Avella, M. (2019). The contribution of washing processes of synthetic clothes to microplastic pollution. *Scientific Reports*, *9*(1), 1–11. https://doi.org/10.1038/s41598-019-43023-x

De Frond, H., Thornton Hampton, L., Kotar, S., Gesulga, K., Matuch, C., Lao, W., Weisberg, S. B., Wong, C. S., & Rochman, C. M. (2022). Monitoring microplastics in drinking water: An interlaboratory study to inform effective methods for quantifying and characterizing microplastics. *Chemosphere*, *298*, 134282. https://doi.org/10.1016/j.chemosphere.2022.134282

Dessì, C., Okoffo, E. D., O'Brien, J. W., Gallen, M., Samanipour, S., Kaserzon, S., Rauert, C., Wang, X., & Thomas, K. V. (2021). Plastics contamination of store-bought rice. *Journal of Hazardous Materials, 416*, 125778. https://doi.org/10.1016/j.jhazmat.2021.125778

De Witte, B., Devriese, L., Bekaert, K., Hoffman, S., Vandermeersch, G., Cooreman, K., & Robbens, J. (2014). Quality assessment of the blue mussel (Mytilus edulis): Comparison between commercial and wild types. *Marine Pollution Bulletin, 85*(1), 146–155. https://doi.org/10.1016/j.marpolbul.2014.06.006

Diaz-Basantes, M. F., Conesa, J. A., & Fullana, A. (2020). Microplastics in honey, beer, milk and refreshments in Ecuador as emerging contaminants. *Sustainability, 12*, 5514. https://doi.org/10.3390/SU12145514

Ding, S., Ng, B. F., Shang, X., Liu, H., Lu, X., & Wan, M. P. (2019). The characteristics and formation mechanisms of emissions from thermal decomposition of 3D printer polymer filaments. *Science of the Total Environment, 692*, 984–994. https://doi.org/10.1016/j.scitotenv.2019.07.257

Domenech, J., & Marcos, R. (2021). Pathways of human exposure to microplastics, and estimation of the total burden. *Current Opinion in Food Science, 39*, 144–151. https://doi.org/10.1016/j.cofs.2021.01.004

Dris, R., Gasperi, J., Mirande, C., Mandin, C., Guerrouache, M., Langlois, V., & Tassin, B. (2017). A first overview of textile fibers, including microplastics, in indoor and outdoor environments. *Environmental Pollution, 221*, 453–458. https://doi.org/10.1016/j.envpol.2016.12.013

Enyoh, C. E., Verla, A. W., Verla, E. N., & Ibe, F. C. (2019). Airborne microplastics: A review study on method for analysis, occurrence, movement and risks. *Environmental Monitoring and Assessment*, 191–668.

Eraslan, F. N., Bhat, M. A., Gaga, E. O., & Gedik, K. (2021). Comprehensive analysis of research trends in volatile organic compounds emitted from building materials: A bibliometric analysis. In J. A. Malik, & S. Marathe (Eds.), *Ecological and Health Effects of Building Materials* (1st ed., pp. 87–109). Springer Nature, Switzerland. https://doi.org/https://doi.org/10.1007/978-3-030-76073-1_6

Eraslan, F. N., Bhat, M. A., Gedik . K., & Gaga, E. O. (2023). The single-use plastic pandemic in the covid-19 era. In M. Vithanage, & M. N. V. Prasad (Eds.), *Microplastics in the Ecosphere: Air, Water, Soil, and Food* (1st ed., pp. 65–76). John Wiley & Sons. https://doi.org/10.1002/9781119879534.ch4

Fang, M., Liao, Z., Ji, X., Zhu, X., Wang, Z., Lu, C., Shi, C., Chen, Z., Ge, L., Zhang, M., Dahlgren, R. A., & Shang, X. (2022). Microplastic ingestion from atmospheric deposition during dining/drinking activities. *Journal of Hazardous Materials, 432*, 128674. https://doi.org/10.1016/j.jhazmat.2022.128674

Gaston, E., Woo, M., Steele, C., Sukumaran, S., & Anderson, S. (2020). Microplastics differ between indoor and outdoor air masses: Insights from multiple microscopy methodologies. *Applied Spectroscopy, 74*, 1079–1098. https://doi.org/10.1177/0003702820920652

Gouin, T., Roche, N., Lohmann, R., & Hodges, G. (2011). A thermodynamic approach for assessing the environmental exposure of chemicals absorbed to microplastic. *Environmental Science and Technology, 45*, 1466–1472. https://doi.org/10.1021/es1032025

Gu, J., Wensing, M., Uhde, E., & Salthammer, T. (2019). Characterization of particulate and gaseous pollutants emitted during operation of a desktop 3D printer. *Environment International, 123*, 476–485. https://doi.org/10.1016/j.envint.2018.12.014

Henry, B., Laitala, K., & Klepp, I. G. (2019). Microfibres from apparel and home textiles: Prospects for including microplastics in environmental sustainability assessment. *Science of the Total Environment, 652*, 483–494. https://doi.org/10.1016/j.scitotenv.2018.10.166

Hernandez, L. M., Xu, E. G., Larsson, H. C. E., Tahara, R., Maisuria, V. B., & Tufenkji, N. (2019). Plastic teabags release billions of microparticles and nanoparticles into tea. *Environmental Science and Technology, 53*(21), 12300–12310. https://doi.org/10.1021/acs.est.9b02540

Hyytiäinen, H. K., Jayaprakash, B., Kirjavainen, P. V., Saari, S. E., Holopainen, R., Keskinen, J., Hämeri, K., Hyvärinen, A., Boor, B. E., & Täubel, M. (2018). Crawling-induced floor dust resuspension affects the microbiota of the infant breathing zone. *Microbiome*, *6*, 1–12. https://doi.org/10.1186/s40168-018-0405-8

Ibrahim, Y. S., Tuan Anuar, S., Azmi, A. A., Wan Mohd Khalik, W. M. A., Lehata, S., Hamzah, S. R., Ismail, D., Ma, Z. F., Dzulkarnaen, A., Zakaria, Z., Mustaffa, N., Tuan Sharif, S. E., & Lee, Y. Y. (2021). Detection of microplastics in human colectomy specimens. *JGH Open*, *5*, 116–121. https://doi.org/10.1002/jgh3.12457

Jenner, L. C., Sadofsky, L. R., Danopoulos, E., & Rotchell, J. M. (2021). Household indoor microplastics within the Humber region (United Kingdom): Quantification and chemical characterisation of particles present. *Atmospheric Environment*, *259*, 118512. https://doi.org/10.1016/j.atmosenv.2021.118512

Karami, A., Golieskardi, A., Keong Choo, C., Larat, V., Galloway, T. S., & Salamatinia, B. (2017). The presence of microplastics in commercial salts from different countries. *Scientific Reports*, *7*(November 2016), 1–11. https://doi.org/10.1038/srep46173

Kashfi, F. S., Ramavandi, B., Arfaeinia, H., Mohammadi, A., Saeedi, R., De-la-Torre, G. E., & Dobaradaran, S. (2022). Occurrence and exposure assessment of microplastics in indoor dusts of buildings with different applications in Bushehr and Shiraz cities, Iran. *Science of the Total Environment*, *829*, 154651. https://doi.org/10.1016/j.scitotenv.2022.154651

Kedzierski, M., Lechat, B., Sire, O., Le Maguer, G., Le Tilly, V., & Bruzaud, S. (2020). Microplastic contamination of packaged meat: Occurrence and associated risks. *Food Packaging and Shelf Life*, *24*, 100489 Contents. https://doi.org/10.1016/j.fpsl.2020.100489

Kevin Luo. (2018). Are you breathing plastic air at home? Here's how microplastics are polluting our lungs. *World Economic Forum Join*, 1–6. https://www.weforum.org/agenda/2018/06/microplastic-pollution-in-air-pollutes-our-lungs/

Klein, M., & Fischer, E. K. (2019). Microplastic abundance in atmospheric deposition within the Metropolitan area of Hamburg, Germany. *Science of the Total Environment*, *685*, 96–103. https://doi.org/10.1016/j.scitotenv.2019.05.405

Knibbs, L. D., He, C., Duchaine, C., & Morawska, L. (2012). Vacuum cleaner emissions as a source of indoor exposure to airborne particles and bacteria. *Environmental Science and Technology*, *46*, 534–542. https://doi.org/10.1021/es202946w

Kole, P. J., Löhr, A. J., Van Belleghem, F. G. A. J., Ragas, A. M. J. (2017 Oct 20). Wear and tear of tyres: A stealthy source of microplastics in the environment. *International Journal of Environmental Research and Public Health*, *14*(10), 1265. https://doi.org/10.3390/ijerph14101265.

Kreider, M. L., Panko, J. M., McAtee, B. L., Sweet, L. I., & Finley, B. L. (2010). Physical and chemical characterization of tire-related particles: Comparison of particles generated using different methodologies. *Science of the Total Environment*, *408*, 652–659. https://doi.org/10.1016/j.scitotenv.2009.10.016

Kutralam-Muniasamy, G., Pérez-Guevara, F., Elizalde-Martínez, I., & Shruti, V. C. (2020). Branded milks – Are they immune from microplastics contamination? *Science of the Total Environment*, *714*, 136823. https://doi.org/10.1016/j.scitotenv.2020.136823

Kwon, J. H., Kim, J. W., Pham, T. D., Tarafdar, A., Hong, S., Chun, S. H., Lee, S. H., Kang, D. Y., Kim, J. Y., Kim, S. Bin, & Jung, J. (2020). Microplastics in food: A review on analytical methods and challenges. *International Journal of Environmental Research and Public Health*, *17*, 1–23. https://doi.org/10.3390/ijerph17186710

Lei, K., Qiao, F., Liu, Q., Wei, Z., Qi, H., Cui, S., Yue, X., Deng, Y., & An, L. (2017). Microplastics releasing from personal care and cosmetic products in China. *Marine Pollution Bulletin*, *123*, 122–126. https://doi.org/10.1016/j.marpolbul.2017.09.016

Leslie, H. A., van Velzen, M. J. M., Brandsma, S. H., Vethaak, A. D., Garcia-Vallejo, J. J., & Lamoree, M. H. (2022). Discovery and quantification of plastic particle pollution in human blood. *Environment International*, *163*, 107199. https://doi.org/10.1016/j.envint.2022.107199

Lewis, R. D., Ong, K. H., Emo, B., Kennedy, J., Kesavan, J., & Elliot, M. (2018). Resuspension of house dust and allergens during walking and vacuum cleaning. *Journal of Occupational and Environmental Hygiene*, *15*, 235–245. https://doi.org/10.1080/15459624.2017.1415438

Li, D., Shi, Y., Yang, L., Xiao, L., Kehoe, D. K., Gun'ko, Y. K., Boland, J. J., & Wang, J. J. (2020). Microplastic release from the degradation of polypropylene feeding bottles during infant formula preparation. *Nature Food*, *1*, 746–754. https://doi.org/10.1038/s43016-020-00171-y

Li, L., Zhao, X., Li, Z., & Song, K. (2021). COVID-19: Performance study of microplastic inhalation risk posed by wearing masks. *Journal of Hazardous Materials*, *411*, 124955. https://doi.org/10.1016/j.jhazmat.2020.124955

Liao, Z., Ji, X., Ma, Y., Lv, B., Huang, W., Zhu, X., Fang, M., Wang, Q., Wang, X., Dahlgren, R., & Shang, X. (2021). Airborne microplastics in indoor and outdoor environments of a coastal city in Eastern China. *Journal of Hazardous Materials*, *417*, 126007. https://doi.org/10.1016/j.jhazmat.2021.126007

Liu, C., Li, J., Zhang, Y., Wang, L., Deng, J., Gao, Y., Yu, L., Zhang, J., & Sun, H. (2019). Widespread distribution of PET and PC microplastics in dust in urban China and their estimated human exposure. *Environment International*, *128*, 116–124. https://doi.org/10.1016/j.envint.2019.04.024

Magnusson, K., Eliasson, K., Fråne, A., Haikonen, K., Hultén, J., Olshammar, M., Stadmark, J., & Voisin, A. (2016). Swedish sources and pathways for microplastics to the marine environment. In *Swedish Environmental Protection Agency* (Vol. 183, Issue C 183). www.ivl.se

Miljöinstitutet, I. S. (2012). *Uppdragsrapport för SDAB, Livscykelanalys på återvinning av däck, Jämförelser mellan däckmaterial och alternativa material I konstgräsplaner, dräneringslager och ridbanor, rapportnummer U3891*.

Mitro, S. D., Dodson, R. E., Singla, V., Adamkiewicz, G., Elmi, A. F., Tilly, M. K., & Zota, A. R. (2016). Consumer product chemicals in indoor dust: A quantitative meta-analysis of U.S. Studies. *Environmental Science and Technology*, *50*, 10661–10672. https://doi.org/10.1021/acs.est.6b02023

Mølhave, L., Schneider, T., Kjærgaard, S. K., Larsen, L., Norn, S., & Jørgensen, O. (2000). House dust in seven Danish offices. *Atmospheric Environment*, *34*, 4767–4779. https://doi.org/10.1016/S1352-2310(00)00104-7

Napper, I. E., Bakir, A., Rowland, S. J., & Thompson, R. C. (2015). Characterisation, quantity and sorptive properties of microplastics extracted from cosmetics. *Marine Pollution Bulletin*, *99*, 178–185. https://doi.org/10.1016/j.marpolbul.2015.07.029

Napper, I. E., & Thompson, R. C. (2016). Release of synthetic microplastic plastic fibres from domestic washing machines: Effects of fabric type and washing conditions. *Marine Pollution Bulletin*, *112*(1–2), 39–45. https://doi.org/10.1016/j.marpolbul.2016.09.025

Nematollahi, M. J., Zarei, F., Keshavarzi, B., Zarei, M., Moore, F., Busquets, R., & Kelly, F. J. (2021). Microplastic occurrence in settled indoor dust in schools. *Science of the Total Environment*, *807*, 150984. https://doi.org/10.1016/j.scitotenv.2021.150984

O'Brien, S., Okoffo, E. D., O'Brien, J. W., Ribeiro, F., Wang, X., Wright, S. L., Samanipour, S., Rauert, C., Toapanta, T. Y. A., Albarracin, R., & Thomas, K. V. (2020). Airborne emissions of microplastic fibres from domestic laundry dryers. *Science of the Total Environment*, *747*, 141175. https://doi.org/10.1016/j.scitotenv.2020.141175

OMEGA. (2017). *Measuring particle pollution*.

Pauly, J. L., Stegmeier, J., Cheney, T., Zhang, P. J., Allaart, A., & Mayer, G. (1998). Inhaled cellulosic and plastic fibers found in human lung tissue. *Cancer Epidemiology, Biomarkers & Prevention*, *7*, 419–428.

Peng, C., Tang, X., Gong, X., Dai, Y., Sun, H., & Wang, L. (2020). Development and application of a mass spectrometry method for quantifying nylon microplastics in environment. *Analytical Chemistry*, *92*, 13930–13935. https://doi.org/10.1021/acs.analchem.0c02801

Pirc, U., Vidmar, M., Mozer, A., & Kržan, A. (2016). Emissions of microplastic fibers from microfiber fleece during domestic washing. *Environmental Science and Pollution Research*, *23*, 22206–22211. https://doi.org/10.1007/s11356-016-7703-0

Potter, P. M., Al-Abed, S. R., Hasan, F., & Lomnicki, S. M. (2021). Influence of polymer additives on gas-phase emissions from 3D printer filaments. *Chemosphere*, *279*, 1–18. https://doi.org/10.1016/j.chemosphere.2021.130543

Prata, J. C. (2018). Airborne microplastics: Consequences to human health? *Environmental Pollution*, *234*, 115–126. https://doi.org/10.1016/j.envpol.2017.11.043

Prata, J. C., Castro, J. L., da Costa, J. P., Duarte, A. C., Rocha-Santos, T., & Cerqueira, M. (2020). The importance of contamination control in airborne fibers and microplastic sampling: Experiences from indoor and outdoor air sampling in Aveiro, Portugal. *Marine Pollution Bulletin*, *159*, 111522. https://doi.org/10.1016/j.marpolbul.2020.111522

Ragusa, A., Svelato, A., Santacroce, C., Catalano, P., Notarstefano, V., Carnevali, O., Papa, F., Rongioletti, M. C. A., Baiocco, F., Draghi, S., D'Amore, E., Rinaldo, D., Matta, M., & Giorgini, E. (2021). Plasticenta: First evidence of microplastics in human placenta. *Environment International*, *146*, 106274. https://doi.org/10.1016/j.envint.2020.106274

Rasmussen, P. E., Levesque, C., Chénier, M., & Gardner, H. D. (2018). Contribution of metals in resuspended dust to indoor and personal inhalation exposures: Relationships between PM10 and settled dust. *Building and Environment*, *143*, 513–522. https://doi.org/10.1016/j.buildenv.2018.07.044

Revel, M., Châtel, A., & Mouneyrac, C. (2018). Micro(nano)plastics: A threat to human health? *Current Opinion in Environmental Science and Health*, *1*, 17–23. https://doi.org/10.1016/j.coesh.2017.10.003

Rodes, C. E., Lawless, P. A., Evans, G. F., Sheldon, L. S., Williams, R. W., Vette, A. F., Creason, J. P., & Walsh, D. (2001). The relationships between personal PM exposures for elderly populations and indoor and outdoor concentrations for three retirement center scenarios. *Journal of Exposure Analysis and Environmental Epidemiology*, *11*, 103–115. https://doi.org/10.1038/sj.jea.7500155

Salthammer, T. (2020). Emerging indoor pollutants. *International Journal of Hygiene and Environmental Health*, *224*, 113423. https://doi.org/10.1016/j.ijheh.2019.113423

Salthammer, T. (2022). Microplastics and their additives in the indoor environment. *Angewandte Chemie – International Edition*, *61*(32), 1–10. https://doi.org/10.1002/anie.202205713

Schneider, M., Stracke, F., Hansen, S., & Schaefer, U. F. (2009). Nanoparticles and their interactions with the dermal barrier. *Dermato-Endocrinol*, *1*, 197–206.

Shen, M., Hu, T., Huang, W., Song, B., Qin, M., Yi, H., Zeng, G., & Zhang, Y. (2021). Can incineration completely eliminate plastic wastes? An investigation of microplastics and heavy metals in the bottom ash and fly ash from an incineration plant. *Science of the Total Environment*, *779*, 146528. https://doi.org/10.1016/j.scitotenv.2021.146528

Simon Magnusson. (2015). *Systemanalys av konstgräsplaner, Miljö-och konstnadsaspekter, Teknisk rapport. Luleå tekniska universitet, Luleå.* https://ltu.diva-portal.org/smash/get/diva2:995183/FULLTEXT01.pdf

Sinclair, R. (2015). Understanding textile fibres and their properties: What is a textile fibre? In *Textiles and Fashion: Materials, Design and Technology*. Elsevier Ltd. https://doi.org/10.1016/B978-1-84569-931-4.00001-5

Soltani, N. S., Taylor, M. P., & Wilson, S. P. (2021). Quantification and exposure assessment of microplastics in Australian indoor house dust. *Environmental Pollution*, *283*, 117064. https://doi.org/10.1016/j.envpol.2021.117064

Tang, C. L., & Seeger, S. (2022). Systematic ranking of filaments regarding their particulate emissions during fused filament fabrication 3D printing by means of a proposed standard test method. *Indoor Air*, *32*, 1–12. https://doi.org/10.1111/ina.13010

Thacharodi, A., Meenatchi, R., Hassan, S., et al. (2024). Microplastics in the environment: A critical overview on its fate, toxicity, implications, management, and bioremediation strategies. *Journal of Environmental Management*, *349*, 119433. https://doi.org/10.1016/j.jenvman.2023.119433

Thatcher, T. L., Layton, D. W. (1995). Deposition, resuspension, and penetration of particles within a residence. *Atmospheric Environment* 29, 1487–1497. https://doi.org/10.1016/1352-2310(95)00016-R.

Tian, L., Skoczynska, E., Siddhanti, D., van Putten, R. J., Leslie, H. A., & Gruter, G. J. M. (2022). Quantification of polyethylene terephthalate microplastics and nanoplastics in sands, indoor dust and sludge using a simplified in-matrix depolymerization method. *Marine Pollution Bulletin*, *175*, 113403. https://doi.org/10.1016/j.marpolbul.2022.113403

Toussaint, B., Raffael, B., Angers-Loustau, A., Gilliland, D., Kestens, V., Petrillo, M., Rio-Echevarria, I. M., & Van den Eede, G. (2019). Review of micro- and nanoplastic contamination in the food chain. *Food Additives and Contaminants – Part A*, *36*(5), 639–673. https://doi.org/10.1080/19440049.2019.1583381

Uddin, S., Fowler, S. W., Habibi, N., Sajid, S., Dupont, S., & Behbehani, M. (2022). A preliminary assessment of size-fractionated microplastics in indoor aerosol – Kuwait's baseline. *Toxics*, *10*, 2–17.

Vianello, A., Jensen, R. L., Liu, L., & Vollertsen, J. (2019). Simulating human exposure to indoor airborne microplastics using a Breathing Thermal Manikin. *Scientific Reports*, *9*, 1–11. https://doi.org/10.1038/s41598-019-45054-w

Wang, B., Tang, Z., Li, Y., Cai, N., & Hu, X. (2021). Experiments and simulations of human walking-induced particulate matter resuspension in indoor environments. *Journal of Cleaner Production*, *295*, 126488. https://doi.org/10.1016/j.jclepro.2021.126488

Winkler, A., Santo, N., Ortenzi, M. A., Bolzoni, E., Bacchetta, R., & Tremolada, P. (2019). Does mechanical stress cause microplastic release from plastic water bottles? *Water Research*, *166*, 115082. https://doi.org/10.1016/j.watres.2019.115082

Wright, S. L., & Kelly, F. J. (2017). Plastic and human health: A micro issue? *Environmental Science and Technology*, *51*(12), 6634–6647. https://doi.org/10.1021/acs.est.7b00423

Wu, T., Täubel, M., Holopainen, R., Viitanen, A. K., Vainiotalo, S., Tuomi, T., Keskinen, J., Hyvärinen, A., Hämeri, K., Saari, S. E., & Boor, B. E. (2018). Infant and adult inhalation exposure to resuspended biological particulate matter. *Environmental Science and Technology*, *52*, 237–247. https://doi.org/10.1021/acs.est.7b04183

Xie, Y., Li, Y., Feng, Y., Cheng, W., & Wang, Y. (2022). Inhalable microplastics prevails in air: Exploring the size detection limit. *Environment International*, *162*(November 2021), 107151. https://doi.org/10.1016/j.envint.2022.107151

Yang, L., Qiao, F., Lei, K., Li, H., Kang, Y., Cui, S., & An, L. (2019). Microfiber release from different fabrics during washing. *Environmental Pollution*, *249*, 136–143. https://doi.org/10.1016/j.envpol.2019.03.011

Yao, Y., Glamoclija, M., Murphy, A., & Gao, Y. (2021). Characterization of microplastics in indoor and ambient air in northern New Jersey. *Environmental Research*, *207*, 112142. https://doi.org/10.1016/j.envres.2021.112142

Zhang, J., Wang, L., & Kannan, K. (2020a). Microplastics in house dust from 12 countries and associated human exposure. *Environment International*, *134*, 105314. https://doi.org/10.1016/j.envint.2019.105314

Zhang, Q., Zhao, Y., Du, F., Cai, H., Wang, G., & Shi, H. (2020b). Microplastic fallout in different indoor environments. *Environmental Science and Technology*, *54*(11), 6530–6539. https://doi.org/10.1021/acs.est.0c00087

Zhu, J., Zhang, X., Liao, K., Wu, P., & Jin, H. (2022). Microplastics in dust from different indoor environments. *Science of the Total Environment*, *833*, 155256. https://doi.org/10.1016/j.scitotenv.2022.155256

14 Microfiber Pollution
Pathways to Health Impact

Disha Katyal
Canadian Institute of Public Health Inspectors (CIPHI)

14.1 INTRODUCTION

Today, the world has become enveloped in plastic, where everything from a baby's bottle to clothes to dentures to coffins is made of plastic. Products manufactured for single use are polluting the environment for hundreds of years, where they enter the food web to return to the descendants of those who disposed of them in the first place. By 2025, it is estimated that approximately 28 million metric tons of plastic will enter the oceans annually from improper waste management in coastal countries (Jambeck et al., 2015).

The release of microfibers from clothing during laundry has been reported as the largest source of microfiber pollution in the oceans by several scientific studies (De Falco et al., 2019; Suaria et al., 2020; Acharya et al., 2021). The floating plastic matter in the world's oceans and seas is primarily made up of microfibers (Green et al., 2018; Mishra et al., 2019; Xu et al., 2021), defined as fibers that are 1 μm–5 mm in length and with a length to diameter ratio greater than 100 (Liu et al., 2019). Microfibers make up 91% of all microplastics found in surface waters worldwide (Barrows et al., 2018; Napper et al., 2021). In deeper water, microfibers have been found to be less abundant (McEachern et al., 2019; Sanchez-Vidal et al., 2018). Nearly 80% of all microfibers are made of cellulose and about 12.9% contain polyester polyethylene terephthalate (PET), which is often used as an indicator of plastic textile microfibers in scientific studies (Sanchez-Vidal et al., 2018). Currently, polyester dominates the textile industry (Schöpel and Stamminger, 2019). Studies in Finland have revealed that polyester microfiber emissions from household laundry practices have been estimated to be 150,000 kg/year (Sillanpää and Sainio, 2017).

Furthermore, the effects of climate change are causing ice sinks in the Arctic sea to melt and release plastic debris from decades before (Obbard et al., 2014). With large volumes of plastic entering the environment, wildlife interaction is unavoidable. Microfibers in the size range of less than 5 mm, are small enough to be ingested by aquatic life, and potentially allowing for the biomagnification of plastic matter and bioaccumulation of toxins carried by particulates (Wang et al., 2018). Bioaccumulation is the process of the physical accumulation of particles, toxins, or pathogens, resulting from the consumption of contaminated prey organisms. The increasing concentration of particles moving up the food chain as a result of bioaccumulation is called biomagnification. The environmental impact of microfibers includes the destruction of habitats (Lamb et al., 2018), ingestion (Setälä et al., 2014),

 DOI: 10.1201/9781003331995-17

and plastic-facilitated transport of pathogens or invasive organisms to new ecosystems (McCormick et al., 2016; Viršek et al., 2017).

The atmospheric environment also plays an important role in the global distribution of microfiber pollution, although more research is required to study airborne microfibers (Akanyange et al., 2021; Xu et al., 2020; Zhang et al., 2020b). The inherent structure of microfibers: low density, medium length, natural or artificial crimp, and larger surface area to volume ratio, allows for long periods of atmospheric suspension and promotes regional transport of up to 1,000 km (Brahney et al., 2020). Dry microfibers with less mass are more susceptible to atmospheric conditions and can travel long distances globally through several cycles of suspension and deposition (Brahney et al., 2020). Airborne microfibers are the main source of contamination of areas that would otherwise be unpolluted, such as the French Pyrenees, the Italian Alps, and Arctic snow (Ambrosini et al., 2019; Bergmann et al., 2019). Most of the airborne microfibers are assumed to originate from textiles as they are released into the air during mechanical drying or friction during wear (Napper and Thompson, 2016; O'Brien et al., 2020); however, there is little to no data available that specifically looks at airborne microfibers independently from other plastic particulate matter.

This chapter focuses on the three major pathways of exposure to microfiber pollution: inhalation, ingestion, and dermal contact, and their potential health impact on humans. The health impact is further discussed in terms of the physical, biological, and chemical implications. This chapter also enlightens the research gaps in this area to provide further directions for mitigative measures.

14.2 PATHWAYS

To understand the health implications of exposure to microfibers, it is critical to study the routes of exposure. The ingestion of physical plastic particulates, inhalation of plastic matter, and dermal contact are the three main pathways discussed below.

14.2.1 INHALATION

Inhalation is a route of exposure to micro- and nanoplastics for the general population. Microfiber inhalation is common in city or urbanized environments due to the high densities of synthetic materials; from the withering of clothes to the degradation of vehicular tires to the breakdown of road paint, etc., and hence microfibers are an emerging concern within the context of air pollution. Aerosolized plastic particulate matter settles and deposits into dust, which is further discussed later in the chapter. A significant correlation between microfiber exposure via inhalation and the higher prevalence of lung disease in textile workers has been noted in several occupational health studies (Too et al., 2016; Hussain et al., 2019; Lai and Christiani, 2013; Karanikas and Hasan, 2022). Prolonged inhalation of aerosolized dust containing hemp and cotton microfibers has been shown to cause byssinosis, respiratory diseases, allergies, and excessive epithelial growth (Van Dijk et al., 2020; Zele et al., 2021). A study conducted in Pakistan revealed that 35.6% of textile workers suffer from byssinosis, a disease of the lungs associated with cotton or other vegetable fiber dust, particularly those working in weaving mills (Memon et al., 2008); the worldwide prevalence of byssinosis among textile workers is up to 40% (Murlidhar et al., 1995). Histopathological analysis of the lungs of individuals working in the plastic and textile industries revealed interstitial fibrosis and granulomatous lesions, postulated to be acrylic, polyester, and nylon dust (Kremer et al., 1994; Muittari and Veneskoski, 1978). Reversible and irreversible obstructive lung diseases, such as asthma and chronic obstructive pulmonary disease, and restrictive lung diseases, such as asbestosis and pulmonary fibrosis, are notably more prevalent in textile workers (Lai and Christiani, 2013). A study conducted in the UK found microfibers in all regions of the human lung samples collected from 13 different thoracic surgical procedures (Jenner et al., 2022). Interestingly, a study compared the number of microfibers absorbed by inhalation and by ingestion and reported that the amount of inhaled microfibers was 3–15 times higher than the ingested ones (Catarino et al., 2018).

14.2.2 Ingestion

14.2.2.1 Food Systems

In order to study the impact of ingesting microfibers on human health, it is essential to understand how microfiber affects aquatic organisms and paves its way into the food web. It has been established by several studies that aquatic life is severely affected by microfibers worldwide and makes its way to humans through food (Sequeira et al., 2020; Kumar et al., 2018; Pazos et al., 2017). A study that looked at aquatic species in the American and Indian markets found microplastic contamination in 28% of fish, 33% of shellfish, and 55% of all species tested (Rochman et al., 2015). Filter-feeding organisms and deposit-eating fish, such as crustaceans, shellfish, zooplankton, and fish, are more susceptible to microfiber ingestion than larger predator fish due to their feeding behaviors (Wesch et al., 2016; Lusher et al., 2020). Interestingly, a study conducted by Mizraji et al. (2017) examined the relationship between feeding patterns and the amount of microfiber consumption in tidal fish and discovered that omnivore fish ingested more microfibers than herbivore and carnivore fish. Organisms that rely on visual cues for feeding are more likely to consume suspended microfibers that resemble natural prey (de Sá et al., 2015). Biomagnification has been reported in a variety of fish and other aquatic species higher up in the food chain (Boerger et al., 2010). It has been observed in bluefin tuna, albacore tuna, and swordfish in the Mediterranean Sea (Romeo et al., 2015). Most of the consumed microfibers remain in the gastrointestinal tract of fish, such as the European anchovy, embryos of bighead carp, hybrid snakehead, and Indian major carp, which poses a risk for humans as these fish and larvae are consumed whole (Capone et al., 2020; Zhang et al., 2021; Prata et al., 2022).

In addition to seafood, microfiber contamination has also been documented in other food products such as salt, honey, and drinks such as water, sodas, and milk. Salt is a staple for the human diet as it provides essential nutritional value and aids food preservation methods, such as for fruits, cheeses, cereals, and drinks. Microfibers have been found in salt samples from 128 different brands from 38 different sources in numerous countries spanning over five continents in the period of 2015–2018 (Peixoto et al., 2019). Sea salt products are known to be the most contaminated with microfibers, followed by lake salt, rock salt, and well salt (Danopoulos et al., 2020b). The amount of contamination is directly correlated to the amount of environmental litter in the area where the salt is sourced (Karami et al., 2017). Particles less than 200 μm are most commonly found in sea salt products (Cabernard et al., 2018).

On the other hand, honey is also affected by microfiber pollution. A study that analyzed 19 samples of honey from different countries revealed high concentrations of microfiber particles (Liebezeit and Liebezeit, 2013). Honey samples from Ecuador were found to contain 54 particles/L in industrial honey and 67 particles/L in craft honey (Diaz-Basantes et al., 2020). The origin of the microfiber contamination of honey is not fully understood. In 2019, a study conducted by the Swiss Beekeepers Association compared microfiber levels in beehives made of wood with those made of polystyrene and found no statistically significant difference (Mühlschlegel et al., 2017). It is predicted that the microfiber contamination of honey is linked to beekeeper activities and the atmosphere rather than the natural manufacturing process (Mühlschlegel et al., 2017). In addition, unrefined cane sugar can also have up to 560 fibers and 540 plastic fragments/kg (Liebezeit and Liebezeit, 2013). Very recently, microfiber contamination has been detected in fruits and vegetables. 52,600–307,750 particles/g were detected in fruits, with apples being the most contaminated, and 72,175–130,500 particles/g in vegetables, with carrots being the most affected (Oliveri Conti et al., 2020). Further research is required for determining the origin and pathways of contamination of produce.

Potable water has also been identified as a potential source of exposure to microfibers for humans by various studies (Danopoulos et al., 2020a; Li et al., 2020; Mason et al., 2018; Mintenig et al., 2019; Oßmann et al., 2018; Schymanski et al., 2018). In Germany, samples including raw water, drinking water, tap water, and bottled water were analyzed, and it was found that the water supply chain had an average of 700 plastic particles/L, ranging between 50 and 150 μm (Mintenig et al., 2019). Other beverages intended for human consumption have also been studied. A study by Shruti et al. (2020) in Mexico analyzed 57 different beverage products, including cold tea, soft drinks, energy drinks, and beers, and found various forms of microplastics, including fibers and fragments, of various sizes, ranging between 0.1 and 3 mm, with different colors (blue, red, brown, black, and green). The chemical nature of the identified microfibers indicated contamination from synthetic textiles and packaging in the beverage products (Shruti et al., 2020). The plastic particulates included polyester, polyamide, acrylonitrile, butadiene, styrene, and PET, which are common raw materials for synthetic and semi-synthetic textiles and packaging materials (Shruti et al., 2020). Microfibers have also been detected in dairy products. All 23 milk samples tested in Mexico, including 5 international and 3 national brands, revealed microfiber contamination (Kutralam-Muniasamy et al., 2020). The same study identified the most common contaminant in milk to be thermoplastic sulfone polymers, which are used in the filtration membrane in dairy processing (Kutralam-Muniasamy et al., 2020). The presence of microfibers in milk raises concerns for infants and young children, who are the primary consumers.

It has repeatedly been shown in several studies that microfibers are ubiquitous and have long entered the food web, particularly through seafood. Microfibers have been found in the digestive tracts, gills, and other internal organs of marine species, such as bivalves, crabs, and fish (Dawson et al., 2021). Microfiber contamination of drinking water (Akhbarizadeh et al., 2020; Gouin et al., 2021; Zhang et al., 2020a) and salt (Peixoto et al., 2019; Zhang et al., 2020a), are other ingestion pathways that expose humans to the pollutant. More research is required to study the extent of biomagnification of microfibers through the food chain and *in vivo* studies are needed to establish the impact on human health.

14.2.2.2 Other Sources

Ingestion and inhalation of dust, both indoors and outdoors, are another pathway of exposure to microfibers in humans (Dris et al., 2017). In a study conducted by Catarino et al. (2016), it was concluded that microfiber ingestion through contaminated mussels is minimal compared to airborne household fibers that fall into food. Indoor dust is made up of up to 88% microfibers, ranging between 1,550 and 120,000 mg/kg (Liu et al., 2019). Adults can ingest up to 64.1 fibers/kg-bw/day, while infants are at more than ten times the risk to adults, up to 889 fibers/kg-bw/day (Liu et al., 2019). Infants and young children are at greater risk likely due to their crawling behavior and higher frequency of nibbling on their hands feeding bottles, toys, and clothes (Liu et al., 2019). Several studies have confirmed the presence of common microfibers, such as PET and polycarbonate (PC), in human stool, placenta, and meconium (Braun et al., 2021; Mohamed Nor et al., 2021; Schwabl et al., 2019). Most recent research has shown that common polymers used in fibrous materials, such as PET and PE, are also found in the human bloodstream within nanosizes; about 1.6 μg/mL of nanoplastics have been recorded (Leslie et al., 2022). Previous studies have shown that microfiber fragments between 1 and 10 μm lead to significant mechanical stretching of the lipid membranes without inflammatory reactions or serious dysfunction of the cell machinery (Fleury and Baulin, 2021).

Plastic products have gained popularity, particularly in the food industry, due to their low production costs, ease of manufacture, and imperviousness to water. Plastic packaging can include substances such as monomers, oligomers, and polymers, as well as additives, including plasticizers, antioxidants, heat stabilizers, and pigments (Meng et al., 2023). Most of these chemicals are not covalently bound to the polymer matrix and therefore leach into the packaged foods; *in vitro* studies have demonstrated how plastic food packaging can leach out more chemicals with increasing temperature, UV degradation, increasing fat and acid contents in the food, and decreasing molecular size (Lithner et al., 2011; Zimmermann et al., 2019). Exposure to leachate from plastic food packaging has been shown to induce reproductive and/or neurobehavioral toxicity in various animal studies, including zebrafish (Lin et al., 2022), sows (Nerin et al., 2014), boars (Schulze et al., 2020; Nerín et al., 2020), mice (Gader Al-Khatim and Galil, 2015), and marine orchid dottybacks (Hamlin et al., 2015).

14.2.3 Dermal Contact

Dermal contact with microfibers is considered a less significant route of exposure. Although the human skin is an efficient barrier against microfibers and other contaminants, some possible entry routes include sweat glands, open skin injuries, and

TABLE 14.1
Summary of Pathways to Exposure and Health Impacts

Mode of Exposure	Pathway	Health Impact
Ingestion	Consumption of contaminated food such as aquatic species, salt, honey, milk, water, fruits, and vegetables.	Swelling and blockage within the gastrointestinal tract, oxidative stress, and cytotoxicity.
Inhalation	Breathing in contaminated city/urbanized dust particles.	Reversible and irreversible obstructive lung diseases, such as asthma and chronic obstructive pulmonary disease, restrictive lung diseases, such as asbestosis, pulmonary fibrosis, and byssinosis.
Dermal contact	Exposure through open wounds, surgical sutures, sweat glands, and hair follicles.	Inflammatory response and oxidative stress.

hair follicles (Schneider et al., 2009). There is some evidence demonstrating dermal uptake of plastic spheres up to 1 μm in rainbow trout from surrounding water (Moore et al., 1998). In humans, surgical sutures in medicine, such as polyester and polypropylene, are known to induce low inflammatory responses (Schirinzi et al., 2017). Human epithelial cells have also been shown to suffer oxidative stress from exposure to microfibers and nanoplastics (Schirinzi et al., 2017). Dermal contact with toxic chemicals carried by microfibers, such as toxic dyes, can produce carcinogenic compounds (Periyasamy, 2023; Athey et al., 2022). The health impact of exposure to plastic particulates through dermal contact has not been thoroughly studied, and there is limited data available to explore the subject. Table 14.1 summarizes the modes of human exposure to microfibers, along with the pathways and potential health impacts.

14.3 POTENTIAL HEALTH IMPACT

14.3.1 Physical

Although there has not been enough research to study the direct pathway of microfibers once they enter the human body, studies with other species have revealed concerns. Swelling and blockage within the gastrointestinal tracts have been noted in fish and crabs (Wang et al., 2016; Wright et al., 2013). Microfibers have been shown to concentrate in the gills, stomachs, and metabolic systems and cause changes in the cellular function of crabs and fish (Karbalaei et al., 2018; Brennecke et al., 2015). Prolonged exposure to microfibers has also been linked to oxidative stress and cytotoxicity in various organisms (Galloway, 2015; Schirinzi et al., 2017; Wright and Kelly, 2017; Anbumani and Kakkar, 2018). Microfibers can induce oxidative stress by releasing chemicals inherently absorbed to their surface and by inducing the release of reactive oxygen species in the host due to a hyperinflammatory response (Kelly and Fussell, 2012; Valavanidis et al., 2013). In laboratory settings, it has been shown how nanoplastic particles can be used to induce inflammatory responses in lung cells and human gastric cancer cells (Forte et al., 2016). Other animal studies have revealed a higher prevalence of chronic discomfort, inflammation, altered cell growth, and immune cell dysfunction in organisms with prolonged exposure to microfibers (Smith et al., 2018).

There are several studies that have shown negative health outcomes of exposure to microfibers in various physiological systems. The incidence of inflammatory bowel disease has been correlated with larger amounts of microfibers found in the gut (Yan et al., 2021). The buildup of microfibers in the digestive tract can introduce them into other physiological systems through the endocytosis and persorption processes (Geiser et al., 2005). The internalization of microfibers in cardiac cells has been demonstrated to cause inflammation, blood cell cytotoxicity, vascular swelling, obstructions, and respiratory high blood pressure in laboratory settings (Wright and Kelly, 2017; Campanale et al., 2020). Exposure to nanoplastic particles can cause red blood cell coagulation and endothelium wall adherence (Barshtein et al., 2016). *In vitro* tests have shown that a buildup of nanoplastic particles in renal epithelial cells impairs healthy renal operation (Monti et al., 2015). Long-term exposure to microfibers can lead to chronic inflammation, impaired organ function, and an increased risk of neoplasia when transferred to distal tissues (Prata et al., 2020). Once microfibers reach the skeletal system, they may stimulate osteoclasts and cause bone loss (Liu et al., 2015; Ormsby et al., 2016). Neurotoxicity has also been documented with persistent exposure to plastic particulate matter, due to immune cell activation and oxidative stress in the brain (MohanKumar et al., 2008). In studies involving mice, microfiber exposure has been linked to changes in neuronal function with alterations in the concentration of neurotransmitters (Deng et al., 2017). Prolonged inflammation caused by microfiber consumption may induce cancer by causing DNA damage (Chang, 2010).

14.3.2 Chemical

Chemicals are extensively used throughout the manufacturing process of synthetic textiles, from formulating raw materials to finishing and storage (Lellis et al., 2019). Semi-synthetic and natural textile fibers also undergo heavy chemical processing for obtaining the right colors and textures and thus cannot be considered inherently natural or environmentally friendly (Islam, 2020). These chemicals can include substances used for cultivating plant and animal fibers, such as herbicides, insecticides, rodenticides, etc., and the chemicals required to recycle or regenerate semi-synthetic fibers (Dolez and Benaddi, 2018). As textiles wear down, microfibers can also serve as vectors for toxic chemicals in the environment. Polycyclic aromatic hydrocarbons, polychlorinated biphenyls, and heavy metals such as lead, nickel, cadmium, and zinc have been found in microfiber samples (Crawford and Quinn, 2017; Wright and Kelly, 2017). Once microfibers enter the human body and transport to deeper tissues or organs, the adhering chemicals can leach out and interfere with endogenous hormones (Cole et al., 2011). Bisphenol A (BPA), triclosan, organotin, and brominated flame retardants are common additives used for manufacturing plastics (Galloway, 2015; Prata et al., 2020).

BPA is a clinically proven toxin that damages the endocrine system by altering hormone levels (Yamamoto et al., 2001; Halden, 2010). BPA poisoning has been noted in tuna fish, pork, and potable water - demonstrating a clear pathway of exposure to humans (Colin et al., 2014). In a study that examined urine samples from 167 men, it was found that BPA levels were negatively proportional to serum levels of inhibin B and the estradiol:testosterone ratio, indicating a critical detrimental effect on men's reproductive health (Meeker et al., 2010). An increase in BPA levels has also been linked to obesity, where it interferes with alpha and beta receptors in fat tissues, causing a hormonal imbalance involving estrogen levels (Vom Saal et al., 2012; Micha łowicz, 2014). There is some evidence of BPA causing breast and prostate cancer in mammals, possibly promoting the same in humans (Michalowicz, 2014). Fetal exposure to BPA has been linked to a greater risk of offspring developing breast cancer (Murray et al., 2007). BPA exposure in animal studies has also been correlated to changes in brain function and behavior in males (Kundakovic et al., 2013) as well as decreased maternal behavior in females (Palanza et al., 2002). BPA has been detected in food and food packaging by several studies, and it has been assumed by regulatory agencies that oral exposures are the main source (Chapin et al., 2008; Vandenberg et al., 2013). With growing evidence highlighting the health impacts of BPA exposure via ingestion, the compound has been subject to restricted use in several countries across the world; the bans are primarily focused on baby bottles, sipper cups, toys, and re-usable water bottles to limit exposure via ingestion for children (FDA, 2012; EPA, 2012; Usman et al., 2022). More recent studies have highlighted concerns over dermal exposures: research has revealed that increased levels of BPA are found in urine samples of individuals that come in direct and prolonged dermal contact with cosmetics and thermal paper, such as individuals who wear or work with makeup and cashiers (Dodson et al., 2012; Mendum et al., 2011; Ehrlich et al., 2014). BPA, when ingested, is absorbed into the mesenteric blood vessels and metabolized by the liver, which converts it into the conjugated form (ex. BPA-glucuronide or BPA-sulfate), and only a small portion of the unconjugated form reaches the bloodstream (Tominaga et al., 2006). However, BPA exposure through other routes, such as dermal contact and inhalation, allows significantly larger amounts of unconjugated BPA to penetrate the bloodstream (Pottenger et al., 2000; Negishi et al., 2004; Tominaga et al., 2006). This is important to note because only the unconjugated form of BPA can bind with estrogen receptors and alter hormonal responses inside the human body and is therefore deemed hazardous (Matthews et al., 2001; Teeguarden et al., 2013). The impact of dermal exposure to BPA and similar endocrine disruptor chemicals has recently gained attention and is an area that requires further research. The dermal contact of textiles contaminated with BPA has not been measured or studied enough to make scientific conclusions. A recent study that analyzed used and washed clothes for BPA levels estimated a dermal exposure dose of 52.1 ng/kg BW/d – indicating a relatively high exposure risk for humans (Wang et al., 2019). In addition, the absorption of BPA through the handling of thermal paper, documented by several studies noted previously, indicates a high risk for textiles that too have direct and prolonged exposures with skin and should be further studied.

Other toxic chemicals that can have an impact on human and environmental health include pigments and dyes, wrinkle-resistant finishing substances, antimicrobial agents, and stain-repellants carried by microfibers originating from textiles (Lacasse and Baumann, 2004; Hill et al., 2017). Textile dyes can act as toxic, mutagenic, and carcinogenic agents (Aquino et al., 2014; Khatri et al., 2018; Sharma et al., 2018). They can also persist in the environment and enter the food web, providing biomagnification (Sandhya, 2010; Newman, 2009). Azo dyes are the most common types of colorants used in textile production, accounting for 60%–70% of the global market (Singh et al., 2014). It has been noted that 15%–50% of azo dyes do not bind to the fabric during the dyeing process – and are therefore released into the environment through microfibers (Rehman et al., 2018). Environmentally, azo dyes and their byproducts negatively impact soil microbial communities and affect the germination and growth of plants (Imran et al., 2014; Rehman et al., 2018). Exposure to azo dyes in humans can lead to allergic reactions, alterations in growth and developmental hormones, as well as carcinogenic effects (Chung, 2016; Sharma et al., 2018). Phthalates are another group of toxic chemicals that are commonly used in textile production, most often used in polyvinyl chloride (PVC) prints and coatings of decorative images (Weil et al., 2006). Phthalates have been well studied and are known to be endocrine-disrupting compounds and have an impact on human reproductive health (Meeker et al., 2009).

14.3.3 Biological

In addition to transporting toxic chemicals, microfibers can carry pathogenic microorganisms and transport them to favorable environments. The hydrophobic surface of plastic waste provides an ideal environment for microbial colonization and biofilm formation (Zettler et al., 2013). Some of the most aggressive pathogens, such as the Vibrio species, can adhere to (and thrive on) microfibers (Kirstein et al., 2016). Vibriosis causes around 80,000 illnesses and 100 deaths in the United States every year (Centers for Disease Control and Prevention [CDC], 2019). The infection is spread through consumption of contaminated seafood or exposing an open wound to contaminated water (CDC, 2019). Pathogenic serotypes of fecal coliforms, such as *Escherichia coli* (Silva et al., 2019), as well as invasive algal species (Casabianca et al., 2019), have been detected on floating microfibers in marine waters. Marine microfibers have also been reported to harbor microbial pathogens causing disease outbreaks in coral reefs (Lamb et al., 2018), fish (Viršek et al., 2017), and shellfish (Amaral-Zettler et al., 2020). Exposure to microfibers carrying microorganisms can affect the gut microbiota, allowing opportunistic pathogens to rapidly grow and cause a hyperinflammatory response, infections, and intoxications (West-Eberhard, 2019).

It is important to note that most of the data available for assessing the health impact of exposure to microfibers is from *in vitro* or laboratory-based studies and, therefore, there is inconsistent and insufficient data available to fully understand or conclude the severity of the health impact in humans. Braeuning (2019) discovered in a literature review that there is a large disparity in the total mass of microfibers found in tissues, the amount ingested, and the amount excreted; there is inadequate research that accounts for microfibers in bodily fluids and tissues as a whole.

14.4 FUTURE MITIGATION STRATEGIES

Limiting consumption of food products that are known to contain high levels of microfibers is the first step to preventing adverse health effects. Currently, microfiber pollutants in food products such as shellfish, fish, bottled water, honey, etc., are not monitored or regulated by food inspection agencies across the world. Consumer awareness and education can play a significant role in decreasing polymer exposure from contaminated food sources.

Reducing the consumption of single-use plastics is a long-term solution for limiting exposure to toxic plastic particulates, which should be mandated with government policies and regulations. Several countries have begun taking legislative initiatives to regulate plastic pollution (Katyal et al., 2020); however, they are noted to be too slow to respond to the rapidly rising levels of plastic waste (Usman et al., 2022). It has also been concluded that there is not enough attention being given to research and development to better understand the impact of microplastics on aquatic organisms, food safety, human health, and for finding plastic alternatives (Usman et al., 2022). Several policies worldwide have been too specific, such as targeting specific toxic materials or plastic products, and have overlooked certain types of plastics, such as microfibers (Usman et al., 2022). Government policies around the world are focused on specific chemicals and specific products to limit exposure to toxins, such as the ingestion of BPA through plastic products (EPA, 2012; FDA, 2012). Most of these policies are targeted toward limiting children's exposure via ingestion, thereby banning BPA in baby products (EPA, 2012; FDA, 2012). Although this is a step in the right direction, significant legislative changes are required to address other routes of exposure to microfibers and chemicals carried by microfibers, such as BPA. In addition to ingestion, other routes of exposure, such as through dermal contact and inhalation, need to be addressed in policies to limit risk and assess exposure through products not monitored for toxin levels, such as textiles. There are huge gaps in knowledge around evaluating the impact of exposure to chemical toxins through microfibers and routine articles. The physical and biological impact of microfibers in humans also requires more data to make concrete scientific conclusions. Further research is required to guide global policy and standards around the use of chemicals in the textile manufacturing industry and regulating microfiber waste. In the meantime, increased awareness about microfibers and their impact on health can help the public make informed decisions and improve regulatory measures worldwide to control microfiber pollution.

14.5 CONCLUSIONS

Plastic waste released into the environment is only expected to increase in the future (Jambeck et al., 2015). Plastic has been shown to negatively affect sensitive ecosystems and offers a medium for opportunistic pathogens to transport to favorable environments and cause disease. Several animal studies have documented cases of entrapment, blockages, and physical harm linked to micro- and macroplastic waste. Epidemiologically, multiple health conditions have been linked to human exposure to polymers and toxic chemicals used for manufacturing plastics. Exposure through contaminated food products, food packaging, inhalation, and ingestion of dust are some of the pathways studied. Once ingested, microfibers can distally travel to other physiological systems through the circulatory system and accumulate in organs such as the liver and kidneys. Laboratory studies have indicated the potential link between exposure to microfibers and the development of cancer. Chemical additives from microfibers have been shown to affect the endocrine system, with alterations in estrogen and testosterone levels and thereby affecting reproductive functions. Furthermore, neurotoxic symptoms, metabolic dysfunction, and an increased risk of obesity have been observed. It is crucial to find mitigation strategies that regulate microfiber contamination of the natural environment and protect food systems, thereby safeguarding human health.

REFERENCES

Acharya, S., Rumi, S. S., Hu, Y., & Abidi, N. (2021). Microfibers from synthetic textiles as a major source of microfibers in the environment: A Review. *Textile Research Journal*, 91(17–18), 2136–2156. https://doi.org/10.1177/0040517521991244

Akanyange, S. N., Lyu, X., Zhao, X., Li, X., Zhang, Y., Crittenden, J. C., Anning, C., Chen, T., Jiang, T., & Zhao, H. (2021). Does microfiber really represent a threat? A review of the atmospheric contamination sources and potential impacts. *The Science of the Total Environment*, 777, 146020. https://doi.org/10.1016/j.scitotenv.2021.146020

Akhbarizadeh, R., Dobaradaran, S., Nabipour, I., Tajbakhsh, S., Darabi, A. H., & Spitz, J. (2020). Abundance, composition, and potential intake of microfibers in canned fish. *Marine Pollution Bulletin*, 160, 111633. https://doi.org/10.1016/j.marpolbul.2020.111633

Amaral-Zettler, L. A., Zettler, E. R., & Mincer, T. J. (2020). Ecology of the plastisphere. *Nature Reviews Microbiology*, 18(3), 139–151. https://doi.org/10.1038/s41579-019-0308-0

Ambrosini, R., Azzoni, R. S., Pittino, F., Diolaiuti, G., Franzetti, A., & Parolini, M. (2019). First evidence of microfiber contamination in the supraglacial debris of an Alpine Glacier. *Environmental Pollution*, 253, 297–301. https://doi.org/10.1016/j.envpol.2019.07.005

Anbumani, S., & Kakkar, P. (2018). Ecotoxicological effects of microfibers on biota: A review. *Environmental Science and Pollution Research*, 25(15), 14373–14396. doi:10.1007/s11356-018-1999-x

Aquino, J. M., Rocha-Filho, R. C., Ruotolo, L. A. M., Bocchi, N., & Biaggio, S. R. (2014). Electrochemical degradation of a real textile wastewater using β-pbo2 and DSA(r) anodes. *Chemical Engineering Journal*, 251, 138–145. https://doi.org/10.1016/j.cej.2014.04.032

Athey, S. N., Carney Almroth, B., Granek, E. F., Hurst, P., Tissot, A. G., & Weis, J. S. (2022). Unraveling physical and chemical effects of textile microfibers. *Water*, 14(23), 3797. https://doi.org/10.3390/w14233797

Barrows, A. P. W., Cathey, S. E., & Petersen, C. W. (2018). Marine environment microfiber contamination: Global patterns and the diversity of microparticle origins. *Environmental Pollution*, 237, 275–284. https://doi.org/10.1016/j.envpol.2018.02.062

Barshtein, G., Livshits, L., Shvartsman, L. D., Shlomai, N. O., Yedgar, S., & Arbell, D. (2016). Polystyrene nanoparticles activate erythrocyte aggregation and adhesion to endothelial cells. *Cell Biochemistry and Biophysics*, 74(1), 19–27. doi:10.1007/s12013-015-0705-6

Bergmann, M., Mützel, S., Primpke, S., Tekman, M. B., Trachsel, J., & Gerdts, G. (2019). White and wonderful? Microfibers prevail in snow from the Alps to the Arctic. *Science Advances*, 5(8). https://doi.org/10.1126/sciadv.aax1157

Boerger, C. M., Lattin, G. L., Moore, S. L., & Moore, C. J. (2010). Plastic ingestion by planktivorous fishes in the North Pacific Central Gyre. *Marine Pollution Bulletin*, 60(12), 2275–2278. doi:10.1016/j.marpolbul.2010.08.007

Braeuning, A. (2019). Uptake of microfibers and related health effects: A critical discussion of Deng et al., Scientific reports 7:46687, 2017. *Archives of Toxicolarchives Toxicology*, 793(1), 46687219–46687220. doi:10.1007/s00204-018-2367-9

Brahney, J., Hallerud, M., Heim, E., Hahnenberger, M., & Sukumaran, S. (2020). Plastic rain in protected areas of the United States. *Science (New York, N.Y.)*, 368(6496), 1257–1260. https://doi.org/10.1126/science.aaz5819

Braun, T., Ehrlich, L., Henrich, W., Koeppel, S., Lomako, I., Schwabl, P., & Liebmann, B. (2021). Detection of microfiber in human placenta and meconium in a clinical setting. *Pharmaceutics*, 13(7), 921. https://doi.org/10.3390/pharmaceutics13070921

Brennecke, D., Ferreira, E. C., Costa, T. M., Appel, D., da Gama, B. A., & Lenz, M. (2015). Ingested microfibers (>100 μm) are translocated to organs of the tropical fiddler crab uca rapax. *Marine Pollution Bulletin*, 96(1–2), 491–495. doi:10.1016/j.marpolbul.2015.05.001

Cabernard, L., Roscher, L., Lorenz, C., Gerdts, G., & Primpke, S. (2018). Comparison of Raman and Fourier transform infrared spectroscopy for the quantification of microfibers in the aquatic environment. *Environmental Science & Technology*, 52(22), 13279–13288. https://doi.org/10.1021/acs.est.8b03438

Campanale, C., Massarelli, C., Savino, I., Locaputo, V., & Uricchio, V. F. (2020). A detailed review study on potential effects of microfibers and additives of concern on human health. *Ijerph*, 17(4), 1212. doi:10.3390/ijerph17041212

Capone, A., Petrillo, M., & Misic, C. (2020). Ingestion and elimination of anthropogenic fibres and microfiber fragments by the European Anchovy (Engraulis Encrasicolus) of the NW Mediterranean Sea. *Marine Biology*, 167(11), 1–15. doi:10.1007/s00227-020-03779-7

Casabianca, S., Capellacci, S., Giacobbe, M. G., Dell'Aversano, C., Tartaglione, L., Varriale, F., Narizzano, R., Risso, F., Moretto, P., Dagnino, A., Bertolotto, R., Barbone, E., Ungaro, N., & Penna, A. (2019). Plastic-associated harmful microalgal assemblages in marine environment. *Environmental Pollution*, 244, 617–626. https://doi.org/10.1016/j.envpol.2018.09.110

Catarino, A. I., Macchia, V., Sanderson, W. G., Thompson, R. C., & Henry, T. B. (2018). Low levels of microfibers (MP) in wild mussels indicate that MP ingestion by humans is minimal compared to exposure via household fibres fallout during a meal. *Environmental Pollution*, 237, 675–684. https://doi.org/10.1016/j.envpol.2018.02.069

Catarino, A. I., Thompson, R., Sanderson, W., & Henry, T. B. (2016). Development and optimization of a standard method for extraction of microfibers in mussels by enzyme digestion of soft tissues. *Environmental Toxicology and Chemistry*, 36(4), 947–951. https://doi.org/10.1002/etc.3608

Centers for Disease Control and Prevention. (2019). Vibrio species causing vibriosis. Retrieved from https://www.cdc.gov/vibrio/faq.html

Chang, C. (2010). The immune effects of naturally occurring and synthetic nanoparticles. *Journal of Autoimmunity*, 34(3), J234–J246. doi:10.1016/j.jaut.2009.11.009

Chapin, R. E., Adams, J., Boekelheide, K., Gray, L. E., Jr, Hayward, S. W., Lees, P. S., McIntyre, B. S., Portier, K. M., Schnorr, T. M., Selevan, S. G., Vandenbergh, J. G., & Woskie, S. R. (2008). NTP-CERHR expert panel report on the reproductive and developmental toxicity of bisphenol A. *Birth Defects Research. Part B, Developmental and Reproductive Toxicology*, 83(3), 157–395. https://doi.org/10.1002/bdrb.20147

Chung, K.-T. (2016). Azo dyes and human health: A Review. *Journal of Environmental Science and Health, Part C*, 34(4), 233–261. https://doi.org/10.1080/10590501.2016.1236602

Cole, M., Lindeque, P., Halsband, C., & Galloway, T. S. (2011). Microfibers as contaminants in the marine environment: A review. *Marine Pollution Bulletin*, 62(12), 2588–2597. doi:10.1016/j.marpolbul.2011.09.025

Colin, A., Bach, C., Rosin, C., Munoz, J.-F., & Dauchy, X. (2014). Is drinking water a major route of human exposure to alkylphenol and bisphenol contaminants in France? *Archives of Environmental Contamination and Toxicology*, 66(1), 86–99. doi:10.1007/s00244-013-9942-0

Crawford, C. B., & Quinn, B. (2017). The interactions of microfibers and chemical pollutants. *Microfiber Pollutants*, 1, 131–157. doi:10.1016/b978-0-12-809406-8.00006-2
Danopoulos, E., Jenner, L. C., Twiddy, M., & Rotchell, J. M. (2020b). Microfiber contamination of seafood intended for human consumption: A systematic review and meta-analysis. *Environmental Health Perspectives*, 128(12), 126002. https://doi.org/10.1289/ehp7171
Danopoulos, E., Twiddy, M., & Rotchell, J. M. (2020a). Microfiber contamination of drinking water: A systematic review. *PLoS One*, 15(7). https://doi.org/10.1371/journal.pone.0236838
Dawson, A. L., Santana, M. F. M., Miller, M. E., & Kroon, F. J. (2021). Relevance and reliability of evidence for microfiber contamination in seafood: A critical review using Australian consumption patterns as a case study. *Environmental Pollution*, 276, 116684. https://doi.org/10.1016/j.envpol.2021.116684
De Falco, F., Di Pace, E., Cocca, M., & Avella, M. (2019). The contribution of washing processes of synthetic clothes to microfiber pollution. *Scientific Reports*, 9(1). https://doi.org/10.1038/s41598-019-43023-x
Deng, Y., Zhang, Y., Lemos, B., & Ren, H. (2017). Tissue accumulation of microfibers in mice and biomarker responses suggest widespread health risks of exposure. *Scientific Reports*, 7(1), 46687–46710. doi:10.1038/srep46687
de Sá, L. C., Luís, L. G., & Guilhermino, L. (2015). Effects of MPs on juveniles of the common goby (pomatoschistus microps): Confusion with prey, reduction of the predatory performance and efficiency, and possible influence of developmental conditions. *Environmental Pollution*, 196, 359–362.
Diaz-Basantes, M. F., Conesa, J. A., & Fullana, A. (2020). Microfibers in honey, beer, milk and refreshments in Ecuador as emerging contaminants. *Sustainability*, 12(14), 5514. https://doi.org/10.3390/su12145514
Dodson, R. E., Nishioka, M., Standley, L. J., Perovich, L. J., Brody, J. G., & Rudel, R. A. (2012). Endocrine disruptors and asthma-associated chemicals in consumer products. *Environmental Health Perspectives*, 120(7), 935–943. https://doi.org/10.1289/ehp.1104052
Dolez, P. I., & Benaddi, H. (2018). Toxicity testing of textiles. In *Advanced Characterization and Testing of Textiles*, 151–188. https://doi.org/10.1016/b978-0-08-100453-1.00008-8
Dris, R., Gasperi, J., Mirande, C., Mandin, C., Guerrouache, M., Langlois, V., & Tassin, B. (2017). A first overview of textile fibers, including microfibers, in indoor and outdoor environments. *Environmental Pollution*, 221, 453–458. https://doi.org/10.1016/j.envpol.2016.12.013
Ehrlich, S., Calafat, A. M., Humblet, O., Smith, T., & Hauser, R. (2014). Handling of thermal receipts as a source of exposure to bisphenol A. *JAMA*, 311(8), 859–860. https://doi.org/10.1001/jama.2013.283735
Environmental Protection Agency [EPA]. (2012). Bisphenol A alternatives in thermal paper. Retrieved from https://www.epa.gov/sites/default/files/2015-08/documents/bpa_final.pdf
Fleury, J.-B., & Baulin, V. A. (2021). Microfibers destabilize lipid membranes by mechanical stretching. *Proceedings of the National Academy of Sciences*, 118(31). https://doi.org/10.1073/pnas.2104610118
Forte, M., Iachetta, G., Tussellino, M., Carotenuto, R., Prisco, M., De Falco, M., et al. (2016). Polystyrene nanoparticles internalization in human gastric adenocarcinoma cells. *Toxicology in Vitro*, 31, 126–136. doi:10.1016/j.tiv.2015.11.006
Gader Al-Khatim, A.-S. A., & Galil, K. A. (2013). Postnatal toxicity in mice attributable to plastic leachables in peritoneal dialysis solution (PDS). *Archives of Environmental & Occupational Health*, 70(2), 91–97. https://doi.org/10.1080/19338244.2013.807760
Galloway, T. S. (2015). Micro- and nano-plastics and human health. In *Marine Anthropogenic Litter* (Cham: Springer), 343–366. doi:10.1007/978-3-319-16510-3_13

Geiser, M., Rothen-Rutishauser, B., Kapp, N., Schürch, S., Kreyling, W., Schulz, H., et al. (2005). Ultrafine particles cross cellular membranes by nonphagocytic mechanisms in lungs and in cultured cells. *Environmental Health Perspectives*, 113(11), 1555–1560. doi:10.1289/ehp.8006

Gouin, T., Cunliffe, D., De France, J., Fawell, J., Jarvis, P., Koelmans, A. A., Marsden, P., Testai, E. E., Asami, M., Bevan, R., Carrier, R., Cotruvo, J., Eckhardt, A., & Ong, C. N. (2021). Clarifying the absence of evidence regarding human health risks to microfiber particles in drinking-water: High quality robust data wanted. *Environment International*, 150, 106141. https://doi.org/10.1016/j.envint.2020.106141

Green, D. S., Kregting, L., Boots, B., Blockley, D. J., Brickle, P., da Costa, M., & Crowley, Q. (2018). A comparison of sampling methods for seawater microfibers and a first report of the microfiber litter in coastal waters of Ascension and Falkland Islands. *Marine Pollution Bulletin*, 137, 695–701. https://doi.org/10.1016/j.marpolbul.2018.11.004

Halden, R. U. (2010). Plastics and health risks. *Annual Review of Public Health*, 31, 179–194. doi:10.1146/annurev.publhealth.012809.103714

Hamlin, H. J., Marciano, K., & Downs, C. A. (2015). Migration of nonylphenol from food-grade plastic is toxic to the coral reef fish species Pseudochromis Fridmani. *Chemosphere*, 139, 223–228. https://doi.org/10.1016/j.chemosphere.2015.06.032

Hill, P. J., Taylor, M., Goswami, P., & Blackburn, R. S. (2017). Substitution of pfas chemistry in outdoor apparel and the impact on repellency performance. *Chemosphere*, 181, 500–507. https://doi.org/10.1016/j.chemosphere.2017.04.122

Hussain, N., Kadir, M. M., Nafees, A. A., Karmaliani, R., & Jamali, T. (2019). Needs assessment regarding occupational health and safety interventions among textile workers: A qualitative case study in Karachi, Pakistan. *JPMA – The Journal of the Pakistan Medical Association*, 69(1), 87–93.

Imran, M., Crowley, D. E., Khalid, A., Hussain, S., Mumtaz, M. W., & Arshad, M. (2014). Microbial biotechnology for decolorization of textile wastewaters. *Reviews in Environmental Science and Bio/Technology*, 14(1), 73–92. https://doi.org/10.1007/s11157-014-9344-4

Islam, S. (2020). Sustainable raw materials: 50 shades of sustainability. *Sustainable Technologies for Fashion and Textiles*, 343–357. https://doi.org/10.1016/c2018-0-00610-6

Jambeck, J. R., Geyer, R., Wilcox, C., Siegler, T. R., Perryman, M., Andrady, A., Narayan, R., & Law, K. L. (2015). Plastic waste inputs from land into the Ocean. *Science*, 347(6223), 768–771. https://doi.org/10.1126/science.1260352

Jenner, L. C., Rotchell, J. M., Bennett, R. T., Cowen, M., Tentzeris, V., & Sadofsky, L. R. (2022). Detection of microfibers in human lung tissue using μFTIR spectroscopy. *The Science of the Total Environment*, 831, 154907. https://doi.org/10.1016/j.scitotenv.2022.154907

Karami, A., Golieskardi, A., Keong Choo, C., Larat, V., Galloway, T. S., & Salamatinia, B. (2017). The presence of microfibers in commercial salts from different countries. *Scientific Reports*, 7(1). https://doi.org/10.1038/srep46173

Karanikas, N., & Hasan, S. M. (2022). Occupational health & safety and other worker well-being areas: Results from labour inspections in the Bangladesh textile industry. *Safety Science*, 146, 105533. https://doi.org/10.1016/j.ssci.2021.105533

Karbalaei, S., Hanachi, P., Walker, T. R., & Cole, M. (2018). Occurrence, sources, human health impacts and mitigation of microfiber pollution. *Environmental Science and Pollution Research*, 25(36), 36046–36063. doi:10.1007/s11356-018-3508-7

Katyal, D., Kong, E., & Villanueva, J. (2020). Microfibers in the environment: Impact on human health and future Mitigation Strategies. *Environmental Health Review*, 63(1), 27–31. https://doi.org/10.5864/d2020-005

Kelly, F. J., & Fussell, J. C. (2012). Size, source and chemical composition as determinants of toxicity attributable to ambient particulate matter. *Atmospheric Environment*, 60, 504–526. doi:10.1016/j.atmosenv.2012.06.039

Khatri, J., Nidheesh, P. V., Anantha Singh, T. S., & Suresh Kumar, M. (2018). Advanced oxidation processes based on zero-valent aluminium for treating textile wastewater. *Chemical Engineering Journal*, 348, 67–73. https://doi.org/10.1016/j.cej.2018.04.074

Kirstein, I. V., Kirmizi, S., Wichels, A., Garin-Fernandez, A., Erler, R., Löder, M., et al. (2016). Dangerous hitchhikers? Evidence for potentially pathogenic vibrio spp. On microfiber particles. *Marine Environmental Research*, 120, 1–8. doi:10.1016/j.marenvres.2016.07.004

Kremer, A. M., Pal, T. M., Boleij, J. S., Schouten, J. P., & Rijcken, B. (1994). Airway hyper-responsiveness and the prevalence of work-related symptoms in workers exposed to irritants. *American Journal of Industrial Medicine*, 26(5), 655–669. https://doi.org/10.1002/ajim.4700260508

Kumar, V. E., Ravikumar, G., & Jeyasanta, K. I. (2018). Occurrence of MPs in fishes from two landing sites in Tuticorin, South East Coast of India. *Marine Pollution Bulletin*, 135, 889–894. doi:10.1016/j.marpolbul.2018.08.023

Kundakovic, M., Gudsnuk, K., Franks, B., Madrid, J., Miller, R. L., Perera, F. P., & Champagne, F. A. (2013). Sex-specific epigenetic disruption and behavioral changes following low-dose in utero bisphenol A exposure. *Proceedings of the National Academy of Sciences*, 110(24), 9956–9961. https://doi.org/10.1073/pnas.1214056110

Kutralam-Muniasamy, G., Pérez-Guevara, F., Elizalde-Martínez, I., & Shruti, V. C. (2020). Branded milks – Are they immune from microfibers contamination? *Science of the Total Environment*, 714, 136823. https://doi.org/10.1016/j.scitotenv.2020.136823

Lacasse, K., & Baumann, W. (2004). Environmental considerations for textile processes and chemicals. In *Textile Chemicals: Environmental Data and Facts*, 484–647. https://doi.org/10.1007/978-3-642-18898-5_7

Lai, P. S., & Christiani, D. C. (2013). Long-term respiratory health effects in textile workers. *Current Opinion in Pulmonary Medicine*, 19(2), 152–157. https://doi.org/10.1097/MCP.0b013e32835cee9a

Lamb, J. B., Willis, B. L., Fiorenza, E. A., Couch, C. S., Howard, R., Rader, D. N., True, J. D., Kelly, L. A., Ahmad, A., Jompa, J., & Harvell, C. D. (2018). Plastic waste associated with disease on coral reefs. *Science*, 359(6374), 460–462. https://doi.org/10.1126/science.aar3320

Lellis, B., Fávaro-Polonio, C. Z., Pamphile, J. A., & Polonio, J. C. (2019). Effects of textile dyes on health and the environment and bioremediation potential of living organisms. *Biotechnology Research and Innovation*, 3(2), 275–290. https://doi.org/10.1016/j.biori.2019.09.001

Leslie, H. A., van Velzen, M. J. M., Brandsma, S. H., Vethaak, A. D., Garcia-Vallejo, J. J., & Lamoree, M. H. (2022). Discovery and quantification of plastic particle pollution in human blood. *Environment International*, 163, 107199. https://doi.org/10.1016/j.envint.2022.107199

Li, Y., Li, W., Jarvis, P., Zhou, W., Zhang, J., Chen, J., Tan, Q., & Tian, Y. (2020). Occurrence, removal and potential threats associated with microfibers in drinking water sources. *Journal of Environmental Chemical Engineering*, 8(6), 104527. https://doi.org/10.1016/j.jece.2020.104527

Liebezeit, G., & Liebezeit, E. (2013). Non-pollen particulates in honey and sugar. *Food Additives & Contaminants: Part A*, 30(12), 2136–2140. https://doi.org/10.1080/19440049.2013.843025

Lin, J., Xiao, Y., Liu, Y., Lei, Y., Cai, Y., Liang, Q., Nie, S., Jia, Y., Chen, S., Huang, C., & Chen, J. (2022). Leachate from plastic food packaging induced reproductive and neurobehavioral toxicity in zebrafish. *Ecotoxicology and Environmental Safety*, 231, 113189. https://doi.org/10.1016/j.ecoenv.2022.113189

Lithner, D., Larsson, Å., & Dave, G. (2011). Environmental and health hazard ranking and assessment of plastic polymers based on chemical composition. *Science of the Total Environment*, 409(18), 3309–3324. https://doi.org/10.1016/j.scitotenv.2011.04.038

Liu, A., Richards, L., Bladen, C. L., Ingham, E., Fisher, J., & Tipper, J. L. (2015). The biological response to nanometre-sized polymer particles. *Acta Biomaterialia*, 23, 38–51. doi:10.1016/j.actbio.2015.05.016

Liu, C., Li, J., Zhang, Y., Wang, L., Deng, J., Gao, Y., Yu, L., Zhang, J., & Sun, H. (2019). Widespread distribution of pet and PC microfibers in dust in urban China and their estimated human exposure. *Environment International*, 128, 116–124. https://doi.org/10.1016/j.envint.2019.04.024

Liu, J., Yang, Y., Ding, J., Zhu, B., & Gao, W. (2019). Microfibers: A preliminary discussion on their definition and sources. *Environmental Science and Pollution Research*, 26(28), 29497–29501. https://doi.org/10.1007/s11356-019-06265-w

Lusher, A. L., Welden, N. A., Sobral, P., & Cole, M. (2020). Sampling, isolating and identifying microfibers ingested by fish and invertebrates. *Analytical Nanoplastics Microfibers Food*, 119–148. doi:10.1201/9780429469596-8

Mason, S. A., Welch, V. G., & Neratko, J. (2018). Synthetic polymer contamination in bottled water. *Frontiers in Chemistry*, 6. https://doi.org/10.3389/fchem.2018.00407

Matthews, J. B., Twomey, K., & Zacharewski, T. R. (2001). In vitro and in vivo interactions of bisphenol A and its metabolite, bisphenol A glucuronide, with estrogen receptors alpha and beta. *Chemical Research in Toxicology*, 14(2), 149–157. https://doi.org/10.1021/tx0001833

McCormick, A. R., Hoellein, T. J., London, M. G., Hittie, J., Scott, J. W., & Kelly, J. J. (2016). Microfiber in surface waters of urban rivers: Concentration, sources, and associated bacterial assemblages. *Ecosphere*, 7(11). https://doi.org/10.1002/ecs2.1556

McEachern, K., Alegria, H., Kalagher, A. L., Hansen, C., Morrison, S., & Hastings, D. (2019). Microfibers in Tampa Bay, Florida: Abundance and variability in estuarine waters and sediments. *Marine Pollution Bulletin*, 148, 97–106. https://doi.org/10.1016/j.marpolbul.2019.07.068

Meeker, J. D., Ehrlich, S., Toth, T. L., Wright, D. L., Calafat, A. M., Trisini, A. T., et al. (2010). Semen quality and sperm DNA damage in relation to urinary bisphenol A among men from an infertility clinic. *Reproductive Toxicology*, 30 (4), 532–539. doi:10.1016/j.reprotox.2010.07.005

Meeker, J. D., Sathyanarayana, S., & Swan, S. H. (2009). Phthalates and other additives in plastics: Human exposure and associated health outcomes. *Philosophical Transactions of the Royal Society B: Biological Sciences*, 364(1526), 2097–2113. https://doi.org/10.1098/rstb.2008.0268

Memon, I., Panhwar, A., Rohra, D. K., Azam, S. I., & Khan, N. (2008). Prevalence of byssinosis in spinning and textile workers of Karachi, Pakistan. *Archives of Environmental & Occupational Health*, 63(3), 137–142. https://doi.org/10.3200/AEOH.63.3.137-142

Mendum, T., Stoler, E., VanBenschoten, H., & Warner, J. C. (2011). Concentration of bisphenol A in thermal paper. *Green Chemistry Letters and Reviews*, 4(1), 81–86. https://doi.org/10.1080/17518253.2010.502908

Meng, W., Sun, H., & Su, G. (2023). Plastic packaging-associated chemicals and their hazards – An overview of reviews. *Chemosphere*, 138795. https://doi.org/10.1016/j.chemosphere.2023.138795

Michałowicz, J. (2014). Bisphenol A – Sources, toxicity and biotransformation. *Environmental Toxicology and Pharmacology*, 37 (2), 738–758. doi:10.1016/j.etap.2014.02.003

Mintenig, S. M., Löder, M. G. J., Primpke, S., & Gerdts, G. (2019). Low numbers of microfibers detected in drinking water from ground water sources. *Science of the Total Environment*, 648, 631–635. https://doi.org/10.1016/j.scitotenv.2018.08.178

Mishra, S., Rath, C. c., & Das, A. P. (2019). Marine microfiber pollution: A review on present status and future challenges. *Marine Pollution Bulletin*, 140, 188–197. https://doi.org/10.1016/j.marpolbul.2019.01.039

Mizraji, R., Ahrendt, C., Perez-Venegas, D., Vargas, J., Pulgar, J., Aldana, M., et al. (2017). Is the feeding type related with the content of MPs in intertidal fish gut? *Marine Pollution Bulletin*, 116(1–2), 498–500. doi:10.1016/j.marpolbul.2017.01.008

Mohamed Nor, N. H., Kooi, M., Diepens, N. J., & Koelmans, A. A. (2021). Lifetime accumulation of microfiber in children and adults. *Environmental Science & Technology*, 55(8), 5084–5096. https://doi.org/10.1021/acs.est.0c07384

MohanKumar, S. M., Campbell, A., Block, M., & Veronesi, B. (2008). Particulate matter, oxidative stress and neurotoxicity. *Neurotoxicology*, 29(3), 479–488. doi:10.1016/j.neuro.2007.12.004

Monti, D. M., Guarnieri, D., Napolitano, G., Piccoli, R., Netti, P., Fusco, S., et al. (2015). Biocompatibility, uptake and endocytosis pathways of polystyrene nanoparticles in primary human renal epithelial cells. *Journal of Biotechnology*, 193, 3–10. doi:10.1016/j.jbiotec.2014.11.004

Moore, J., Ototake, M., & Nakanishi, T. (1998). Particulate antigen uptake during immersion immunisation of fish: The effectiveness of prolonged exposure and the roles of skin and Gill. *Fish & Shellfish Immunology*, 8(6), 393–408. https://doi.org/10.1006/fsim.1998.0143

Mühlschlegel, P., Hauk, A., Walter, U., & Sieber, R. (2017). Lack of evidence for microfiber contamination in Honey. *Food Additives & Contaminants: Part A*, 34(11), 1982–1989. https://doi.org/10.1080/19440049.2017.1347281

Muittari, A., & Veneskoski, T. (1978). Natural and synthetic fibers as causes of asthma and rhinitis. *Annals of Allergy*, 41(1), 48–50.

Murlidhar, V., Murlidhar, V. J., & Kanhere, V. (1995). Byssinosis in a Bombay textile mill. *The National Medical Journal of India*, 8(5), 204–207.

Murray, T. J., Maffini, M. V., Ucci, A. A., Sonnenschein, C., & Soto, A. M. (2007). Induction of mammary gland ductal hyperplasias and carcinoma in situ following fetal bisphenol A exposure. *Reproductive Toxicology*, 23(3), 383–390. https://doi.org/10.1016/j.reprotox.2006.10.002

Napper, I. E., Baroth, A., Barrett, A. C., Bhola, S., Chowdhury, G. W., Davies, B. F. R., Duncan, E. M., Kumar, S., Nelms, S. E., Hasan Niloy, M. N., Nishat, B., Maddalene, T., Thompson, R. C., & Koldewey, H. (2021). The abundance and characteristics of microfibers in surface water in the transboundary Ganges River. *Environmental Pollution*, 274, 116348. https://doi.org/10.1016/j.envpol.2020.116348

Napper, I. E., & Thompson, R. C. (2016). Release of synthetic microfiber plastic fibres from domestic washing machines: Effects of fabric type and washing conditions. *Marine Pollution Bulletin*, 112(1–2), 39–45. https://doi.org/10.1016/j.marpolbul.2016.09.025

Negishi, T., Tominaga, T., Ishii, Y., Kyuwa, S., Hayasaka, I., Kuroda, Y., & Yoshikawa, Y. (2004). Comparative study on toxicokinetics of bisphenol A in F344 rats, monkeys (Macaca fascicularis), and chimpanzees (Pan Troglodytes). *Experimental Animals*, 53(4), 391–394. https://doi.org/10.1538/expanim.53.391

Nerín, C., Su, Q.-Z., Vera, P., Mendoza, N., & Ausejo, R. (2020). Influence of nonylphenol from multilayer plastic films on artificial insemination of sows. *Analytical and Bioanalytical Chemistry*, 412(24), 6519–6528. https://doi.org/10.1007/s00216-020-02698-2

Nerin, C., Ubeda, J. L., Alfaro, P., Dahmani, Y., Aznar, M., Canellas, E., & Ausejo, R. (2014). Compounds from multilayer plastic bags cause reproductive failures in artificial insemination. *Scientific Reports*, 4(1). https://doi.org/10.1038/srep04913

Newman, M. C. (2009). *Fundamentals of Ecotoxicology*. CRC Press.

Obbard, R. W., Sadri, S., Wong, Y. Q., Khitun, A. A., Baker, I., & Thompson, R. C. (2014). Global warming releases microfiber legacy frozen in Arctic Sea Ice. *Earth's Future*, 2(6), 315–320. https://doi.org/10.1002/2014ef000240

Oßmann, B. E., Sarau, G., Holtmannspötter, H., Pischetsrieder, M., Christiansen, S. H., & Dicke, W. (2018). Small-sized microfibers and pigmented particles in bottled mineral water. *Water Research*, 141, 307–316. https://doi.org/10.1016/j.watres.2018.05.027

O'Brien, S., Okoffo, E. D., O'Brien, J. W., Ribeiro, F., Wang, X., Wright, S. L., Samanipour, S., Rauert, C., Toapanta, T. Y., Albarracin, R., & Thomas, K. V. (2020). Airborne emissions of microfiber fibres from domestic laundry dryers. *Science of the Total Environment*, 747, 141175. https://doi.org/10.1016/j.scitotenv.2020.141175

Oliveri Conti, G., Ferrante, M., Banni, M., Favara, C., Nicolosi, I., Cristaldi, A., Fiore, M., & Zuccarello, P. (2020). Micro- and nano-plastics in edible fruit and vegetables. The first diet risks assessment for the general population. *Environmental Research*, 187, 109677. https://doi.org/10.1016/j.envres.2020.109677

Ormsby, R. T., Cantley, M., Kogawa, M., Solomon, L. B., Haynes, D. R., Findlay, D. M., et al. (2016). Evidence that osteocyte perilacunar remodelling contributes to polyethylene wear particle induced osteolysis. *Acta Biomaterialia*, 33, 242–251. doi:10.1016/j.actbio.2016.01.016

Palanza, P. L., Howdeshell, K. L., Parmigiani, S., & vom Saal, F. S. (2002). Exposure to a low dose of bisphenol A during fetal life or in adulthood alters maternal behavior in mice. *Environmental Health Perspectives*, 110(suppl 3), 415–422. https://doi.org/10.1289/ehp.02110s3415

Pazos, R. S., Maiztegui, T., Colautti, D. C., Paracampo, A. H., & Gómez, N. (2017). MPs in gut contents of coastal freshwater fish from Río de la Plata estuary. *Marine Pollution Bulletin*, 122(1–2), 85–90. doi:10.1016/j.marpolbul.2017.06.007

Peixoto, D., Pinheiro, C., Amorim, J., Oliva-Teles, L., Guilhermino, L., & Vieira, M. N. (2019). Microfiber pollution in commercial salt for human consumption: A Review. *Estuarine, Coastal and Shelf Science*, 219, 161–168. https://doi.org/10.1016/j.ecss.2019.02.018

Periyasamy, A. P. (2023). Microfiber emissions from functionalized textiles: Potential threat for human health and environmental risks. *Toxics*, 11(5), 406. https://doi.org/10.3390/toxics11050406

Pottenger, L. H., Domoradzki, J. Y., Markham, D. A., Hansen, S. C., Cagen, S. Z., & Waechter, J. M., Jr (2000). The relative bioavailability and metabolism of bisphenol A in rats is dependent upon the route of administration. *Toxicological Sciences: An Official Journal of the Society of Toxicology*, 54(1), 3–18. https://doi.org/10.1093/toxsci/54.1.3

Prata, J. C., da Costa, J. P., Duarte, A. C., & Rocha-Santos, T. (2022). Suspected microfibers in Atlantic horse mackerel fish (Trachurus trachurus) captured in Portugal. *Marine Pollution Bulletin*, 174, 113249. doi:10.1016/j.marpolbul.2021.113249

Prata, J. C., da Costa, J. P., Lopes, I., Duarte, A. C., & Rocha-Santos, T. (2020). Environmental exposure to MPs: An overview on possible human health effects. *Science of the Total Environment*, 702, 134455. doi:10.1016/j.scitotenv.2019.134455

Rehman, K., Shahzad, T., Sahar, A., Hussain, S., Mahmood, F., Siddique, M. H., Siddique, M. A., & Rashid, M. I. (2018). Effect of reactive black 5 azo dye on soil processes related to C and N cycling. https://doi.org/10.7287/peerj.preprints.26557v1

Rochman, C. M., Tahir, A., Williams, S. L., Baxa, D. V., Lam, R., Miller, J. T., Teh, F.-C., Werorilangi, S., & Teh, S. J. (2015). Anthropogenic debris in seafood: Plastic debris and fibers from textiles in fish and bivalves sold for human consumption. *Scientific Reports*, 5(1). https://doi.org/10.1038/srep14340

Romeo, T., Pietro, B., Pedà, C., Consoli, P., Andaloro, F., & Fossi, M. C. (2015). First evidence of presence of plastic debris in stomach of large pelagic fish in the Mediterranean sea. *Marine Pollution Bulletin*, 95(1), 358–361. doi:10.1016/j.marpolbul.2015.04.048

Sanchez-Vidal, A., Thompson, R. C., Canals, M., & de Haan, W. P. (2018). The imprint of microfibres in Southern European Deep Seas. *PLoS One*, 13(11). https://doi.org/10.1371/journal.pone.0207033

Sandhya, S. (2010). Biodegradation of azo dyes under anaerobic condition: Role of Azoreductase. In *The Handbook of Environmental Chemistry*, 39–57. https://doi.org/10.1007/698_2009_43

Schirinzi, G. F., Pérez-Pomeda, I., Sanchís, J., Rossini, C., Farré, M., & Barceló, D. (2017). Cytotoxic effects of commonly used nanomaterials and MPs on cerebral and epithelial human cells. *Environmental Research*, 159, 579–587. doi:10.1016/j.envres.2017.08.043

Schneider, M., Stracke, F., Hansen, S., & Schaefer, U. F. (2009). Nanoparticles and their interactions with the dermal barrier. *Dermato-Endocrinology*, 1(4), 197–206. https://doi.org/10.4161/derm.1.4.9501

Schöpel, B., & Stamminger, R. (2019). A comprehensive literature study on microfibres from washing machines. *Tenside Surfactants Detergents*, 56(2), 94–104. https://doi.org/10.3139/113.110610

Schulze, M., Schröter, F., Jung, M., & Jakop, U. (2020). Evaluation of a panel of spermatological methods for assessing reprotoxic compounds in multilayer semen plastic bags. *Scientific Reports*, 10(1). https://doi.org/10.1038/s41598-020-79415-7

Schwabl, P., Köppel, S., Königshofer, P., Bucsics, T., Trauner, M., Reiberger, T., & Liebmann, B. (2019). Detection of various microfibers in human stool. *Annals of Internal Medicine*, 171(7), 453–457. https://doi.org/10.7326/m19-0618

Schymanski, D., Goldbeck, C., Humpf, H.-U., & Fürst, P. (2018). Analysis of microfibers in water by micro-raman spectroscopy: Release of plastic particles from different packaging into mineral water. *Water Research*, 129, 154–162. https://doi.org/10.1016/j.watres.2017.11.011

Sequeira, I. F., Prata, J. C., da Costa, J. P., Duarte, A. C., & Rocha-Santos, T. (2020). Worldwide contamination of fish with microfibers: A brief global overview. *Marine Pollution Bulletin*, 160, 111681. doi:10.1016/j.marpolbul.2020.111681

Setälä, O., Fleming-Lehtinen, V., & Lehtiniemi, M. (2014). Ingestion and transfer of microfibers in the planktonic food web. *Environmental Pollution*, 185, 77–83. https://doi.org/10.1016/j.envpol.2013.10.013

Sharma, B., Dangi, A. K., & Shukla, P. (2018). Contemporary enzyme based technologies for bioremediation: A Review. *Journal of Environmental Management*, 210, 10–22. https://doi.org/10.1016/j.jenvman.2017.12.075

Shruti, V. C., Pérez-Guevara, F., Elizalde-Martínez, I., & Kutralam-Muniasamy, G. (2020). First study of its kind on the microfiber contamination of soft drinks, cold tea and energy drinks – Future research and environmental considerations. *Science of the Total Environment*, 726, 138580. https://doi.org/10.1016/j.scitotenv.2020.138580

Sillanpää, M., & Sainio, P. (2017). Release of polyester and cotton fibers from textiles in machine washings. *Environmental Science and Pollution Research*, 24(23), 19313–19321. https://doi.org/10.1007/s11356-017-9621-1

Silva, M. M., Maldonado, G. C., Castro, R. O., de Sá Felizardo, J., Cardoso, R. P., Anjos, R. M., & Araújo, F. V. (2019). Dispersal of potentially pathogenic bacteria by plastic debris in Guanabara Bay, RJ, Brazil. *Marine Pollution Bulletin*, 141, 561–568. https://doi.org/10.1016/j.marpolbul.2019.02.064

Singh, S., Chatterji, S., Nandini, P. T., Prasad, A. S., & Rao, K. V. (2014). Biodegradation of azo dye direct orange 16 by Micrococcus luteus strain SSN2. *International Journal of Environmental Science and Technology*, 12(7), 2161–2168. https://doi.org/10.1007/s13762-014-0588-x

Smith, M., Love, D. C., Rochman, C. M., & Neff, R. A. (2018). MPs in seafood and the implications for human health. *Current Environmental Health Reports*, 5(3), 375–386. doi:10.1007/s40572-018-0206-z

Suaria, G., Achtypi, A., Perold, V., Lee, J. R., Pierucci, A., Bornman, T. G., Aliani, S., & Ryan, P. G. (2020). Microfibers in oceanic surface waters: A global characterization. *Science Advances*, 6(23). https://doi.org/10.1126/sciadv.aay8493

Teeguarden, J., Hanson-Drury, S., Fisher, J. W., & Doerge, D. R. (2013). Are typical human serum BPA concentrations measurable and sufficient to be estrogenic in the general population? *Food and Chemical Toxicology: An International Journal Published for the British Industrial Biological Research Association*, 62, 949–963. https://doi.org/10.1016/j.fct.2013.08.001

Tominaga, T., Negishi, T., Hirooka, H., Miyachi, A., Inoue, A., Hayasaka, I., & Yoshikawa, Y. (2006). Toxicokinetics of bisphenol A in rats, monkeys and chimpanzees by the LC-MS/MS method. *Toxicology*, 226(2–3), 208–217. https://doi.org/10.1016/j.tox.2006.07.004

Too, C. L., Muhamad, N. A., Ilar, A., Padyukov, L., Alfredsson, L., Klareskog, L., Murad, S., Bengtsson, C., & MyEIRA Study Group (2016). Occupational exposure to textile dust increases the risk of rheumatoid arthritis: Results from a Malaysian population-based case-control study. *Annals of the Rheumatic Diseases*, 75(6), 997–1002. https://doi.org/10.1136/annrheumdis-2015-208278

U.S. Food and Drug Administration [FDA]. (2012). *Bisphenol A (BPA): Use in Food Contact Application*. U.S. Food and Drug Administration. https://www.fda.gov/food/food-additives-petitions/bisphenol-bpa-use-food-contact-application#:~:text=FDA%20has%20amended%20its%20regulations,(ACC)%20%5B9%5D.

Usman, S., Abdull Razis, A. F., Shaari, K., Azmai, M. N., Saad, M. Z., Mat Isa, N., & Nazarudin, M. F. (2022). The burden of microfibers pollution and contending policies and regulations. *International Journal of Environmental Research and Public Health*, 19(11), 6773. https://doi.org/10.3390/ijerph19116773

Valavanidis, A., Vlachogianni, T., Fiotakis, K., & Loridas, S. (2013). Pulmonary oxidative stress, inflammation and cancer: Respirable particulate matter, fibrous dusts and ozone as major causes of lung carcinogenesis through reactive oxygen species mechanisms. *International Journal of Environmental Research and Public Health*, 10(9), 3886–3907. doi:10.3390/ijerph10093886

Vandenberg, L. N., Hunt, P. A., Myers, J. P., & Vom Saal, F. S. (2013). Human exposures to bisphenol A: Mismatches between data and assumptions. *Reviews on Environmental Health*, 28(1), 37–58. https://doi.org/10.1515/reveh-2012-0034

van Dijk, F., Van Eck, G., Cole, M., Salvati, A., Bos, S., Gosens, R., & Melgert, B. (2020). Exposure to textile microfiber fibers impairs epithelial growth. *Airway Cell Biology and Immunopathology*. https://doi.org/10.1183/13993003.congress-2020.1972

Viršek, M. K., Lovšin, M. N., Koren, Š., Kržan, A., & Peterlin, M. (2017). Microfibers as a vector for the transport of the bacterial fish pathogen species aeromonas salmonicida. *Marine Pollution Bulletin*, 125(1–2), 301–309. https://doi.org/10.1016/j.marpolbul.2017.08.024

Vom Saal, F. S., Nagel, S. C., Coe, B. L., Angle, B. M., & Taylor, J. A. (2012). The estrogenic endocrine disrupting chemical bisphenol A (BPA) and obesity. *Molecular and Cellular Endocrinology*, 354(1–2), 74–84. doi:10.1016/j.mce.2012.01.001

Wang, F., Wong, C. S., Chen, D., Lu, X., Wang, F., & Zeng, E. Y. (2018). Interaction of toxic chemicals with microfibers: A critical review. *Water Research*, 139, 208–219. https://doi.org/10.1016/j.watres.2018.04.003

Wang, J., Tan, Z., Peng, J., Qiu, Q., & Li, M. (2016). The behaviors of MPs in the marine environment. *Marine Environmental Research*, 113, 7–17. doi:10.1016/j.marenvres.2015.10.014

Wang, L., Zhang, Y., Liu, Y., Gong, X., Zhang, T., & Sun, H. (2019). Widespread occurrence of bisphenol A in daily clothes and its high exposure risk in humans. *Environmental Science & Technology*, 53(12), 7095–7102. https://doi.org/10.1021/acs.est.9b02090

Weil, E. D., Levchik, S., & Moy, P. (2006). Flame and smoke retardants in vinyl chloride polymers – Commercial usage and current developments. *Journal of Fire Sciences*, 24(3), 211–236. https://doi.org/10.1177/0734904106057951

Wesch, C., Bredimus, K., Paulus, M., & Klein, R. (2016). Towards the suitable monitoring of ingestion of MPs by marine biota: A review. *Environmental Pollution*, 218, 1200–1208. doi:10.1016/j.envpol.2016.08.076

West-Eberhard, M. J. (2019). Nutrition, the visceral immune system, and the evolutionary origins of pathogenic obesity. *Proceedings of the National Academy of Sciences*, 116(3), 723–731. doi:10.1073/pnas.1809046116

Wright, S. L., & Kelly, F. J. (2017). Plastic and human health: A micro issue? *Environmental Science & Technology*, 51(12), 6634–6647. doi:10.1021/acs.est.7b00423

Wright, S. L., Thompson, R. C., & Galloway, T. S. (2013). The physical impacts of MPs on marine organisms: A review. *Environmental Pollution*, 178, 483–492. doi:10.1016/j.envpol.2013.02.031

Xu, C., Zhang, B., Gu, C., Shen, C., Yin, S., Aamir, M., & Li, F. (2020). Are we underestimating the sources of microfiber pollution in terrestrial environment? *Journal of Hazardous Materials*, 400, 123228. https://doi.org/10.1016/j.jhazmat.2020.123228

Xu, Y., Chan, F. K., Stanton, T., Johnson, M. F., Kay, P., He, J., Wang, J., Kong, C., Wang, Z., Liu, D., & Xu, Y. (2021). Synthesis of dominant plastic microfibre prevalence and pollution control feasibility in Chinese freshwater environments. *Science of the Total Environment*, 783, 146863. https://doi.org/10.1016/j.scitotenv.2021.146863

Yamamoto, Y., Kobayashi, Y., & Matsumoto, H. (2001). Lipid peroxidation is an early symptom triggered by aluminum, but not the primary cause of elongation inhibition in pea roots. *Plant Physiology*, 125(1), 199–208. doi:10.1104/pp.125.1.199

Yan, Z., Liu, Y., Zhang, T., Zhang, F., Ren, H., & Zhang, Y. (2021). Analysis of microfibers in human feces reveals a correlation between fecal microfibers and inflammatory bowel disease status. *Environmental Science & Technology*, doi:10.1021/acs.est.1c03924

Zele, Y. T., Kumie, A., Deressa, W., Bråtveit, M., & Moen, B. E. (2021). Registered health problems and demographic profile of integrated textile factory workers in Ethiopia: A cross-sectional study. *BMC Public Health*, 21(1), 1526. https://doi.org/10.1186/s12889-021-11556-4

Zettler, E. R., Mincer, T. J., & Amaral-Zettler, L. A. (2013). Life in the "plastisphere": Microbial communities on plastic marine debris. *Environmental Science & Technology*, 47(13), 7137–7146. https://doi.org/10.1021/es401288x

Zhang, C., Wang, J., Zhou, A., Ye, Q., Feng, Y., Wang, Z., & Zou, J. (2021). Species-specific effect of microfibers on fish embryos and observation of toxicity kinetics in larvae. *Journal of Hazardous Materials*, 403, 123948. doi:10.1016/j.jhazmat.2020.123948

Zhang, Q., Xu, E. G., Li, J., Chen, Q., Ma, L., Zeng, E. Y., & Shi, H. (2020a). A review of microfibers in table salt, drinking water, and air: Direct human exposure. *Environmental Science & Technology*, 54(7), 3740–3751. https://doi.org/10.1021/acs.est.9b04535

Zhang, Y., Kang, S., Allen, S., Allen, D., Gao, T., & Sillanpää, M. (2020b). Atmospheric microfibers: A review on current status and Perspectives. *Earth-Science Reviews*, 203, 103118. https://doi.org/10.1016/j.earscirev.2020.103118

Zimmermann, L., Dierkes, G., Ternes, T. A., Völker, C., & Wagner, M. (2019). Benchmarking the in vitro toxicity and chemical composition of plastic consumer products. *Environmental Science & Technology*, 53(19), 11467–11477. https://doi.org/10.1021/acs.est.9b02293

Section 4

Potential Mitigation Strategies and Awareness

15 Cross-Disciplinary Collaboration Challenging Microfibre Fragmentation in the Fashion and Textile Industries

Alana M. James, Nkumbu Mutambo, and Miranda Prendergast-Miller
Northumbria University

Kelly J. Sheridan
Northumbria University
The Microfibre Consortium

Abbie Rogers
Northumbria University

15.1 INTRODUCTION

The fashion and textile value chain is a long and complex system which relies heavily on globalised logistics to facilitate sourcing, manufacture, and processing, prior to the product arriving at the point of sale. Operating in a linear format, each individual garment has a unique journey from cradle to grave, making transparency and traceability of the supply chain challenging and, in some cases, impossible. Every stage across the value chain creates an environmental impact, negatively affecting the natural world through the use of resources and the generation of emissions, waste, and pollutants. The measurement of this impact has in recent years become a priority across the industry but remains largely voluntary and unregulated by legislation or governing bodies. Common impact measures include raw materials and resources; energy use; water use; and emissions to land, sea, and air and are often categorised by the stage of the value chain (raw material extraction, manufacturing and processing,

DOI: 10.1201/9781003331995-19

transportation, usage and retail, and waste disposal) (Bjørn et al., 2016). These measures are integrated into lifecycle analysis (LCA) methods, which aim to provide an overall data set of the environmental impact caused by any product. However, LCA tools have recently been heavily criticised due to their reliance on methods of self-reporting and their siloed approach to how metrics are implemented across the supply chain (Gonçalves and Silva, 2021). Furthermore, the authenticity and possibilities of biases from the vested interest of funders or board members have also been brought into question (Britten, 2022; Donald, 2022, Northam, 2023).

Despite LCA tools encapsulating the whole of the value chain, there are impacts that are increasingly being overlooked, including the prevalence of microfibres as pollutants through shedding and fibre fragmentation from textiles. Microfibres are ubiquitous and have been detected across all environments, in our waters (Andrady, 2011; Kechi-Okafor et al., 2023), coastlines, land (Forster et al., 2023), and air (Dris et al., 2016) we breathe. Microfibres, alongside microplastics, are regarded as emerging contaminants and derive predominately from the loss of textile fibres. Although not exclusively, textile fibre loss occurs mainly as a result of washing and drying of clothing using domestic washing machines and driers. Knowledge of microfibres as environmental pollutants remains fairly limited across multiple stakeholders including manufacturers, brands, and consumers. Where awareness exists, the translation of this into preventative action is often stunted due to the issue being low on sustainability agendas in the absence of global standards and policies. Where positive action is being implemented, this is often concentrated in the consumer use phase of the value chain, concentrating predominantly on garment care practices. Further exploration of fibre fragmentation across all stages of the fashion and textile value chain is needed, considering textile and garment manufacturing processes and end-of-life waste streams. Furthermore, the consideration of preventative measures in the early stages of the value chain, through variables such as fibre type, yarn, textile construction, and fabric finishes, is yet to be explored in depth. The integration of such considerations, however, will rely on stakeholder collaboration, collegiate knowledge and awareness, and effective communication channels across the sector. These strategies are yet to be evidenced, indicating a significant gap in knowledge, with a focus on bringing together multiple disciplinary knowledge and multi-stakeholder voices being drastically needed throughout the value chain.

15.1.1 Research Aim and Objectives

This research explores microfibres as emerging pollutants caused by the fashion and textile industries from a multi-stakeholder perspective, focusing specifically on the attitudes and actions of these industries. Acknowledging the need for a breadth of perspectives, the research engaged with actors across the value chain, considering both upstream and downstream approaches. By identifying drivers and barriers to change across the fashion lifecycle and framing these within the parameters of policy and legislation, the research aimed to co-create a targeted directive to inform future industry practices.

This was achieved through the following objectives:

- To conduct an extensive review of industry- and academic-based literature, and policy and legislative documentation, to establish parameters and governance across the sector.
- To engage with industry stakeholders from across the fashion and textiles value chain, adopting both an upstream (textile design) and downstream (fibre waste management) approach through qualitative, semi-structured interviews.
- To develop a stakeholder roadmap, informed by the insights generated during the qualitative interview series, and to inform future practices towards the reduction and mitigation of microfibre loss in the fashion and textile industries.

15.2 CONTEXTUAL REVIEW

Sustainability has become an area of focus globally over the last two decades, with the UN's Sustainable Development Goals standing out as one of the most influential and salient examples of how these values are being universally implemented. As a result, regional, national, and international authorities have begun to apply increasing legislative pressure on several industries that are viewed as contributors to some of the greatest negative impacts. While their efficacy may be questioned, a growing number of policies aimed at mitigating the social, environmental, and economic impacts of the fashion and textile industries have emerged. These policies cover a range of areas including labour conditions and worker rights, reducing over-production and consumption, a reduction in carbon emissions, and mitigation against pollution across the lifecycle of garments.

15.2.1 Governance and Legislation

As an area of growing concern, microfibre pollution has become the subject of a growing number of policy initiatives. For instance, couched within larger ambitions and initiatives around the circular economy, the European Union has released policy strategy documents detailing proposed and soon-to-be-implemented legislation. For example, the EU Strategy for Sustainable and Circular Textiles (2022), which is set to be achieved by 2030, will introduce Extended Producer Responsibility (EPR), which would incentivise eco-design measures like recycling, repair, and reuse. It will also see the introduction of digital product passports to increase transparency and traceability. Additionally, the strategy will include design requirements aimed at increasing the lifespan of garments and reducing the unintentional release of harmful substances, including microplastics. Though the design stage of the value chain is recognised as a critical point for effective mitigation against pollution in this strategy, it is not clear what evidence would be used to inform design requirements or in fact how adherence to these measures would be assessed. Similarly, at a national level, the Waste & Resources Action Programme (WRAP), a British charity working with businesses, individuals, and communities to achieve a circular economy, has set ambitions through the development of the Textiles 2030 roadmap. This initiative is aimed at achieving water, carbon, and circularity targets in the fashion and textile industries, as well as informing legislation (WRAP, 2021). Funded in part by its signatories and the UK government, Textiles 2030 is a voluntary agreement that outlines a pathway towards a circular future, including the launch of circular design standards in 2025 that encourage the use of recyclable materials and longer product lifecycles. However, it is unclear how the issue of microfibre pollution will be addressed if at all in this strategy. Moreover, it remains to be seen how effective such a voluntary agreement can actually be.

Another example of a country-level policy is France's eco-labelling regulation that came into effect in January 2023 and requires companies with a certain level of market share (currently, a turnover higher than 50 million euros and responsibility for at least 25,000 "waste generating products" on the French market (Recover, 2023)) to add labels detailing environmental information about their products. In addition to information about a product's recyclability, reusability, and origins, such labels will have to indicate the presence of plastic microfibres if an item is composed of 50% or more of synthetic fibres. Interestingly, this law also prohibits the use of terms such as "respectful of the environment" or "biodegradable" (Recover, 2023). While providing consumers with more information about products can be a useful tool in positively shifting their decision-making, this law lacks any ability in and of itself to change the way in which the goods themselves are designed and produced.

While the increase in legislation and initiatives addressing the environmental impact of the fashion and textile industries is certainly an important step in the right direction, some aspects of these policies need to be critically re-examined. For instance, the French labelling law is a notable example of how much a current policy has been influenced by prevailing narratives that centre plastics and synthetic fibres as the main pollutants while natural fibres are tacitly framed as harmless due to their biodegradability (Anscombe, 2019). However, this has been called into question by evidence of natural fibres persisting in the environment (Liu et al., 2023; Stanton et al., 2019). Moreover, though the consistent focus on the policy of increasing the use of recycled materials in textiles may appear to be a common-sense way of closing the loop, it could be questioned whether the increased use of recycled fibres would be net positive. For instance, there is a need to consider the impact that an increase in demand for recycled polyester (rPET) in the fashion and textiles industries would have especially, considering the fact that 99% of rPET used in garments is not from post-consumer textile waste but single-use polyethylene terephthalate (PET) bottles (Pointing, 2023). As the recycling of PET bottles is a good example of a closed loop, diverting these materials to produce rPET textiles negatively impacts the balance of the PET bottle circularity system and may result in greater waste of materials as some of those recycled textiles will end up in landfills and out of the loop (Bryce, 2021). It could also be questioned whether there is adequate textile recycling infrastructure in countries like the UK to meet the fast-approaching deadlines of 2030. Furthermore, there is also a need for more data on what impacts, if any, using recycled fibres would have on fibre fragmentation. This is evidenced by recent studies with divergent viewpoints. For example, a study by Özkan and Gündoğdu (2021) found a 2.3 times increase in fibre loss from rPET knitted fabrics compared with those made from virgin polyester, while a report from TMC found no significant increase in shed by rPET fabrics but also calling for more investigation of chemically rPET (Bland and Sheridan, 2023).

15.2.2 The Prevalence of Microfibres

The presence of microfibres in the atmosphere and household dust (Dris et al., 2017) suggests that garment laundering is not the only source of shed fibres. Indeed, evidence that fibres are also lost to the environment from general wear and use (Carr, 2017) indicates that the source of fibres needs further investigation. While the focus has been on fashion textiles, recent evidence highlights the potential input of other fibre sources from personal care products, including wet wipes and face masks (Kaur et al., 2023), cigarette butts (Shah et al., 2023), and other household textiles (e.g. carpets) (Gavigan et al., 2020). Once shed, microfibres are affected by abiotic and biotic factors prevalent in the receiving environment, meaning that the characteristics and properties of the fibre particles are likely to change over time. Consequently, their longer-term impacts are also likely to change over time, e.g. ingestion and bioaccumulation; further fragmentation into smaller particles; sorption or leaching of chemicals; growth of microbial films, etc. (Galloway et al., 2017; Horton et al., 2017).

Microfibres have been detected on high mountain peaks, deep ocean troughs, farmland, urban areas, indoor air, and the atmosphere. Due to their environmental presence, microplastics and microfibres have also been detected in tissues and organs of a variety of organisms, from small marine plankton (size) to seabirds and sea mammals (Rebelein et al., 2021). Controlled laboratory studies have shown that soil-dwelling organisms are unable to discern between fibres and soil particles, with microfibres being ingested by earthworms, snails, isopods, and enchytraeids. While microfibres have been detected in soils, there is little evidence of their uptake into plant tissues. However, microplastics enter the human food chain as they have been detected in the human placenta, breastmilk, and stool. However, the prevalence of microfibres in the environment is variable, and comparison between published studies is challenging due to the range of methods and strategies used to sample, count, identify, and report fibres as there are currently no agreed international protocols in place, and this needs to be addressed (Debraj and Lavanya, 2023). Nevertheless, there is a general trend in that microfibres tend to be the dominant form of microplastics found in the environment. In a global study, Gavigan et al. (2020) estimated that between 1950 and 2016, 2.9 Mt of synthetic microfibres entered waterbodies (freshwater and marine combined), while 1.9 Mt entered soils as biosolid applications; 0.6 Mt entered landfills and 0.3 Mt were incinerated. They estimated that polyester fibres were the dominant fibres lost to the environment. Gavigan et al. (2020) concluded that with the increase in fast fashion, synthetic fibre production, and improvements to sanitation (e.g. access to wastewater treatment plants, washing machine ownership), the loss of microfibres to the wider environment is likely to increase in the future. Therefore, there is a need to consider mitigation options upstream of fibre losses to the environment.

Soil environments are regarded as an important sink for microfibres because while wastewater treatment plants are 98% efficient at removing fibres from waste effluent, these fibres are retained in sewage sludge (biosolids) which are then applied as a soil amendment to agricultural land. Biosolids are an important source of nutrients and organic matter and help recycle nutrients and build organic matter. Previous studies have shown the persistence of synthetic fibres in soil environments, where fibres were still detected 15 years after the last known biosolid application (Zubris and Richards, 2005). Therefore, it is considered that synthetic fibres are not biodegradable and remain highly persistent in all environments. However, more recently, there have been studies reporting the presence of naturally derived fibres as microfibre particles (Liu et al., 2023). These have been observed in air and rivers (Stanton et al., 2019), and indoor air. This finding has implications for the persistence of microfibres in the environment as naturally derived fibres (e.g. cellulose-based) are not biodegradable as previously assumed. This is perhaps unsurprising because naturally derived textile fibres bear little resemblance to the original fibre. For example, the manufacturing of cotton fibre results in a significant change to the chemical fibre structure, from cellulose I to cellulose II (Sawada et al., 2022), with the addition of a myriad of chemical additives. While additives improve the look, texture, and durability of clothing, they mean that natural fibres bear little resemblance to their natural state due to extensive fibre processing, dyeing, and finishing (e.g. fire retardants; antimicrobials; water repellents; anti-wrinkle) and mechanical, chemical, and biotechnological treatments. Therefore, it is now argued that natural and synthetic fibres should be considered as anthropogenic fibres (Athey and Erdle, 2022), and there are now emerging data comparing the impacts of both fibre types.

15.2.3 Measuring Environmental Impact

In addition to challenges surrounding the methods used to sample, count, and identify fibres in the environment, there exist similar issues in measuring the environmental impact with a distinct lack of standardisation evidenced across the sector. Consequently, a wide inventory of environmental impact assessment (EIA) metrics, methods and tools are increasingly being devised and utilised. The challenges that this poses for comparability and traceability of EIA are further complicated by complex globalised supply chains that are typified by practices of out-sourcing to second and third-tier factories. Furthermore, the understanding of the extent to which fashion impacts the natural world remains limited due to the absence of baselines and standardised impact assessment methods. Progress to date with EIA has been constrained with little policy to govern and verify action, resulting in the parameters being subjective and diverse measures being applied haphazardly across the breadth of the sector. Existing metrics generally aim to locate, identify, analyse, and address environmental impacts resulting from a linear, extractive, and resource-dependent production system and a waste-generating, disposal-oriented consumption model (Moldan et al., 2012). This includes greenhouse gas emissions; water consumption (Bailey et al., 2022); use of fossil fuel-derived materials; chemical pollution from manufacturing processes and natural fibre production; the release of microfibres into natural environments; growing levels of pre- and post-consumer waste (Ellen

MacArthur Foundation, 2017). These approaches are premised on the need to replace a linear take-make-dispose approach, with a regenerative circular economy (CE) to challenge unlimited growth.

There are four main evaluative methods used to measure and manage environmental sustainability: lifecycle assessment accounts for the total lifecycle of a product from the original extraction of raw materials to waste disposal representing the end of garment life (Bjørn et al., 2016) and requires intensive data sampling at the midpoint of the production process and at the endpoint of resources, eco-systems, and human health; environmental footprint modelling relies on the quantification of an output or the use of a natural resource, i.e. carbon emissions or the impact of chemical pollutants; eco-efficiency modelling offsets economic performance with environmental effect with an aim of creating more value with less impact, thereby acknowledging and rewarding impact reduction; and the Higg index is a compendium of metrics devised by the Sustainable Apparel Coalition (SAC) specifically to enable the textile and garment industries to numerically calculate ratings about their production processes and products. It also enables consumers to access data regarding the environmental impact and the sustainability level of outputs from fashion suppliers and brands. Overall, these existing metrics are largely acknowledged by researchers to display shortcomings. For example, the Higg index has been criticised for focusing only on a specific section of the product value chain, rather than adopting a whole-systems approach. Further critique of the Higg Index included companies self-reporting impact data and claims of greenwashing through the communication of inaccurate data cumulating in the SAC temporarily suspending their product labelling tool. Furthermore, microfibres as pollutants in the environment are not considered, nor are they formally measured. Many of these challenges are common to assessment metrics being utilised across the sector, causing brands to be reluctant to place their trust in third-party organisations and have begun favouring the development of their own EIA measures. This has expanded the suite of tools available and has further hindered comparability and reliability between measures and brands.

15.2.4 Industry Action

As concerns over microfibre pollution have increased in the scientific and legislative arena, there has also been a rise in discourse and action from businesses and organisations in or related to the fashion and textiles industries. For instance, several NGOs and initiatives have been developed with the intention of better understanding the impacts of microfibre pollution. A notable example is The Microfibre Consortium (TMC), which is working to reduce the impact of fibre fragmentation by connecting practices in the industry with academic research. At the time of writing, they have developed a test method with plans of developing a global rating system that would be implemented by signatories by 2030. The TMC's goal of developing baselines and measurement tools that will inform textile and garment production results from a view that mitigation against fibre shedding upstream (i.e. during R&D of fibres and textiles) is a more effective means of addressing the issue (TMC, 2023). Furthermore, Forum for the Future (FFTF), an international sustainability NGO working in partnership with a range of stakeholders including businesses, released a report called Tackling Microfibres at Source (2023). Conducted in partnership with the textile manufacturers Ramatex Group, Nanyang Environment and Water Resources Institute (NEWRI) and Universiti Teknologi Malaysia (UTM), the project brought to light some key insights particularly from the perspective of suppliers whose views and very role in addressing sustainability concerns are often obfuscated by the larger focus placed on brands. In addition to more research on drivers of fibre fragmentation during production as well as impacts on human health, collaboration between brands and their suppliers was noted as a key to driving real change. Moreover, the study also highlighted the supplier perspective that implementing change is often a question of financial viability. Ultimately, there needs to be a willingness from both brands and consumers to share the economic cost of sustainable innovations with suppliers. Finally, the report highlighted that rather than focusing only on shedding which occurs during consumer use (including wear and washing), reducing fragmentation during production would have a great impact as it would translate to lowered shed during use. This view is echoed in a 2022 report by First Sentier MUFG Sustainable Investment Institute, which points out that it is estimated that "10–15% of a textile mass is lost in production in the form of microfibres" (Harvey et al., 2022: 4).

While this approach targeted at reducing fragmentation earlier in the product lifecycle has led to some textile innovations like Polartec's "Shed Less" technology, which purports to cut shed by 85% during washing (Polartec, 2023), it is interesting that many of the products being developed are focused primarily on mitigating impacts after garments and textiles are already in use by reducing shed or capturing fibres during washing. These include, for example, a detergent launched by fast fashion giant Inditex and the chemical company BASF claiming to reduce shedding by 80% (BASF, 2022), or the growing number of washing machine filters by Gulp, PlantCare, or Grundig to name a few. Furthermore, performance-wear brand Patagonia and electronics giant Samsung have joined forces towards the "goal of cleaner oceans" (Samsung, 2022) and developed a washing machine purporting to reduce microplastics by 54% (Fashion United, 2023). While cross-sector collaborations like this are a welcome step in the right direction, it is apparent from this and the other examples above that there is currently a marked focus on aquatic environments as well as microplastics/synthetic fibres in discourse and action related to the issue. Moreover, the prominence of such "solutions" has the effect of implicitly framing the issue as one that is primarily the consumer's responsibility to solve.

15.2.5 Future Directions

It is apparent that there are still large knowledge gaps that would increase the efficacy of efforts towards the mitigation of pollutants like microfibres. For instance, as mentioned earlier in Section 15.2.3, there is still a need for a more robust understanding of the sources as well as factors involved in the fragmentation and shedding of all fibre types if there is to be any real progress towards combating the issue at the source. Furthermore, the impacts of microfibre pollution on all environments, including the soil and the atmosphere as well as human health, need to be better understood. This is particularly important because when microfibres and other contaminants are filtered from wastewater, they end up in biosolids/sewage sludge, which is then applied to agricultural land as fertilisers. In the UK, for example, 96% of sewage sludge is used in farming according to the Environment Agency (2020). The need for more understanding of microfibres cannot be overstressed as it is critical that all sustainability-focused policies are both informed by evidence and responsive to the globally distributed nature of production and consumption networks.

While there is no single solution to microfibre pollution and certainly a good deal of value in the washing-focused approach, it is critical that this (or any other approach) is not treated as a panacea for more complex environmental issues. Moreover, it is important to ensure that these solutions, with their implicit focus on consumers and aquatic impacts, do not detract from the development of solutions that tackle other points of the value chain (especially at the R&D and design phase) or reduce this very multifaceted problem to one that can be solved via domestic filtration systems alone (which also pose an issue of how filtered waste is to be safely disposed of, considering that most household filter waste ends up in the bin or flushed down the toilet), thus detracting from the investigation of solutions that would address other aspects of the problem like atmospheric shedding.

15.3 METHODOLOGY

This research aimed to gain insights into the awareness of microfibre fragmentation from multiple stakeholders across the fashion and textile value chain. To achieve the breadth of perspectives required, participants from across the sector were approached to share their views and to report on any action being adopted as a result. The identification of barriers and challenges regarding the engagement with microfibre pollution remained the focus, aiming to build a holistic understanding of the awareness and consequential action across the industry. These insights were generated to form the basis of a roadmap towards future-directed action, leading to recommendations and future research opportunities being scoped.

Reflective of the value chain, nine stakeholders engaged in a series of semi-structured, informal interviews, all of which were carried out online and recorded through audio transcript software. This method was selected as the most appropriate due to the nature of the information required, with opinions, knowledge, and expertise being shared. Alternative qualitative methods, such as focus groups, were considered; however, due to commercially sensitive material potentially being shared, individual interviews were favoured to encourage the sharing of rich data. Ethical approval was granted, with all participants providing their informed consent to their data being utilised to inform the research. Participants included: materials manager (outdoor apparel); sustainability manager (mid-market sportswear brand); corporate social responsibility manager (global fashion brand); product developer (high-end outerwear brand); dyes/chemical consultant; waste management compliance specialist (UK government); soils specialist (UK government); wastewater specialist (regional water service company); chief technical development and innovation officer (textile recycler). It was anticipated that these participants would provide a fair and equitable overview of the industry, considering all stages of the product value chain. Where engagement with certain stakeholders wasn't achieved, as with textile manufacturers for example, the methodology was designed to compensate for this to ensure this knowledge was encompassed within the study. In this case, engagement with a materials manager was initiated, who was best placed to be able to discuss their working relationship with the textile sourcing and development sector. All interviews lasted between 45 and 60 minutes and followed a similar questioning sequence for analysis and comparability purposes, with the semi-structured nature of the interview allowing for follow-up questions to be posed. Furthermore, it was agreed on behalf of the research team to share a summary of findings upon completion of the project to facilitate findings to inform future practice.

In preparation of the interviews, language and terminology were carefully considered to gain the most benefit from the primary data collection and to ensure a common understanding between the interviewer and interviewee. The researchers acknowledged a potential bias in their questioning regarding microfibres and aimed to overcome this through their subtle questioning regarding sustainability agendas and measurement of environmental impact prior to direct questioning regarding microfibres. This was to provide the participant with the opportunity to share this knowledge without prompt, indicating the importance of microfibre pollution within their role. Additionally, the cross-disciplinary nature of the research team (forensic science, environmental science, and design) was matched with the most appropriate to the interviewees to again enable the freedom of conversation and nuances in answers to be addressed through additional, unplanned lines of inquiry.

Audio recordings were transcribed utilising transcription software and then checked for accuracy by the research team. This facilitated the qualitative data to be scrutinised utilising content analysis methods, where insights were highlighted in relation to their relevance to the aims of the study. Thematic analysis was then conducted, firstly with each individual interview data set and then across data sets through cross-analysis, allowing for similarities, differences, and anomalies in the data to be identified. Codes across the data were then established, leading to the development of prominent findings and conclusions. These analysis methods were selected due to the qualitative nature of the data collected, ensuring a rigorous and in-depth exploration of the data collected.

15.4 FINDINGS AND DISCUSSION

The goal of this study was to begin to build an overview of the breadth of agendas and action across multiple actors and sectors in the fashion lifecycle in relation to microfibre pollution. As a reflection of the thematic analysis methods utilised and of the participants in relation to the prominent sectors of the value chain, results were segmented by fashion and textile industries; waste management; and environmental sector. While this study did not engage directly with consumers as a stakeholder group, all participants discussed the prominence of their role in enacting change both across the broader sustainability agenda and with microfibre pollution. For this reason, consumers were included in the key findings and also featured in the conclusions of this study. The themes discovered were comparable across the three sectors, with a lack of awareness, baselines, and legislation reoccurring frequently during the interview series.

15.4.1 Key Findings: Fashion & Textiles Industries

Several key insights emerged from discussions with stakeholders from the fashion and textile industries. Firstly, mitigating microfibre pollution is not a high-priority area of action for many in these sectors. Indeed, some participants pointed to other initiatives like revamping care labels and establishing repair services as current areas of focus. This was driven by several interrelated factors that feed into and off each other, which include but are not limited to: lack of awareness; the absence of baselines, insufficient legislation, and the scale of a business.

15.4.1.1 Lack of Awareness

While all our participants were aware in a general sense of microfibres and some of the discourse about them, most acknowledged critical gaps in their own knowledge citing, for instance, the need for more data on factors involved in fibre fragmentation as well as a better grasp on the social and environmental impacts of microfibre pollution.

In addition to the participants' own gaps in understanding, it was also apparent from our discussions that there is still a fundamental lack of understanding in the industry about the issue. This was evidenced in part by the intertwining of terminology (i.e. microplastics and microfibres used to mean the same thing). Moreover, participants also alluded to the prevalent view that natural fibres are safer/better than synthetic fibres due to their biodegradability. They noted that many consumers as well as industry professionals (such as designers, buyers, and marketers) believe that replacing polyester in favour of natural fibres is a conclusive solution to the problem. Such misconceptions are driven by the dominant narrative centred on microplastics and their impacts on aquatic environments. Furthermore, considering this persistent narrative as well as acknowledged gaps in understanding, it was unsurprising that many stakeholders seemed to lack an awareness of the potential of microfibre pollution in soil environments via sewage sludge or atmospheric shedding. Finally, while participants admitted to the issue currently being low on brand sustainability agendas, they also expressed openness and intention to engage as more information becomes available.

15.4.1.2 The Absence of Baselines

Another theme that emerged from our discussion with fashion and textile industry stakeholders was the lack of clear and reliable baselines. While some expressed an awareness of some of the emerging research about fibre fragmentation, there was still a lack of clarity about what these data translated to in terms of practical action. Indeed, all of our participants indicated that it was difficult for them to act in the absence of any quantifiable/ evidence-based benchmarks, especially in the light of the high cost associated with many paths of action (such as textile innovation or revamping an entire range). There was also concern from some about their inability to verify or contextualise claims being made by some fabric suppliers with regard to low-shed materials. Furthermore, the lack of standard test methods to assess fibre shedding (which contributes to a lack of baselines) was also referenced. Although there are a few test methods, one participant pointed out that it is not enough that some companies are doing tests and there needs to be a consensus on the most appropriate method/set of methods for the industry to develop reliable standards. Lastly, participants also expressed the view that the lack of clear and reliable data contributed to scepticism among some in the industry about the reality of the problem.

15.4.1.3 Insufficient Legislation

In addition to a lack of incentive for brands to act, the lack of baselines and testing standards has contributed in part to the lack of unified/clear legislation at the regional and global levels. While voluntary commitments certainly have a level of utility, legislated policy is a powerful tool because non-compliance generally results in huge financial losses. This view was echoed by participants who all pointed to legislation as a major driver for action, noting for instance the high priority given to existing legislation such as the ban on perfluorochemicals and polyfluoroalkyl substances, which are commonly used in coatings for protective/outerwear. As brands have finite resources to respond to a growing number of sustainability and social responsibility concerns, it is unsurprising that priority is given to areas that have a clear legislative push behind them. One participant emphasised this as being the root cause of the problem and the rationale for microfibre pollution not yet being a sustainability priority for their business. In addition to incentivising action, legislation also provides clear guidelines. Moreover, for such guidelines to be effective, it was pointed out by stakeholders that policy needs to be drafted with an understanding and consideration of how the industry works (which some believed was currently lacking).

15.4.1.4 Scale of Business

Lastly, it emerged from our discussions that the size of a brand can make it challenging to engage in innovation at earlier stages of the value chain such as R&D/design due to the high financial and temporal cost associated with developing textiles. Moreover, the high cost of certifications of performance standards (i.e. waterproofing) was also referenced as a possible reason for inaction. This point was especially salient for brands in the performance category as their products tend to require more certifications. Moreover, the lack of access to/the cost associated with conducting testing of shedding was referenced as potentially prohibitive for smaller brands. Our discussions revealed that such barriers lead some to focus on interventions that are easier to implement such as providing comprehensive care instructions for laundering and lengthening product or repair and recycling initiatives. While the size of a brand should not preclude it entirely from any responsibility, it is important (especially for policymakers) to be aware of the differing levels of capacity that various brands must engage in certain types of action.

15.4.2 Key Findings: Waste Management & Environmental Sectors

Interestingly, discussions with stakeholders from the waste management and environmental sectors revealed themes similar to those detailed above. For instance, though plastic pollution was mentioned as an area of growing concern, participants from this stakeholder group also viewed microfibres as a low-priority area in their respective sectors. Identified interrelated factors include but are not limited to: lack of awareness, absence of baselines, and insufficient collaboration as drivers for the current state of things.

15.4.2.1 Lack of Awareness

Discussions with participants revealed that though there was some recognition of microfibres as an emerging pollutant in both the waste management and environmental sectors, there is still a fundamental lack of understanding about the issue. In addition to a general need for more data and evidence, one reason cited by many participants (particularly in waste management) for this lack of awareness is the current focus in the sector on plastics and microplastics. Indeed, microfibre pollution was viewed by some as a subcategory of microplastic pollution. Fuelled in part by public outcry over visible pollution spilling over from the waste stream as well as dominant narratives centred on microplastics in aquatic environments, microfibre pollution remains low on the agenda. Moreover, interviews indicated that there was a lack of nuanced information regarding both the sources and impacts of different fibre types in the waste stream. The view that natural fibres are safer/better due to biodegradability was also noted.

While some stakeholders were aware of the potential for microfibre pollution in soil environments via sewage sludge or other solid waste, there was still a lot of uncertainty about the impacts of fibres on soil health and function or about how/if fibres could be removed from the sludge. As nearly all sewage sludge in the UK is used as agricultural fertiliser, it is critical that there is an understanding not only of the environmental impacts but also the harm fibres may pose to human health. Finally, all stakeholders expressed an interest in increasing their understanding of the issue; however, it was also apparent that microfibres were currently overshadowed by several other pollutants which either have more data about their impact or are addressed in a regulatory framework. Therefore, significant action in the near future is not very likely.

15.4.2.2 Absence of Baselines

The lack of awareness outlined above is undoubtedly related to the lack of baselines. As there is still a dearth of understanding about the sources and impacts of microfibre pollution that enters the environment via waste streams, there are no clear standards/benchmarks to aim towards. This, coupled with the high economic and time involved in change, makes it difficult for organisations to act. In addition to an evidentiary base, participants also noted that there was a need for standard testing methods.

All of these issues contribute to a lack of unified/clear legislation and standards at the regional and global levels. Similar to these findings from the fashion and textile industries, compliance with legislation was referenced as a key driver of action. However, it was also noted that legislation often takes a long time to be developed and implemented. As such, there was concern expressed that policy could lag behind developments in the fashion and textile industries, potentially resulting in ineffective legislation and wasted resources.

15.4.2.3 Insufficient Collaboration

In addition to baselines and legislation, stakeholders also pointed to the need for cross-sector collaboration for action to be most effective. Pointing to successful strides made through collaborations with the packaging industry for instance, one participant felt that the waste management sector's knowledge and understanding of the end-of-life fate of textiles could be used to inform the design of material and garment innovation earlier in the fashion value chain. Furthermore, knowledge from the fashion and textiles industries could be used to inform the design of better waste management systems. While a lack of collaboration was noted, these suggestions provided by the stakeholders interviewed highlight the interest and willingness to engage with other sectors to find solutions. Another stakeholder suggested that instead of the typical approach focused on enforcement, policymakers could shift to a collaborative way of working with stakeholders.

15.4.3 Key Findings: Consumers from the Perspective of the Stakeholders

Although this phase of investigations did not include the perspectives of consumers, they were referenced by all stakeholders interviewed. Therefore, the following section highlights some of the key perceptions participants shared about consumers in relation to microfibre pollution. Many of the stakeholders believed that there was some level of awareness about fibre shedding among some consumers. However, they asserted that this awareness was often undergirded by a lack of understanding of the issue as well as a poor understanding of how clothing or the fashion and textile industries work. While many brands acknowledged a rise in engagement from consumers on sustainability issues, they felt that consumers were often misinformed, leading to unrealistic expectations of what could be done. For instance, some felt that consumers demand change without taking into account the differences in capacity in relation to the scale of brands.

Moreover, participants felt that many consumers are currently focused on the dangers of synthetic fibres like polyester while viewing natural fibres as a harmless alternative due to their biodegradability. This may be another example of the consequence of the dominant narrative centred on aquatic environments and microplastics. Furthermore, many stakeholders expressed the view that despite the seeming increase in awareness, price is still an important driver of consumer behaviour. Therefore, there is a general lack of willingness on the part of consumers to bear the economic cost of sustainable changes. Interestingly, consumers are viewed as the main source of microfibre pollution by those in the waste management sector. This could explain the focus in that sector on interventions related to the use and maintenance stage of the value chain like washing machine filters and detergents and changing consumer behaviour around cloth washing routines.

Finally, consumers were also framed as an important driver for change (especially by those in the waste management sector) as they have the ability to move the needle by holding organisations accountable. For instance, public outcry about visible plastic contaminants that had leaked out of the waste stream was cited as contributing to the growing pressure to tackle plastic pollution. Though these insights are not directly from consumers, they do offer some useful insights into the steps that may need to be taken in order to better equip consumers for the role they have to play in mitigating microfibre pollution.

15.5 CONCLUSIONS

This study was conducted to investigate the issue of microfibre pollution from a multi-stakeholder perspective, using the insights gathered to develop a clear directive for future action. Therefore, the findings expounded upon above have been examined not only to establish an understanding of the contextual factors currently shaping engagement, but also as a basis to inform future efforts towards the reduction and mitigation of microfibre loss in the fashion and textile industries. This concluding discussion does not offer simple or singular solutions to the issue; however, it does present useful suggestions for areas that need to be addressed and further investigated by a range of actors, including legislators, industry stakeholders, and researchers. Looking across the data set and at the review of the wider context, the areas that require focus are awareness, governance, and stakeholder collaboration.

15.5.1 AWARENESS

Though the issue of knowledge levels and awareness has received some level of attention, it was clear from both the interviews and review of literature that there is still a lack of awareness about microfibre pollution throughout the value chain. In addition to the explicit acknowledgment of gaps in understanding, the lack of awareness was also evidenced implicitly in the prevalence of confusing or biased narratives at all levels. For instance, the framing (in policy literature and interview data) of synthetic fibres as the dominant polluting fibre type while perceiving, and even marketing, naturally derived fibres as harmless, points to a need for greater awareness and careful consideration of messaging. Moreover, the intertwining of terminology between microfibres and microplastics demonstrates a fundamental lack of understanding and indicates a need for better communication strategies. The efficacy of action for all stakeholders is predicated in large part on the accuracy of the information they draw on when making decisions. The better informed stakeholders across the value chain are, the better equipped actors are in making better decisions and directing future action.

15.5.2 GOVERNANCE

An important finding of the study is the significance of legislation as a driver for stakeholders both upstream and downstream. While it is encouraging to see a growing acknowledgment of microfibre pollution by several legislative authorities, public bodies, and NGOs, as pointed out in out in the discussion above, there is still a need for further investigation of the social and environmental impact of the issue. Such understandings would help to ensure the robustness of legislation as well as the overall messaging to stakeholders. It is critical not only for there to be policies in place but also for greater consideration to be given to how such laws are designed.

For instance, policymakers must consider the direct and indirect consequences their initiatives will have (i.e. the possible issues relating to the rPET or the disposal of filter waste). Moreover, policies need to include robust EIA mechanisms.

Additionally, as action in the absence of baselines and thresholds is challenging for businesses and organisations (both upstream and downstream in the value chain), there is a need for legislation and standards that help establish clear and consistent guidelines.

A cost–benefit analysis of textile policy published by WRAP (2022) intensified the need for a range of complementary approaches that consider every stage in the textile and garment lifecycle in tackling sustainability concerns. Therefore, rather than focusing on one part of the value chain like consumer use and washing, there is a need for a range of policy instruments aimed at mitigating microfibre pollution throughout the garment lifecycle, including the design phase. Though developing such a multimodal yet unified set of standards and policies will certainly pose its own challenges (for instance, how can such policy be implemented globally), there is an opportunity for policymakers to lead the way by facilitating the creation of transformational legislation through cross-sector collaboration and knowledge sharing.

15.5.3 Stakeholder Collaboration

Like any sector where there is a competitive advantage in protecting unique intellectual property, knowledge sharing is still a challenge in the fashion and textile industries. However, tackling microfibre pollution (and arguably, all sustainably concerns) will require a great deal of collaboration within and beyond the industry. As such, there is an opportunity to develop tools, systems, and forums that best facilitate and support boundary-crossing collaborations aimed at maximising the impact and efficacy of action. For instance, there is a need to establish shared spaces of understanding through which multi-disciplinary knowledge from stakeholders involved in the product's beginning and end of life can be shared and co-created. Furthermore, as the earlier stages of the values chain have been identified as key points in reducing fibre fragmentation, it is essential for connections to be made between the design practice, and the evidence and data being gathered by scientists.

REFERENCES

Andrady, A. L. (2011). Microplastics in the marine environment. *Marine Pollution Bulletin*, *62*(8), 1596–1605.

Anscombe, C. (2019, March 22). Are natural fibres really better for the environment than microplastic fibres? *Phys Org*. https://phys.org/news/2019-03-natural-fibres-environment-microplastic.html

Athey, S. N., & Erdle, L. M. (2022). Are we underestimating anthropogenic microfiber pollution? A critical review of occurrence, methods, and reporting. *Environmental Toxicology and Chemistry*, *41*(4), 822–837.

Bailey, K., Basu, A., & Sharma, S. (2022). The environmental impacts of fast fashion on water quality: a systematic review. *Water*, *14*(7), 1073.

BASF. (2022, November 22). Inditex and BASF develop the first detergent designed to reduce microfiber release from textiles during washing [Press release]. https://miro.com/app/board/uXjVPpGTcqU=/?moveToWidget=3458764549059765377&cot=14

Bjørn, A., Margni, M., Roy, P. O., Bulle, C., & Hauschild, M. Z. (2016). A proposal to measure absolute environmental sustainability in life cycle assessment. *Ecological Indicators*, *63*, 1–13.

Bland, E., & Sheridan, K. (2023). Technical research report: Recycled polyester within the context of fibre fragmentation. In *The Microfibre Consortium*. Retrieved August 21, 2023, from https://www.microfibreconsortium.com/resources-1

Britten, F. (2022, June 28). Fashion brands pause use of sustainability index tool over greenwashing claims. *The Guardian*. https://www.theguardian.com/fashion/2022/jun/28/fashion-brands-pause-use-of-sustainability-index-tool-over-greenwashing-claims

Bryce, E. (2021, November 6). Are clothes made from recycled materials really more sustainable? The Guardian. Retrieved August 7, 2023, from https://www.theguardian.com/environment/2021/nov/06/clothes-made-from-recycled-materials-sustainable-plastic-climate

Carr, S. A. (2017). Sources and dispersive modes of micro-fibers in the environment. *Integrated Environmental Assessment and Management*, *13*(3), 466–469.

Debraj, D., & Lavanya, M. (2023). Microplastics everywhere: A review on existing methods of extraction. *Science of the Total Environment*, 164878.

Directorate-General for Environment. (2022). EU strategy for sustainable and circular textiles. In *European Commission*. European Commission. Retrieved August 4, 2023, from https://environment.ec.europa.eu/publications/textiles-strategy_en

Donald, R. (2022, June 3). Industry-linked sustainability standard allows clothing giants to ramp up emissions. *The Intercept*. https://theintercept.com/2022/06/03/sustainable-fashion-greenwashing-higg/

Dris, R., Gasperi, J., Mirande, C., Mandin, C., Guerrouache, M., Langlois, V., & Tassin, B. (2017). A first overview of textile fibers, including microplastics, in indoor and outdoor environments. *Environmental Pollution*, *221*, 453–458.

Dris, R., Gasperi, J., Saad, M., Mirande, C., & Tassin, B. (2016). Synthetic fibers in atmospheric fallout: a source of microplastics in the environment? *Marine Pollution Bulletin*, *104*(1–2), 290–293.

Environment Agency. (2020). Environment Agency strategy for safe and sustainable sludge use. In *Environment Agency*. https://www.gov.uk/government publications/environment-agency-strategy-for-safe-and-sustainable-sludge-use/environment-agency-strategy-for-safe-and-sustainable-sludge-use#why-change-is-needed

Fashion United. (2023, January). Patagonia and Samsung developed a washing machine that reduces microplastics. *Fashion United*. Retrieved August 5, 2023, from https://fashionunited.in/news/fashion/patagonia-and-samsung-developed-a-washing-machine-that-reduces-microplastics/2023011337991

Forster, N. A., Wilson, S. C., & Tighe, M. K. (2023). Microplastic pollution on hiking and running trails in Australian protected environments. *Science of the Total Environment*, *874*, 162473.

Forum for the Future. (2023). Tackling microfibres at source. In *Forum for the Future*. Retrieved August 4, 2023, from https://www.forumforthefuture.org/tackling-microfibres-at-source

Galloway, T. S., Cole, M., & Lewis, C. (2017). Interactions of microplastic debris throughout the marine ecosystem. *Nature Ecology & Evolution*, *1*(5), 0116.

Gavigan, J., Kefela, T., Macadam-Somer, I., Suh, S., & Geyer, R. (2020). Synthetic microfiber emissions to land rival those to waterbodies and are growing. *PLoS One*, *15*(9), e0237839.

Gonçalves, A., & Silva, C. (2021). Looking for sustainability scoring in apparel: A review on environmental footprint, social impacts and transparency. *Energies*, *14*, 3032. https://doi.org/10.3390/en14113032

Harvey, E., Sullivan, R., & Amos, N. (2022). *Microfibres: The Invisible Pollution from Textiles*. First Sentier MUFG Sustainable Investment Institute.

Horton, A. A., Walton, A., Spurgeon, D. J., Lahive, E., & Svendsen, C. (2017). Microplastics in freshwater and terrestrial environments: Evaluating the current understanding to identify the knowledge gaps and future research priorities. *Science of the Total Environment*, *586*, 127–141.

Kaur, M., Ghosh, D., Guleria, S., Arya, S. K., Puri, S., & Khatri, M. (2023). Microplastics/nanoplastics released from facemasks as contaminants of emerging concern. *Marine Pollution Bulletin*, *191*, 114954.

KeChi-Okafor, C., Khan, F. R., Al-Naimi, U., Béguerie, V., Bowen, L., Gallidabino, M. D., … & Sheridan, K. J. (2023). Prevalence and characterisation of microfibres along the Kenyan and Tanzanian coast. *Frontiers in Ecology and Evolution*, *11*, 1020919.

Liu, J., Zhu, B., An, L., Ding, J., & Xu, Y. (2023). Atmospheric microfibers dominated by natural and regenerated cellulosic fibers: Explanations from the textile engineering perspective. *Environmental Pollution*, *317*, 120771.

MacArthur, E. (2017). *Foundation a New Textiles Economy: Redesigning Fashion's Future*. London, UK.

Moldan, B., Janoušková, S., & Hák, T. (2012). How to understand and measure environmental sustainability: Indicators and targets. *Ecological Indicators*, *17*, 4–13.

Northam, L. (2023, April 9). Pulling a fast one – How sustainability programmes can enable throwaway fashion – Real Leather. *Stay Different. Real Leather. Stay Different.* https://choosereallleather.com/education/pulling-a-fast-one-how-sustainability-programmes-can-enable-throwaway-fashion/

Özkan, İ., & Gündoğdu, S. (2021). Investigation on the microfiber release under controlled washings from the knitted fabrics produced by recycled and virgin polyester yarns. *Journal of the Textile Institute*, *112*(2), 264–272. https://doi.org/10.1080/00405000.2020.1741760

Pointing, C. (2023, January 19). Recycled polyester doesn't fix fast fashion's over-production problems. *Good on You*. Retrieved August 7, 2023, from https://goodonyou.eco/recycled-polyester-fast-fashion/

Polartec. (2023, February). Polartec introduces shed less technology that dramatically reduces fiber fragmentation in laundering tests [Press release]. https://www.polartec.com/news/polartec-power-shield-press-release

Rebelein, A., Int-Veen, I., Kammann, U., & Scharsack, J. P. (2021). Microplastic fibers-underestimated threat to aquatic organisms? *Science of the Total Environment*, *777*, 146045.

Recover. (2023, January 19). France's new eco-labeling law. *Recover*. Retrieved August 4, 2023, from https://recoverfiber.com/newsroom/france-new-eco-labeling-law

Samsung. (2022, January 25). Samsung collaborates with Patagonia to keep microplastics out of our oceans [Press release]. https://news.samsung.com/global/samsung-collaborates-with-patagonia-to-keep-microplastics-out-of-our-oceans

Sawada, D., Nishiyama, Y., Shah, R., Forsyth, V. T., Mossou, E., O'Neill, H. M., … & Langan, P. (2022). Untangling the threads of cellulose mercerization. *Nature Communications*, *13*(1), 6189.

Shah, G., Bhatt, U., & Soni, V. (2023). Cigarette: An unsung anthropogenic evil in the environment. *Environmental Science and Pollution Research*, 1–12.

Stanton, T., Johnson, M., Nathanail, P., MacNaughtan, W., & Gomes, R. L. (2019). Freshwater and airborne textile fibre populations are dominated by 'natural', not microplastic, fibres. *Science of the Total Environment*, *666*, 377–389.

TMC. (2023). The microfibre roadmap. Retrieved August 7, 2023, from https://www.microfibreconsortium.com/roadmap

WRAP. (2021). Textiles 2030 circularity pathway. In *WRAP*. Retrieved August 4, 2023, from https://www.wrap.org.uk/textiles2030

WRAP. (2022). *Textiles policy CBA*. Retrieved August 5, 2023, from https://wrap.org.uk/resources/report/textiles-policy-options-and-cost-benefit-analysis

Zubris, K. A. V., & Richards, B. K. (2005). Synthetic fibers as an indicator of land application of sludge. *Environmental Pollution*, *138*(2), 201–211.

16 Mitigation Strategies for Microfiber Pollution

Erika Iveth Cedillo-González
University of Modena and Reggio Emilia

16.1 INTRODUCTION

It is known that most of the microfibers present in the environment come from the laundering of synthetic textiles. In 1998, Habib et al. first noted that microfibers from synthetic textiles were abundant in sewage sludge, sludge-derived products, and effluents from primary/secondary sewage treatment plants (Habib et al., 1998). They related the presence of such microfibers in the municipal sewage sludge to the washing of synthetic clothes in washing machines, explaining that during the wash and rinse cycles, broken, worn, and abraded microfibers detach from textiles by mechanical action and become suspended in water. Because of their poor biodegradability, microfibers cannot be decomposed by aerobic or anaerobic bacteria in sewage treatment plants and are eventually concentrated in sewage sludge or discharged in effluents. Microfiber contamination in sludge was so constant that, using microfibers as an indicator, the authors proposed a simple and rapid test procedure to determine whether commercial soil conditioners and fertilizers contained municipal sewage sludge or its derived products (biosolids). In 2015, Zubris and Richards confirmed the long-term presence of synthetic microfibers in sludge (from washing machines) using the same method (Zubris & Richards, 2005). In their study of soils and field sites, they used microfibers as a semi-quantitative indicator of previous sludge applications and found that microfibers were still detectable in sludge products and soil columns even 5 years after application. In the case of field site soils, microfibers were also detected 15 years after application. Interestingly, in both cases, it was found that microfibers held their original characteristics observed in the applied sludge.

However, the first study to demonstrate that synthetic clothing washing could promote marine microfiber pollution was conducted by Browne et al. in 2011. In this study that carried out the analysis of shorelines at 18 global sites in Oceania, Asia, North America, South America, Africa, and Europe from the poles to the equator, the authors found that an important origin of microplastics seems to be sewage containing microfibers from laundering. The forensic analysis of microplastics from sediments demonstrated that the fractions of polyester (PES) and acrylic microfibers utilized in clothing contributed to microfibers present in habitats where sewage and sewage effluents are discharged. Furthermore, analysis of wastewater samples from domestic washing machines revealed that a single garment could generate more than 1,900 microfibers per wash cycle. Thus, the findings of their study showed that

DOI: 10.1201/9781003331995-20

the source of many microfibers present in the marine environment may be sewage contamination due to laundering rather than fragmentation or cleaning products.

Since the publication of such findings, many research groups have focused on developing different mitigation strategies to reduce microfiber pollution. Among them, setting laundering parameters (such as washing temperature, type and dosage of detergent, use of softener, type of washing machine, washing load, and sequential washings) and modifications to the composition and structure of synthetic textiles can avoid damage to garments and decrease microfiber release. Other strategies related to laundering practices include the use of devices (Cora Ball, bags, filters, and ceramic membranes) to capture microfibers released from synthetic textiles during domestic and industrial laundering. Besides performing laundering using specific washing parameters, other strategies for mitigating microfiber pollution include developing finishing treatments for synthetic fabrics that avoid fabric damage and implementing education and awareness campaigns. It is worth noting that neither of such strategies can prevent microfiber release by 100%, but conventional wastewater treatment plants (WWTPs) can be used to capture microfibers that escape these mitigation strategies. Further use of recovered microfibers, such as raw materials for the building or energy sector, can increase the mitigation impact of WWTPs or any other strategies in microfiber capture (i.e., domestic and industrial filters). The following sections explain and analyze each one of these strategies.

16.2 SETTING DOMESTIC LAUNDERING PRACTICES FOR REDUCED MICROFIBER RELEASE

Several research groups have focused their efforts on the evaluation of the effect of domestic laundering practices (washing temperature, use of detergent and softener, type of washing machine, washing load, and sequential washings) on the release of microfibers to advise final users on how to set washing programs to mitigate microfiber pollution. For instance, findings made by several research groups suggest that synthetic fabric textiles such as PES, polyamide, and acetate should be washed at low temperatures (≤30°C) because shedding increases at high laundry temperatures (De Falco et al., 2018b; Kelly et al., 2019; Rathinamoorthy & Raja Balasaraswathi, 2022; Yang et al., 2019). This phenomenon has been related to the increase in the surface hydrolysis of synthetic fabrics due to the alkaline environment promoted by the use of detergent (De Falco et al., 2018b). De Falco et al. found that the release of microfibers from PES textiles washed at 65°C was higher than the release observed at 40°C (De Falco et al., 2018b). A similar trend was found by Yang et al., who found that PES, polyamide, and acetate textiles released more microfibers when increasing the washing temperature from 30°C to 60°C (Yang et al., 2019). On the other hand, low washing temperatures (≤30°C) have no significant impact on microfiber release from synthetic garments (Kelly et al., 2019; Yang et al., 2019), suggesting that high temperatures during laundering should be avoided.

Washing natural/synthetic blends or pure synthetic fabrics with specific dosages and types of detergents and softeners can help mitigate microfiber pollution. Generally speaking, detergents negatively impact microfiber release, while softeners decrease the quantity of microfiber detached from synthetic garments. In the case of detergents, Napper and Thompson found that the number of microfibers released from a 65% PES/35% cotton blend fabric can be reduced if it is washed without detergent (Napper & Thompson, 2016). Other research groups later confirmed the same trend. Using controlled laboratory conditions, Hernández et al. reported quantitative data regarding two characteristics of microfibers (size and mass) released from 100% PES textiles during simulated home washing (Hernandez et al., 2017). Their findings indicated that performing laundering with detergent (either liquid or powder) released a larger mass of microfibers into wastewater than laundering with deionized water alone. Carney Almroth et al. obtained a similar result when performing laundering cycles of synthetic fabrics, finding a significant increase in the number of microfibers released for three out of four PES and polyacrylic textiles when a commercial detergent was present during washing (Carney Almroth et al., 2018). De Falco et al. reported that the amount of microfibers released from PES and polypropylene fabrics decreased when these textiles were washed with water alone compared to those washed using liquid or powder detergents (De Falco et al., 2018b). Their findings indicate that powder detergents favor microfiber shedding more than liquid ones. The same behavior was further reported for other synthetic textiles such as PES, polyamide, acetate, and rayon fabrics (Yang et al., 2019; Zambrano et al., 2019), and the negative impact of detergents on the release of microfibers from synthetic textiles has also been indirectly observed in other studies focusing on the development of methods for microfiber quantification instead of microfiber release. For example, Corami et al. found that washing 90% PES/10% polyamide towels with a detergent increases the release of microfibers while investigating an analytical methodology for the simultaneous quantification and identification of microfibers via micro-FTIR (Corami et al., 2020). Such a negative impact on microfiber release is likely related to damage to microfibers. According to Yang et al., during laundering, surfactants and other detergent components are fixed on the microfiber surface, reducing frictional forces and leading to damage that generates and releases microfibers (Yang et al., 2019). However, some contrasting findings reported by other research groups show a different trend on the effect of detergents on microfiber release. For instance, some studies found that using detergents in washing procedures is not a major factor involved in microfiber release from PES textiles (Kelly et al., 2019; Pirc et al., 2016). Additionally, Cesa et al. found that using detergents in the washing practices of acrylic, PES, and polyamide textiles significantly decreased the mass of microfibers emitted from the garments (Cesa et al., 2020). Such contrasting findings make one suppose that the magnitude of the negative impact of detergents on microfiber release can also be related to other factors, such as the kind of washing machine used to perform the tests. According to Kelly et al., steel balls in the laboratory washing machines used in some of the previous studies (Carney Almroth et al., 2018; De Falco et al., 2018b; Hernandez et al., 2017; Zambrano et al., 2019) could interact with the detergent, leading to unrealistic microfiber release. These authors hypothesized that steel balls may aggravate the effects of the detergent, possibly by forcing it into the textile weave or promoting the surfactant's agitation, which leads to increased production of bubbles, thus increasing microfiber release when detergents are present during laundering (Kelly et al., 2019).

Since a detergent's primary function is dirt removal and keeping textiles in hygienic conditions, avoiding surfactants during garment washing procedures becomes a measure that society will hardly adopt to mitigate microfiber pollution. However, some mitigation grade can still be achieved by using softeners when washing synthetic textiles. The main role of softeners is to improve the handling of washed textiles by making the fabrics smooth, flexible, and soft. They act as lubricants, and several investigations have found that fabric softeners should reduce microfiber shedding (De Falco et al., 2018b; Rathinamoorthy & Balasaraswathi, 2022; Rathinamoorthy & Raja Balasaraswathi, 2022; Weis & De Falco, 2022).

For instance, De Falco et al. found that the softening agent has a mitigating effect on the number of microfibers released from synthetic PES fabrics (De Falco et al., 2018b). They estimated that using a softener during the domestic laundering of PES fabrics (5 kg load) could decrease the release of microfibers by more than 35% compared with the amount released under the same laundering conditions but using a liquid detergent. This positive effect on microfiber release reduction was explained by the ability of the softener to decrease the friction between fibers, facilitating microfibrils to lay parallel to the fiber bundle and thus reducing damage and breaking (De Falco et al., 2018b). In agreement with the previous trend, Rathinamoorthy and Balasaraswathi found a significant decrease in microfiber shedding from a 100% PES fabric when using high dosages of softener (between 1 and 5 mL) (Rathinamoorthy & Raja Balasaraswathi, 2022). These authors found that the microfiber count was reduced 2.8 times when 5 mL of softener was used during the laundering process. Even if the evidence shows that a high dosage of softener decreases microfiber shedding, lower softener dosages should be used because a high softener concentration negatively affects fabrics. However, at low softener dosages (<3 g/L), a poor correlation between the microfiber shedding and concentration was observed, and this is in good agreement with findings reported by other research groups, which show that softener does not affect microfiber release (Lant et al., 2020; Pirc et al., 2016).

Even if a softener mainly tends to decrease microfiber release, Rathinamoorthy and Balasaraswathi found that washing conditions are essential to prevent shedding through softeners (Rathinamoorthy & Raja Balasaraswathi, 2022). Independent of the softener brand used by the authors for washing treatments (three different commercially available brands were investigated), shedding was significantly reduced only when softeners were added during washing. At the same time, no significant reduction was noted when using softener after washing. The authors related this behavior to the fabric's zeta potential. In the presence of a detergent, i.e., in an alkaline medium, PES fabrics present a negative zeta potential as a result of the dissociation of acidic groups on the surface. Softeners are substances of cationic nature, and consequently, they adsorb on the fabric surface, restricting external forces from damaging the fibers. Thus, the softener can behave as a protective layer during washing by adsorbing on the fabric's surface, reducing microfiber shedding (Rathinamoorthy & Raja Balasaraswathi, 2022).

Water hardness plays a crucial role in the laundry process by affecting the efficiency of detergents, thus playing a significant role in the quantity of detergents used for laundry. Depending on the specific washing conditions and the type of detergent used, water hardness can make the addition of higher dosages necessary (Cameron,

2011), promoting the release of higher quantities of microfibers as previously discussed. Additionally, the results reported by De Falco et al. suggest that water hardness may also affect microfiber release independent of the detergent type (De Falco et al., 2018b). They found that washing a plain weave PES fabric with hard water (27° d or 481 ppm $CaCO_3$) released more microfibers than washing the same fabric with distilled water. As explained by the authors, the increased water hardness could promote fabric abrasion, similar to the accelerated abrasive damage of cotton fabrics when laundered in hard water (De Falco et al., 2018b). However, this effect likely depends on the hardness level of the water, as Corami et al. found that hard tap water (30°F or 300 ppm $CaCO_3$) did not affect shedding and the release of microfibers in 90% PES and 10% polyamide towels (Corami et al., 2020). Therefore, either by its single effect or by its effect combined with detergent dosage, water hardness should also be considered during the development of microfiber mitigation purposes from a laundering perspective.

Strategies to mitigate microfiber release from fabrics should also consider the type of washing machine and washing loads. For example, Yang et al. found that a higher quantity of microfibers was released from synthetic garments after washing with a pulsator laundry machine than using a platen laundry machine (Yang et al., 2019), but the authors did not offer any explanation for this finding. Additionally, it has been reported that top-load machines release more microfibers than front-load machines (Hartline et al., 2016; Zambrano et al., 2019). Hartline et al. hypothesized that a top-load washing machine with its central agitator might be more abrasive than a front-load machine that works through drum rotation, boosting microfiber release. However, Kelly et al. showed that this difference could be because front-load washing machines use a much lower amount of water than top-load machines (Kelly et al., 2019). In independent studies, Kelly et al. and Volgare et al. reported that a high water-volume-to-fabric ratio is one of the most influential factors for microfiber release from PES textiles (Kelly et al., 2019; Volgare et al., 2021). A high water-volume-to-fabric ratio is characteristic of European "delicate" washing cycles. Delicate cycles may boost microfiber release because of their higher overall hydrodynamic pressure on the textile weave. Individual microfibers are characterized by a large surface area-to-volume ratio and low Reynolds numbers. As water passes through and over the fabric during laundering, each microfiber will experience huge viscous forces. Such forces could extract small fibers from the main textile weave. This correlation between the water-volume-to-fabric ratio and microfiber release simplifies the setting of washing parameters for microfiber mitigation purposes. Instead of changing washing machines (a choice that should consider users' economic possibilities), people can avoid performing washing cycles with low washing loads (i.e., delicate cycles).

Finally, using garments as long as possible and avoiding purchasing new (synthetic) ones can also be considered a strategy for mitigating microfiber pollution. Several research groups have reported a decrease in the number of microfibers released from laundering synthetic fabrics between subsequent washing cycles until a constant level is reached. Such a trend has been observed in fabrics composed of PES (Belzagui et al., 2019; Carney Almroth et al., 2018; Cesa et al., 2020; De Falco et al., 2019b; Kärkkäinen & Sillanpää, 2021; Kelly et al., 2019; Napper & Thompson, 2016; Pirc et al., 2016), acrylic (Cesa et al., 2020; Kärkkäinen & Sillanpää, 2021; Napper & Thompson, 2016), and polyamide (Cesa et al., 2020; Kärkkäinen & Sillanpää, 2021).

The information presented in this section showed that the impact of washing parameters on microfibers calls for setting washing parameters to reduce microfiber shedding. A relatively straightforward strategy that can be implemented to reduce microfiber release from synthetic textiles is using front-loading washing machines or, if not possible, performing washing cycles of synthetic garments using full wash loads. Low temperatures, reduced quantities of liquid detergents, and adding a softener during washing will also decrease the number of microfibers released. Additionally, avoiding delicate cycles can further prevent microfiber release. Such parameter settings should not strongly influence the quality of the washing process, provided that caution is considered. According to the BOSCH washing machine producer, a cold wash temperature won't dramatically affect the cleaning performance of a washing machine used for washing clothes that are only lightly soiled and stained (Bosch UK, 2023).

Additionally, most modern detergents are effective in cold water (ARIEL UK, 2023). However, to not compromise the hygienic characteristics of garments after washing, the dosages recommended by detergent producers must be followed. However, microfiber release can still be minimized if liquid detergents are preferred and used with a softener during washing.

16.3 TECHNOLOGIES ALREADY PRESENT IN THE MARKET

Several devices have been developed to capture microfibers released from textiles during laundering practices. Some are designed to capture microfibers during domestic laundering, while others are appropriate for industrial laundering procedures. In the case of domestic devices, they can be either placed inside the washing machine's drum during laundering (in-drum devices) or externally installed in the drain pipe to filter the effluent wastewater (external devices) (Figure 16.1). The aim of in-drum devices is to capture microfibers released from textiles during laundry, while the aim of domestic external devices (which work similarly to industrial ones) is to capture microfibers present in wastewater.

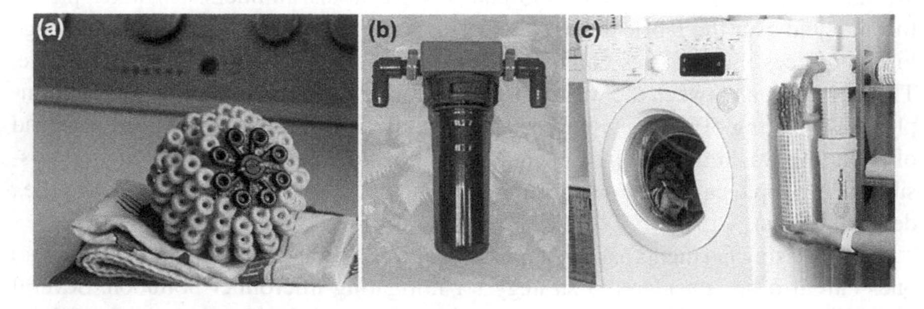

FIGURE 16.1 Examples of three microfiber-capture technologies already present in the market. (a) Cora Ball© in-drum device, (b) Microplastics LUV-R© domestic filter, and (c) PlanetCare© domestic filter. (*Sources*: (a) Photo courtesy of Cora Ball, VT, USA, (b) Photo courtesy of Environmental Enhancements, NS, Canada, and (c) Photo courtesy of PlanetCare Limited, U.K.)

16.3.1 In-Drum Devices

In-drum devices are microfiber traps and include washing bags or washing balls. Both devices have been designed to be user-friendly and without the need for installation.

Washing bags are mesh bags that reduce microfiber release into effluent wastewater from washing machines. Garments are placed inside the bag, and the bag is placed into the washing machine drum. The trapped fibers are removed by hand after washing. Several washing bags are commercially available, and research groups have tested some specialized ones that prevent microfiber pollution. For instance, Napper et al. (2020) investigated the mass of microfibers captured by the GUPPYFRIEND® washing bag (Langbrett, Germany) and the prototype washing bag from Fourth Element (Fourth Element, U.K.) (Napper et al., 2020). According to its website, the GUPPYFRIEND® washing bag protects clothes and reduces fiber shedding. It filters the released microfibers and does not lose any fibers (Global, 2022). On the other side, the Fourth Element washing bag prototype was explicitly designed for Fourth Element clothing, particularly thermal undergarments mainly made from fleece-type fabrics that are expected to release more microfibers than conventional garments (Napper et al., 2020). To evaluate the microfiber retention efficiency of both washing bags, the authors simulated the washing of a typical mixed load using three different synthetic fabrics made of either 100% PES, 100% acrylic, or 60% PES/40% cotton blend. A front-loading washing machine and a 30°C washing cycle of 45-minute synthetic wash at 1,000 RPM without detergent or conditioner were used, which showed that the washing bag decreased the number of released microfibers in wastewater. Laundering the mixed garments without any device released an average of 0.44 ± 0.04 g (mean + S.E) microfibers per wash. The GUPPYFRIEND® washing bag significantly reduced that value by $54\% \pm 14\%$, while the Fourth Element washing bag was least practical, reducing the quantity of microfibers present in wastewater by $21\% \pm 9\%$ only. The efficiency of the GUPPYFRIEND® washing bag was also investigated by Kärkkäinen and Sillanpää (2021). In this case, the authors performed five sequential laundering cycles using two different and new technical sports t-shirts composed of 100% PES. The GUPPYFRIEND® bag was tested for both microfiber-catching efficiency and cleaning. For the cleaning test, the bag was used to wash the textiles stained with blackcurrant juice and cream cheese. Laundering was performed in a front-load washing machine with a 75-minute "Mix" program, which performs washing at 40°C and uses a spin-dry rate of 1,200. A liquid detergent was also used. Each textile went through five sequential washing–drying cycles (Kärkkäinen & Sillanpää, 2021). In this second study, the authors reported that washing the synthetic textiles with no microfiber-capture device releases approximately 2.39×10^6 microfibers/kg (Figure 16.2a), but the GUPPYFRIEND® washing bag reduced microfiber emissions during washing by 39% (the release was reduced to 1.46×10^6 microfibers/kg). This difference between the studies of Kärkkäinen and Sillanpää (2021) and Napper et al. (2020) could be related to the differences between the fabrics or the washing conditions. Additionally, the authors observed that the GUPPYFRIEND® does not interfere with cleaning, as the stains of blackcurrant juice and cream cheese were cleaned away from the textile. Thus, this washing bag can be used to reduce fiber emissions without compromising cleaning efficiency (Kärkkäinen & Sillanpää, 2021).

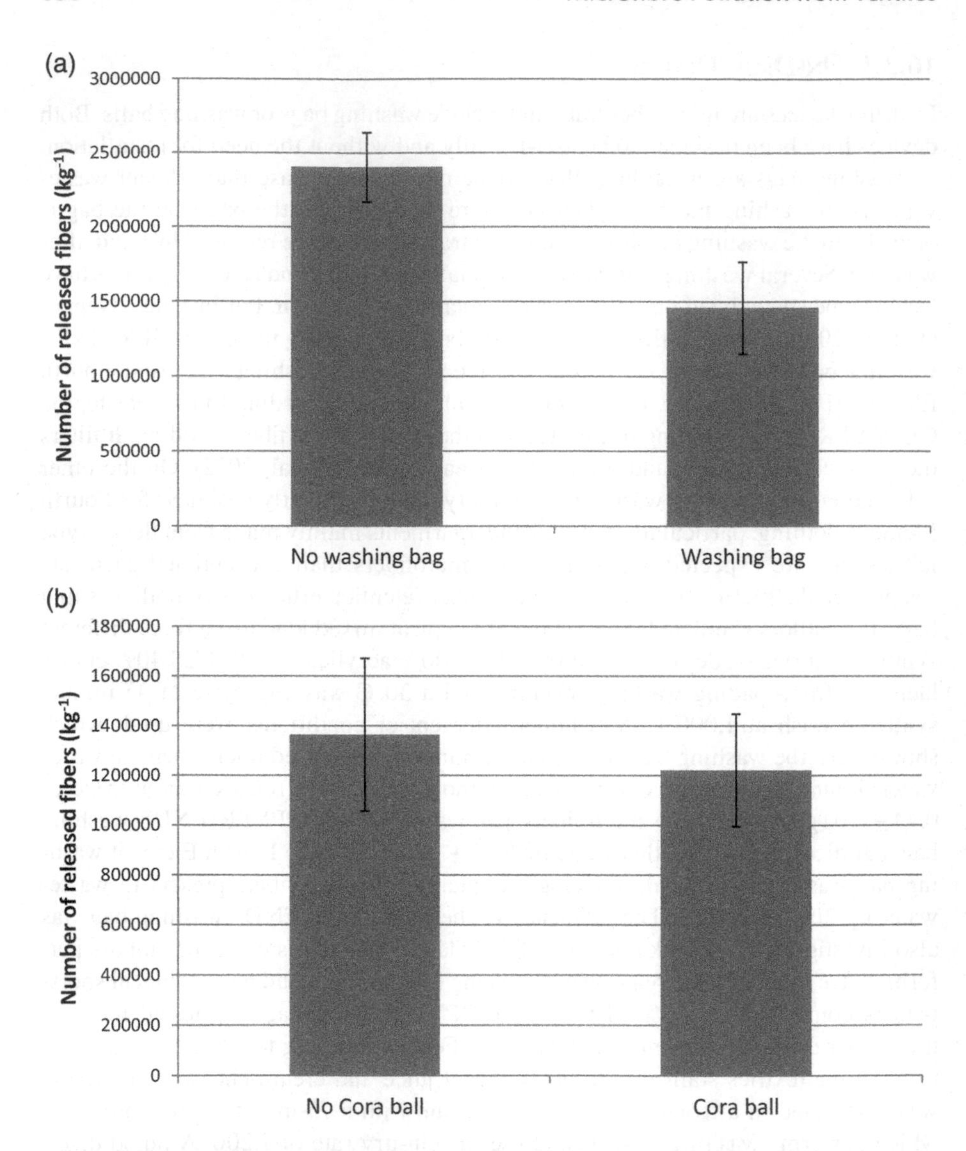

FIGURE 16.2 The number of the released polyester fibers with and without the GUPPYFRIEND® washing bag and the Cora Ball© device. Mean ± SD of three replicate samples. (*Source*: Reprinted from "Quantification of different microplastic fibres discharged from textiles in machine wash and tumble drying" by Kärkkäinen, N. and Sillanpää, M., 2021, Environmental Science and Pollution Research, 28, 16253–16263. CC BY 4.0 (http://creativecommons.org/licenses/by/4.0/).)

Cora Ball© is a microfiber-capture device produced by Cora Ball, VT, USA (*Cora Ball-The Laundry Ball Protecting the Ocean and Your Clothes*, 2022). As stated on the product's website, this item is an in-drum ball-shaped device of 13 cm in diameter made of soft, recycled plastic material. Cora Ball© tangles microfibers when added to the washing machine drum and reduces the amount of fiber breaking off from clothes, helping them last longer and protecting the aquatic ecosystem from microfiber pollution (*Cora Ball – The Laundry Ball Protecting the Ocean and Your Clothes*, 2022). Figure 16.1a shows that it has stalks with small hooks on the end. The filtering system of coral reefs inspires the design. The ball should be placed in the washing machine with clothing, where the stalks with hooks capture microfibers (*Cora Ball – The Laundry Ball Protecting the Ocean and Your Clothes*, 2022). After washing, the stuck fibers are removed by hand (Kärkkäinen & Sillanpää, 2021).

At least three research groups have also tested this commercial device's efficiency in mitigating microfiber marine pollution. For instance, in 2019, McIlwraith et al. first independently tested the efficiency of Cora Ball© in reducing microfiber emissions from washing machines (McIlwraith et al., 2019). They investigated the differences in the number and length of microfibers released from a 100% PES fleece blanket washed on an SDL Atlas M6 Vortex top-loading washing machine. Regarding the number of microfibers per liter of wastewater effluent, the Cora Ball© reduced the quantity of released microfibers by an average of 26%. The similarities in the microfiber lengths obtained in the control test and those obtained in the test with the Cora Ball© device suggest that this device can capture microfibers over a broad range of lengths.

A year later, Napper et al. (2020) also reported the mass of microfibers captured by the same Cora Ball© in-drum device. To perform their tests, they followed the same experimental procedure that was used to test the GUPPYFRIEND® washing bag. They found that washing a mixed load of clothes with Cora Ball reduced the number of microfibers being released into wastewater by 31% ± 8% (Napper et al., 2020) compared to washing the same load without any device (control testing). The efficiency of Cora Ball© was further investigated by Kärkkäinen and Sillanpää (2021) also, in this case, using the same experimental procedures used to test the GUPPYFRIEND® washing bag. The authors reported that washing synthetic textiles with no microfiber-capture device releases approximately 1.37×10^6 microfibers/kg (Figure 16.2b). Cora Ball© caught 10% of the short PES microfibers investigated (the release was reduced to 1.23×10^6 microfibers/kg). The difference between the studies of Kärkkäinen and Sillanpää (2021) and those performed by McIlwraith et al. (2019) and Napper et al. (2020) can be explained by different lengths of the studied microfibers since the trapping efficiency of Cora Ball© increases with the size of microfibers (Kärkkäinen & Sillanpää, 2021).

16.3.2 External Devices

Filters for domestic laundering machines are designed to be fitted externally to wastewater pipes, and they filter microfibers from the effluent wastewater discharge. Several filter devices are already commercially available. An example of such a device is the MicroPlastics LUV-R© filter, produced by Environmental Enhancements, NS, Canada (Figure 16.1b). MicroPlastics LUV-R© is a stainless-steel mesh filter with a pore size of 150 μm attached to the wastewater pipe (Napper et al., 2020). The product's website states that "MicroPlastics LUV-R© avoids septic system failure by removing lint and untreatable synthetic solids from washing machine discharge" (*MicroPlastics LINT LUV-R*, 2022). Filter action is dynamic, i.e., as the filter collects lint, it increases its efficiency (87% initial efficiency capture of microfibers and 100%

at filter saturation). The trapped lint acts as a filter itself. It also needs cleaning after 2–3 loads of laundry (*MicroPlastics LINT LUV-R*, 2022).

In 2019, McIlwraith et al. first tested the efficiency of MicroPlastics LUV-R© filter in reducing microfiber emissions from washing machines. Their results demonstrated that the MicroPlastics LUV-R© filter significantly reduced the released microfibers from a 100% PES fleece blanket per liter of effluent by an average of 87%. Additionally, a 72% reduction in fiber length was reached compared to no mitigation strategy, indicating that big fibers are captured by the MicroPlastics LUV-R© filter.

Different results were reported for the same filter by Napper et al. in 2020. A washing cycle of 45 minutes at 30°C and 1,000 RPM without detergent or conditioner was applied. The authors found that the MicroPlastics LUV-R© filter had no significant effect in decreasing the quantity of microfibers being released into wastewater (it only reduced by 29% ± 15% the average microfiber release derived from washing a mixed load of clothes without any device of control) (Napper et al., 2020). The differences in the results of microfiber retention between the works of McIlwraith et al. (2019) and Napper et al. (2020) may have resulted from the different methodologies used in these studies. For instance, McIlwraith et al. used 100% PES fleece blankets, a kind of textile that has been reported to show high shedding rates.

XEROS Technology Group, U.K., has also developed a washing machine filtration device called XFilter (*Filtration–Xeros Technology*, 2022). XFilter is designed to be integrated into any domestic washing machine during manufacture to help trap the microfibers that clothes release. The company's website states that its filtration technology captures over 99% of microplastics. They also stated that currently their technology is ready for application. Working alongside the industry, they have engineered the filtration solution to work with any washing machine model for their partners to scale. Two models are available: XF¹ (domestic) and XF² (commercial) (*Filtration–Xeros Technology*, 2022). The XF¹ filtration device captures over 99% of microplastics and over 80% of cellulosic microfibers. To achieve the lowest LCA impact, minimizing its impact on the planet, XFilter is designed to last the lifetime of a washing machine with no replacement cartridges. It works in the detergent drawer and can be easily removed to empty the trapped fibers into household waste, making it as simple as emptying the lint from a tumble dryer (*Filtration–Xeros Technology*, 2022). XF² has the same effective capture rates as XF¹, but is designed to be compatible at a commercial scale. XF² is ready for commercial scale by integrating the filter directly into a commercial washing machine. It can also be placed as a stand-alone unit attached to a series of machines or even into a whole laundry. The system is designed with a self-cleaning mechanism to last 60 wash cycles before emptying. It only takes a minute to dispose of the fibers from the collection tray, which is then put back into XFilter to continue to collect further fiber fragments (*Filtration-Xeros Technology*, 2022).

In the same study in 2020, Napper et al. investigated the microfiber retention efficiency of the XFiltra prototype using the same testing conditions used in the MicroPlastics LUV-R© tests. The authors found that the XFiltra prototype significantly decreased the quantity of microfibers released into wastewater. Washing a mixed load of clothes with XFiltra reduced the number of microfibers released into wastewater by 78% ± 5% (compared to a washing cycle without any device) (Napper et al., 2020). The high microfiber retention efficiency of the XFiltra prototype was related to its mesh pore size (60 μm). Additionally, it is an "active device" that uses a motor-powered centrifugal separator that needs an external electrical supply to facilitate the flow of wastewater through the filtration mesh (Napper et al., 2020).

The same study included a third filter, the PlanetCare© washing machine filter (Figure 16.1c), produced by PlanetCare Limited, U.K. (Napper et al., 2020). As stated on the company's website, "the PlanetCare© washing machine filter is a solution that catches 90% of microfibers and takes care of the whole cycle for a completely closed loop." Each filter is provided with reusable cartridges that are replaced once they are full of microfibers and sent back to the company, which can adequately refurbish them for future use (PlanetCare, 2022). Napper et al. (2020) found that the PlanetCare© filter had no significant effect in reducing the number of microfibers being released into wastewater (it only reduced by 25% ± 20% the average microfiber release of 0.44 ± 0.04 g (mean + S.E) derived from washing a mixed load of clothes without any device of control).

Industrial laundering is also a source of microfiber pollution. Microfiltration (MF) or ultrafiltration (UF) devices can be a valuable strategy to reduce microfiber release from industrial laundering practices. To examine the use of such devices in the mitigation of microfiber pollution from industrial washing practices, Luogo et al. (2022) investigated the microfiber retention of two ceramic membranes from LiqTech Ceramics A/S (Ballerup, Denmark) after the first washing cycle of a polyvinyl chloride (PVC) tent (Hvalsø Teltudlejning ApS, Hvalsø, Denmark) using an industrial tent washing machine (Luogo et al., 2022). Industrial washing of fabrics was investigated because, considering the washing capacity specifications of their industrial tent washing machine, the authors estimated that an average of 1.08 kg/week of PVC microfibers would be released from one industrial washing machine (Luogo et al., 2022). The investigated ceramic membranes are commercially available and can be classified as MF or UF membranes. Both are formed by highly porous multichannel ceramic membranes of 30 cylindrical channels. Each channel is 3 mm in diameter and 305 mm long, with a 0.09 m^2 effective membrane area. The devices use silicon carbide (SiC) as support, and SiC constitutes the membrane component in the case of the MF membrane and monoclinic zirconia (ZrO_2) in the case of the UF membrane. The tests were performed on a commercial pilot-scale filtration setup from LiqTech LabBrain (LiqTech Ceramics A/S, Ballerup, Denmark). The experiment consisted of the filtration of wastewater from a single washing cycle of the PVC tent in the industrial washing machine (Luogo et al., 2022). By visually counting the average number of microfibers found in 20 mL of wastewater using light microscopy, the authors estimated a microfiber concentration of 22.6 mg/L in the wastewater feed. After filtration with the MF membrane, the concentration of microfibers decreased to 0.3 mg/L, showing that the MF membrane could remove 98.5% of microfibers from laundering wastewater.

On the other hand, filtration with the UF membrane resulted in a concentration of 0.135 mg/L. This low value, relative to the smaller pore size of the UF membrane, means that the microfiber removal percentage for this membrane is around 99.2%. Additionally, the authors concluded that using the UF will provide better performance during filtration and less back-flushing, with less permeability loss and avoiding membrane clogging during a more extended period. Instead, the MF membrane will need to undergo cleaning in place more often than the UF membrane, and its operational cost will be higher (Luogo et al., 2022).

Even if the removal efficiencies of the commercially available devices may seem low, such devices can still contribute to mitigating microfiber pollution derived from domestic laundering practices. For example, as a case study that presents a simplified idea of the effectiveness of commercially available mitigation strategies, McIlwraith et al. (2019) approximated their maximum potential to divert microfibers from treated wastewater in a single region (Toronto). Based on their findings, it was estimated that in the City of Toronto, every year, between 23 and 36 trillion microfibers are released into wastewater from washing machines. If, on a municipal level, Cora Ball© or MicroPlastics LUV-R© is placed in all households in Toronto, then the quantity of microfibers released into wastewater from laundering could be reduced by up to 6–9 or 20–31 trillion fibers, respectively, every year based on their data. Assuming 99% capture in the sludge of WWTPs, this would translate to ~61–92 billion fewer microfibers being directly released into the environment annually with Cora Ball© and 204–309 billion with LUV-R© (McIlwraith et al., 2019).

The overall effectiveness of any microfiber trap to mitigate microplastic pollution is significantly affected by its user-friendliness (Kärkkäinen & Sillanpää, 2021). For example, the filters will require potential space for installation in washing machines (Napper et al., 2020), while other devices require cleaning by the consumer to remove fibers, except those made by PlanetCare, which are returned to the manufacturer for disposal and replacement (Napper et al., 2020). Regarding the disposal of the caught microfibers, more research is needed to establish disposal and device cleaning procedures by users to avoid unintentional incorrect disposal of the microfibers during the cleaning of the devices.

Table 16.1 presents an overview of the commercially available devices examined in this section.

16.4 DESIGN OF TEXTILES WITH REDUCED MICROFIBER RELEASE

Reducing the release of microfibers from textiles may depend on fabric structures and their composition. Currently, no precise patrons are found in the literature. Moreover, differences between the fabrics (and their textile structure) used in the investigations make it difficult to perform comparisons. For instance, Napper and Thompson found that textiles made of natural/synthetic blends tend to release fewer microfibers than pure synthetic fabrics. They also observed that a piece of fabric made of a 65% PES/35% cotton blend shed fewer microfibers than 100% synthetic fabrics made of PES or acrylic (Napper & Thompson, 2016). Yang et al. investigated microfibers released from three typical synthetic textile fabrics (PES, polyamide, and acetate) and found that PES and polyamide (23,094 ± 1,812 and 69,723 ± 40,773 microfibers/m^2 per wash, respectively) fabrics released fewer microfibers than acetate fabrics (74,816 ± 10,656 microfibers/m^2 per wash) (Yang et al., 2019). Zambrano et al. found that PES (0.1–1.0 mg/g fabric) fabrics released fewer microfibers than cellulose-based fabrics (0.2–4 mg/g fabric) with the same fabric structures (Zambrano et al., 2019). In agreement with such results, Cesa et al. found that textile characteristics affect microfiber release from cotton, acrylic, PES, and polyamide textiles (Cesa et al.,

TABLE 16.1
Commercially Available Devices for Mitigating Microfiber Pollution from Domestic and Industrial Laundering.

Type of Device	Domestic Laundering	Industrial Laundering	Advantages	Disadvantages	Reference
In-drum GUPPYFRIEND ® washing bag device	✓		Easy to use Does not interfere with stain removal from garments It can be used in any washing machine	Made of polyester, it can release microfibers itself Users should dispose of the collected microfibers Regarding disposal of the caught microfibers, more research is needed to establish disposal and device cleaning procedures by users to avoid unintentional incorrect disposal of the microfibers during the cleaning of the devices	Global (2022), Kärkkäinen and Sillanpää (2021), Napper et al. (2020)
In-drum Cora Ball© device	✓		Easy to use It can be used in any washing machine Captures microfibers over a wide range of lengths	Made of recycled plastic material, it can release microplastics After washing, the stuck fibers are removed by hand, thus presenting the same advantages of microfiber disposition as washing bags	Cora Ball–The Laundry Ball Protecting the Ocean and Your Clothes (2022), Kärkkäinen and Sillanpää (2021), Napper et al. (2020)
MicroPlastics LUV-R© External device	✓		Removes microfibers from wastewater It can be used in any washing machine	Requires installation as it is added after the wastewater pipe Requires cleaning after 2–3 loads of laundry Efficiency varies according to washing conditions	McIlwraith et al. (2019), *MicroPlastics LINT LUV-R* (2022), Napper et al. (2020)

(*Continued*)

TABLE 16.1 (*Continued*)
Commercially Available Devices for Mitigating Microfiber Pollution from Domestic and Industrial Laundering.

Type of Device	Domestic Laundering	Industrial Laundering	Advantages	Disadvantages	Reference
XEROS XF[1] external device	✓		Designed to be integrated into any domestic washing machine No need for the replacement of cartridges	Integration of the filter is during the manufacturing stage of washing machines	*Filtration-Xeros Technology* (2022), Napper et al. (2020)
PlanetCare© external device	✓		Each filter includes reusable cartridges that are replaced once they are full of microfibers and sent them back to the company to be adequately refurbished for future use	Users must collect the cartridges Currently, cartridges are being only stored by the producer, and reuse strategies are being investigated	PlanetCare (2022)
XEROS XF[2] external device		✓	It can be directly integrated into a commercial washing machine. It can also be placed as a stand-alone unit attached to a series of machines It has a self-cleaning mechanism to last 60 wash cycles before it needs to be emptied		Filtration-Xeros Technology (2022)
LiqTech Ceramics microfiltration (MF) and ultrafiltration (UF) membranes		✓	Prevent microfiber release from industrial washing procedures	MF will need cleaning more often than the UF, with higher operational costs	Luogo et al. (2022)

2020). They reported that textiles with irregular shorter fibers and lower tenacities resulted in higher microfiber releases. Thus, composition and textile structure may play an important role in microfiber production. Fabrics with higher abrasion resistance, low hairiness, and higher yarn-breaking strength have a lower tendency to form fuzz or release microfibers during the mechanical action of washing (Cesa et al., 2020; Zambrano et al., 2019).

Yang et al. related microfiber release to the weight per unit area of a textile, which is dependent on yarn diameter, the thickness of the fabric, and linear density. The number of microfibers will increase with the yarn count due to the presence of more fiber per cross-section (Yang et al., 2019). In contrast, a higher number of yarns per unit length (unit length of textile) promotes a tighter structure with a lower probability of microfiber release. Thus, these parameters should be considered during product design to create yarn and textiles that release only a few microfibers during laundering. Moreover, textile geometry is another critical factor that according to Yang and co-workers affects microplastic release during washing (Yang et al., 2019). For example, knitted fabrics could release more microfibers than woven fabrics due to their less compact structure. Moreover, a fabric made of short-staple fibers shows higher microfiber release than fabrics made with continuous filaments. This might be one of the reasons why polyamide fabric and acetate fabrics released more microfibers than PES fabrics in the experiment. Additionally, mechanical factors during yarn production, such as the air-textured yarn and dying process (cone-dyed ones or mass-cone-dyed ones) can influence microfiber release.

Reducing the release of microfibers from textiles may depend on the fabric polymer and structure and their surface, as well as bulk composition. For this reason, modification of synthetic fabrics is being investigated as an innovative strategy to mitigate microfiber release not only during the washing of garments but also during use throughout their lifecycle. As an example, De Falco et al. proposed an innovative finishing treatment of raw 100% polyamide-6,6 (PA) woven fabrics based on the use of pectin modified with glycidyl methacrylate to mitigate the microplastic impact of synthetic fabrics (De Falco et al., 2018a). The modified pectin was grafted on the fabric by a crosslinking reaction. After wash trials performed using a laboratory simulator of real washing machines (Linitest apparatus, URAI S.p.A., Assago, Italy) according to the ISO 105-C06:2010 standard method, the authors found that the treatment reduced about 90% the number of microfibers released by fabrics. The same research group investigated the ElectroFluidoDynamic (EFD) method as a finishing treatment by applying homogeneous coatings of biodegradable poly (lactic acid) (PLA) and poly (butylene succinate-co-butylene adipate) (PBSA) on PA fabrics to reduce the release of microfibers (De Falco et al., 2019a). The authors found that the treatments led to a reduction of more than 80% of the number of microfibers released during laundering tests of the treated fabrics. In addition, the coatings proved to endure washing cycles, with PLA being more durable than PBSA coatings, showing a promising application as a mitigation strategy for microfiber release from synthetic textiles. Chapter 18 presents a complete overview of finishing treatments for synthetic fabrics to reduce microfiber release.

Microfiber pollution mitigation through specific fabric structures, composition (in terms of polymer), or finishing treatments seems to be a valid alternative with a high potential for reduction in microfiber release. For instance, textiles made of natural/

synthetic blends should release lower quantities of synthetic microfibers. Although several research groups have found differences in microfiber release between garments of different polymers, microfiber detachment likely depends more on fabric structure than on composition. Therefore, fabrics with a lower tendency to form fuzz (i.e., textiles with high abrasion resistance, low hairiness, and high yarn-breaking strength) should be preferred to decrease the release of microfibers during the mechanical action of washing.

16.5 WASTEWATER TREATMENT PLANTS (WWTPS)

Currently, most WWTPs are not optimized as a mitigation strategy for microfiber pollution. Most of the research in the literature indicates that, due to the high volumes of wastewater influents managed by WWTPs, these act more as a source of microfibers to the environment (through effluents and sludges) than as a removal strategy. However, according to a review recently published by Le et al., WWTPs can reach a high removal rate of microfibers through primary, secondary, and tertiary treatments (Le et al., 2022). As illustrated in the Sankey diagram proposed by the authors (Figure 16.3), primary treatment, which typically includes screening, grit and grease removal, and primary settlement, can remove between 40% and 65% of the microfibers initially present in the influents. The removal is promoted by the characteristics of microfibers (long and light plastic particles that easily float on water), which lead to their effective capture by screening and floatation for oil and grease removal (Le et al., 2022).

Secondary treatment can add about 20%–60% of microfiber removal on top of the efficiency of primary treatment (Le et al., 2022). In this treatment, microfibers can be caught by activated sludge flocs and removed from the secondary clarification tank. Finally, sand filtration and MF (tertiary treatment) can remove any remaining

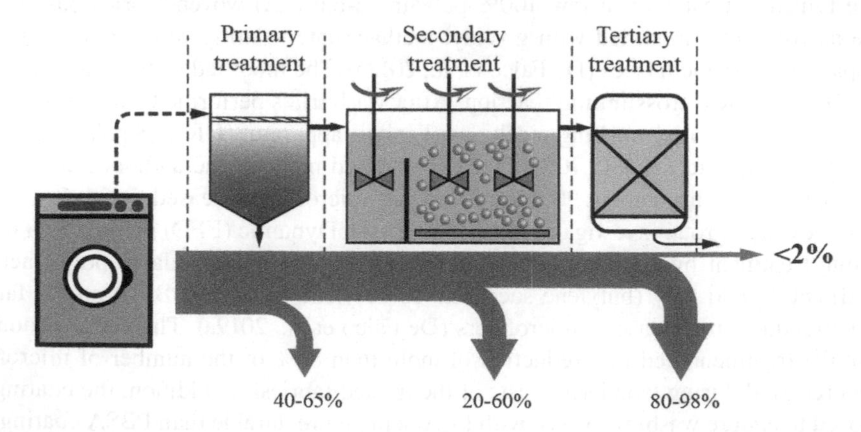

FIGURE 16.3 A Sankey diagram showing the removal of microfibers from WWTPs. (*Source:* Reprinted from Science of the Total Environment, 852 /15 December 2022, Linh-Thy Le, Kim-Qui N. Nguyen, Phuong-Thao Nguyen, Hung C. Duong, Xuan-Thanh Bui, Ngoc Bich Hoang, Long D. Nghiem, "Microfibers in laundry wastewater: problem and solution", Article number 158412, Copyright (2022), with permission from Elsevier.)

microfiber before effluent discharge (Le et al., 2022). In tertiary treatment, an alternative technique, the membrane bioreactor, can help achieve an excellent removal performance (90%–99%) compared to others, e.g., disc filter (80%–98%), granular filter (80%–98%), and dissolved air flotation (60%–85%) (Le et al., 2022). Even if not all WWTPs include all three treatment stages, it is relevant to highlight that WWTPs can offer effective and satisfactory microfiber removal performance from wastewater.

16.6 ENVIRONMENTAL EDUCATION STRATEGIES AND AWARENESS CAMPAIGNS

In 2019, Grelaud and Ziveri (2020) carried out awareness campaigns at beach sites to reduce the amount of all kinds of plastic waste that enters the environment through beaches. The activities consisted of awareness campaigns directly on sites or through social media or radio, new trash bins for mixed or recyclable waste, or new signs on existing bins (CoNISMa, 2020; Grelaud & Ziveri, 2020). The authors' results show that awareness campaigns could be efficient in decreasing the amount of litter from recreational use of the Mediterranean island beaches.

In 2021, Oliani and Cedillo-González carried out an awareness campaign specifically focused on microfiber marine pollution by publishing a seven-chapter column in Lo Spunk, an Italian newspaper for children aged 6 to 11 years (Oliani & Cedillo-González, 2021). The column, called "Pelucco's Diary," was focused on microfiber marine pollution and was specially designed to teach children how microfibers can be released through washing cycles of synthetic textiles and travel through pipes and rivers to reach the ocean. The main character, named Pelucco (a PES microfiber), was used to teach children the environmental impacts of microfiber pollution. The same story was further developed by the same authors in a book called "Pelucco: A Microplastic's journey" (Oliani & Cedillo-González, 2023) and has been tested with Italian children aged 3 to 6 years (group = 115 children) and 6 to 11 years (group = 42). The authors found that before telling the story, almost 94% of children were unaware of the contribution of plastic materials to humanity's well-being and viewed plastic as a synonym for pollution. Furthermore, they were unaware of microfiber pollution. After hearing Pelucco's story, 100% of children recognized the marvels of plastic, described the generation of microfibers from synthetic textiles and other plastic wastes, and suggested strategies to stop the ocean's microfiber pollution (Oliani & Cedillo-González, 2022a, b).

16.7 REUSE OF MICROFIBERS

All the previous strategies were based on avoiding microfiber release from synthetic textiles or capturing them before their introduction into aquatic environments. However, most current strategies to mitigate microfiber pollution do not consider the fate of the captured microfibers. This aspect is highly relevant, as adequate disposal of any waste material is a crucial factor for the success of any mitigation strategy. In this regard, some research groups propose using microfibers derived from synthetic textiles as secondary raw materials for fabricating building materials. For instance, Malchiodi et al. (2022) investigated the direct reuse of microfibers in green thermal-insulating and mechanical-performing composite construction materials. They first collected blended finishing textile waste microfibers (61% pure cotton, 29% cotton blend (cotton and synthetics), and 10% synthetics) and then reused them in designing thermal-insulating and mechanical-performing fiber-reinforced cementitious composites (FRCs) (Malchiodi et al., 2022). Portland cement-based FRCs were prepared, including up to 4wt% of blended textile waste microfibers in untreated, water-saturated, and NaOH-treated conditions. By increasing the microfiber content from 0 to 4wt%, it was observed up to a +320% maximum bending load, +715% toughness, −80% linear shrinkage, and double-insulating power of Portland cement. It was also found that toughness and linear shrinkage reduction were better enhanced in microfibers treated with NaOH and water. The authors also found that including textile waste microfibers also improved Portland cement's thermal-insulating and mechanical properties. Remarkably, considering the composition of the microfibers, this suggested solution could promote the removal of at least 4kg of microfibers (almost the same number of microfibers collected in a day in Paris (Dris et al., 2016; Zhang et al., 2022)) per ton of cement paste. Hence, mitigating microfibers from the environment and producing green and optimized FRCs were validated (Malchiodi et al., 2022).

Belzagui et al. proposed a novel and sustainable method to immobilize microfibers in a polymeric matrix, turning them into a composite. To prepare the composites, different proportions of PES (5%, 10% and 15% /v/v)) microfibers were mixed with low-density polyethylene, which is the material proposed for the immobilization of microfibers (Belzagui et al., 2022). The authors found that the tensile stress and Young's modulus of the thermoplastic polymer matrix improved at the expense of reducing the maximum deformation achievable when including microfibers at a concentration of 10%. It was also found that the composites behaved homogeneously with the microfiber's concentration up to 10%. Even if the authors found that a lower microfiber concentration (5%) also worked fine, their objective was to treat the current "fibres' microplastic pollution from laundering," trying to include the highest quantity of microfibers in the material. As no microfibers were detached from the final composite, it was found that they were fully contained in the matrix. Although scanning electron microscopy images showed low microfiber–matrix compatibility due to the smooth surface of the PES microfibers, the authors stated that in real-world conditions, other types of microfibers with higher roughness than PES will be included, increasing the adhesion between the pollutants and the recycled polymer.

Research has also been made to use microfibers as a renewable energy source. Yousef et al. used lint microfibers generated during the drying of clothes, mainly cotton, PES, and lignin, to generate three main energy products (oil, gas, and char) using a pyrolysis treatment in a pilot plant (Yousef et al., 2021b). They found that lint microfibers could thermally decompose to generate energy products with conversion rates of 65% (at 400°C and 600°C) and 78.4% (at 500°C). Toluene was the main component of the obtained biogas and bio-oil at 500°C with 93% and 33% concentrations, respectively. The preliminary environmental assessment (in terms of Greenhouse gas emissions) of the suggested conversion strategy was expressed in kg CO_2-eq/t and was performed based on the obtained biogas, bio-oil, and char. A CO_2-eq is a metric that compares the emissions from several greenhouse gases based on their global-warming potential. It converts the amounts of other gases to the equivalent amount of CO_2 with the same global warming potential. The authors found that almost 45 tons of lint microfibers can be generated by 1 million individuals annually, and this amount is enough to produce 13.8 tons of oil (31%), 21.5 tons of gas (47.7%), and 9.7 tons of char (21.6%). Those energy products have an estimated profitability of 120,400$ and a reduction in the carbon footprint of −42,039,000 kg CO_2-eq/t of lint microfibers. A deeper insight into the generation of energy products by lint microfibers in the dryer was further investigated by the same research group (Yousef et al., 2021a) using a thermogravimetric Fourier transform infrared coupled to a chromatography–mass spectrometry system to study the thermal and chemical decomposition of the collected lint samples and to simulate the pyrolysis process. Results showed that the lint decomposes in three stages, with a mass loss of 58% in the maximum decomposition zone. It was also found that isobutane (10.77%), furan, 3-methyl- (4.78%), furfural (6.48%), CO_2 (11.68%), and other compounds were strongly present in the volatile products, which suggests that lint could be a new sustainable biomass source for renewable energy.

16.8 CONCLUSION AND FUTURE SCOPE

As most of the microfibers in the environment come from the domestic laundering of synthetic garments, mitigation strategies such as setting washing programs with reduced releases or designing microfiber-capture devices have explicitly been investigated to stop microfiber release from laundering practices. Moreover, modifications in the composition and structure of synthetic textiles and implementing education and awareness campaigns constitute alternatives to reduce microfiber release into the environment. Although neither of these strategies can prevent 100% of the microfiber release, WWTPs can catch the microfibers that escape other mitigation strategies through primary, secondary, and tertiary treatments. Those recovered microfibers can be reused as secondary raw materials in the building sector or as raw materials for producing oil, gas, and char through pyrolysis. However, it is essential to highlight that as discussed in this chapter, each step of the overall laundering practice (setting of a washing load, temperature, detergent dosage, use of softener, or even the use of a commercially available microfiber-capture device) strongly influences the amounts of synthetic microfibers that may enter the environment. As most individuals who carry out such practices are unaware of the influence of their specific washing practices on environmental protection, future research can focus on a deeper investigation of the mechanisms that promote microfiber release during washing, taking into account both microfiber release and cleaning efficiency. This information can then be used to elaborate washing guides that can contribute to minimizing microfiber release without compromising the hygienic characteristics of garments.

REFERENCES

ARIEL UK. (2023). How to use and dose our washing gel correctly. https://www.ariel.co.uk/en-gb/how-to-wash/how-to-dose/how-to-dose-ariel-gel

Belzagui, F., Crespi, M., Álvarez, A., Gutiérrez-Bouzán, C., & Vilaseca, M. (2019). Microplastics' emissions: Microfibers' detachment from textile garments. *Environmental Pollution*, *248*, 1028–1035. https://doi.org/10.1016/j.envpol.2019.02.059

Belzagui, F., Gutiérrez-Bouzán, C., & Carrillo-Navarrete, F. (2022). Novel treatment to immobilize and use textiles microfibers retained in polymeric filters through their incorporation in composite materials. *Polymers*, *14*(15), Article 15. https://doi.org/10.3390/polym14152971

Bosch UK. (2023). *Is it better to wash my clothes in cold or hot water?* https://www.bosch-home.co.uk/customer-service/get-support/washing-machines/is-it-better-to-wash-my-clothes-in-cold-or-hot-water

Browne, M. A., Crump, P., Niven, S. J., Teuten, E., Tonkin, A., Galloway, T., & Thompson, R. (2011). Accumulation of microplastic on shorelines woldwide: Sources and sinks. *Environmental Science & Technology*, *45*(21), 9175–9179. https://doi.org/10.1021/es201811s

Cameron, B. A. (2011). Detergent considerations for consumers: Laundering in hard water-How much extra detergent is required. *Journal of Extension*, *49*, 1–11.

Carney Almroth, B. M., Åström, L., Roslund, S., Petersson, H., Johansson, M., & Persson, N.-K. (2018). Quantifying shedding of synthetic fibers from textiles; a source of microplastics released into the environment. *Environmental Science and Pollution Research*, *25*(2), 1191–1199. https://doi.org/10.1007/s11356-017-0528-7

Cesa, F. S., Turra, A., Checon, H. H., Leonardi, B., & Baruque-Ramos, J. (2020). Laundering and textile parameters influence fibers release in household washings. *Environmental Pollution*, *257*, 113553. https://doi.org/10.1016/j.envpol.2019.113553

CoNISMa. (2020). *D4.2.2 pilot activities implementation*. https://blueislands.interreg-med.eu/fileadmin/user_upload/Sites/Sustainable_Tourism/Projects/BLUEISLANDS/D4.2.2_Pilot_Activities_Implementation_CoNISMa.pdf

Cora Ball-The Laundry Ball Protecting the Ocean and Your Clothes. (2022). Cora ball. https://www.coraball.com/

Corami, F., Rosso, B., Bravo, B., Gambaro, A., & Barbante, C. (2020). A novel method for purification, quantitative analysis and characterization of microplastic fibers using micro-FTIR. *Chemosphere*, *238*, 124564. https://doi.org/10.1016/j.chemosphere.2019.124564

De Falco, F., Cocca, M., Guarino, V., Gentile, G., Ambrogi, V., Ambrosio, L., & Avella, M. (2019a). Novel finishing treatments of polyamide fabrics by electrofluidodynamic process to reduce microplastic release during washings. *Polymer Degradation and Stability*, *165*, 110–116. https://doi.org/10.1016/j.polymdegradstab.2019.05.001

De Falco, F., Di Pace, E., Cocca, M., & Avella, M. (2019b). The contribution of washing processes of synthetic clothes to microplastic pollution. *Scientific Reports*, *9*(1), Article 1. https://doi.org/10.1038/s41598-019-43023-x

De Falco, F., Gentile, G., Avolio, R., Errico, M. E., Di Pace, E., Ambrogi, V., Avella, M., & Cocca, M. (2018a). Pectin based finishing to mitigate the impact of microplastics released by polyamide fabrics. *Carbohydrate Polymers*, *198*, 175–180. https://doi.org/10.1016/j.carbpol.2018.06.062

De Falco, F., Gullo, M. P., Gentile, G., Di Pace, E., Cocca, M., Gelabert, L., Brouta-Agnésa, M., Rovira, A., Escudero, R., Villalba, R., Mossotti, R., Montarsolo, A., Gavignano, S., Tonin, C., & Avella, M. (2018b). Evaluation of microplastic release caused by textile washing processes of synthetic fabrics. *Environmental Pollution*, *236*, 916–925. https://doi.org/10.1016/j.envpol.2017.10.057

Dris, R., Gasperi, J., Saad, M., Mirande, C., & Tassin, B. (2016). Synthetic fibers in atmospheric fallout: A source of microplastics in the environment? *Marine Pollution Bulletin*, *104*(1), 290–293. https://doi.org/10.1016/j.marpolbul.2016.01.006

Filtration-Xeros Technology. (2022, mayo 12). https://www.xerostech.com/filtration/, https://www.xerostech.com/filtration/

Global, G. S. | E. |. (2022). *GUPPYFRIEND washing bag online shop | STOP! Micro waste*. Guppyfriend Shop | Europe | Global. https://en.guppyfriend.com/

Grelaud, M., & Ziveri, P. (2020). The generation of marine litter in Mediterranean island beaches as an effect of tourism and its mitigation. *Scientific Reports*, *10*(1), Article 1. https://doi.org/10.1038/s41598-020-77225-5

Habib, D., Locke, D. C., & Cannone, L. J. (1998). Synthetic fibers as indicators of municipal sewage sludge, sludge products, and sewage treatment plant effluents. *Water, Air, and Soil Pollution*, *103*(1), 1–8. https://doi.org/10.1023/A:1004908110793

Hartline, N. L., Bruce, N. J., Karba, S. N., Ruff, E. O., Sonar, S. U., & Holden, P. A. (2016). Microfiber masses recovered from conventional machine washing of new or aged garments. *Environmental Science & Technology*, *50*(21), 11532–11538. https://doi.org/10.1021/acs.est.6b03045

Hernandez, E., Nowack, B., & Mitrano, D. M. (2017). Polyester textiles as a source of microplastics from households: A mechanistic study to understand microfiber release during washing. *Environmental Science & Technology*, *51*(12), 7036–7046. https://doi.org/10.1021/acs.est.7b01750

Kärkkäinen, N., & Sillanpää, M. (2021). Quantification of different microplastic fibres discharged from textiles in machine wash and tumble drying. *Environmental Science and Pollution Research*, *28*(13), 16253–16263. https://doi.org/10.1007/s11356-020-11988-2

Kelly, M. R., Lant, N. J., Kurr, M., & Burgess, J. G. (2019). Importance of water-volume on the release of microplastic fibers from laundry. *Environmental Science & Technology*, *53*(20), 11735–11744. https://doi.org/10.1021/acs.est.9b03022

Lant, N. J., Hayward, A. S., Peththawadu, M. M. D., Sheridan, K. J., & Dean, J. R. (2020). Microfiber release from real soiled consumer laundry and the impact of fabric care products and washing conditions. *PLoS One*, *15*(6), e0233332. https://doi.org/10.1371/journal.pone.0233332

Le, L.-T., Nguyen, K.-Q. N., Nguyen, P.-T., Duong, H. C., Bui, X.-T., Hoang, N. B., & Nghiem, L. D. (2022). Microfibers in laundry wastewater: Problem and solution. *Science of the Total Environment*, *852*, 158412. https://doi.org/10.1016/j.scitotenv.2022.158412

Luogo, B. D. P., Salim, T., Zhang, W., Hartmann, N. B., Malpei, F., & Candelario, V. M. (2022). Reuse of water in laundry applications with micro- and ultrafiltration ceramic membrane. *Membranes*, *12*(2), Article 2. https://doi.org/10.3390/membranes12020223

Malchiodi, B., Cedillo-González, E. I., Siligardi, C., & Pozzi, P. (2022). A practical valorization approach for mitigating textile fibrous microplastics in the environment: Collection of textile-processing waste microfibers and direct reuse in green thermal-insulating and mechanical-performing composite construction materials. *Microplastics*, *1*(3), Article 3. https://doi.org/10.3390/microplastics1030029

McIlwraith, H. K., Lin, J., Erdle, L. M., Mallos, N., Diamond, M. L., & Rochman, C. M. (2019). Capturing microfibers - marketed technologies reduce microfiber emissions from washing machines. *Marine Pollution Bulletin*, *139*, 40–45. https://doi.org/10.1016/j.marpolbul.2018.12.012

MicroPlastics LINT LUV-R. (2022). https://environmentalenhancements.com/store/index.php/products/products-micro-plastics

Napper, I. E., Barrett, A. C., & Thompson, R. C. (2020). The efficiency of devices intended to reduce microfibre release during clothes washing. *Science of the Total Environment*, *738*, 140412. https://doi.org/10.1016/j.scitotenv.2020.140412

Napper, I. E., & Thompson, R. C. (2016). Release of synthetic microplastic plastic fibres from domestic washing machines: Effects of fabric type and washing conditions. *Marine Pollution Bulletin*, *112*(1), 39–45. https://doi.org/10.1016/j.marpolbul.2016.09.025

Oliani, P., & Cedillo-González, E. I. (2021, junio). Il Diario di Pelucco. *Lo Spunk, il giornale delle bambine e dei bambini*, *IV*(5), 16.

Oliani, P., & Cedillo-González, E. I. (2022a, septiembre 20). Pelucco, the microplastic journey: A book that teaches children the wonders of plastic and the environmental impacts of microplastic pollution. In *7th International Marine Debris Conference*, Busan, South Korea.

Oliani, P., & Cedillo-González, E. I. (2022b, noviembre 15). *THE MICROPLASTICS JOURNEY: TEACHING TO THE YOUNGEST MINDS THE WONDERS OF PLASTIC AND THE ENVIRONMENTAL IMPACT OF MICROPLASTIC POLLUTION THROUGH A STORY FOR CHILDREN*. MICRO 2022 Online Atlas Edition, Lanzarore, Spain. https://doi.org/10.5281/zenodo.7216636

Oliani, P., & Cedillo-González, E. I. (2023). *Pelucco: A microplastic's journey*. Amazon KDP. ISBN: 979-8860263468. https://a.co/d/j43b5t1

Pirc, U., Vidmar, M., Mozer, A., & Kržan, A. (2016). Emissions of microplastic fibers from microfiber fleece during domestic washing. *Environmental Science and Pollution Research*, *23*(21), 22206–22211. https://doi.org/10.1007/s11356-016-7703-0

PlanetCare. (2022). *A fully circular, closed-loop solution*. PlanetCare. https://planetcare.org/pages/how-it-works

Rathinamoorthy, R., & Balasaraswathi, S. R. (2022). *Microfiber Pollution* (1st ed. 2022 edition). Springer.

Rathinamoorthy, R., & Raja Balasaraswathi, S. (2022). Investigations on the impact of handwash and laundry softener on microfiber shedding from polyester textiles. *The Journal of the Textile Institute*, *113*(7), 1428–1437. https://doi.org/10.1080/00405000.2021.1929709

Volgare, M., De Falco, F., Avolio, R., Castaldo, R., Errico, M. E., Gentile, G., Ambrogi, V., & Cocca, M. (2021). Washing load influences the microplastic release from polyester fabrics by affecting wettability and mechanical stress. *Scientific Reports*, *11*(1), Article 1. https://doi.org/10.1038/s41598-021-98836-6

Weis, J. S., & De Falco, F. (2022). Microfibers: Environmental problems and textile solutions. *Microplastics*, *1*(4), Article 4. https://doi.org/10.3390/microplastics1040043

Yang, L., Qiao, F., Lei, K., Li, H., Kang, Y., Cui, S., & An, L. (2019). Microfiber release from different fabrics during washing. *Environmental Pollution*, *249*, 136–143. https://doi.org/10.1016/j.envpol.2019.03.011

Yousef, S., Eimontas, J., Striūgas, N., Mohamed, A., & Abdelnaby, M. A. (2021a). Morphology, compositions, thermal behavior and kinetics of pyrolysis of lint-microfibers generated from clothes dryer. *Journal of Analytical and Applied Pyrolysis*, *155*, 105037. https://doi.org/10.1016/j.jaap.2021.105037

Yousef, S., Eimontas, J., Zakarauskas, K., Striūgas, N., & Mohamed, A. (2021b). A new strategy for using lint-microfibers generated from clothes dryer as a sustainable source of renewable energy. *Science of the Total Environment*, *762*, 143107. https://doi.org/10.1016/j.scitotenv.2020.143107

Zambrano, M. C., Pawlak, J. J., Daystar, J., Ankeny, M., Cheng, J. J., & Venditti, R. A. (2019). Microfibers generated from the laundering of cotton, rayon and polyester based fabrics and their aquatic biodegradation. *Marine Pollution Bulletin*, *142*, 394–407. https://doi.org/10.1016/j.marpolbul.2019.02.062

Zhang, Y.-Q., Lykaki, M., Markiewicz, M., Alrajoula, M. T., Kraas, C., & Stolte, S. (2022). Environmental contamination by microplastics originating from textiles: Emission, transport, fate and toxicity. *Journal of Hazardous Materials*, *430*, 128453. https://doi.org/10.1016/j.jhazmat.2022.128453

Zubris, K. A. V., & Richards, B. K. (2005). Synthetic fibers as an indicator of land application of sludge. *Environmental Pollution*, *138*(2), 201–211. https://doi.org/10.1016/j.envpol.2005.04.013

17 Together We Go Further and Faster

A Review of Existing Policy Plans for Microfibre Pollution

Songyi Yan
The University of Manchester

17.1 INTRODUCTION

There is growing concern and interest surrounding the issue of microfibre pollution (MFP) globally, as a result of increased research and media attention (Amos, 2020; Misund et al., 2020; Athey and Erdle, 2022), along with the efforts made by non-governmental organisations (NGOs) to tackle this issue (Mendenhall, 2018). Within the academic remit, most research centres are investigating the pathways, microfibre shedding rate, and material parameters that affect microfibre release (Gaylarde et al., 2021; Liu et al., 2021). Whilst various industrial practitioners are developing feasible mitigation solutions to tackle MFP for different entry points (i.e. design and manufacturing, use and maintenance, end-of-life, end-of-pipe), policymakers are increasingly looking for policy options at local, national, and supranational levels to better oversee current and future environmental challenges and all creatures' health risks associated with microfibres.

When 'microfibre' is considered a pollutant, it refers to three different types of microfibres – synthetic, natural, and regenerated cellulosic microfibres (Henry et al., 2019). Past studies and media predominantly referred to synthetic microfibres when it comes to MFP; one of the reasons could be they are closely linked to a wider and well-known disclosure – microplastic pollution, yet without clear explanation, it can distract from clarity about what constitutes microfibres (Yan et al., 2020a). Although synthetic fibres still dominate the global textile market, the production of natural and regenerated cellulosic fibres is also continuously growing (Sandin et al., 2019), which can further have a significant implication on MFP. To reiterate this further, studies indicated that a single 6-kg domestic wash could release as many as 700,000 microfibres (Napper and Thompson, 2016). Depending on different types of fabric and washing variables, from 6,490 to 87,165 tonnes of microfibres could be released per year in the UK, which is the equivalent of around 600–7,500 double-decker buses (Hazlehurst et al., 2023). When taking a closer look at these released microfibres, it is highlighted that cellulose-based fabrics (which include natural and regenerated cellulosic fibres) shed more microfibres than polyester fabrics – twice as abundant

DOI: 10.1201/9781003331995-21

as synthetic microfibres that can be found in deep-sea sediments (Zambrano et al., 2019). Furthermore, cellulose-based fabric often contains chemical additives (e.g. colourants and finishes) to provide desirable properties; these additives could be toxic and potentially affect the biodegradability of released cellulose-based microfibres (Athey and Erdle, 2022). Therefore, although most research and communication centre on synthetic microfibres, natural and regenerated cellulosic microfibres warrant further investigation. The term 'microfibre' in this chapter refers to synthetic, natural, and regenerated cellulosic microfibres.

MFP is a complex issue that not only poses multiple challenges, but also lacks in-depth research from both (hard) science and social science. For example, there remain knowledge gaps (e.g. terminology, definition, sources, pathways, impacts of microfibres, etc.), low accessibility and availability of knowledge and existing mitigation solutions from the public/consumers' point of view, the lack of consistency and clarity within the current communication (Yan et al., 2020b) and the lack of best practices to inform policy plans (OECD, 2021; De Falco and Cocca, 2022). Microfibres are released at any stage of the garment lifecycle from manufacturing to end of life, which means that responsibilities and actions should be shared among different stakeholders (e.g. textile/garment manufacturers, brands, washing machine companies, wastewater treatment companies, etc.), which needs to be integrated and reflected in potential mitigation solutions as well as policy plans. What seems evident from past studies is that, whilst we have understood the need for multiple stakeholders to tackle the issue of MFP collaboratively (Yan et al., 2020b; De Falco and Cocca, 2022), we have not yet sufficiently explored whether and how a multi-stakeholder approach for MFP connects existing policy plans and objectives. Thus, this chapter sets out to first review some prominent formal government-led and voluntary research and/or industry-led policy options in tackling MFP and to discuss their associated benefits and challenges. Moreover, existing formal and voluntary policy options share common objectives to enhance research, strengthen the facts, bridge the knowledge gaps, develop harmonised test methods, and foster innovative solutions (Legifrance, 2020; EC, 2022; Euratex, 2022). However, what currently has not been fully explored and discussed within many existing policy options echoes the challenges aforementioned – knowledge sharing, public communication, and education. This chapter will further discuss this aspect and other opportunities and challenges to broaden and deepen the scope of the existing policy plans to better tackle the MFP issue.

17.2 REVIEW OF PROMINENT FORMAL POLICY PLANS

Formal policy plans that are governed by governments and authorities are arising at local, national, and supranational levels. The existing recommended and implemented policy actions centre on targeting synthetic microfibres and focus on identifying and assessing potential and feasible mitigation actions at different entry points from garment manufacturing, use, and maintenance, to end-of-life and end-of-pipe to quantify and remove microfibre pollutants from wastewater streams. Table 17.1 highlights some proposed and implemented policy plans at local, national, and supranational levels. These proposed and/or implemented formal policy plans target different

TABLE 17.1
Prominent Examples of Formal Policies Targeting MFP

Description	Fostering Research (Addressing Scientific Knowledge Gaps)	Standardising Test Methods	Developing Mitigation Solutions	Public Education
National				
	France			
All new industrial, commercial, and household washing machines are required to install a **microfibre filter** by 1 January 2025 (Legifrance, 2020)			X	
	Australia			
Australian Government requires **microfibre filters** to be installed on new household and commercial washing machines by 1 July 2030 (DAWE, 2021)			X	
	Canada			
Under its Zero Plastic Waste strategy, Canada is aiming to **close research gaps** on macro- and microplastics. The strategy includes researching synthetic microfibre shedding during washing and designing **test methods** (Government of Canada, 2020)	X	X		
	Sweden			
The Swedish Environmental Protection Agency has financed research grants to **bridge the knowledge gaps** on microfibre pollution and create potential **mitigation measures** (Swedish EPA, 2019)	X		X	

(Continued)

TABLE 17.1 (*Continued*)
Prominent Examples of Formal Policies Targeting MFP

Description	Fostering Research (Addressing Scientific Knowledge Gaps)	Standardising Test Methods	Developing Mitigation Solutions	Public Education
UK				
The UK government supported research projects to investigate the **definition** of MFP, develop **test methodologies**, **quantify** synthetic microfibres in wastewaters, and test the effectiveness of **removal** treatment processes	X	X	X	
An All-Party Westminster Parliamentary Group on Microplastics ('APPG') has suggested an extended producer responsibility scheme ('**EPR**') for textiles from 2023. Requiring **microfibre filters** to be fitted into all new washing machines from 2025; washing machine and/or filter manufacturers should **communicate** to the public how collected microfibre waste should be correctly recycled or disposed of; increasing public awareness through **communication campaign**; introducing incentives to encourage the establishment of **recycling technology for microfibres** (APPG, 2021)			X	X

(Continued)

TABLE 17.1 (*Continued*)
Prominent Examples of Formal Policies Targeting MFP

Description		Fostering Research (Addressing Scientific Knowledge Gaps)	Standardising Test Methods	Developing Mitigation Solutions	Public Education
Local		**United States**			
	New York State introduced a bill requiring all garments containing more than 50% synthetic fibres to include the message '*This garment sheds plastic microfibers when washed*' on their clothing **label** (New York State Assembly, 2018)				X
	In 2019, California introduced state legislation, requiring **microfibre filters** installed in domestic, commercial, and industrial washing machines, adopting a **methodology** to quantify the amount of released microfibres, and establishing infrastructure and providing **public knowledge and educational campaign** (California Legislative Information, 2019)		X	X	X
	Connecticut passed a state-wide bill aiming to increase consumer awareness, by establishing an **education** and **clothing labelling** program (The State of Connecticut, 2018)				X
Supranational		**EU**			
	The European Commission aims to address the unintentional release of synthetic microfibres, including standardising **test methods**; closing scientific **knowledge gaps**; and developing **labelling** and **certification** measures (EC, 2022)	X	X		X
	The European Environmental Agency (EEA) aims to support tackling microplastic pollution by strengthening the **knowledge base** of synthetic microfibre releases (sources, pathways, and impacts), creating **mitigation solutions** targeting different stages of garment lifecycle (e.g. design, use, end-of-life), and **raising public awareness** and promoting sustainable behaviour in addressing MFP (EEA, 2022)	X		X	X

policy objectives (Table 17.1), which are categorised into (i) fostering research to strengthen knowledge foundation, (ii) standardising testing methods, (iii) providing feasible mitigation solutions, and (iv) public education.

Seeing as the public plays a significant part in 'creating' MFP, as we wear and wash our clothes; it is not surprising that there are policies geared at increasing public awareness and providing public education (The State of Connecticut, 2018; California Legislative Information, 2019; Zarghamee et al., 2022; EC, 2022). One approach that appears repeatedly within different policy plans is the clothing labelling scheme, which is considered an effective method to educate the public about MFP. Currently, most clothing labelling schemes are only designed to educate consumers to maintain clothing quality and longevity by informing them about appropriate washing methods in the form of care labels. This, however, does not reflect the environmental impact these garments may have, such as shedding of microfibres during the laundering process (ISO, 2012). Thus, disclosing information on microfibres and appropriate washing instructions on clothing care labels, product packaging, and other garment tags can potentially enhance consumers' knowledge and awareness (Yan et al., 2020a; OECD, 2020; EC, 2022). Table 17.1 outlines that public education initiatives mainly centre on local (US-focused) and supranational (EU-focused) levels. This might raise the question of how to bridge national and/or global solutions with local practices and how to ensure clarity and consistency within public education and communication among different levels.

Aside from care labels, the media has spotlighted the fact that 87% of treated sewage sludge was sold to farmers to use it as a fertiliser. This, however, can be seen as problematic, as the sewage sludge is contaminated with the microfibres that are filtered from wastewater during treatment, which are then sent back to the environment and jeopardise the initial good intention (The Guardian, 2022). Among the policies outlined in Table 17.1, only the UK government proposed a plan addressing the end-of-pipe stage, thereby focusing on wastewater treatment and proper disposal of wastewater sludge. The APPG (2021) has introduced incentives to encourage the establishment of recycling technologies for collected microfibre wastes. Although Table 17.1 is not exhaustive, to the author's knowledge, there are only very limited policies available that centre their attention on the end-of-pipe stage. Thus, future policy plans need to consider the collection, management, and other feasible solutions for captured microfibre wastes.

What needs to be highlighted is the fact that these policy actions and plans proposed and/or implemented by local, national, and supranational governance will not solve the problem alone, but they can act as a foundation for more comprehensive strategies and solutions (OECD, 2020). However, changes made to existing policies and/or creating new ones are complex and thus take a long time, as proposals need to be reviewed, amended, and/or agreed upon by multiple parties in multiple rounds before they can be passed. Yet, the scale of the threat posed by microfibres requires immediate action, which is a further challenge (Thodges, 2019). Moreover, formal governance adopts a top-down approach, with politicians, governments, and authorities initiating the changes without the involvement of other relevant key stakeholders and society (Garcia-Vazquez and Garcia-Ael, 2021). As highlighted earlier, multi-stakeholder collaboration in the context of MFP is essential and beneficial for generating more efficient solutions and policy plans (Misund et al., 2020; Yan et al., 2020b; Garcia-Vazquez and Garcia-Ael, 2021). Therefore, a private, voluntary governance approach is suggested, which often offers a variety of voluntary initiatives involving public and private actors to address social and environmental problems (Rasche, 2012).

17.3 REVIEW OF EXISTING VOLUNTARY POLICY GOVERNANCE FOR TACKLING MFP

In parallel with formal government-led initiatives, several voluntary research and/or industry-led governances have emerged to enforce microfibre mitigation. This indirect form of governance, which is also referred to as a multi-stakeholder initiative (MSI), is adapting a collaborative form of governance, involving different actors (public and private) in society to define the problems, implement and evaluate solutions, and enforce rules that aim to foster sustainable behaviour changes (Rasche, 2012; Moog et al., 2015). This form of governance contributes to accelerating the research process, extending opportunities, and deepening the knowledge of the subject (i.e. MFP). Table 17.2 outlines four prominent examples of voluntary governance for MFP, their involved stakeholders, and the policy plans that are proposed and/or implemented.

MSI governance approaches have been gradually created globally. In the UK, The Microfibre Consortium (TMC) has been established with the aim of bridging the gap between academic research and practices in the textile supply chain, involving stakeholders from research institutes, textile retailers and manufacturers, and NGOs. In the European context, a number of industry associations (International Association for Soaps, Detergents and Maintenance Products, Comité International de la Rayonne et des Fibres Synthétiques, European Outdoor Group, Euratex, and the Federation of the European Sporting Goods Industry) have formed a voluntary Cross Industry Agreement (Euratex, 2022). Today, the Cross Industry Agreement is reaching its final stage of establishing a harmonised test method, following extensive stakeholder engagement. The established Cross Industry Agreement test method will be handed over to the European Committee for Standardization (CEN) for final assessment and developed into an official CEN standard. The established testing standard will not only enable global testing and data analysis, but also help evaluate the effectiveness of mitigation solutions, in addition to developing formal governance processes. The Outdoor Industry Association and European Outdoor Group recognised the Outdoor Industry's potential contribution to MFP; their 5-year action plans (by 2023) outlined a variety of tasks; yet there is no further information on the current progress and updates.

Some fashion and apparel brands have been actively working on reducing the widest-reaching and invisible impacts caused by MFP. Brands such as Patagonia have been tackling this issue since 2014 by collaborating with research institutes, domestic and industrial washing machine companies, and creative communicators to establish test methods for microfibre shedding rate, develop effective filters and washing methods for domestic and industrial uses, and raise public awareness through creative communication. Whilst some brands are able to initiate the

TABLE 17.2
Prominent Examples of Voluntary Research and/or Industry-Led Policy for MFP

Description	Stakeholders	Fostering Research (Addressing Scientific Knowledge Gaps)	Standardising Test Methods	Developing Mitigation Solutions	Knowledge Sharing among Stakeholders	Public Education
		UK				
The Microfibre Consortium (TMC) To date, the action plans included standardising test methods, investigating various production parameters on shedding behaviours, providing mitigation solutions from the textile manufacturing stage, and establishing knowledge-sharing platforms for stakeholders	Research institutions, textile and apparel brands, suppliers, and manufacturers, NGOs	X	X	X	X	

(Continued)

TABLE 17.2 (*Continued*)
Prominent Examples of Voluntary Research and/or Industry-Led Policy for MFP

Description	Stakeholders	Fostering Research (Addressing Scientific Knowledge Gaps)	Standardising Test Methods	Developing Mitigation Solutions	Knowledge Sharing among Stakeholders	Public Education
		European				
The Cross Industry Agreement The partnership aims to develop and standardise test methods; to foster knowledge sharing among stakeholders; to address knowledge gaps; and to develop feasible solutions through industrial collaborations	Cross industry suppliers and manufacturers (white goods, outdoor apparel, sportswear), research institutions	X	X	X	X	

(Continued)

TABLE 17.2 (*Continued*)

Prominent Examples of Voluntary Research and/or Industry-Led Policy for MFP

Description	Stakeholders	Fostering Research (Addressing Scientific Knowledge Gaps)	Standardising Test Methods	Developing Mitigation Solutions	Knowledge Sharing among Stakeholders	Public Education
		US				
Outdoor Industry Association and **European Outdoor Group** This partnership established a 5-year plan: by 2023 to drive the data collection to better understand the sources and pathways of microfibre release, to establish consistent test methodologies, to implement best practices to reduce microfibre emission on the supply chain, to invest in innovative solutions, and to enhance knowledge sharing among stakeholders	Collectively representing over 1,500 outdoor companies; other industries (i.e. the fashion and textile industries, chemical manufacturers, the home appliance industry, and water treatment facilities); research institutions	X	X	X	X	

(*Continued*)

TABLE 17.2 (*Continued*)
Prominent Examples of Voluntary Research and/or Industry-Led Policy for MFP

Description	Stakeholders	Fostering Research (Addressing Scientific Knowledge Gaps)	Standardising Test Methods	Developing Mitigation Solutions	Knowledge Sharing among Stakeholders	Public Education
Surfrider Foundation Study the latest science, facilitate coordinating solutions, create campaigns to raise public awareness, and promote existing policies and legislation (e.g. California legislation)	Outdoor apparel companies (e.g. Patagonia), outdoor industry associations, NGOs	X		X		X

changes by themselves, others might require to be a part of a collective force. Thus, it is unsurprising that various MSIs have successfully attracted fashion and apparel brands' attention to engage with tackling the issue of MFP. Many well-known brands (e.g. Nike, Boohoo, Next, M&S, etc.) are acting as signatories to commit to the action plans and goals (e.g. TMC, 2022); however, it is currently unclear how and to what extent these brands have contributed to the initiatives. Whilst it is great for brands to demonstrate their motivation and commitment to tackle the issue of MFP by signing up with different MSIs (e.g. Boohoo), without disclosing transparently what has been done, it entails a risk of being seen as greenwashing and taking a 'too easy' approach.

Furthermore, compared with formal governance, MSI governance involves a wider array of stakeholders across different sectors/industries at national and international levels to develop initiatives and form policy plans. Because of the multi-actor nature of this approach, it places knowledge/information sharing among different stakeholders as one of the most significant working tasks. This, however, also seemingly implies that public education and communication seem to take a backseat in current plans (See Table 17.2).

17.3.1 Benefits and Challenges of Voluntary Industry and/or Research-Led Policy and Governance

Voluntary joint basis policy and governance (MSI governance) comprises several benefits: Firstly, it can facilitate information/data gathering and sharing among various stakeholders across industries at national and international levels. For instance, the Microfibre Consortium has created the 'Microfibre Data Portal', where signatories can voluntarily submit their product data links to microfibre shedding (using established the TMC test method – ISO 105-C06). A rich data portal can facilitate data analysis, identify priorities for research actions, and strengthen the understanding and knowledge of the mechanisms and production parameters that influence microfibre shedding behaviour. The TMC claimed that the knowledge will be further disseminated to stakeholders' networks (e.g. textile and fashion brands, suppliers, and manufacturers) and to the public in 2023 through the TMC's 'Microfibre Knowledge Hub' to help different actors act effectively on MFP. Having an information/knowledge sharing system is essential as currently there is a lack of consistency over different testing methodologies, which makes it difficult to compare and synthesise findings and provide potential solutions. Secondly, involving multiple stakeholders across different sectors in the governance can potentially help determine and/or create the best available practices in different sectors (e.g. white goods industry, apparel industry, wastewater treatment, etc.). It can also contribute to accelerating industrial and cross-industrial research and development (R&D) (OECD, 2021). Thirdly, as demonstrated in Table 17.2, MSIs share similar policy plans and objectives with formal government-led initiatives. As such, the former can support, promote, and complete government-led policy actions and contribute to accelerating the completion of the common policy objectives.

Although a multi-stakeholder, multinational, and cross-industrial initiative has been amplified, it is not always easy to implement, since different stakeholders (private, public, and third-party sectors) might have different agendas and goals in mind

(Matzembacher et al., 2021; Rathinamoorthy and Raja Balasaraswathi, 2022). For instance, tensions and doubts can emerge when organisations seek to address their financial needs for competing in the same market, whilst the public, regulatory institutions, and NGOs add pressure on firms to improve their sustainability performance and disclosure transparency (Matzembacher et al., 2021). Moreover, the agreed-upon regulations can be too rigorous and strict, which in turn can pose barriers for other stakeholders to follow, considering they might be technically and financially incapable. As such, it is necessary to consider all stakeholders' abilities and aim for a good balance between the restrictiveness and feasibility of implemented rules. For example, the presented prominent voluntary agreements are considered in favour of textile/garment retailers and manufacturers, whilst primary textile suppliers/producers that are based in developing countries (global south) and are often technically and financially restrained are currently underrepresented. Many proposed mitigation solutions need to take place in supply chains, yet solely pressuring suppliers to become more sustainable and putting the burden of cost on them simply reinforces the unbalanced powers between brands and suppliers. The complex issue of MFP requires collaboration between stakeholders, and how stakeholders interact with each other (especially for brands and their suppliers) remains a research gap and a barrier for going forward.

Whilst data/information sharing among stakeholders can enhance the research progress as aforementioned, challenges remain. Using textile and fashion supply chains as an example, they are complex and globally dispersed, which makes it difficult to monitor fabric and garment manufacturing practices, as well as their environmental footprints. As a result, textile and garment manufacturers, fashion brands, and other downstream stakeholders often have inconsistent, incomplete, and/or inadequate information on products (OECD, 2021). Whilst this challenge could be addressed by blockchain technology to ensure transparency, voluntary or mandatory certification schemes can also be introduced as an alternative solution. Furthermore, past studies indicated that third-party accreditations may be more reliable and could increase consumers' trust towards the product/service (Turunen and Halme, 2021). Other studies suggest that macro-environmental factors such as political status need to be considered; seeing the certificated label is more effective in countries with low political trust (e.g. Portugal) (Misund et al., 2020). To explain, an individual's pro-environmental behaviour can be influenced by the level of trust in the country's institutions and authorities; as such, the intervention should be designed ad hoc for each culture and region (Misund et al., 2020).

17.4 FUTURE POLICY ACTION AREAS FOR IMPROVEMENT

17.4.1 Action Plans for Knowledge Sharing and Public Education

Knowledge and communication play significant roles in tackling MFP; knowledge has been qualitatively and quantitatively explored, and it has been proven in many studies that it has a direct and significant relationship with behaviour change intentions (Yan et al., 2020b; Ojinnaka and Aw, 2020; Herweyers et al., 2020; Garcia-Vazquez and Garcia-Ael, 2021). This being said, in order to effectively communicate and educate about the issue of MFP, it is essential to understand how knowledge is currently communicated and what are its associated opportunities and barriers. A few studies investigated

MFP from a knowledge and communication perspective; for instance, Yan et al. (2020b) and Garcia-Vazquez and Garcia-Ael's (2021) indicated that there is a lack of public (social) knowledge of MFP and its solutions and legislation. The former further emphasised that public knowledge (social knowledge) can further lead to individual learning and behaviour changes (Yan et al., 2020b). However, creating public knowledge and finding best practices for MFP communication remain a not fully explored area within research as well as existing policy plans. Although disseminating knowledge and raising public awareness are included in both formal and voluntary policy plans, it is unclear what messages/knowledge is embedded and how related stakeholders (e.g. public, media, communicators, researchers, etc.) are involved in the policy creation and implementation processes. Previous studies shed some light on these questions, indicating that there is a need to establish a knowledge/communication foundation for MFP, by identifying key knowledge types, addressing main knowledge gaps, and designing ad-hoc educational campaigns or interventions for the public (Yan et al., 2020b; Garcia-Vazquez and Garcia-Ael, 2021). To achieve these, it is crucial to have standardised terminologies/definitions and test methods. The former is crucial, as MFP is currently being referred to through different terminologies such as 'microplastic pollution' and 'fibre fragmentation' within research and policy plans (TMC, 2022; Liu et al., 2021). Furthermore, whilst establishing labelling and certification schemes is necessary as it is an alternative way to raise public awareness, it is also vital to have researchers, mass media, NGOs, educators, and creative communicators involved to create understandable science communication to a wider audience and stimulate discussion.

Moreover, the messages/knowledge of MFP can also be embedded in broader communication schemes that aim at fostering sustainable behaviour changes (e.g. climate change, circular economy, sustainable consumption), which has the potential to be a cost-effective communication option (OECD, 2021; EEA, 2022). An example of this is the #What's in my clothes# campaign in the UK, which encourages consumers to read their clothing labels and be aware of the materials they purchased; the campaign further provided information on synthetic microfibre shedding and prevention measures that consumers can implement at home (during laundering) (Fashion Revolution, 2019). However, again this can only be effectively communicated when we have agreed-upon definitions addressing MFP and harmonised test methods for microfibre shedding. To explain, widely agreed information and methods can further lead to measuring evaluating, and communicating the effectiveness of mitigation solutions, since the information on product effectiveness (effectiveness knowledge) can affect consumers' uptake of the solutions and whether they will implement it for a long term (Yan et al., 2020b).

17.4.2 Explore Synergies with Other Environmental Policy Objectives

As mentioned above, the communication of MFP can be a part of wider disclosure; similarly, preventative mitigation actions can also be integrated into wider policy plans for social and environmental protection. For instance, it is noticed that there are several potential synergies between actions to mitigate microfibre releases and actions targeting promoting a transition towards a more circular textile sector (OECD, 2020; EEA, 2022). One example of an existing synergy within policy is the

EU circular economy action plan, which envisages a comprehensive EU strategy for sustainable textiles, encompassing the strategies to tackle synthetic microfibres. Its proposed actions include reducing the production and consumption of fast fashion, promoting upcycling and repairing to extend the lifetime of the garment and prevent landfilling garments, and establishing eco/circular design practices to enable the reuse, repair, and recycling of the existing materials (EC, 2020; EEA, 2022). How do these approaches help address MFP? To explain, landfilled textiles/garments can release microfibres via waterways and/or being airborne (breaking off the garment); in this sense, policies aimed at preventing the mismanagement of waste textiles (e.g. through 'binding them' as reuse and recycling) and reducing the production of fast fashion (mainly made from synthetic fabrics) could contribute to reducing microfibre shedding through eliminating release in raw material manufacturing, production, and end-of-life stages. However, more concrete studies investigating the relationship between microfibre shedding and reusing and recycling of garments are needed. Whilst policy actions potentially reduce the negative effect of microfibre release, their rebound effects (trade-off) of any other environmental risks are worth considering. For instance, whilst eliminating microfibre release through implementing eco/circular design and/or through manufacturing innovative materials, water and energy consumption, and economic and social implications should be investigated.

17.5 CONCLUSIONS AND FUTURE DIRECTIONS

Textiles release microfibres throughout their lifecycle, as shown in Figure 17.1, from production to use, including wearing, washing, and drying (Athey et al., 2020; De Falco et al., 2020; Kapp and Miller, 2020). Textiles then continue to release microfibres at their end-of-life stage by being reused, recycled, or landfilled (Gavigan et al., 2020). Thus, MFP needs to be tackled with multiple stakeholders in mind and

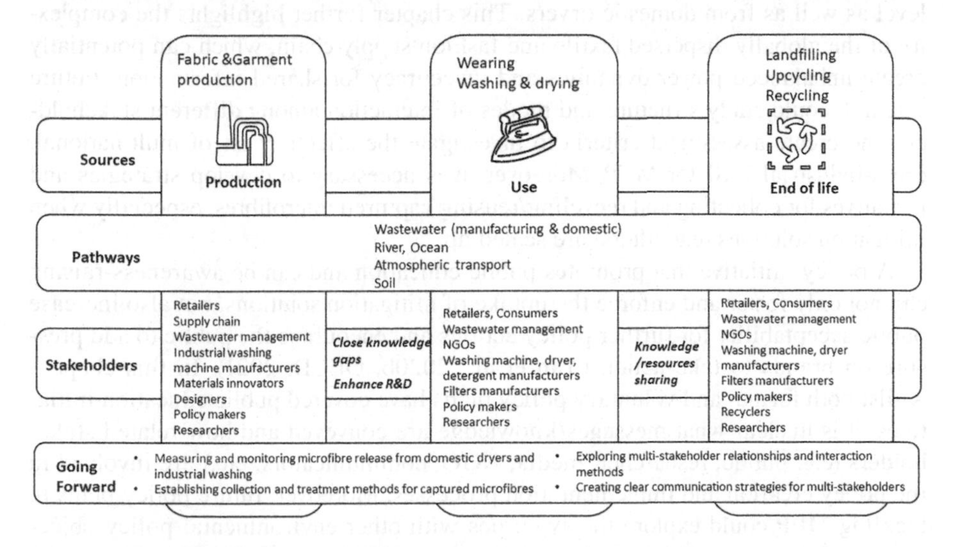

FIGURE 17.1 MSIs for MFP – Sources, Pathways, Multi-stakeholder, and Going Forward visualised as a fabric for life.

being involved at multiple stages. Figure 17.1 identifies the sources and pathways of MFP, its relevant stakeholders that should be involved in MSIs to achieve the identified objectives, and areas for future consideration and investigation.

This chapter set out to explore whether and how a multi-stakeholder approach for MFP connects existing policy plans. These questions are addressed by reviewing prominent examples of formal government-led and voluntary research and/or industry-led policy plans in relation to MFP. It is noted that formal governance takes a top-down centralised approach, in which decisions are taken and policies are formulated by the government and other authorities, leaving the other actors to obey without initial participation and collaboration. This approach is further seen as slow to tackle MFP, which requires immediate action. Thus, the promotion of multi-stakeholder, multinational, and cross-industry (interdisciplinary) collaboration within governance can be a key enabler for achieving policy objectives, in particular to accelerate research, close knowledge gaps, establish standardised test methods, identify best practices, enhance R&D within and across sectors, and create platforms for knowledge/resources sharing (OECD, 2021). More inclusive and comprehensive research evidence across different sectors can accelerate solutions, and it is the foundation for the determination and implementation of further policy measures. Therefore, platforms such as the TMC's 'Microfibre Data Portal' and 'Microfibre Knowledge Hub' should be created, promoted, and sustained to facilitate the dissemination of information and knowledge and enhance multi-stakeholder dialogue in the long term (OECD, 2021). Although the existing MSIs have clearly identified and have been working on several shared underpinning objectives, the challenges or areas for future consideration remain. Through the review, it is noted that most of the current research and policy plans predominantly focus on standardising test methods for microfibre shedding for domestic washing. It leaves a knowledge gap such as harmonised measurements and monitoring microfibre release on an industrial (washing) level as well as from domestic dryers. This chapter further highlights the complexity of the globally dispersed textile and fashion supply chain, which can potentially create unbalanced power dynamics and inaccuracy for shared information. Future research could study structure and modes of interaction among different stakeholders and create assessment criteria to investigate the effectiveness of multinational, cross-industrial MSI for MFP. Moreover, it is necessary to develop strategies and initiatives for collecting and recycling/reusing captured microfibres, especially when mitigation solutions (e.g. filters) are scaled up.

A policy initiative that promotes public education and can be awareness-raising can not only foster and enforce the uptake of mitigation solutions, but also increase public acceptability for further policy actions and encourage the public to add pressure on brands to take action (Yan et al., 2020b; OECD, 2021). In this chapter, whilst both formal and voluntary policy plans have covered public education initiatives, it is unclear what messages/knowledge are conveyed and how related stakeholders (e.g. public, researcher, media, NGO, communicators, etc.) are involved in the policy creation and implementation processes. Moreover, future policy plans in tackling MFP could explore the synergies with other environmental policy objectives, such as sustainable/circular textile plans. Whilst synthetic microfibres take the central stage within existing policy plans, other types of microfibres (i.e. natural and regenerated cellulosic microfibres) remain understudied (Yan et al., 2020a;

Finnegan et al., 2022). The main challenge associated with the latter types is their biodegradability and toxicity, which could be potentially addressed in synergy with other wider policy approaches (e.g. textile toxicity). This chapter identified and discussed existing prominent formal and voluntary policy plans and objectives to tackle MFP. Future research could tackle one of the aspects that have been identified within the Going Forward section in Figure 17.1.

REFERENCES

Amos, J. (2020). *Plastic pollution: Washed clothing's synthetic mountain of 'fluff*. BBC. https://www.bbc.co.uk/news/science-environment-54182646 (accessed October 3, 2022)

APPG. (2021). *Microplastics policies for the government*. All-Party Parliamentary Group on Microplastics. https://www.thewi.org.uk/__data/assets/pdf_file/0003/550038/WI_APPGMicroplastics_Report.pdf (accessed September 11, 2022)

Athey, S. N., Adams, J. K., Erdle, L. M., Jantunen, L. M., Helm, P. A., Finkelstein, S. A., & Diamond, M. L. (2020). The widespread environmental footprint of indigo denim microfibers from blue jeans. *Environmental Science & Technology Letters*, *7*(11), 840–847.

Athey, S. N., & Erdle, L. M. (2022). Are we underestimating anthropogenic microfiber pollution? A critical review of occurrence, methods, and reporting. *Environmental Toxicology and Chemistry*, *41*(4), 822–837.

California Legislative Information. (2019). *AB-129 microfiber pollution*. https://leginfo.legislature.ca.gov/faces/billTextClient.xhtml?bill_id=201920200AB129 (accessed August 21, 2022)

DAWE. (2021). *National plastics plan 2021*. https://www.dcceew.gov.au/sites/default/files/documents/national-plastics-plan-2021.pdf (accessed August 26, 2022)

De Falco, F., & Cocca, M. (2022). Innovative approaches to mitigate microfibre pollution. In S. W. Judith, M. Cocca, and F. De Falco (Eds), *Polluting Textiles* (pp. 245–264). Routledge, London.

De Falco, F., Cocca, M., Avella, M., & Thompson, R. C. (2020). Microfiber release to water, via laundering, and to air, via everyday use: A comparison between polyester clothing with differing textile parameters. *Environmental Science & Technology*, *54*(6), 3288–3296.

Euratex. (2022). *Cross industry agreement*. https://euratex.eu/cia/ (accessed August 27, 2022)

European Commission. (2022). *EU strategy for sustainable and circular textiles*. https://eur-lex.europa.eu/legal-content/EN/TXT/?uri=CELEX%3A52022DC0141 (accessed September, 21, 2022)

European Environmental Agency. (2022). *Microplastics from textiles: Towards a circular economy for textiles in Europe*. https://www.eea.europa.eu/publications/microplastics-from-textiles-towards-a (accessed August 24, 2022)

Fashion Revolution. (2019). *#whatsinmyclothes: The truth behind the label*. https://www.fashionrevolution.org/whatsinmyclothes-the-truth-behind-the-label/ (accessed October 04, 2022)

Finnegan, A. M. D., Süsserott, R. C., Gabbott, S. E., & Gouramanis, C. (2022). Man-made natural and regenerated cellulosic fibres greatly outnumber microplastic fibres in the atmosphere. *Environmental Pollution*, 119808.

Garcia-Vazquez, E., & Garcia-Ael, C. (2021). The invisible enemy. Public knowledge of microplastics is needed to face the current microplastics crisis. *Sustainable Production and Consumption*, *28*, 1076–1089.

Gavigan, J., Kefela, T., Macadam-Somer, I., Suh, S., & Geyer, R. (2020). Synthetic microfiber emissions to land rival those to waterbodies and are growing. *PLoS One*, *15*(9), e0237839.

Gaylarde, C., Baptista-Neto, J. A., & da Fonseca, E. M. (2021). Plastic microfibre pollution: How important is clothes' laundering? *Heliyon*, *7*(5), e07105.

Government of Canada. (2020). *A proposed integrated management approach to plastic products: Discussion paper.* https://www.canada.ca/en/environment-climate-change/services/canadian-environmental-protection-act-registry/plastics-proposed-integrated-management-approach.html (accessed August 24, 2022)

Hazlehurst, A., Tiffin, L., Sumner, M., & Taylor, M. (2023). Quantification of microfibre release from textiles during domestic laundering. *Environmental Science and Pollution Research*, 1–18.

Henry, B., Laitala, K., & Klepp, I. G. (2019). Microfibres from apparel and home textiles: Prospects for including microplastics in environmental sustainability assessment. *Science of the Total Environment*, *652*, 483–494.

Herweyers, L., Catarci Carteny, C., Scheelen, L., Watts, R., & Du Bois, E. (2020). Consumers' perceptions and attitudes toward products preventing microfiber pollution in aquatic environments as a result of the domestic washing of synthetic clothes. *Sustainability*, *12*(6), 2244.

ISO 3758:2012. (2012) *Textiles - Care labelling code using symbols.* https://www.iso.org/standard/42918.html (accessed September 04, 2022)

Kapp, K. J., & Miller, R. Z. (2020). Electric clothes dryers: An underestimated source of microfiber pollution. *PLoS One*, *15*(10), e0239165.

Legifrance. (2020). *LOI n° 2020-105 du 10 février 2020 relative à la lutte contre le gaspillage et à l'économie circulaire* (1), https://www.legifrance.gouv.fr/jorf/id/JORFTEXT000041553759 (accessed August 23, 2022)

Liu, J., Liang, J., Ding, J., Zhang, G., Zeng, X., Yang, Q., … & Gao, W. (2021). Microfiber pollution: An ongoing major environmental issue related to the sustainable development of textile and clothing industry. *Environment, Development and Sustainability*, *23*, 11240–11256.

Matzembacher, D. E., Vieira, L. M., & de Barcellos, M. D. (2021). An analysis of multi-stakeholder initiatives to reduce food loss and waste in an emerging country-Brazil. *Industrial Marketing Management*, *93*, 591–604.

Mendenhall, E. (2018). Oceans of plastic: A research agenda to propel policy development. *Marine Policy*, *96*, 291–298.

Misund, A., Tiller, R., Canning-Clode, J., Freitas, M., Schmidt, J. O., & Javidpour, J. (2020). Can we shop ourselves to a clean sea? An experimental panel approach to assess the persuasiveness of private labels as a private governance approach to microplastic pollution. *Marine Pollution Bulletin*, *153*, 110927.

Moog, S., Spicer, A., & Böhm, S. (2015). The politics of multi-stakeholder initiatives: The crisis of the Forest Stewardship Council. *Journal of Business Ethics*, *128*, 469–493.

Napper, I. E., & Thompson, R. C. (2016). Release of synthetic microplastic plastic fibres from domestic washing machines: Effects of fabric type and washing conditions. *Marine Pollution Bulletin*, *112*(1–2), 39–45.

New York State Assembly. (2018). *Bill no. A01549.* https://www.nysenate.gov/legislation/bills/2019/A1549 (accessed August 28, 2022)

OECD. (2020). *Workshop on microplastics from synthetic textiles: Knowledge, mitigation, and policy.* https://www.oecd.org/water/Workshop_MP_Textile_Summary_Note_FINAL.pdf (accessed August 21, 2022)

OECD. (2021). OECD iLibrary. https://www.oecd-ilibrary.org/sites/6e06626c-en/index.html?itemId=/content/component/6e06626c-en#snotes-d7e28026 (accessed August 21, 2022)

Ojinnaka, D., & Aw, M. (2020). Micro and nano plastics: A consumer perception study on the environment, food safety threat and control systems. *Biomedical Journal of Scientific & Technical Research*, *32*(2).

Rasche, A. (2012). Global policies and local practice: Loose and tight couplings in multi-stakeholder initiatives. *Business Ethics Quarterly*, *22*(4), 679–708.

Rathinamoorthy, R., & Raja Balasaraswathi, S. (2022). Microfiber pollution prevention-mitigation strategies and challenges. In *Microfiber Pollution* (pp. 205–243). Springer Nature, Singapore.

Sandin, G., Roos, S., & Johansson, M. (2019) *Environmental impact of textile fibres: What we know and what we don't know*. Mistra Future Fashion (online): https://mistrafuture-fashion.com/ download-publications-on-sustainable-fashion (accessed August 24, 2022)

Swedish EPA. (2019). *Microplastics*. https://www.naturvardsverket.se/amnesomraden/plast/om-plast/mikroplast/ (accessed August 27, 2022)

The Guardian. (2022). *Microplastics in sewage: A toxic combination that is poisoning our land.* https://www.theguardian.com/commentisfree/2022/may/26/microplastics-sewage-poison-land-britain-waterways-chemicals?CMP=share_btn_tw (accessed November 05, 2022)

The Microfibre Consortium. (2022). https://www.microfibreconsortium.com/ (accessed August 20, 2022)

The State of Connecticut. (2018). *An act concerning clothing fiber pollution*. https://www.cga.ct.gov/2018/TOB/s/2018SB-00341-R00-SB.htm (Accessed August 21, 2022)

Thodges. (2019). *Bills and best practices for microfiber pollution solutions*. Surfrider. https://www.surfrider.org/coastal-blog/entry/bills-and-best-practices-for-microfiber-pollution-solutions (accessed October 20, 2022)

Turunen, L. L. M., & Halme, M. (2021). Communicating actionable sustainability information to consumers: The Shades of Green instrument for fashion. *Journal of Cleaner Production, 297*, 126605.

Yan, S., Henninger, C. E., Jones, C., & McCormick, H. (2020b). Sustainable knowledge from consumer perspective addressing microfibre pollution. *Journal of Fashion Marketing and Management: An International Journal, 24*(3), 437–454.

Yan, S., Jones, C., Henninger, C. E., & McCormick, H. (2020a). Textile industry insights towards impact of regenerated cellulosic and synthetic fibres on microfibre pollution. In *Sustainability in the Textile and Apparel Industries: Sourcing Synthetic and Novel Alternative Raw Materials*, (pp.157–171). Springer, Cham.

Zambrano, M. C., Pawlak, J. J., Daystar, J., Ankeny, M., Cheng, J. J., & Venditti, R. A. (2019). Microfibers generated from the laundering of cotton, rayon and polyester based fabrics and their aquatic biodegradation. *Marine Pollution Bulletin, 142*, 394–407.

Zarghamee, R., Fowler, S. L., & Myers, A. (2022). *California announces first statewide microplastics strategy*. https://www.pillsburylaw.com/en/news-and-insights/ca-first-microplastics-strategy.html (accessed August 21, 2022)

18 Sustainable Approaches in Textile Finishing to Control Microfiber Releases

Aravin Prince Periyasamy
Aalto University & VTT Technical Research Center of Finland

18.1 INTRODUCTION

In most cases, the origin of the microplastics that are released into the aquatic environment may be differentiated into two categories: primary and secondary. Primary microplastics are microplastics that are directly released into the environment as small plastic particles (less than 5 mm in size). This category of microplastics includes the microplastics that are generated during the process of home washing. Other sources of the principal microplastics include things like road markings, the natural wear and tear of tires, marine coatings, and cosmetics. When larger pieces of plastic are released into the environment, the natural weathering process breaks them down over time, resulting in the production of secondary microplastics (Andrady, 2011; Periyasamy, 2021). Among the sources that have been discussed, garments are responsible for 35% of the release in the amount of microplastics, primarily during washing cycles (*Microplastics: Sources, Effects, and Solutions*, 2018). This chapter will cover the many sustainable and environmental methods that can be used to reduce the amount of microplastics produced by domestic washing.

18.2 REDUCTION TECHNIQUES FOR FIBER EMISSION

The generation and release of microplastics are influenced by many factors during domestic washing, including the type of fabric, the geometry of the fabric (woven, knit, or nonwoven), the type of yarn (twist, evenness, hairiness, staple fiber, filament, and the number of fibers per cross-section), the processing history of the fabric (spinning, knitting, weaving, scouring, bleaching, dyeing, finishing particularly different functional finishes (Periyasamy et al., 2020), and drying processes), and the physicochemical properties of the fibers themselves (Bishop, 1995; Hernandez et al., 2017; Periyasamy, 2021).

The shattering of the short-staple fibers is primarily responsible for the creation of microfibers and their subsequent discharge (Acharya et al., 2021; Belzagui et al., 2019). Due to the fact that the staple length of natural fibers can vary, it can be challenging to produce evenness in the yarn made from these fibers. The natural fibers include both short-staple fibers, whose length ranges from 2.5 to 6.0 cm, and long-staple fibers, whose lengths are greater than 6.0 cm (Babu, 2015; Elhawary, 2015; Negm &

 DOI: 10.1201/9781003331995-22

Sanad, 2020). However, since most of the short-staple fibers are eliminated during the combing process, combed yarn unquestionably offers consistency. Staple fibers of the proper length are often cut from synthetic fibers like polyethylene terephthalate (PET), polyamide (PA), polyacrylonitrile (PAN), and regenerated fibers like Lyocell and Viscose rayon for use in staple yarn manufacturers (typically less than 32 mm) (Bhat & Kandagor, 2014; Elhawary, 2015; El-Sayed & Sanad, 2010; Erdumlu & Ozipek, 2008; Hagewood, 2014). Fuzz is the result of various factors such as the friction of fibers, their cross-sectional shape, thickness, stiffness, yarn hairiness, and abrasion resistance. Short-staple fibers (also called protruding fibers) and increased hairiness in the yarn are responsible for a significant quantity of microplastic emissions after washing (Ratnam, 2010), which may then create fuzz (Figure 18.1).

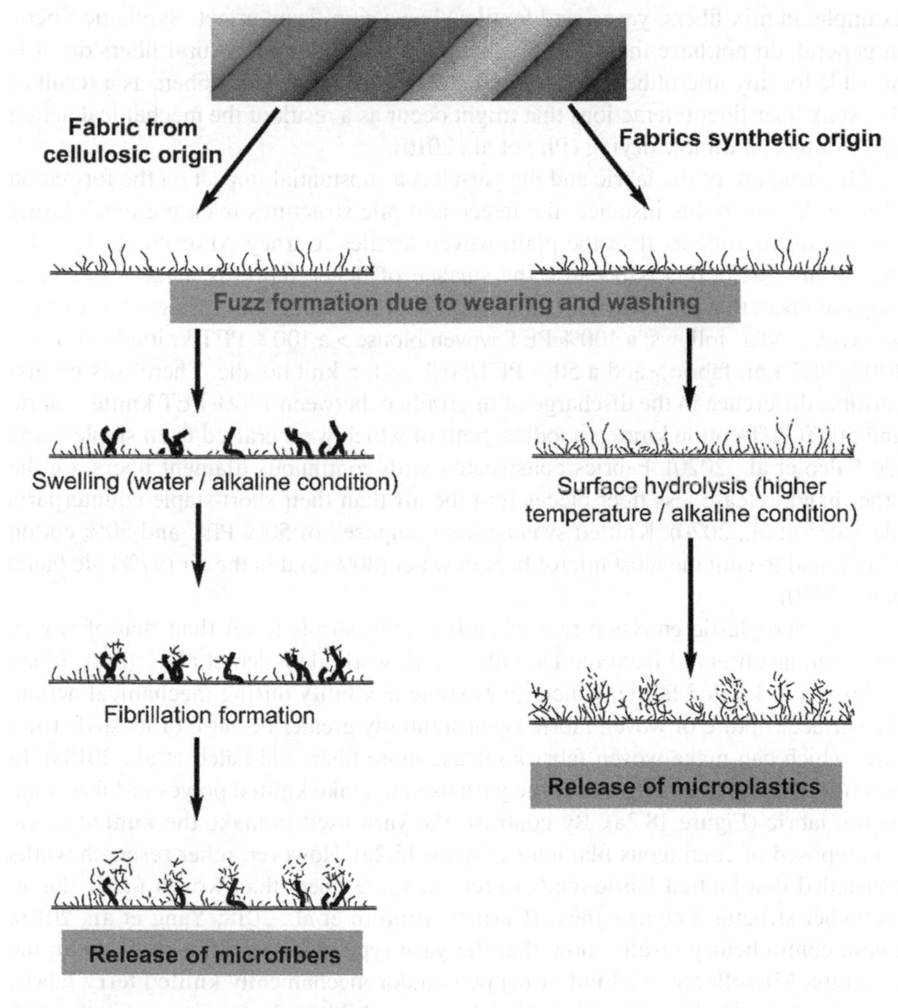

FIGURE 18.1 A depiction in the form of a schematic illustrating the process by which microfibers are generated in textiles of cellulosic origin and polyester fabrics. (*Source:* Modified and reused from Periyasamy & Tehrani-Bagha, 2022 under C.C 4.0 license.)

The majority of the time, pilling is caused by staple fibers in the fabric breaking apart at the surface as a result of being subjected to mechanical stress. The development of fuzz is the initial stage of the pilling process, which is characterized by the gradual separation of fibers from a surface owing to the mechanical action of the washing and wearing cycles (Muthusamy et al., 2021; Periyasamy, 2020b). Microfiber development is expected to follow a similar mechanism (seen in Figure 18.1) for both synthetic fibers like polyester and cellulosic fibers like cotton and lyocell during pill formation. According to the cited paper (Chiweshe & Crews, 2000), the use of fabric conditioners during the washing process of synthetic clothes increases the likelihood that the garments may pill.

When it comes to the formation and release of microplastic during laundry, the percentage of various components used in the construction of the yarn or fabric (for example, in mix fibers, yarns, and textiles) has a significant effect. Synthetic fibers, in general, do not have the natural convolution or crimp that natural fibers do. It is possible for tiny microfibers to become detached from the major fibers as a result of the weak fiber-fiber interactions that might occur as a result of the mechanical action of washing and tumble drying (Pirc et al., 2016).

The structure of the fabric and the yarn has a substantial impact on the formation of microfibers; in this instance, the fleece and pile structures release a much larger number of microfibers than the plain-woven textiles (Carney Almroth et al., 2018; Pirc et al., 2016), this is because the surface of fleece fabric is made up of open loops or fibers that protrude out (Lant et al., 2020; Wilson, 2012). Fragments of fiber were released as follows: a 100% PET woven blouse > a 100% PET knitted t-shirt > a 100% PET knit fabric > and a 50% PET/50% cotton knit hoodie. There was no discernible difference in the discharge of microfibers between 100% PET knitted fabric and 50/50 PET/cotton knitted hoodies, both of which were crafted from staple yarns (de Falco et al., 2020). Fabrics constructed with continuous filament fibers, on the other hand, release less fiber pieces into the air than their short-staple counterparts (de Falco et al., 2020). Knitted sweatshirts composed of 50% PET and 50% cotton were found to emit the most microfibers in water (80%) and in the air (97%) (de Falco et al., 2020).

The microplastic emission rate of knitted polyester is lower than that of woven polyester, as observed from the De Falco et al. work (de Falco et al., 2018a). When compared to knitted fabric, which gives some flexibility during mechanical action, the surface rupture of woven fabric is substantially greater because of its stiff structure, which can make woven fabrics release more fibers (de Falco et al., 2018a). In addition, there is less hairiness in the yarn used to make knitted polyester fabric than woven fabric (Figure 18.2a). By contrast, the yarn used to make the knitted fabric is composed of continuous filaments (Figure 18.2a). However, other research works concluded that knitted fabric tends to release more fibers than woven fabric due to its higher structural compactness (Carney Almroth et al., 2018; Yang et al., 2019). These contradictory results show that the yarn type plays an influential role in the structure. Microfleece is a kind of napped and/or mechanically knitted terry fabric. In contrast, doubled yarn is utilized for the warp while single yarn is used in the weft (Figure 18.2b and c). A greater number of microfibers were discharged from these microfleeces than from conventional jersey knits (Carney Almroth et al., 2018).

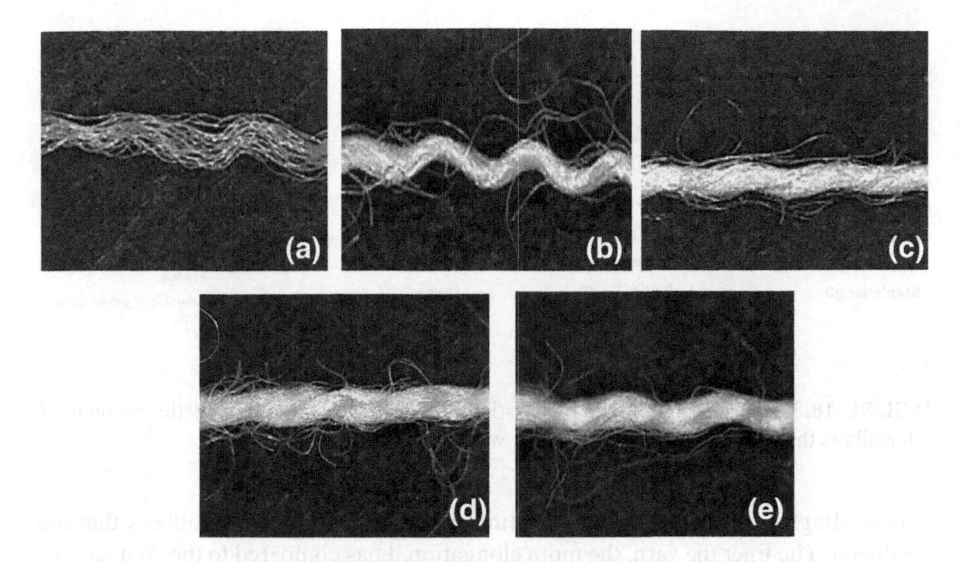

FIGURE 18.2 Optical microscope images of (a) continuous polyester yarn from double knit jeresy; (b) a staple polyester weft yarn; (c) a staple polyester warp yarn from plain weave polyester; (d) a staple polypropylene weft yarn; and (e) a staple polypropylene warp yarn from plain weave polypropylene. (*Source:* Reprinted from de Falco et al., 2018b with kind permission of Elsevier publications.)

The amount of microfibers released by old clothes is around 25% more than by new clothes (Hartline et al., 2016). Over time, the mechanical structure of fibers degrades owing to factors such as exposure to sunlight, wear, and cleaning. Sunlight has a wide range of wavelengths that can damage fabrics in a variety of ways. The fabric's exposure to UV, visible, and infrared light can generate more heat, speed up the oxidation of fibers, and hasten their progressive disintegration (Gogotov & Barazov, 2014). In addition to these, the wear and tear on fibers are accelerated by things like CO_2, NO_2, and NO_3 gas exposure, perspiration, abrasion, friction during wear, and frequent washings and dryings (Yousif & Haddad, 2013).

The mechanical actions that take place during the washing process have the potential to break staple fibers down and detach them from the yarn (Zambrano et al., 2019). Since the emission of microfibers is heavily influenced by factors such as the twist of yarns, breaking strength, elongation, bending stiffness, friction, and the softness of the yarn (Carney Almroth et al., 2018; de Falco et al., 2019b, 2020; Zambrano et al., 2019). The hairiness of the yarn contributes to the production of fuzz as it also contributes to the formation of pilling in the structure of the fabric, which ultimately leads in the discharge of microfibers.

Figure 18.3 presents a comprehensive overview of the characteristics of textiles and their impact on the reduction of microfibers. In fact, fiber is the raw material for textiles, and the characteristics of textile products, including how they behave with regard to the shedding of microfibers, are dependent on the properties of fibers. When selecting fibers, it is better to choose those that have a longer staple length and continuous filaments that have a greater tenacity and higher elongation (Periyasamy &

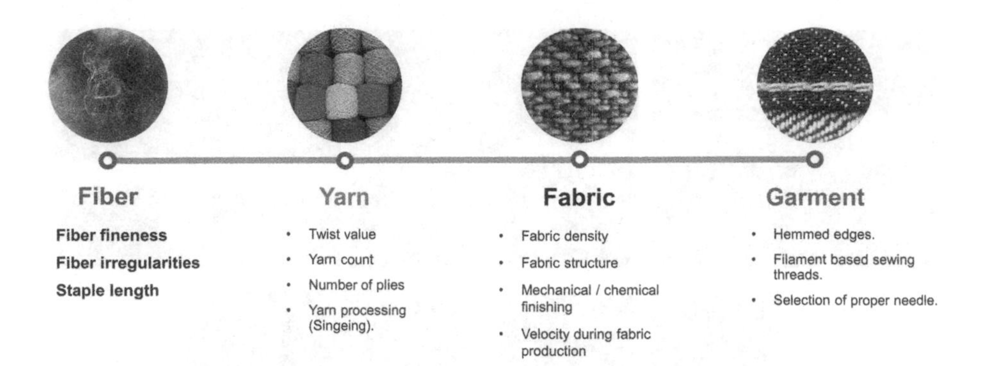

FIGURE 18.3 Various textile characteristics are desirable for reducing the amount of microfibers that are released throughout the washing and drying processes.

Tehrani-Bagha, 2022). This helps to minimize the amount of microfibers that are produced. The finer the yarn, the more elongation it has compared to the coarser it is, and it also has fewer short fibers, also known as protruding fibers. As a result, there are fewer opportunities for the release of microfibers. Fabrics manufactured with combed yarns are recommended for use in the production of textiles based on the origin of natural fibers since these yarns have a more consistent appearance than carded and open-end spun yarns do. Additionally, the combed yarn has reduced hairiness, also known as hairy fibers, which results in a reduction in the amount of microfibers that are emitted. The higher twist per meter results in increased yarn strength, which in turn leads to a decrease in the amount of generations of microfibers. A fabric with a higher floating structure, such as satin or 3/1 twill, produces more microfibers than a fabric with a lower floating structure because the higher floating structure provides more opportunities for the short fibers in the fabric and yarn structure to get dislodged (Periyasamy & Tehrani-Bagha, 2022). It's possible that the fabric has a larger cover factor, and its compact structure helps to limit the emission of microfibers due to the increased fiber-to-fiber interactions. Some of the fabric's physical attributes have a directionally proportionate relationship to the amount of microfibers that are released. For example, the greater the abrasion resistance of the fabric, the less microfibers are released with regular home washing (Periyasamy & Tehrani-Bagha, 2022). On the other hand, some of the finishes enhance the emission, which is something that can be further addressed in Section 18.2.2. There are just a few chemical and enzymatic finishes on the textiles that have the potential to reduce the release of microfibers. During the preparation of the garment, sealing the edges offers more potential for reducing the emission of microfibers. However, the friction that occurs between the sewing thread and needle operation and the fabric, as well as the mechanical damage that occurs when the needle comes into contact with the fibers, may lead to an increase in the amount of microfibers that are released (Periyasamy & Tehrani-Bagha, 2022).

Figure 18.4 explains the significant criteria that need to be taken into account in order to reduce the amount of microfibers and microfibers that are released throughout the washing and drying processes. During an extended period of washing, the

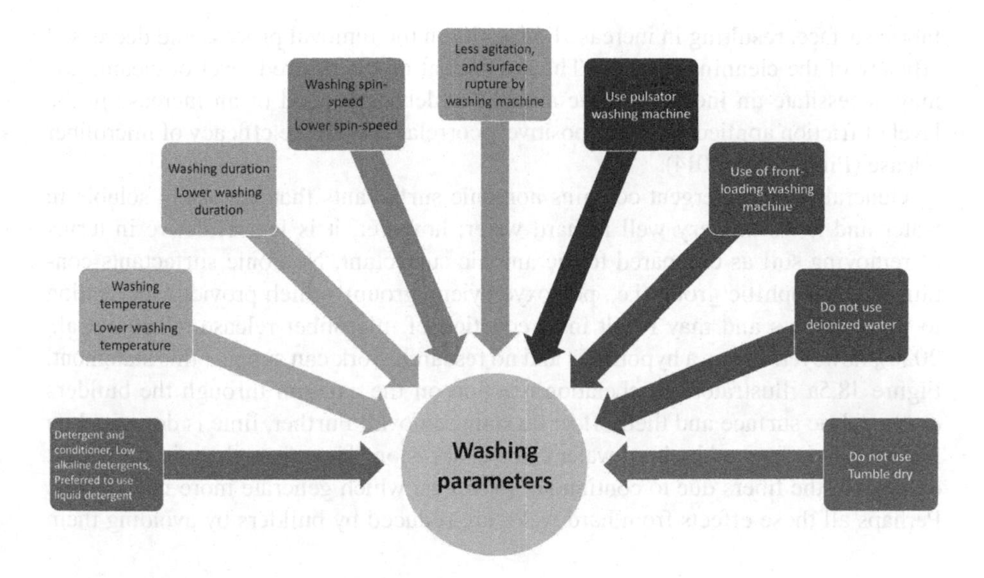

FIGURE 18.4 Factors to consider in order to minimize the microfibers/microplastics throughout the laundering and drying processes.

temperature causes the fiber to deteriorate owing to the oxidation of the fibers derived from cellulosic-based materials and the hydrolysis of the fibers that are derived from polyester (Napper & Thompson, 2016). As temperatures rise, detergents moisten surfaces, and fibers slip more easily from yarns and fabric structures. It is better to use formulations of liquid detergents that are based on nonionic detergents rather than formulations that have an anionic nature and inorganic salts. The formulation ought to demonstrate its better washing efficiency at lower temperatures and with higher levels of water hardness (Winkler et al., 2019). Increased friction between the clothing and the machine drum, caused by increased washing rotational speed and agitation, leads to an increased amount of microfibers being released. Therefore, the optimal speed will lower the total microfiber release rate.

Moreover, the ratio of the garment to washing liquor in the washing machine plays a vital role in the emission of microfibers (Kelly et al., 2019). The authors observed 65 mg of microfibers per kg of garment, and it increased to 125 mg of microfibers when the wash water was used twice the amount. In addition, the higher water volume increases the mechanical stress on the garments during the washing process. New additives (i.e., builders and softeners) for washing liquor were studied by Piñol et al. (2014) to reduce microfiber emissions. The primary role of builders is to sequester the cations (such as Ca^{2+} and Mg^{+}) present in hard water during the laundering procedure. Anionic surfactants have been found to effectively remove soil and stains. However, it is important to note that the efficacy of these surfactants may be compromised in the presence of multivalent ions commonly found in hard water. The precipitation of anionic surfactants occurs due to their bonding with positive ions, resulting in their deposition on the surface of the fabric. Moreover, the calcium ion exhibits the characteristic of serving as an adhesive agent between the stain and the fabric surface, resulting in increased difficulty in the removal process and decreased efficacy of the cleaning process. The attainment of the desired level of cleanliness may necessitate an increase in the amount of detergent used or an increase in the level of friction applied, which is positively correlated with the efficacy of microfiber release (Piñol et al., 2014).

Generally, the detergent contains nonionic surfactants that are easily soluble in water and it works very well in hard water; however, it is less effective in terms of removing soil as compared to the anionic surfactant. Nonionic surfactants contain the hydrophilic group (i.e., polyoxyethylene group), which provides lubrication to the garments and may result in a reduction of microfiber release (Cheng et al., 2020); however, this is a hypothesis and no research work can support this statement. Figure 18.5a illustrates the chelation reaction on the calcium through the builders on the fabric surface and their effect on stain removal. Further, lime is deposited on fabrics as it is formed by hard water causing abrasion. Therefore, there is significant damage to the fibers due to continuous washings, which generate more microfibers. Perhaps all these effects from hard water are reduced by builders by avoiding them

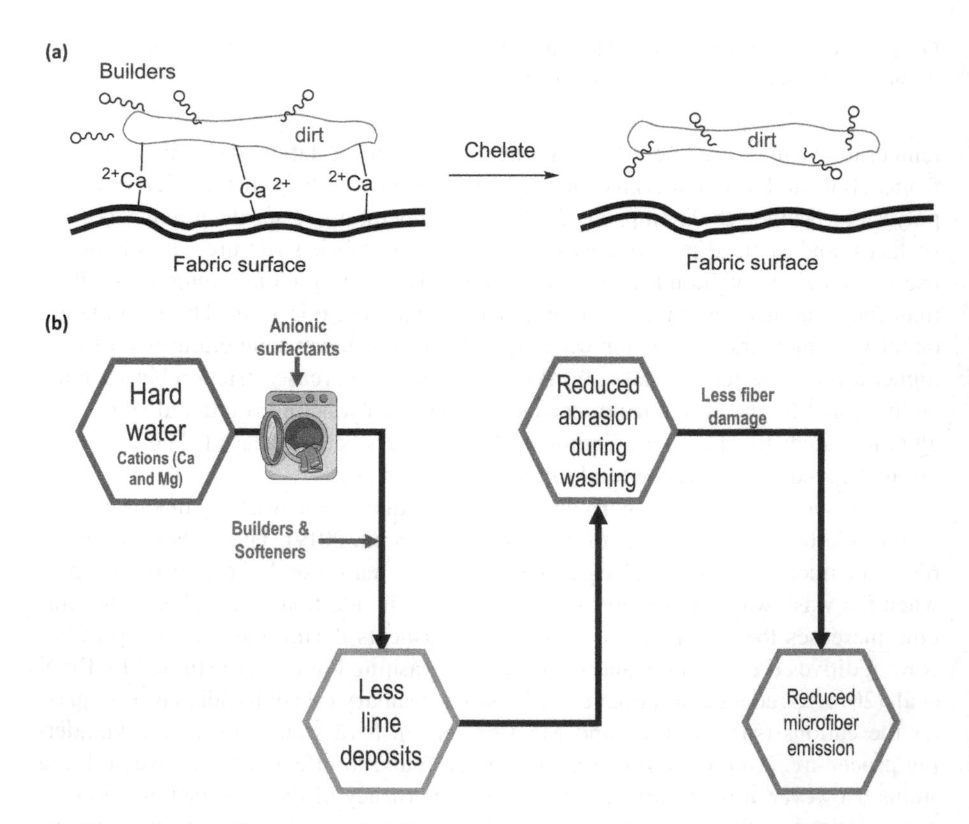

FIGURE 18.5 Builders and their impacts on stain removal under hard water (a) and action on microfiber reduction (b). (*Source:* (a) adopted from Mermaids (2015) Study of new additives for washing liquor and their contribution to the reduction of fiber breakage, 2015 (b) created by the author.)

creating problems (Figure 18.5b). On the other hand, builders also deposit, yet the quantity is minimum during the washing cycle without using builders.

Enzymes, including cellulases, proteases, lipases, and amylases, were added to detergents to increase their effectiveness in removing stains and reduce the time it takes to wash the laundry (Niyonzima, 2019). To convert the cellulosic fiber into monosaccharides (Lee et al., 1983), enzymes like cellulase just hydrolyze the -1,4-linkages found in the cellulose fiber. Consequently, the cellulase-containing detergents cause the fibers to become detached from the surface of the fabric, which leads to an enhanced release of microfibers. In general, the enzymes that are linked with detergents are costlier. Although they are effective for washing and removing stains, they also cause more microfibers to be released during the washing process.

18.2.1 Mechanical Finishing

Mechanical finishing, including singeing and calendaring, are used to control the dispersion of fiber particles. Singeing is the practice of using a carefully directed open flame to burn away the coarse hairs that have worked their way to the surface of a fabric. Calendaring is the process of compressing the fabric by passing it between two or more rollers while maintaining precise control over the time, temperature, and pressure (Hossain et al., 2021). The fabric is subjected to sanforizing and heat-setting procedures so that it may be shrunk and fixed in both the warp and the weft directions. The durability of mechanical finishes is often just average, and they are unable to protect the fabric from the release of microfibers after only a few washing cycles have been completed. After just a few washes, the surface of the fabric will begin to exhibit protrusions of the fabric's threads (Senthil Kumar & Sundaresan, 2013).

The surface of the fabric is given a mechanical treatment by shearing and brushing, which results in the cloth being soft and thick. During the shearing process, the surface fibers, also known as projecting fibers, are chopped to the required length in a consistent fashion. The brushing process contributes to the fiber being pulled away from the structure of the yarn. There is a considerable physical change in the surface structure of the fabric that occurs in both of the finishes, and this shift facilitates the release of microfibers (Figure 18.6). In most cases, the increased release of microfibers that occurs during the first few washes is caused by the release of entrapped microfibers that were formed during the mechanical finishing step (Cai et al., 2020).

18.2.2 Chemical Finishing

18.2.2.1 Enzymatic Treatment

Surface modifications made to textiles may be used to help tackle the problem of microfibers being released into the environment. Biopolishing, which is used on cellulosic clothing and involves the enzyme cellulase, is described as the process of removing fibrils or microfibrils, from the surfaces of cellulosic fabrics. The development of pilling and the emergence of color fading are both indicators that there are fibrils present in the final products, which is a problem since fibrils cause problems with the finished goods. The removal of fibrils by enzymatic means leads to the production of garments/textiles that are not only softer but also cleaner. Because of this,

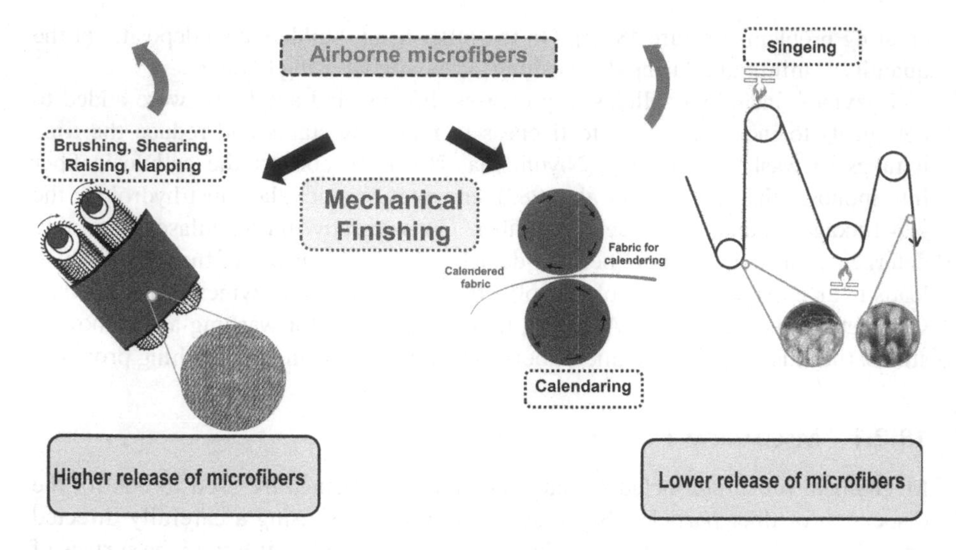

FIGURE 18.6 Various mechanical finishings and their impact of microfiber emission during domestic washing. (*Source:* Adapted from Periyasamy, A. P., Tehrani-Bagha, A. (2022). A review on microplastic emissions from textile materials and their reduction techniques. *Polymer Degradation and Stability*, 199, 109901. https://doi.org/10.1016/j.polymdegradstab.2022.10990, 2022.)

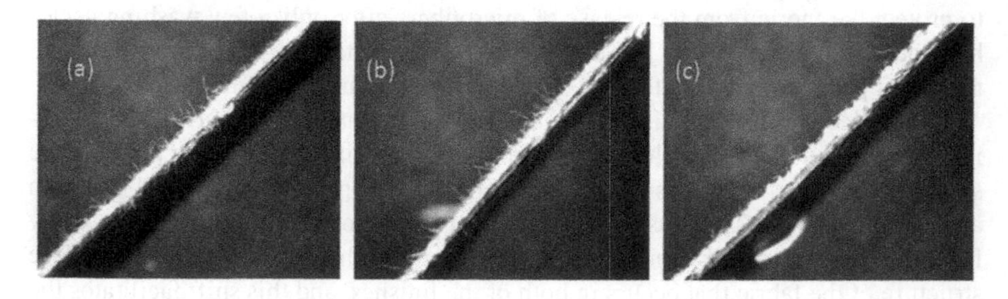

FIGURE 18.7 Microscopical image of protruding fibers on a pristine fabric surface (a), fabric treated with soluble cellulase (b), and immobilized cellulase-treated fabric (c). (*Source:* Reprinted from Sankarraj & Nallathambi, 2018, with kind permission of Elsevier publications.)

the original color is not altered in any way throughout the procedure (Andreaus et al., 2019; Arja, n.d.). According to Nisa's findings (Sankarraj & Nallathambi, 2018), the immobilized cellulase has a powerful effect on the projecting threads that may be seen on the surface of cotton garments (see Figure 18.7). Following the application of the enzymatic treatment, there is a discernible decrease in the number of projecting fibers from the surface of the fabric as compared to the untreated fabric. Due to the free mobilization assault of the fabric in the interior, the soluble enzyme treatment is more effective than the immobilized enzyme treatment in comparison. This causes the fibers to increase on the surface of the fabric. On the other hand, there is no information on the protrusion of fibers that develop after wash and wear cycles in comparison to the endurance of the enzymatic treatment. In another research study, the microfiber discharging problem was approached using an environmentally friendly method that involved the enzyme lipase and the surface modification of polyester fabric, as a result, the microfiber shedding has been reduced as a result of the surface modification (Ramasamy & Subramanian, 2022).

18.2.2.2 Coating and Chemical Finishing

Finishing and coating textiles with a variety of chemicals, such as silicone, polyurethane, and acrylic polymers, may help to minimize the number of microfibers that are released. A layer is formed by these polymers on the surface of the fibers, which serves to protect the fibers from abrasion and mechanical strain. Because of the abrasion that occurs during use and washing, these polymeric films have the potential to release microparticles in addition to microfibers. Numerous research has been conducted in order to evaluate the mitigating impact that several typical finishing procedures have had on microfibers. The chemical treatment may lessen the number of protruding fibers thanks to its strong attachment to the surface of the textile, which in turn guarantees a decrease in the number of microfibers that are released (Zambrano et al., 2021a, b). There are different resins, namely, acrylic resins, polyurethane resins, and silicone emulsions were applied to the different synthetic fabrics through padding and exhaustion methods (Piñol et al., 2014). This pattern was seen in woven and knitted textiles made of PET, polypropylene (PP), PAN, and PA. Acrylic resins had the best pilling resistance out of the chosen auxiliaries for all of the materials. On the other hand, the microplastic emission from the acrylic resin finished fabric is less compared to polyurethane resin finished fabrics. In a similar vein, the fabric that had been treated with silicon emulsion was noticeably softer than the untreated PET fabric, despite the fact that it pilled more than the latter. According to the results of the laundry tests, the number of microfibers released by fabrics treated with silicone emulsion was much lower than that released by fabrics done with other chemical finishes. The silicone finish results in a softer surface on PET fabric, which reduces the amount of friction that occurs during the washing process between the fabric and the detergent. This is the primary source of the problem. However, it is unable to locate any reports that discuss the durability of these chemical finishes in order to determine whether or not they are short-term or long-term remedies for the release of microfibers (Periyasamy & Tehrani-Bagha, 2022).

18.2.2.3 Finishing with Pectin

Using a basic finishing process that was based on pectin, de Falco et al. (2018c) were able to limit the number of microfibers that were released from the PA fabric. It was accomplished via two separate procedures, the first of which was the synthesis of PEC-glycidyl methacrylate (GMA, 97%) and the second of which was the grafting of PEC-GMA onto the fabric (PEC-GMA-PA). This produced an emulsion mixture. Because of the issues over sustainability, pectin was chosen for further consideration because it is readily available in large quantities and has the potential to react with fabrics. The chemistry that occurs between PEC and GMA was articulated very well by Maior et al. (2008). In accordance with the ISO 105-C06:2010 standard test procedure, the pristine-PA and PEC-GMA-PA textiles were washed for 45 minutes at a temperature of 40°C using a commercial detergent called ariel. According to the findings, the pristine-PA fabric releases fiber pieces with dimensions of 12 ± 222 μm in length, 18 ± 3 μm in mean diameter, and 0.359 g of microfibers for every kilogram of wash. In contrast, the PEC-GMA-PA treatment reduces the number of microfibers to 550 ± 384 μm in length, 16 ± 4 μm in mean diameter, and 0.058 g of microfibers per kilogram of fabric. This represents a decrease of microfibers of almost 90%.

18.2.2.4 Finishing with Biodegradable Polymers

Electrofluidodynamic coating, also known as EFD coating, was used to coat the PA fabric with biodegradable polymers such as polylactic acid (PLA) and poly (butylene succinate-co-butylene adipate) (PBSA), respectively (de Falco et al., 2019a). The resultant textile exhibits a significant decrease of 90% in the discharge of microplastics when compared to the original fabric. The EFD treatment offers several benefits, including the uniformity of coating, the ability to coat particles at the nanoscale level, and the preservation of the fabric's fundamental properties during surface finishing. The microfiber release per gram of washed fabric for PLA and PBSA-finished PA fabric was found to be 428 ± 92 and 456 ± 120, respectively. The pristine PA fabric may emit approximately 3,966 ± 1,425 microfibers per gram. Further, a 90% reduction in microfiber shedding from the PA fabric is reported by this research rather than the previous research (pectin-based finishing) (de Falco et al., 2018c).

Throughout the finishing process, a homogeneous surface coating is applied to the fabric. This coating serves to protect the material from the effects of mechanical and chemical reactions that occur during the washing process. An unwashed piece of PA fabric has released an approximate weight of 0.35 g of microfibers/kg of fabric after it has been laundered. In addition, the coating is long-lasting enough to withstand at least five washing cycles, and the PLA coatings are more long-lasting than the PBSA coatings.

18.2.2.5 Finishing with Chitosan

Piñol et al. conducted an investigation on the microfibers that were released from the chitosan-finished PET fabric (Piñol et al., 2014). There are 1,726, 2,497, and 2,237 microfibers per gram of fabric that were emitted by chitosan-finished PET fabric for 1%, 2%, and 3% chitosan concentrations, respectively. The pristine PET fabric emitted 3,047 microfibers per gram of fabric. When compared to the untreated sample, the textiles that had been treated with chitosan at a concentration of 3% emitted 27% fewer microfibers. However, when the findings of the tests are duplicated on several occasions, there is never a consistent outcome. This may be because of the low affinity that chitosan has with PET fabric. When compared to pectin (de Falco et al., 2018c) and PLA/PBSA (de Falco et al., 2019a), this treatment is seen as being non-durable and less effective than the former two. The possible important results of different mechanical and chemical finishing methods on the reduction of fiber pieces in the laundry are summarized in Table 18.1, which may be seen here.

Additional finishes, such as compacting, in addition to singeing and calendaring, bring a decrease in the number of fiber pieces that are shed (Piñol et al., 2014). On the other hand, the findings are encouraging with regard to polymeric chemical finishes. A layer of protection against abrasion that is provided by a polymeric film that

TABLE 18.1
Influence of Mechanical and Chemical Finishing on Microfibers Releases

Finishing Techniques	Results	References
Finishing with chitosan	A reduction of 27% in the release of microfibers from polyester fabric has been observed. Poor durability and inconsistent results	Piñol et al. (2014)
	18%–43% reduction of microfibers	Mossotti et al. (2018)
	48% reduction of microfibers.	Kang et al. (2021)
Finishing with silicone and acrylic resin	A notable decrease in the amount of microfibers emitted from PET fabric. There is no measurement of durability available for either of the finishes.	Piñol et al. (2014)
Finishing with PLA	63.4% reduction of microfibers released from the treated PA fabric	de Falco et al. (2019a)
Finishing with PBSA	90% reduction of microfibers released from the treated PA fabric	de Falco et al. (2019a)
Finishing with pectin	90% reduction of microfibers released from the treated PA fabric	de Falco et al. (2018c)
Mechanical finishing: singeing and calendering	Achieve a substantial reduction in the emission of microfibers. However, the durability of the finishing is non-satisfactory.	Periyasamy and Tehrani-Bagha (2022)
Shearing and brushing	Both finishings release a lot of microfibers during domestic washing, creating fuzz fibers on the fabric's surface for comfort.	Periyasamy and Tehrani-Bagha (2022)

Source: Adapted from Periyasamy et al. (2022).

is covering the fibers may help to limit the amount of microfibers that are released. However, the polymeric film that serves as the protective layer has the potential to shed plastic microparticles into the environment. Utilization of acrylic silicone emulsion, pectin, PLA, and PBSA-based finishes indicates effectiveness. However, in order to prevent further pollution of the environment and to complete the process of sustainable finishing using chemicals and polymers that are safe for the environment, it is very necessary. However, there are a great many alternatives, such as enzyme treatments, plasma treatments, and the use of different biopolymers, that have never been tested before and have offered researchers potential avenues of investigation.

18.2.2.6 Finishing on Denim Garments

In order to remove the dyes from the surface of the fabric, a variety of denim finishing techniques were used. These techniques included stone washing, sandblasting, scraping, whiskers, tacking, grinding (also known as the demolish look), and ripping cuts (Periyasamy, 2020a; Periyasamy et al., 2019; Periyasamy & Militky, 2017). In most cases, the denim fabric has been colored with indigo using the ring dyeing technique (Chakraborty, 2014; Periyasamy et al., 2021). The fading effect is accentuated whenever the abrasion occurs on the fabric's surface. Since that opens up the structure of the yarn and creates a soft, fluffy, flannel-like appearance, each of these finishing techniques has increased the potential for higher fiber release during household washing. It is possible that the removal of fuzz fibers from denim during the finishing process uses cellulase enzymes (Choudhury, 2017). There are no studies to back up this claim; however, there is a hairiness on the cloth surface that considerably increases the release of microfibers throughout the washing cycles (Figure 18.8).

18.2.3 Positive Actions toward Avoidance and Reduction of Microfibers

The most important things to do to minimize pollution caused by microfibers are to take preventative measures to avoid and restrict the development of microfibers from the very beginning stages of the manufacture of textiles and the usage phase. Positive activities that may be taken toward the avoidance and minimization of the generations of microfibers are depicted in Figure 18.9.

18.2.3.1 Production Phase

Due to the enormous amount of synthetic textiles that are produced each year, the microfibers that are produced from synthetic textiles need to have proper management across the whole of the supply chain, beginning with the creation of the fabric (Truscott L, 2022). One potential solution is the development of fibers and textiles that are more durable, allowing for longer periods of use and reuse, and that reduce shedding during activities such as wearing and washing. Alterations to the way knitting

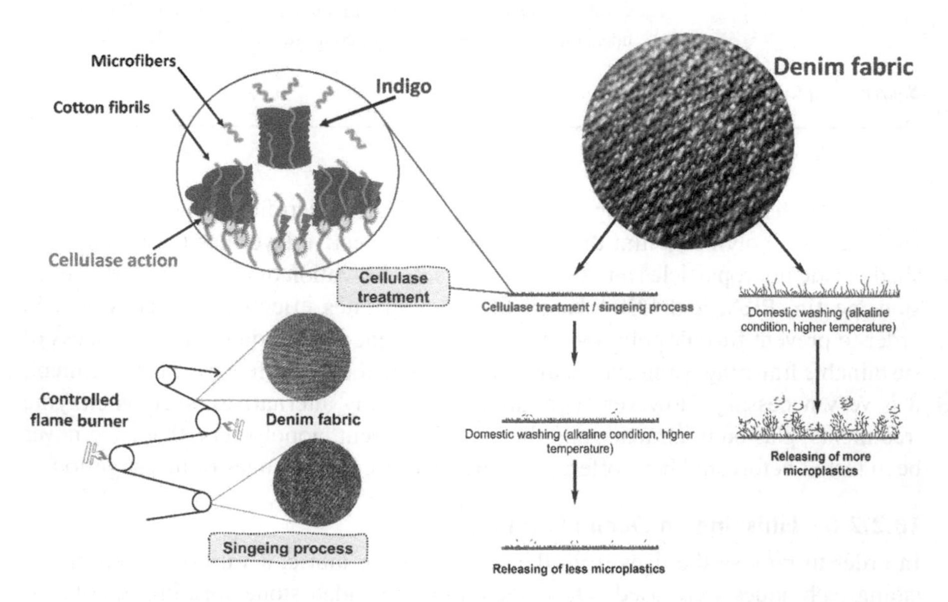

FIGURE 18.8 The mechanism for the microplastic generations from denim and their possible finishing techniques to reduce the microplastic generations. (*Source:* Copyright ACS (Periyasamy & Periyasami, 2023) under CC-BY license.)

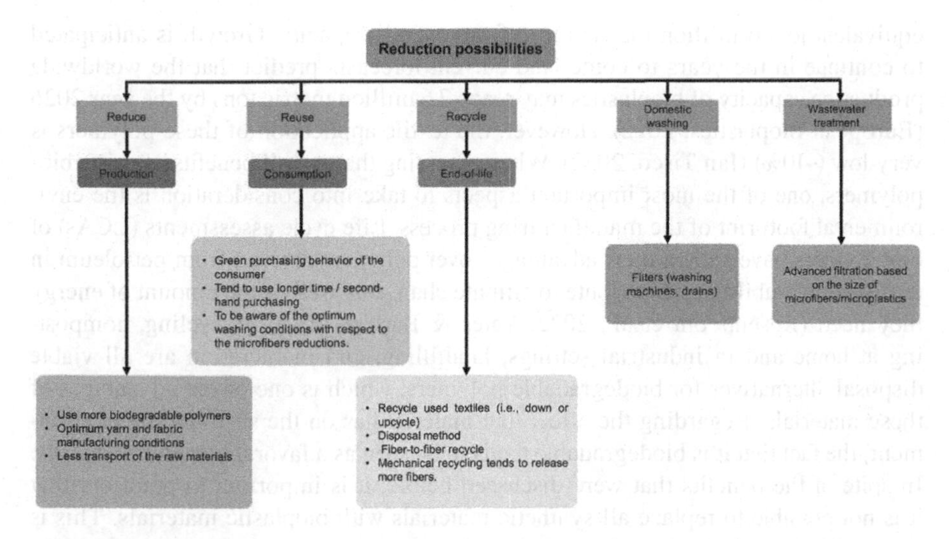

FIGURE 18.9 Positive actions toward the avoidance and reduction of microfibers and microfiber generations.

and weaving are done, such as making the fibers more durable and preventing or reducing the amount of shedding that occurs, may also be advantageous. Dyeing, bleaching, and finishing are other processes that need to be modified in order to reduce their negative impacts on the shedding of fibers and the amount of harmful chemicals that are released into the environment (Periyasamy & Militky, 2020).

Utilizing natural cellulosic materials and developing alternative textile materials, such as those made of bio-based and/or biodegradable polymers, may both be potential solutions for achieving the goal of establishing a bioeconomy that is both completely sustainable and circular. The most commonly discussed alternatives are PLA (Reddy et al., 2015), polyhydroxyalkanoate (PHA) (Degeratu et al., 2019), polycaprolactone (PCL) (Meng et al., 2007), polybutylene succinate (PBS) (Savitha et al., 2022), polyhydroxybutyrate (PHB) (Acevedo et al., 2018), Ioncell (Sixta et al., 2015), BioCelsol (Vehviläinen et al., 2020), Infinna (*Infinna Fiber: It's Time for the Textile Industry to Lose Its Virginity*, 2022), and Renewcell (*Renewcell: We Make Fashion Circular*, n.d.). PLA, PHA, and PHB are all bio-based and biodegradable fibers derived from renewable sources; Ioncell, BioCelsol, Infinna, and Renewcell are considered regenerated cellulosic fibers. PCL, PHA, and PBS are all plastics. However, at the moment, it seems as though not all of the proposed alternatives are appropriate for use in textiles or for resolving the microfibers problem. This is due to the fact that biodegradable plastics do not degrade in the same way under all environmental conditions, which is a topic that will require additional research in the near future. PLA is the alternative to PET that has been researched the most. When used in textile applications or processing, PLA has several drawbacks, such as poor heat resistance; nevertheless, research is still underway to find solutions to these drawbacks. In 2021, the total manufacturing capacity of bioplastics on a global scale rose by 16%, reaching 2.4 million metric tons. In 2021, the capacity for biodegradable bioplastics was equivalent to 1.6 million metric tons of the overall capacity. Growth is anticipated to continue in the years to come, and current forecasts predict that the worldwide production capacity of bioplastics may reach 7.6 million metric tons by the year 2026 (European Bioplastics, 2022). However, the textile application of these polymers is very low (~10%) (Ian Tiseo, 2022). When assessing the overall benefits of using biopolymers, one of the most important aspects to take into consideration is the environmental footprint of the manufacturing process. Life cycle assessments (LCAs) of biopolymers have shown their advantages over polymers derived from petroleum in terms of their ability to contribute to climate change as well as the amount of energy they need (Rosenboom et al., 2022; Yates & Barlow, 2013). Recycling, composting at home and in industrial settings, landfilling, and incineration are all viable disposal alternatives for biodegradable polymers, which is one of the advantages of these materials. Regarding the effect that material has on the surrounding environment, the fact that it is biodegradable tends to be seen as a favorable material quality. In spite of the benefits that were discussed before, it is important to point out that it is not possible to replace all synthetic materials with bioplastic materials. This is especially true when taking into account the enormous manufacturing volume that is necessary.

18.2.3.2 Consumption Phase

In 2022, the average person in Europe will have created more than 15kg worth of waste from textiles. Clothing and other textiles used in the house that are no longer recognized by their users are the primary contributors to textile waste, accounting for around 85% of the total waste (Hedrich et al., 2022). A short lifespan for a clothing item is not always the result of poor quality or durability; rather, it may be attributed to the behavior of consumers. Textiles are important items that, in general, have the potential to leave a considerable ecological footprint in terms of the additional resources used, the amount of carbon dioxide emissions, the amount of chemical emissions, and the amount of water consumed. Considering that the majority of microfibers/microplastics are generated in the first few washes of new textiles, educational initiatives are encouraging longer usage by customers to prevent so-called fast fashion. These ideas should also be supported by textile manufacturing initiatives to make clothing of greater quality and increased durability. Creating a streamlined system for the distribution and sale of pre-owned garments is yet another step. Because microfibers and microplastics shed less with each wear and wash cycle, increasing the sale and usage of previously owned items may help restrict their release to a minimum, as well as lower export rates and littering. In addition, optimizing of washing procedures to reduce the abrasion of synthetic textiles can help reduce shedding. This optimization can include reductions in the amount of detergent used, the speed at which the washing is done, and the volume of water used.

18.2.3.3 End of Life, Recycling, and Disposal Phase

It is necessary to cut down on the emission of microfibers and microplastics materials (Quartinello et al., 2017). These kinds of applications make it feasible to recycle fiber back into regular fiber (Sun, 2019). The different blending of materials, coatings, dyes, and so on offers the primary challenge that must be overcome by the textile recycling processes that are now available; yet, these methods are still restricted in their scope. The chemical recycling of plastic polymers also requires further improvement in order to obtain a greater rate of recovery and conversion efficiency as well as to decrease influences on the properties and quality of the materials (Mohsin et al., 2018; Molnar & Ronkay, 2019; Wang et al., 2019; Wu et al., 2019). The linked difficulties with pollution and energy consumption, however, cannot be ignored. Fabrics that include a combination of natural and synthetic polymers (for example, cotton, wool, and silk with PET or PA) may be recycled to remove the natural fibers using enzymatic digestion. The remaining refractory PET or PA can then be recycled to make PET or PA fibers and then made into yarn; however, the mechanical and other properties depend on the type of recycling. Because it is an upscaling many regenerated cellulose fibers are developed in Europe, particularly Finland and Sweden. This is accomplished by easily converting the paper and pulp industry into the regenerated cellulose fiber industry. Research on regenerated cellulose fibers has recently shown a significant amount of interest, and the market is anticipated to expand at a compound yearly growth rate of 10.5% by 2026, reaching 7.1 million metric tons by 2021 (Määttänen et al., 2021; Vehviläinen et al., 2020; Truscott L, 2022).

18.2.3.4 Domestic Washing and Wastewater Treatment Phase

Fiber pollution might also be avoided by upgrading or installing better filters in washing machines (de Falco et al., 2018b; Hernandez et al., 2017). As another preventative measure, installing a detachable stainless steel filter at the drain connection point or using an external filtering system at bespoke pipe connections are also viable options. A filter with an installed stainless steel filter with a mesh of 150–200 μm was recently discovered to retain 87% of the number of microfibers/microplastics in washing effluent (Pirc et al., 2016). In addition, it has been observed that using laundry balls made to collect fibers during washing may reduce fiber counts in washing effluent by 26%. It has also been claimed that the use of washing bags to encapsulate synthetic fabrics during laundry may lower microplastic emissions by capturing fibers from the hems. It is important to consider the mesh or pore size of the filters, as well as their filtering and capture capacity, when evaluating the overall efficacy of various methods for reducing fibers. By incorporating an additional end filtering step into wastewater treatment facilities, such as sand filtration, membrane bioreactor (MBR) treatment, or pile fabric filtration, microfiber and microplastic emissions may be mitigated. The Stockholm wastewater treatment plants in Sweden are now renovating a portion of their existing sludge system by installing a MBR, which will make it the biggest MBR facility in the world. In addition, granular activated carbon and other methods typically employed for eliminating micropollutants in water treatment may lower microfibers (especially 50 μm) by up to around 61% (Sachidhanandham & Periyasamy, 2021; Wang et al., 2020). Yet, this kind of technological development may need more resources, like electricity, chemicals, and so on, as well as entail substantial upfront investment and ongoing upkeep expenditures. To be clear, the aforementioned upgrades to wastewater treatment facilities may not be an option in many areas of the globe due to a lack of necessary infrastructure. In addition, it is helpful to enhance sludge procedures, such as pre-treatment to reduce microfiber content. Also, clean-up technology, such as plastic collectors, has been used in recent years to collect plastic debris from the water's surface and prevent it from being carried farther downstream.

18.3 CONCLUSIONS AND RECOMMENDATIONS FOR FUTURE REMARKS

Concern is rising over the environmental impact of textile items that produce microfibers. Faster abrasion, distortion, fiber tangling, and surface fuzz production are all brought on by the mechanical agitation used in automated washing machines and dryers. The emitted microplastics are significantly influenced by the world's population, as it will reach 7.9 billion by 2021 (*Current World Population*, 2023). Up to 95% of microfibers may be filtered out by the various stages of wastewater treatment facilities. However, the 5% that aren't immediately absorbed by wildlife make their way to waterways. Since textile production and consumption are expected to rise in the years ahead, ignoring this issue or responding too slowly would only make the situation worse. The public has to be made aware of the gravity of this issue, and stakeholders should be compelled to take action via rising environmental consciousness. Last but not least, lawmakers and environmental authorities should swiftly implement new environmental legislation to reduce the severity of this issue.

This chapter covered the many sustainable finishing procedures, including chemical and mechanical finishing on textile products, as well as the influence that these finishing methods have on the release of microfibers. For example, mechanical finishes such as brushing produce a greater number of microfibers, while finishes like singeing and calendaring produce a smaller number of microfibers. In the chemical finishing process, bio-washing, PLA/PBSA displays results that are promising on the microfiber reductions; however, it still needs to be used at the pilot level in order to verify their efficiency and confirm it.

The approach in which customers make their purchases has a significant impact on the reduction of microfibers. The current patterns of consumption lead to harm done to the environment, which poses a threat to our health as well as the health of future generations. The elimination of microfibers could be helped by adopting more environmentally responsible shopping practices. A green shopping habit includes the acquisition of sustainable fashion since it is dependent on socioeconomic and environmentally favorable characteristics. Consumers should adopt environmentally conscious shopping practices and show a genuine interest in learning about the reduction of microfibers that occurs as a result of household washing procedures. It is recommended that warning signs be included on garments that are manufactured from fleece, fuzz, and pile materials. In the meanwhile, businesses should also enhance their understanding of the mitigation of microfibers in relation to different types of fabrics.

REFERENCES

Acevedo, F., Villegas, P., Urtuvia, V., Hermosilla, J., Navia, R., & Seeger, M. (2018). Bacterial polyhydroxybutyrate for electrospun fiber production. *International Journal of Biological Macromolecules*, *106*, 692–697. https://doi.org/10.1016/j.ijbiomac.2017.08.066

Acharya, S., Rumi, S. S., Hu, Y., & Abidi, N. (2021). Microfibers from synthetic textiles as a major source of microplastics in the environment: A review. *Textile Research Journal*, *91*(17–18), 2136–2156. https://doi.org/10.1177/0040517521991244

Andrady, A. L. (2011). Microplastics in the marine environment. *Marine Pollution Bulletin*, *62*(8), 1596–1605. https://doi.org/10.1016/j.marpolbul.2011.05.030

Andreaus, J., Colombi, B. L., Gonçalves, J. A., & Alves dos Santos, K. (2019). Processing of cotton and man-made cellulosic fibers. In *Advances in Textile Biotechnology* (pp. 185–238). Elsevier. https://doi.org/10.1016/B978-0-08-102632-8.00009-8

Arja, M.-O. (n.d.). Cellulases in the textile industry. In *Industrial Enzymes* (pp. 51–63). Springer: Netherlands. https://doi.org/10.1007/1-4020-5377-0_4

Babu, K. M. (2015). Natural textile fibres. In *Textiles and Fashion* (pp. 57–78). Elsevier. https://doi.org/10.1016/B978-1-84569-931-4.00003-9

Belzagui, F., Crespi, M., Álvarez, A., Gutiérrez-Bouzán, C., & Vilaseca, M. (2019). Microplastics' emissions: Microfibers' detachment from textile garments. *Environmental Pollution*, *248*, 1028–1035. https://doi.org/10.1016/j.envpol.2019.02.059

Bhat, G., & Kandagor, V. (2014). 1 – Synthetic polymer fibers and their processing requirements. In D. Zhang (Ed.), *Advances in Filament Yarn Spinning of Textiles and Polymers* (pp. 3–30). Woodhead Publishing. https://doi.org/10.1533/9780857099174.1.3

Bishop, D. P. (1995). Physical and chemical effects of domestic laundering processes. In C. M. Carr (Ed.), *Chemistry of the Textiles Industry* (pp. 125–172). Springer: Netherlands. https://doi.org/10.1007/978-94-011-0595-8_4

Cai, Y., Mitrano, D. M., Heuberger, M., Hufenus, R., & Nowack, B. (2020). The origin of microplastic fiber in polyester textiles: The textile production process matters. *Journal of Cleaner Production*, *267*, 121970. https://doi.org/10.1016/j.jclepro.2020.121970

Carney Almroth, B. M., Åström, L., Roslund, S., Petersson, H., Johansson, M., & Persson, N.-K. (2018). Quantifying shedding of synthetic fibers from textiles; a source of microplastics released into the environment. *Environmental Science and Pollution Research*, *25*(2), 1191–1199. https://doi.org/10.1007/s11356-017-0528-7

Chakraborty, J. N. (2014). Dyeing with indigo. In *Fundamentals and Practices in Colouration of Textiles* (pp. 106–120). Woodhead Publishing Limited. https://doi.org/10.1533/9780857092823.106

Cheng, K. C., Khoo, Z. S., Lo, N. W., Tan, W. J., & Chemmangattuvalappil, N. G. (2020). Design and performance optimisation of detergent product containing binary mixture of anionic-nonionic surfactants. *Heliyon*, *6*(5), e03861. https://doi.org/10.1016/j.heliyon.2020.e03861

Chiweshe, A., & Crews, P. C. (2000). Influence of household fabric softeners and laundry enzymes on pilling and breaking strength. *Textile Chemist and Colorist and American Dyestuff Reporter*, *32*(9), 41–47. https://www.scopus.com/inward/record.uri?eid=2-s2.0-0033777771&partnerID=40&md5=ffe5437819ee720aaa8935d615156783

Choudhury, A. K. R. (2017). Environmental impacts of denim washing. In *Sustainability in Denim* (pp. 49–81). Elsevier. https://doi.org/10.1016/B978-0-08-102043-2.00003-4

Current World Population. (2023). https://www.worldometers.info/world-population/#ref-1

de Falco, F., Cocca, M., Avella, M., & Thompson, R. C. (2020). Microfiber release to water, via laundering, and to air, via everyday use: A comparison between polyester clothing with differing textile parameters. *ACS Applied Materials and Interfaces*, *54*(6), 3288–3296. https://doi.org/10.1021/acs.est.9b06892

de Falco, F., Cocca, M., Guarino, V., Gentile, G., Ambrogi, V., Ambrosio, L., & Avella, M. (2019a). Novel finishing treatments of polyamide fabrics by electrofluidodynamic process to reduce microplastic release during washings. *Polymer Degradation and Stability*, *165*, 110–116. https://doi.org/10.1016/j.polymdegradstab.2019.05.001

de Falco, F., di Pace, E., Cocca, M., & Avella, M. (2019b). The contribution of washing processes of synthetic clothes to microplastic pollution. *Scientific Reports*, *9*(1), 6633. https://doi.org/10.1038/s41598-019-43023-x

de Falco, F., Gentile, G., Avolio, R., Errico, M. E., di Pace, E., Ambrogi, V., Avella, M., & Cocca, M. (2018c). Pectin based finishing to mitigate the impact of microplastics released by polyamide fabrics. *Carbohydrate Polymers*, *198*, 175–180. https://doi.org/10.1016/j.carbpol.2018.06.062

de Falco, F., Gullo, M. P., Gentile, G., di Pace, E., Cocca, M., Gelabert, L., Brouta-Agnésa, M., Rovira, A., Escudero, R., Villalba, R., Mossotti, R., Montarsolo, A., Gavignano, S., Tonin, C., & Avella, M. (2018a). Evaluation of microplastic release caused by textile washing processes of synthetic fabrics. *Environmental Pollution*, *236*, 916–925. https://doi.org/10.1016/j.envpol.2017.10.057

de Falco, F., Gullo, M. P., Gentile, G., di Pace, E., Cocca, M., Gelabert, L., Brouta-Agnésa, M., Rovira, A., Escudero, R., Villalba, R., Mossotti, R., Montarsolo, A., Gavignano, S., Tonin, C., & Avella, M. (2018b). Evaluation of microplastic release caused by textile washing processes of synthetic fabrics. *Environmental Pollution*, *236*, 916–925. https://doi.org/10.1016/j.envpol.2017.10.057

Degeratu, C. N., Mabilleau, G., Aguado, E., Mallet, R., Chappard, D., Cincu, C., & Stancu, I. C. (2019). Polyhydroxyalkanoate (PHBV) fibers obtained by a wet spinning method: Good in vitro cytocompatibility but absence of in vivo biocompatibility when used as a bone graft. *Morphologie*, *103*(341), 94–102. https://doi.org/10.1016/j.morpho.2019.02.003

Elhawary, I. A. (2015). Fibre to yarn. In R. Sinclair (Ed.), *Textiles and Fashion* (pp. 191–212). Elsevier. https://doi.org/10.1016/B978-1-84569-931-4.00009-X

El-Sayed, M. A. M., & Sanad, S. H. (2010). Compact spinning technology. In C. A. Lawrence (Ed.), *Advances in Yarn Spinning Technology* (pp. 237–260). Woodhead Publishing. https://doi.org/10.1533/9780857090218.2.237

Erdumlu, N., & Ozipek, B. (2008). Investigation of regenerated bamboo fibre and yarn characteristics. *Fibres and Textiles in Eastern Europe*, *16*(4), 43–47.

European Bioplastics. (2022). *Global bioplastics production will more than triple within the next five years*. https://www.european-bioplastics.org/global-bioplastics-production-will-more-than-triple-within-the-next-five-years/

Gogotov, I. N., & Barazov, S. Kh. (2014). The effect of ultraviolet light and temperature on the degradation of composite polypropylene. *International Polymer Science and Technology*, *41*(3), 55–58. https://doi.org/10.1177/0307174X1404100313

Hagewood, J. (2014). Technologies for the manufacture of synthetic polymer fibers. In D. Zhang (Ed.), *Advances in Filament Yarn Spinning of Textiles and Polymers* (pp. 48–71). Woodhead Publishing. https://doi.org/10.1533/9780857099174.1.48

Hartline, N. L., Bruce, N. J., Karba, S. N., Ruff, E. O., Sonar, S. U., & Holden, P. A. (2016). Microfiber masses recovered from conventional machine washing of new or aged garments. *Environmental Science & Technology*, *50*(21), 11532–11538. https://doi.org/10.1021/acs.est.6b03045

Hedrich, S., Janmark, J., Langguth, N., Magnus, K.-H., & Strand, M. (2022). Scaling textile recycling in Europe-turning waste into value. McKinsey. https://www.mckinsey.com/industries/retail/our-insights/scaling-textile-recycling-in-europe-turning-waste-into-value

Hernandez, E., Nowack, B., & Mitrano, D. M. (2017). Polyester textiles as a source of microplastics from households: A mechanistic study to understand microfiber release during washing. *Environmental Science & Technology, 51*(12), 7036–7046. https://doi.org/10.1021/acs.est.7b01750

Hossain, Md. S., Islam, M. D. M., Dey, S. C., & Hasan, N. (2021). An approach to improve the pilling resistance properties of three thread polyester cotton blended fleece fabric. *Heliyon, 7*(4), e06921. https://doi.org/10.1016/j.heliyon.2021.e06921

Ian Tiseo. (2022). *Production capacity of bioplastics worldwide from 2020 to 2026*. Statista. https://www.statista.com/statistics/678684/global-production-capacity-of-bioplastics-by-type/

Infinna Fiber: It's Time for the Textile Industry to Lose Its Virginity. (2022). Infinited fiber. https://infinitedfiber.com

Kang, H., Park, S., Lee, B., Ahn, J., & Kim, S. (2021). Impact of chitosan pretreatment to reduce microfibers released from synthetic garments during laundering. *Water, 13*(18), 2480. https://doi.org/10.3390/w13182480

Kelly, M. R., Lant, N. J., Kurr, M., & Burgess, J. G. (2019). Importance of water-volume on the release of microplastic fibers from laundry. *Environmental Science & Technology, 53*(20), 11735–11744. https://doi.org/10.1021/acs.est.9b03022

Lant, N. J., Hayward, A. S., Peththawadu, M. M. D., Sheridan, K. J., & Dean, J. R. (2020). Microfiber release from real soiled consumer laundry and the impact of fabric care products and washing conditions. *PLoS One, 15*(6), e0233332. https://doi.org/10.1371/journal.pone.0233332

Lee, S. B., Kim, I. H., Ryu, D. D. Y., & Taguchi, H. (1983). Structural properties of cellulose and cellulase reaction mechanism. *Biotechnology and Bioengineering, 25*(1), 33–51. https://doi.org/10.1002/bit.260250105

Määttänen, M., Gunnarsson, M., Wedin, H., Stibing, S., Olsson, C., Köhnke, T., Asikainen, S., Vehviläinen, M., & Harlin, A. (2021). Pre-treatments of pre-consumer cotton-based textile waste for production of textile fibres in the cold NaOH(aq) and cellulose carbamate processes. *Cellulose, 28*(6), 3869–3886. https://doi.org/10.1007/s10570-021-03753-6

Maior, J. F. A. S., Reis, A. V., Muniz, E. C., & Cavalcanti, O. A. (2008). Reaction of pectin and glycidyl methacrylate and ulterior formation of free films by reticulation. *International Journal of Pharmaceutics, 355*(1–2), 184–194. https://doi.org/10.1016/j.ijpharm.2007.12.006

Meng, Q., Hu, J., Zhu, Y., Lu, J., & Liu, Y. (2007). Polycaprolactone-based shape memory segmented polyurethane fiber. *Journal of Applied Polymer Science, 106*(4), 2515–2523. https://doi.org/10.1002/app.26764

MERMAIDS EU Life+ (2015). Report on localization and estimation of laundry microplastics sources and on micro and nanoplastics present in washing wastewater effluents (A1).

Microplastics: Sources, Effects and Solutions. (2018). https://www.europarl.europa.eu/news/en/headlines/society/20181116STO19217/microplastics-sources-effects-and-solutions

Mohsin, M. A., Alnaqbi, M. A., Busheer, R. M., & Haik, Y. (2018). Sodium methoxide catalyzed depolymerization of waste polyethylene terephthalate under microwave irradiation. *Catalysis in Industry, 10*(1), 41–48. https://doi.org/10.1134/S2070050418010087

Molnar, B., & Ronkay, F. (2019). Effect of solid-state polycondensation on crystalline structure and mechanical properties of recycled polyethylene-terephthalate. *Polymer Bulletin, 76*(5), 2387–2398. https://doi.org/10.1007/s00289-018-2504-x

Mossotti, R., Montarsolo, A., Patrucco, A., Zoccola, M., Caringella, R., Pozzo, P. D., & Tonin, C. (2018). *Mitigation of the Impact Caused by Microfibers Released during Washings by Implementing New Chitosan Finishing Treatments* (pp. 223–229). https://doi.org/10.1007/978-3-319-71279-6_31

Muthusamy, L. P., Periyasamy, A. P., & Govindan, N. (2021). Prediction of pilling grade of alkali-treated regenerated cellulosic fabric using fuzzy inference system. *The Journal of the Textile Institute*, 1–11. https://doi.org/10.1080/00405000.2021.2012909

Napper, I. E., & Thompson, R. C. (2016). Release of synthetic microplastic plastic fibres from domestic washing machines: Effects of fabric type and washing conditions. *Marine Pollution Bulletin*, *112*(1–2), 39–45. https://doi.org/10.1016/j.marpolbul.2016.09.025

Negm, M., & Sanad, S. (2020). Cotton fibres, picking, ginning, spinning and weaving. In *Handbook of Natural Fibres* (pp. 3–48). Elsevier. https://doi.org/10.1016/B978-0-12-818782-1.00001-8

Niyonzima, F. N. (2019). Detergent-compatible bacterial cellulases. *Journal of Basic Microbiology*, *59*(2), 134–147. https://doi.org/10.1002/jobm.201800436

Periyasamy, A. P. (2020a). Environmental hazards of denim processing-I. *Asian Dyer*, *17*(1), 56–60.

Periyasamy, A. P. (2020b). Effects of alkali pretreatment on lyocell woven fabric and its influence on pilling properties. *The Journal of the Textile Institute*, *111*(6), 846–854. https://doi.org/10.1080/00405000.2019.1665294

Periyasamy, A. P. (2021). Evaluation of microfiber release from jeans: The impact of different washing conditions. *Environmental Science and Pollution Research*. https://doi.org/10.1007/s11356-021-14761-1

Periyasamy, A. P., & Militky, J. (2017). Denim processing and health hazards. In *Sustainability in Denim* (pp. 161–196). Elsevier. https://doi.org/10.1016/B978-0-08-102043-2.00007-1

Periyasamy, A. P., & Militky, J. (2020). *Sustainability in Textile Dyeing: Recent Developments* (pp. 37–79). https://doi.org/10.1007/978-3-030-38545-3_2

Periyasamy, A. P., Militky, J., Sachinandham, A., & Duraisamy, G. (2021). Nanotechnology in textile finishing: Recent developments. In *Handbook of Nanomaterials and Nanocomposites for Energy and Environmental Applications* (pp. 1–31). Springer International Publishing. https://doi.org/10.1007/978-3-030-11155-7_55-1

Periyasamy, A. P., & Periyasami, S. (2023). Critical review on sustainability in denim: A step toward sustainable production and consumption of denim. *ACS Omega*. https://doi.org/10.1021/acsomega.2c06374

Periyasamy, A. P., Ramamoorthy, S. K., & Lavate, S. S. (2019). Eco-friendly denim processing. In *Handbook of Ecomaterials* (Vol. 3, pp. 1559–1579). https://doi.org/10.1007/978-3-319-68255-6_102

Periyasamy, A. P., & Tehrani-Bagha, A. (2022). A review on microplastic emission from textile materials and its reduction techniques. *Polymer Degradation and Stability*, *199*, 109901. https://doi.org/10.1016/j.polymdegradstab.2022.109901

Periyasamy, A. P., Venkataraman, M., Kremenakova, D., Militky, J., & Zhou, Y. (2020). Progress in sol-gel technology for the coatings of fabrics. *Materials*, *13*(8), 1838. https://doi.org/10.3390/ma13081838

Piñol, L., Rodriguez, L., Brouta, M., Escamilla, M., & Rosa Escudero. (2014). *Mapping and estimation of laundry effluent microplastics release*. https://static1.squarespace.com/static/5afae80b7c93276139def3ec/t/5b06e71c70a6ad631bc540ee/1527179088560/OCEAN+CLEAN+WASH+GUIDE.pdf

Pirc, U., Vidmar, M., Mozer, A., & Kržan, A. (2016). Emissions of microplastic fibers from microfiber fleece during domestic washing. *Environmental Science and Pollution Research*, *23*(21), 22206–22211. https://doi.org/10.1007/s11356-016-7703-0

Quartinello, F., Vajnhandl, S., Volmajer Valh, J., Farmer, T. J., Vončina, B., Lobnik, A., Herrero Acero, E., Pellis, A., & Guebitz, G. M. (2017). Synergistic chemo-enzymatic hydrolysis of poly(ethylene terephthalate) from textile waste. *Microbial Biotechnology*, *10*(6), 1376–1383. https://doi.org/10.1111/1751-7915.12734

Ramasamy, R., & Subramanian, R. B. (2022). Enzyme hydrolysis of polyester knitted fabric: A method to control the microfiber shedding from synthetic textile. *Environmental Science and Pollution Research*, *29*(54), 81265–81278. https://doi.org/10.1007/s11356-022-21467-5

Ratnam, T. V. (2010). *SITRA Norms for Spinning Mills*. South India Textile Research Association.

Reddy, N., Yang, Y., Reddy, N., & Yang, Y. (2015). Polylactic acid (PLA) fibers. In N. Reddy & Y. Yang (Eds.), *Innovative Biofibers from Renewable Resources* (pp. 377–385). Springer: Berlin Heidelberg. https://doi.org/10.1007/978-3-662-45136-6_66

Renewcell: We Make Fashion Circular. (n.d.). Renewcell. Retrieved October 9, 2022, from https://www.renewcell.com/en/

Rosenboom, J.-G., Langer, R., & Traverso, G. (2022). Bioplastics for a circular economy. *Nature Reviews Materials*, *7*(2), 117–137. https://doi.org/10.1038/s41578-021-00407-8

Sachidhanandham, A., & Periyasamy, A. P. (2021). Environmentally friendly wastewater treatment methods for the textile industry. In *Handbook of Nanomaterials and Nanocomposites for Energy and Environmental Applications* (pp. 2269–2307). Springer International Publishing. https://doi.org/10.1007/978-3-030-36268-3_54

Sankarraj, N., & Nallathambi, G. (2018). Enzymatic biopolishing of cotton fabric with free/immobilized cellulase. *Carbohydrate Polymers*, *191*, 95–102. https://doi.org/10.1016/j.carbpol.2018.02.067

Savitha, K. S., Ravji Paghadar, B., Senthil Kumar, M., & Jagadish, R. L. (2022). Polybutylene succinate, a potential bio-degradable polymer: Synthesis, copolymerization and bio-degradation. *Polymer Chemistry*, *13*(24), 3562–3612. https://doi.org/10.1039/D2PY00204C

Senthil Kumar, R., & Sundaresan, S. (2013). Mechanical finishing techniques for technical textiles. In *Advances in the Dyeing and Finishing of Technical Textiles* (pp. 135–153). Elsevier. https://doi.org/10.1533/9780857097613.2.135

Sixta, H., Michud, A., Hauru, L., Asaadi, S., Ma, Y., King, A. W. T., Kilpeläinen, I., & Hummel, M. (2015). Ioncell-F: A high-strength regenerated cellulose fibre. *Nordic Pulp & Paper Research Journal*, *30*(1), 43–57. https://doi.org/10.3183/npprj-2015-30-01-p043-057

Sun, D. (2019). An investigation into the performance viability of recycled polyester from recycled polyethylene terephthalate (R-PET). *Journal of Textile Science & Fashion Technology*, *2*(4). https://doi.org/10.33552/JTSFT.2019.02.000543

Truscott L. (2022). *Preferred Fiber & Materials Market Report (2022)*. Textile Exchange. https://textileexchange.org/knowledge-center/reports/preferred-fiber-and-materials/

Vehviläinen, M., Määttänen, M., Grönqvist, S., & Harlin, A. (2020). Sustainable continuous process for cellulosic regenerated fibers. *Chemical Fibers International*, *70*(4). https://www.genios.de/fachzeitschriften/artikel/CFI/20201216/sustainable-continuous-process-for-/20201216553737.html

Wang, Z., Lin, T., & Chen, W. (2020). Occurrence and removal of microplastics in an advanced drinking water treatment plant (ADWTP). *Science of the Total Environment*, *700*, 134520. https://doi.org/10.1016/j.scitotenv.2019.134520

Wang, Y., Zhang, Y., Song, H., Wang, Y., Deng, T., & Hou, X. (2019). Zinc-catalyzed ester bond cleavage: Chemical degradation of polyethylene terephthalate. *Journal of Cleaner Production*, *208*, 1469–1475. https://doi.org/10.1016/j.jclepro.2018.10.117

Wilson, J. (2012). Woven structures and their characteristics. In *Woven Textiles* (pp. 163–204). Elsevier. https://doi.org/10.1533/9780857095589.2.163

Winkler, A., Santo, N., Ortenzi, M. A., Bolzoni, E., Bacchetta, R., & Tremolada, P. (2019). Does mechanical stress cause microplastic release from plastic water bottles? *Water Research*, *166*, 115082. https://doi.org/10.1016/j.watres.2019.115082

Wu, H., Lv, S., He, Y., & Qu, J.-P. (2019). The study of the thermomechanical degradation and mechanical properties of PET recycled by industrial-scale elongational processing. *Polymer Testing*, *77*, 105882. https://doi.org/10.1016/j.polymertesting.2019.04.029

Yang, L., Qiao, F., Lei, K., Li, H., Kang, Y., Cui, S., & An, L. (2019). Microfiber release from different fabrics during washing. *Environmental Pollution*, *249*, 136–143. https://doi.org/10.1016/j.envpol.2019.03.011

Yates, M. R., & Barlow, C. Y. (2013). Life cycle assessments of biodegradable, commercial biopolymers-A critical review. *Resources, Conservation and Recycling*, *78*, 54–66. https://doi.org/10.1016/j.resconrec.2013.06.010

Yousif, E., & Haddad, R. (2013). Photodegradation and photostabilization of polymers, especially polystyrene: Review. *SpringerPlus*, *2*(1), 398. https://doi.org/10.1186/2193-1801-2-398

Zambrano, M. C., Pawlak, J. J., Daystar, J., Ankeny, M., Cheng, J. J., & Venditti, R. A. (2019). Microfibers generated from the laundering of cotton, rayon and polyester based fabrics and their aquatic biodegradation. *Marine Pollution Bulletin*, *142*, 394–407. https://doi.org/10.1016/j.marpolbul.2019.02.062

Zambrano, M. C., Pawlak, J. J., Daystar, J., Ankeny, M., & Venditti, R. A. (2021a). Impact of dyes and finishes on the microfibers released on the laundering of cotton knitted fabrics. *Environmental Pollution*, *272*, 115998. https://doi.org/10.1016/j.envpol.2020.115998

Zambrano, M. C., Pawlak, J. J., Daystar, J., Ankeny, M., & Venditti, R. A. (2021b). Impact of dyes and finishes on the aquatic biodegradability of cotton textile fibers and microfibers released on laundering clothes: Correlations between enzyme adsorption and activity and biodegradation rates. *Marine Pollution Bulletin*, *165*, 112030. https://doi.org/10.1016/j.marpolbul.2021.112030

19 Microfiber from Textiles – The New Task of Standardization

Edith Classen
Hohenstein Institut für Textil Innovation GmbH

19.1 INTRODUCTION

The topic of microfiber is critical for all stakeholders in the textile sector (scientists, industry, business, brands, and non-governmental organizations (NGOs)). In the last few years, a lot of information and knowledge have been generated in various research groups all over the world and will be generated in running investigations. At present, much data from different studies are available. However, the comparability of different studies is hindered by the diversity of analytical technologies and pre-treatment procedures, as well as the diversity of selected textiles and applied washing processes (Cesa et al., 2017; Haap et al., 2019). The use of standardized analytical methods is beneficial to presenting comparable data. The use of standardized textile materials or methods is helpful in developing a more comprehensive understanding of the generation of microfibers from textiles and in developing solutions to avoid or minimize fiber release. A critical cause for fiber abrasion during use is the washing process. Harmonized protocols could help to make a better comparability of studies. There is a growing need to generate a comprehensive understanding of fiber abrasion during washing. This was the starting point of different activities in the national, European, and international standardization organizations of the stakeholders in the textile world. This chapter will overview the activities and update the already-reached results until March 2023 in the aspects of standardization.

19.2 STANDARDIZATION

A standard is a technical document designed to be used as a rule, guideline, or definition developed with all interested parties. All stakeholders benefit from standardization through increased product safety and quality, as well as lower transaction costs and prices. The work is done in national standardization bodies (e.g., German Institute for Standardization (Deutsche Institut für Normung (DIN)), French Standardization Association (Association Françai se de Normalisation (AFNOR)), Italian National Unification Body (Ente nationale Italiano di Unificaziono (UNI), British Standards Institution (BSI), the Bureau of Indian Standards (BIS), the Standardization

DOI: 10.1201/9781003331995-23

Administration of China (SAC), American National Standards Institute (ANSI), and others), in the European Committee for Standardization (CEN), and the International Organization for Standardization (ISO).

19.2.1 ISO Standards

The International Electrotechnical Commission (IEC) and the ISO are the publishers of standards for mechanical engineering. IEC deals with electrical and electronic issues and ISO deals with mechanical issues. ISO, founded in 1946, is a worldwide federation of national standards bodies. ISO standards set industry standards, simplify technical rules, and make them internationally comparable. ISO standards can be product standards, test methods, codes of practice, guideline standards, management system standards, etc. ISO standards are prepared by ISO technical committees. Everybody interested in a subject for which a technical committee has been established has the right to be represented on the committee. ISO is not responsible for identifying any patent rights. Details of any patent rights identified during the development of the standard are mentioned in the introduction and/or on the ISO list of patent declarations received. Any trade name used in a standard is information given for the convenience of users and does not constitute an endorsement.

Developing a new ISO standard starts with the new work item proposal (NWIP) and developing a Committee draft (CD). If the CD is accepted, the draft international standard (DIS) work starts. The DIS has to be accepted by the national standard bodies (NSBs). After acceptance of DIS, the final draft international standard (FDIS) development starts and is finished by a favorable vote of the NSBs. The immediate steps are the completion and publication of the standard. ISO standards can stand alone or be listed as originally European or national standards with the ISO suffix (e.g., DIN ISO). This only means that a German DIN standard corresponds to international requirements (ISO, 2023).

ISO international standards and other normative ISO deliverables (technical specification (TS), technical report (TR), publicly available specification (PAS), and international workshop agreements (IWA)) are voluntary. They do not include contractual, legal, or statutory requirements. Voluntary standards do not replace national laws, with which users are understood to comply and which take precedence (ISO, 2023).

- A **Technical Specification (ISO TS)** addresses work still under technical development or where it is believed that there will be a future, but not an immediate, possibility of agreement on an international standard. The ISO TS is published for immediate use, but it also provides a means to obtain feedback. The aim is that it will be transformed and republished as an international standard.
- A **Technical Report (ISO TR)** contains information and can include data obtained from a survey, an informative report, or information on the perceived “state of the art”.
- A **Publicly Available Specification (ISO PAS)** is published to respond to an urgent market need, representing either the consensus of the experts

within a working group or a consensus in an organization external to ISO. ISO PAS are published for immediate use and serve to obtain feedback for an eventual transformation into an international standard. ISO PAS has a maximum life of 6 years, after which it can be transformed into an international standard or withdrawn.

- **International Workshop Agreements (ISO IWA)** are prepared through a workshop outside of ISO committee structures to respond to urgent market requirements. If there is an existing ISO committee whose scope covers the topic, the published ISO IWA is automatically allocated to this committee for maintenance. An IWA is reviewed 3 years after publication and can be further processed to become an ISO PAS, ISO TS, or ISO standard. An ISO IWA can exist for a maximum of 6 years.

19.2.2 CEN Standards

European Standards (EN) are technical standards drafted and maintained by CEN (European Committee for Standardization), CENELEC (European Committee for Electrotechnical Standardization), and ETSI (European Telecommunications Standards Institute). Standards for textiles and textile machines are developed in CEN. The members of the CEN National Body committees develop a EN. In some cases, the request for a standard comes from the European Commission or other stakeholders to produce European harmonized standards supporting EU legislation and policies (CEN, 2023).

A EN is a document for common and repeated use that provides rules, guidelines, or characteristics for activities or their results and is based on the consolidated results of science, technology, and experience. Experts from industry, associations, the public, administration, science, and societal organizations develop ENs and other deliverables together. EN should help build the European internal market for goods and services, remove trade barriers, and strengthen Europe's position in the global economy.

The national CEN members from 34 countries implement a EN as a national standard. The work is assigned to a CEN Technical Committee (CEN/TC) in the field concerned, and a "standstill" is enforced on all national work surrounding the same topic. After establishing the CEN/TC, mirror committees of national-level stakeholders decide on the national contributions regarding the development of the standard. The experts develop the proposal for a standard. The proposal goes on to the drafting stage, which is based on consensus-building. After finalizing the draft standard (prEN (pr = preliminary)), it goes up for public inquiry and is open to all interested parties. After inquiry, the votes and comments on the standard are evaluated, and depending on the result, the draft standard is either published or worked upon and submitted to a formal vote.

A TS is a normative document developed when various alternatives are not consensual, pending future harmonization, or parallel deployment. The specifications and/or evolving technologies must exist. A TS is approved through a weighted vote by the CEN national members. The TS shall be announced at the national level and may be adopted as a national standard, but conflicting national standards may continue to exist. A TS may, however, not conflict with a EN. If a conflicting EN is subsequently published, the TS must be withdrawn. TSs have no time limit and shall be reviewed at intervals of not more than 3 years, starting from their publication date.

A TR is an informative document that provides information on the technical content of standardization work. TR is approved through a simple majority vote by the CEN national members and involves no obligation at the national level. A TR has no specified time limit, but a TR is regularly reviewed by the responsible technical body to ensure that it remains valid (CEN, 2023).

19.2.3 Cooperation between CEN and ISO

The Vienna Agreement regulated technical cooperation between the ISO and the CEN and was signed in 1991. The Vienna Agreement between ISO and CEN aims to carry out standardization work at only one level, if possible, but to bring about simultaneous recognition as an international standard and EN through suitable coordination procedures. The main objectives of this agreement are to provide a framework for the optimal use of resources and expertise available for standardization work, prevent duplication of effort, reduce time when preparing standards, and provide a mechanism for information exchange between international and European Standardization Organizations (ESOs) to increase the transparency of ongoing work at international and European levels.

Cooperation with ISO is prioritized if international standards correspond to European legislative and market requirements, and non-European global players also implement these standards. With every new standardization project accepted in CEN, it should be checked whether it can be developed at the ISO level. The work programs will be coordinated jointly via international and European draft standards.

19.2.4 DIN Standard/A National Standard

National standards can be developed by national standardization bodies in different countries (e.g., Germany – DIN, France - AFNOR, USA - ANSI, India - BIS, China – SAC, and others). National standards are valid in the country and can be developed further to a EN or ISO standard. DIN standards are the results of work at national. After acceptance of a proposal for a new standard, the standards project is carried out according to set rules of procedure by the relevant DIN standards committee. All stakeholders, including manufacturers, consumers, businesses, research institutes, public authorities, and testing bodies, can participate in this work. The standard is developed in one of the 70 standards committees, each responsible for a specific subject area. Standards are developed with complete consensus; that is, they are developed by experts to arrive at a common standpoint, considering the state of the art.

The DIN SPEC is a kind of preliminary stage to the DIN NORM and has the advantage that it can be implemented more quickly. It can be launched within a few months since it is created in smaller working groups, and there is no obligation to reach a consensus. It can be the basis for a DIN standard and helps the initiators prepare the ground for later standardization.

19.3 DEVELOPMENT OF THE ISO 4484 SERIES

Since May 2020, different standard drafts dealing with textile microplastics (MPs) have been under development (ISO 4484-1, ISO DIS 4484-2, and ISO FDIS 4484-3). The work is done by the ISO/TC 38 WG 34 (textiles) in collaboration with the CEN/TC 248 WG 37 (textiles and textile products) under the Vienna Agreement. Part 1 and Part 2 are developed under the CEN lead and Part 3 is developed under the ISO lead.

19.3.1 ISO 4484-1

The ISO 4484-1: textiles and textile products — MPs from textile sources — Part 1: determination of material loss from fabrics during washing (ISO 4484-1, 2023), published in January 2023, describes a method for systematically collecting material loss from fabrics under laundering test conditions to achieve comparable and accurate results. There is no direct correlation to material loss during domestic and commercial laundering. The method assesses material loss of all fiber types (synthetic and natural fibers). The laboratory washing procedure used is a modification of ISO 105-C06 (ISO 105-CO6, 1994). Any collected debris is assumed to bc fiber fragments and the weight of the collected debris is determined. For the identification of the nature/composition of this debris, the methods described in the Draft ISO 4484-2 (ISO 4484-2, 2023) can be used.

For testing, test specimens are prepared by cutting a rectangle from the fabric roll, folding it, and seaming it. The test specimens are dried in an oven and cooled in a desiccator to reach a dry, constant mass, and the dry mass of each specimen is determined. A closed canister with water and steel balls is used for the laboratory washing process and the test specimen is added to the preheated canister. The canister with the test specimen is put back into the accelerated laundering device, and the washing process starts under rotation. After washing, the wash liquor is filtrated in a filtration device with a glass fiber filter (1.6 μm) with a vacuum pump; the wash liquor of the various containers is separately filtered. If the filter has become clogged, a second conditioned, dried filter can be used, and the results are combined. Any remaining fibers from the bottom edge of the funnel are rinsed with water into the specimen tray. Filters with fibers are dried in an oven and the weight is determined. Contamination from the lab clothes and from outside must be avoided during the process.

This simple method can be performed everywhere, and no specific analytical equipment is necessary. To avoid residues of detergents on the filters, the test is done without detergents. However, the test can be carried out with liquid detergents. Powder detergents should not be used because of insoluble residues.

19.3.2 ISO DIS 4484-2

The ISO DIS 4484-2: textiles and textile products— MPs from textile sources — Part 2: qualitative and quantitative evaluation of MPs (ISO DIS 4484-2, 2023) describes a qualitative-quantitative analytical evaluation of MPs. For assessing the potential impact of MP from textiles, their number, shape, and size are relevant parameters, and these parameters can be analyzed with microscopic, molecular spectroscopic, and thermo-analytical methods. Both molecular spectroscopic methods, micro-FTIR and micro-Raman, are powerful tools for the identification and counting of plastic particles up to submicron size. Various samples, e.g., solid samples from textile production processes, water samples from the textile production process and/or from the washing of clothing, and air samples for testing the air quality in the workplace of textile companies, can be investigated. Depending on the matrix, pre-treatment of the sample is necessary to concentrate the MPs and eliminate inorganic and organic contaminants that can interfere with their identification.

With the mentioned methods, information about size, shape, surface, and estimated mass can be provided, which can be used for ecotoxicological assessments. The results of the investigation can be given in terms of the estimated surface area and mass of MPs per unit sample, regardless of the origin of the investigated sample. With this method, only one single filter can be analyzed. That can lead to errors in the qualitative and quantitative determination of MPs, and it is necessary to investigate different filters to be more accurate and precise.

The document describes in detail the sample preparation, the purification procedures, the preparation of standardized MPs, and the analytical procedures of the different analytical techniques.

Samples of the textile sector can be contaminated; therefore, contamination has to be investigated as a first step to avoid the wrong interpretation of the data. This Part 2 uses high instrumental equipment, and the investigation needs more time than the testing, according to Part 1. The DIS was positively voted for by ISO in 2022. However, by CEN, the second vote was also negative. Now, the standard will be developed only as an ISO standard, and the publication is expected at the end of 2023.

19.3.3 ISO FDIS 4484-3

The ISO FDIS 4484-3: textiles and textile products—MPs from textile sources—Part 3: measurement of collected material mass released from textile end products by domestic washing method (ISO FDIS 4484-3, 2022), a third approach to investigate the fiber debris from textiles.

The test method collects all materials, including fiber fragments, from washing. There are many types of washing machines in the world available; therefore, the standard washing machines in ISO 6330 are used (ISO 6330, 2021). The washing conditions of textile end products are indicated by the care labeling according to ISO 3758 (2012). The standard can be used for textile end products such as clothing (e.g., fleece, shirts, trousers, and blouses) and home textile end products (e.g., blankets, rugs, and curtains). These products are composed of all fibers (natural or synthetic fibers), including blends, and can be washed in a domestic washing machine. A given mass of textile end products is washed in a domestic washing machine under specific conditions of temperature and mechanical conditions for a given number of washing cycles. The textile material sheds through the washing cycles and is collected using a standardized filter bag attached to the outlet hose of the washing machine.

The collected materials by the filter bag are transferred to the membrane filter and then weighed. The mass of the collected materials can be calculated from the weight of the original membrane filter and the weight of the membrane filter with shed material.

Cleaning the washing machine before the test is critical to avoid contamination of the washing machine. Therefore, checking methods and washing machine requirements are included in the standard. The identification and quantification of components contained in the collected materials can be determined by applying Part 2. The standard can be used in the textile industry to develop textile end products to reduce or minimize shedding materials from the textile product (e.g., garment) through washing. The final vote of the standard starts in March 2023 and will end in April 2023. The standard is expected to be published by the end of 2023.

19.3.4 DIN SPEC 4872

A German development is the DIN SPEC 4872:2023 test method for textiles – determination of fiber release during washing and aerobic degradation level in an aqueous medium in consideration of ecotoxicity (DIN SPEC 4872, 2023). DIN SPEC 4872 has been developed according to the PAS procedure in a DIN SPEC consortium and does not require the participation of all stakeholders. The members of the DIN SPEC consortium (Hohenstein Laboratories GmbH & Co. KG, FREUDENBERG Performance Materials Apparel SE & Co. KG, Intrinsic Advanced Materials, TRIGEMA Inh. W. Grupp e.K., Paradies GmbH, and the DBL ITEX Gaebler Industrie-Textilpflege GmbH & Co. KG) started with the development in 2022. The DIN SEPC 4872 was published in February 2023. DIN SPEC 4872 is a standardized test method for determining and classifying the environmental impact of textiles during washing. The DIN SPEC 4872 analyzes fiber release after a laboratory washing procedure with dynamic image analysis (DIA). The biodegradability of the fiber abrasion in wastewater is defined according to ISO 14851 with some modifications (ISO 14851, 2019) and determines the degree of degradation within a defined period. Finally, an ecotoxicity test according to ISO 20079:2006–12 (ISO 20079, 2006) is conducted to determine the toxicity of the fiber residues after the biodegradation process. The ecotoxicological effects are determined with a duckweed growth inhibition assay by comparing aquatic media obtained with and without the addition of textile residues after the biodegradation test. The method of quantitative determination of the textile discharge and the ability to biodegrade in the aqueous medium is coupled with an ecotoxicological assessment, able to point out their potential impact on the environment. The biodegradation test provides information on the percentage by which a material degrades under specific conditions in an aqueous medium. The ecotoxicological assessment of the influence of the textile residues is limited after the biodegradation test. Once the test procedure has been completed, a classification code is assigned that reflects the degree of fiber discharge, the biodegradation rate, and the ecotoxicological potential of the textile product. This classification aims to provide reliable data for optimizing the product portfolio, and targeted product development enables active and conscious control or prevention of environmental pollution.

Textiles made of natural or synthetic fibers and mixtures of the fibers can be tested. The fiber release of the textile products in the wash water is determined with the DIA which can provide data about the particle shape and size distribution of the fibers (Haap et al., 2019). The results of the DIA support the classification of textile products in terms of fiber release during washing. For biodegradation testing, the required amount of fiber charge is produced by beating under a nitrogen atmosphere. The ecotoxicological assessment constitutes a screening and does not include a full consideration of the ecotoxicological impact. A suitable ecotoxicity test is carried out to determine the toxicity of residues after the biodegradation process. DIN SPEC 4872 is available free of charge by DIN (DIN SPEC 4872, 2023).

19.3.5 AATCC TM212–2021

The AATCC TM 212–2021: test method for fiber fragment release during home laundering was developed by the American Association of Textile Chemists and Colorists (AATCC) Committee RA100 in 2021 (AATCC TM 212, 2021). This test determines the mass of fiber fragments released in an accelerated laundering testing machine. Accelerated laundering is expected to provide a relative approximation of fiber fragment release in full-scale home laundering, but an exact correlation has not been determined. This test is applicable to textiles that are expected to withstand home laundering. The test method is similar to the test setup in ISO 4484-1 and optionally uses detergent to simulate the washing condition of textiles.

19.3.6 Other Relevant Documents for Microplastic, Not Primarily for Textiles

The ISO/TR 21960:2020 (en) plastics — environmental aspects — state of knowledge and methodologies is a technical report developed under the scope of ISO/TC 61 plastics (ISO/TR 21960, 2020). This technical report summarizes the current scientific literature on the occurrence of macroplastics and MPs in the environment and biota. The report gives an overview of testing methods, including sampling from various environmental matrices, sample preparation, and analysis, and describes chemical and physical testing methods for identifying and quantifying plastics. The report gives recommendations for the three essential steps, sampling, sample preparation, and analysis, which are necessary for the standardization of methods toward harmonized procedures.

The technical report follows the resulting requirements. Independent from these, terms are used in the text that are in the scope of other ISO/TCs, such as ISO/TC 38 textiles, ISO/TC 45 rubber and rubber products, and ISO/TC 217 cosmetics (ISO/TR 21960, 2020). Table 19.1 gives an overview of standards in the development of just published microfiber from textiles (as of March 2023).

19.4 OTHER ACTIVITIES IN EUROPE

In Europe, the EU and numerous initiatives are active on the topic of MPs and standardization. The activities of the environmental coalition on standards (ECOS),

TABLE 19.1
Summary of the Different Test Methods Developed for the Fiber Release from Textile Material

Number of the Standard	Material	Methods	Characteristic Aspects
ISO 4484-1	Textiles	Quantitative analysis Gravimetric analysis	Textile material Washing in laboratory scale Filtration
ISO DIS 4484-2	Textiles	Quantitative and qualitative analysis Determination of the number, shape, and size Micro-FTIR and micro-Raman (microscopic molecular spectroscopy	Textile material Washing in Laboratory scale Filtration
ISO FDIS 4484-3	Textiles	Quantitative analysis Gravimetric analysis	Garment Household washing machine Filtration
DIN SPEC 4872	Textiles	Quantitative analysis with dynamic image analysis (DIA) (size, shape, and amount) Biodegradation test Ecotoxicological test	Textile material Washing in laboratory scale Grind of textile material
AATCC TM 212	Textiles	Quantitative analysis Gravimetric method	Samples of textiles Washing in laboratory scale
ISO/TR 21960	Plastic	Summary of analytical methods	

microfiber consortium (TCM), and trade association for the home appliance industry in Europe (APPLiA) as examples and describe the various activities.

Textiles and plastics were among the key value chains in the EU circular economy action plan (CEAP) in March 2020 (CEAP, 2020). The wearing and washing of textiles made from synthetic (plastic) fibers is one recognized source of MPs in the environment (EU, 2022). The EU Commission addresses the unintentional release of MPs from synthetic textiles (EU Textile Strategy, 2022).

The ECOS is an international NGO with a network of members and experts advocating for environmentally friendly technical standards, policies, and laws. ECOS has shared recommendations for the upcoming measures meant to curb MP pollution with the European Commission. The new initiative aims to tackle MPs unintentionally released into the environment, and it will focus on labeling, standardization, certification, and regulatory measures for the primary sources of emitting these plastics, including pellets, tires, and textiles (ECOS, 2023).

The TCM aims to develop practical solutions for the textile industry to minimize fiber fragmentation and release to the environment from manufacturing and the product life cycle. TCM's vision is to work toward zero impact from fiber fragmentation from textiles to the environment. The Microfiber 2030 Commitment presents the opportunity to align globally as an industry and has a real impact on microfiber release by bringing all stakeholders together. The cross-industry connection expedites understanding and reduces the possibility of duplication of research and resources (TCM, 2023).

APPLiA is a trade association for the home appliance industry in Europe. It represents the interests of manufacturers of large appliances (e.g., refrigerators, ovens, washing machines, and dryers), of small appliances (e.g., vacuum cleaners, irons, toasters, and toothbrushes), and heating, ventilation, and air conditioning appliances (e.g., air conditioners, heat pumps, and local space heaters). APPLiA is the coordinator of the Consortium on MPs release during household washing processes. The Consortium comprises manufacturers (e.g., washing machines, detergent, and filters) and test houses and is working on a standardized testing methodology assessing MP release during the household washing cycle (APPLiA, 2023).

19.5 CONCLUSIONS

The standardization and establishment of test methods for textile microfibers have been under development in the last 3 years. Because the MP topic is a topic worldwide, the activities were based on ISO and CEN. ISO 4484-1 is the first published standard developed under CEN lead. This method determines the fiber debris from fabrics on the laboratory scale. This method is simple and needs no high-tech equipment. The comparison of the different textile materials is possible. The results are essential for the manufacturer to improve materials and select the material with the lowest fiber debris for a product. The second one is the ISO DIS 4484-2, which was developed under CEN's lead and gave qualitative and quantitative information about the fiber release from textiles. This method needs a highly equipped laboratory, which is time-consuming. The results give much information about the fiber debris and are essential for the qualitative analysis of the fiber debris. Because of the negative vote in CEN, the standard will be only published as an ISO standard. The third part, the ISO FDIS 4484-3, is developed under ISO lead. This method determines the fiber a garment releases in a home washing machine. This method needs a proper cleaning of the washing machine before testing to avoid contamination from the washing machine. This method covers the fiber released from the garment; all additional dirt and contaminants from the environment that can be collected on the surface of the textile material during use can influence the results.

AATCC TM 212, an American standard, is similar to the ISO 4484-1 method. These described four methods can be used to investigate MP from textiles, and the methods can be selected depending on which information is requested and necessary.

The DIN SPEC 4872 is developed in Germany, which determines the microfiber discharge from textile materials with the DIA in the first step. In the second step, the biodegradation of this fiber release is investigated, and the third step is the determination of the ecotoxicological effect. The textiles can be classified under DIN Spec 4872.

The projects show that developing the field of the analysis of microfibers from textiles needs time. The industry wishes to have methods for determining MP from textiles; however, they wish to have cheap methods with moderate technical equipment. Scientists develop new analytical methods based on new analytical methods, which can sometimes be time-consuming and expensive. The industry needs reproducible test methods that should give results during the production process of the textiles. To evaluate test methods, reference test fibers should be available to compare and evaluate the analysis method. For the development of standardized methods, round-robin tests are necessary to guarantee that the test methods are working well in the comparison of different laboratories around the world. Finding a consensus among all stakeholders in the working groups needs a lot of discussions, standard test materials, and reproducible test methods.

The following work for the working groups will be the development of standardized methods to determine microfibers from textiles in the air. A lot of investigations are running, and results will be coming in the following years.

REFERENCES

AATCC TM 212. (2021). Test method for fibre fragment release during home laundering. https://members.aatcc.org/store/tm212/3573/

APPLiA. (2023). APPLiA homepage. https://www.applia-europe.eu/topics

CEAP. (2020). https://eur-lex.europa.eu/legal-content/EN/TXT/?qid=1583933814386&uri=COM:2020:98:FIN

CEN. (2023). European Committee of standardization (CEN). https://www.cencenelec.eu/about-cen/

Cesa, F.S. et al. (2017). Synthetic fibers as microplastics in the marine environment: A review from textile perspective with a focus on domestic washings. *Sci. Total Environ.* 598, 1116–1129, doi:10.1016/j.scitotenv.2017.04.172.

DIN SPEC 4872. (2023). Test method for textiles – Determination of fiber release during washing and aerobic degradation level in aqueous medium in consideration of ecotoxicity

ECOS. (2023). Environmental coalition of standards (ECOS). https://ecostandard.org/work_area/textiles/

EU. (2022). Microplastics from textiles: Towards a circular economy for textiles in Europe. https://www.eea.europa.eu/publications/microplastics-from-textiles-towards-a, published 10 Feb 2022. Last modified 17 Mar 2022

EU Textile Strategy. (2022). EU textile strategy. https://environment.ec.europa.eu/strategy/textiles-strategy_en

Haap, J. et al. (2019). Fibers released by textile laundry: A new analytical approach for the determination of fibers in effluents. *Water*. 11, 2088

ISO. (2023). ISO International standard organization. www.iso.org

ISO 105-C06. (1994). Textiles – Tests for color fastness – Part C06: Color fastness to domestic and commercial laundering

ISO 14851. (2019). Determination of the ultimate aerobic biodegradability of plastic materials in an aqueous medium – Method by measuring the oxygen demand in a closed respirometer

ISO 20079. (2006-12). Water quality – Determination of the toxic effect of water constituents and waste water on duckweed (Lemna minor) – Duckweed growth inhibition test (ISO 20079:2005).

ISO 3758. (2012). Textiles – Care labelling code using symbols
ISO 4484-1. (2023). (Textiles and textile products – Microplastics from textile sources – Part 1: Determination of material loss from fabrics during washing)
ISO 6330. (2021). ISO 6330 textiles – Domestic washing and drying procedures for textile testing
ISO DIS 4484-2. (2022). Textiles and textile products – Microplastics from textile sources – Part 2: Qualitative and quantitative evaluation of microplastics
ISO FDIS 4484-3. (2022). Textiles and textile products – Microplastics from textile sources – Part 3: Measurement of collected material mass released from textile end products by domestic washing method
ISO TR 21960. (2020). https://www.iso.org/obp/ui/#iso:std:iso:tr:21960:ed-1:v1:en
TCM. (2023). The microfiber consortium (TCM). https://www.microfibreconsortium.com

20 Aligning the Textile Industry towards the Mitigation of Microfibre Pollution

Kelly J. Sheridan
The Microfibre Consortium
Northumbria University

Anna Bateman, Elliot Bland, Charline Ducas, and Sophie Mather
The Microfibre Consortium

20.1 INTRODUCTION

Microfibres are widespread environmental pollutants, dispersed across land, in oceans, waterways and the atmosphere, gaining the attention of industry, government, media, NGOs and consumers. Predominately originating from textiles, fibre fragments shed (are lost) from textiles at all stages of the product life cycle; during manufacture; consumer use, including washing, drying and general wear; and as a result of disposal (Weis and De Falco, 2022) (Figure 20.1). Synthetic microfibres (e.g. polyester, nylon) do not biodegrade naturally, and although natural fibres (e.g. cotton) may degrade faster, chemicals used in the dyeing and finishing processes of these materials delay their degradation, allowing them to persist and accumulate in the environment like synthetic microfibres (Athey et al., 2020). The sorbent properties of cellulosic fibres, due to the presence of carbonyl and hydroxyl functional groups in their chemical composition, allow them to be carriers of environmentally detrimental hydrophobic compounds (Ali et al., 2018).

Despite their anthropogenic existence and posed risk being known by some industries for many years, microfibres as an environmental concern has, until recently, gained little attention. The first documentation of microfibres in the environment came in 1954, when jute fibres were observed in suspended sea water (Atkins et al., 1954). Over 20 years later, synthetic microfibres were found in samples taken from coastal sediment in the UK (Buchanan, 1971). For decades, the omnipresence of textile fibres in domestic environments has been exploited by forensic scientists to solve crimes (Grieve, 2000). It has led to a significant body of research on their prevalence (Jones and Johansson, 2023; Palmer and Burch, 2009; Watt et al., 2005;

DOI: 10.1201/9781003331995-24

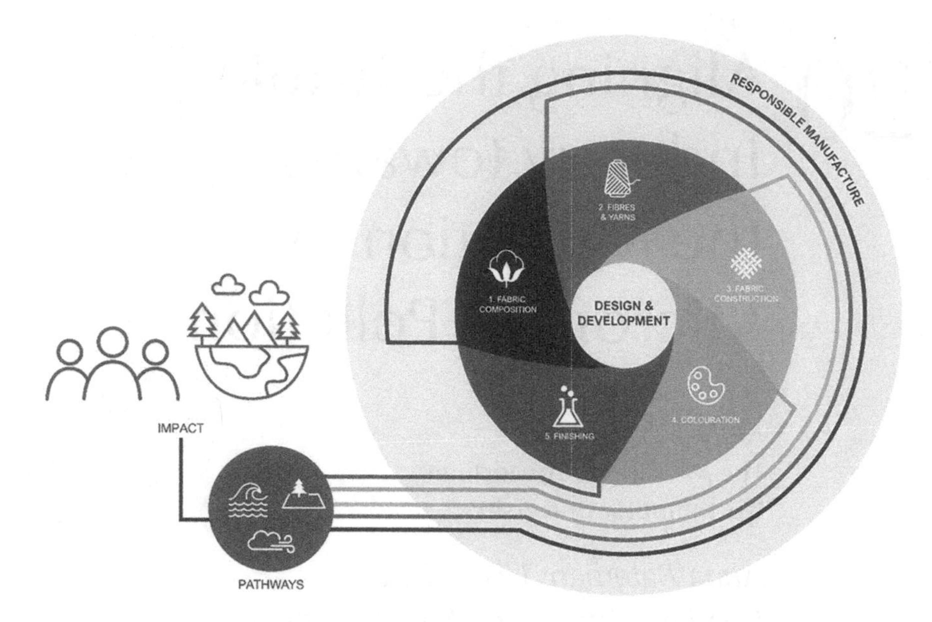

FIGURE 20.1 Microfibre pathways to the environment.

Cantrell et al., 2001; Grieve and Biermann, 1997), characterisation and discrimination (Grieve et al., 2005; Palmer et al., 2015; Lunstroot et al., 2016) and the factors that influence their propensity to shed from textiles alongside their subsequent transfer and persistence properties (Pounds and Smalldon, 1975a–c). Moreover, the presence of lint in manufacturing facilities as a consequence of textile processing has been recognised as potentially hazardous to human health (Athey et al., 2022).

There are numerous barriers that hamper collective efforts to raise awareness and address microfibres as environmental pollutants, such as their minute size, making them very difficult to see without the aid of magnification (KeChi-Okafor et al., 2023), rendering it a largely invisible problem; the general lack of awareness and misunderstanding that it is a synthetic-only issue (Athey and Erdle, 2021); the lack of alignment and sometimes misuse of terminology; limited data evidencing the full impact of the physical presence and toxicological impact of microfibres on human and animal health and the environment; and lack of accountability, with shifting of responsibilities between industries and stakeholders. Nevertheless, their persistence, high emission rates to the environment and projected increase due to predicted growth in global fibre production, stresses the urgency of the challenge. It is vital that the textile industry, alongside other key stakeholders – science, government, policy makers, and citizens, take collaborative action now and adopt a no-regrets approach to find solutions to mitigate microfibre pollution.

Microfibres found in the environment come from a range of sources, which originate predominately from apparel. Other microfibre sources include home textiles (e.g. carpets, curtains, upholstery and bedding), non-wovens (e.g. wet wipes, cigarette filters, surgical masks) and non-textile sources such as fishing gear (Henry et al., 2019; Hu et al., 2022; Wu et al., 2023). Fundamentally, microfibres are textile fibres, which can be defined as a unit of matter, characterised by their flexibility, fineness and a high ratio of length to thickness. Textile fibres can be synthetic (e.g. acrylic, nylon, polyester, polypropylene), man-made cellulosic (e.g. lyocell, modal, viscose), or naturally sourced (e.g. cotton, flax, jute, silk, wool) and are typically 5× thinner than the diameter of a human hair.

The emerging concern about the presence of microfibres in the environment is relatively recent, primarily since the pioneering work of Thompson et al. (2004), who reported their ubiquity in coastal sediments and waters in the UK. Since then, their anthropogenic presence has been well documented. They have been found in almost every environment on Earth: in marine and freshwater environments, wastewater, stormwater, terrestrial environments and the urban atmosphere (Athey et al., 2020; Napper et al., 2023; Prata, 2018; Stanton et al., 2019). They are easily transported and accumulate in the natural environment, demonstrated by their presence in remote regions far from urban areas, such as the Arctic and Mount Everest (Napper et al., 2020; Pakhomova et al., 2022).

However, there remains much misunderstanding when it comes to the types of microfibres that dominate in the environment and the raw materials from which they originate. Largely driven by the visible prevalence of plastic pollution, many environmental studies are focussed on the detection of 'microplastic' matter (Athey and Erdle, 2021). Non-synthetic material is often overlooked and undocumented, either because it is out of scope or due to a lack of appropriate knowledge, skills and/or appropriate instrumentation to characterise non-synthetic fibres. Not only has this resulted in a general underestimation of microfibre concentrations, but it also gives the wrong impression that all microfibres are microplastic. These findings and the manner in which they are often subsequently reported in the media create a misleading narrative that synthetic clothing is the sole cause of microfibre pollution. This narrative ignores countless forensic studies that provide unequivocal evidence to the contrary. Forensic studies consistently find approximately 70% or more of all fibres are non-synthetic, with the vast majority originating from a natural source (Jones and Johansson, 2023; Palmer and Burch, 2009; Watt et al., 2005; Cantrell et al., 2001; Grieve and Biermann, 1997). Thankfully, a more holistic understanding of the topic is growing with recent microfibre studies recognising the importance of full characterisation and identification of all encountered microfibres, providing a more complete picture of the situation (Stanton et al., 2019; KeChi-Okafor et al., 2023). The latest findings from these studies are congruent with on-land forensic studies, supporting the growing body of evidence that microfibre pollution is not isolated to synthetic fibre fragments.

Whilst microfibre loss occurs throughout the textile lifecycle, from production to disposal, pathways are required to facilitate their release to the environment. There are three major pathways that result in microfibre release to the natural environment, and when occurring in conjunction, the intensity of the problem grows:

- **Water** – release from textile manufacturing and laundering; run-off from land.
- **Air** – release into the atmosphere via textile manufacturing, drying and consumer use.
- **Terrestrial** – end-of-use disposal of textiles, application of microfibre containing sewage sludge to agricultural land.

Wastewater effluent has been identified as a major pathway contributing to microfibre pollution, primarily through the laundering of apparel. Studies have shown that the concentration of microfibres is greater downstream of a wastewater treatment plant than upstream (Dalu et al., 2021). Wastewater treatment systems, if efficient and working correctly, can remove up to 99% of microfibres in some cases, but are not 100% efficient in their removal (Gies et al., 2018). Removed microfibres are often sequestered in sewage sludge (biosolids), a by-product of the wastewater treatment process. Sludge can be incinerated, though it is often applied to agricultural land as fertiliser. Consequently, microfibres find their way to terrestrial environments where they can persist in soil for up to 15 years or through water run-off can find their way to rivers and subsequently back to oceans (Naderi Beni et al., 2023).

Early on, domestic laundering was identified as the major source of microfibre pollution (Hartline et al., 2016). As the most documented pathway to microfibre pollution, laundering has captured the attention of researchers and policymakers alike. The focus has been so great that policymakers in France have passed legislation that as of 2025, washing machine manufacturers will be mandated to fit filters which capture microfibres in effluent and block their pathway to wastewater treatment plants (WWTPs). The USA, Australia and the UK will seemingly following suit (Weis and De Falco, 2022). However, a significant proportion of laundering around the world is done by hand. Whilst some effluent may be processed by WWTPs (if the infrastructure is in place), only around 20% of global wastewater is treated by WWTPs (UNESCO, 2018).

Less attention has been given to air as a major pathway, despite microfibres being found in remote, non-urban locations and scientific demonstration of their ability to naturally move through the air in the absence of external agitation forces (Mbachu et al., 2020). The use of electric clothing dryers also offers a direct pathway to the atmosphere (Kapp and Miller, 2020). Studies on atmospheric microfibre concentrations have found that concentrations are much greater in indoor air than outdoor air, which increases the potential for human exposure (Dris et al., 2017).

Once released, microfibres have the potential to harm the environment, wildlife and people, as negative health effects have been documented in a wide range of aquatic and terrestrial organisms (de Sá et al., 2018). Microfibres have been found in human intestines, placental tissue, blood and lung tissue (Ibrahim et al., 2021; Leslie et al., 2022; Pauly et al., 1998; Ragusa et al., 2021), but the exact effect on humans is unknown. Understanding of the toxicological impact of microfibres is incomplete, but there is sufficient evidence that microfibres pose physical, chemical and biological risks when they enter the environment. Physical risks are caused due to their shape; chemical risks from the leaching of chemicals from manufacturing processes or absorption into the environment; and biological risks as a result of the transport of pathogens or microbes (Henry et al., 2019).

Recent studies have demonstrated the negative impacts of microfibres on aquatic organisms, including tissue damage, reduced growth, body condition and mortality (Rebelein et al., 2021). The negative health effects of microfibres could have large-scale implications on full ecosystems, due to their impact on lower tropic levels (Rebelein et al., 2021). Once in the environment, they are ingested by organisms and can move through the food chain to other trophic levels (Belzagui et al., 2020).

Yet, the root cause of microfibre pollution is the textile itself – either during production, use or end of use. All types of fibres have a propensity to shed to some extent, and thus, they can be lost from all textile materials, regardless of the source of the fibres (Zambrano et al., 2019). By seeking to understand the drivers of fibre fragmentation, science-led collective action can be taken to mitigate loss. Opportunities exist to address the problem through a range of interventions in design and development, material strategies, circular design and environmental management approaches. Within design and development, textile learnings can be embedded across all process steps in the value chain to minimise fibre shedding in materials. Within manufacturing, guidance can align the sector to ensure clear and measurable action on managing microfibre loss in production.

As research continues to emerge from marine and environmental scientists and the environmental impact of microfibre pollution is subject to further examination, engagement with the textile industry remains nascent. Without the support of the industry, progress would be invariably harder, and proposed mitigation actions may be impractical. Harnessing the experience and knowledge of the industry, alongside science and policymakers, is the key to developing practical, actionable solutions at source. Nevertheless, it is critical that industry takes a 'no regrets' approach and adopts scientific concern about the risks posed by microfibre pollution in accepting that it is not an isolated issue. The textile industry must embed fibre fragmentation within broader sustainability strategies, integrating strategies for mitigation through a multifaceted approach:

1. **Circularity Strategies**
   - Preventing fibres from leaving the technosphere and entering the biosphere (oceans, fresh water, drinking water, air, soil, biota) and
   - Designing out pollution and ensuring the materials used to produce textiles are safe for human health and ecosystems.
2. **Preferred Material Strategies**
   - Ensuring the alternative materials brought to the market do not have unintentional consequences and increased shedding.
3. **Environmental Strategies**
   - Ensuring environmental responsibility in the supply chain does not overlook microfibres and the risks posed by this pollutant in wastewater treatment, sludge management, etc.

20.2 THE MICROFIBRE CONSORTIUM

Recognising the need to connect and translate deep academic research with the reality of commercial supply chain production, TMC was formed in 2018 by Founder, Sophie Mather, as a stand-alone organisation. It followed the initial formation of The Outdoor Microfibre Consortium by the European Outdoor Group, the previous year.

TMC is a non-profit, science-led, organisation with a vision to work towards zero impact from fibre fragmentation from textiles to the natural environment. TMC is a membership organisation with a signatory base comprising global brands and retailers, textile product manufacturers, chemical and dye companies, industry associations, researchers and affiliated organisations. Driven by research, with industry change at its core, TMC cultivates a collaborative approach to improve collective understanding of fibre fragmentation in regard to microfibre pollution. Knowledge gained on the topic is used to provide a broad network of stakeholders with practical solutions for reducing the formation and release of microfibres into the environment from textile manufacturing and the product lifecycle. By doing so, TMC acts as a central point for coordinated action.

The remainder of this chapter will focus on the progress of TMC, how it has brought the textile industry together to work towards one common goal, the tools it has generated and the impact made to date.

20.2.1 The Microfibre 2030 Commitment and Roadmap

In 2021, TMC launched The Microfibre 2030 Commitment (The Microfibre Consortium, 2021a), a framework with a common ambition to work towards zero impact from fibre fragmentation from textiles to the natural environment by 2030. Signatories of the 2030 commitment embed fibre fragmentation into their sustainability agenda and become publicly accountable for addressing microfibre pollution. They commit to actively contributing to research and data collection, closing the knowledge gap, identifying the root causes of fibre fragmentation and contributing to the development of innovative solutions.

Supporting The Microfibre 2030 Commitment, The Microfibre Roadmap (The Microfibre Consortium, 2021b) (Figure 20.2) sets the strategic direction, pace of

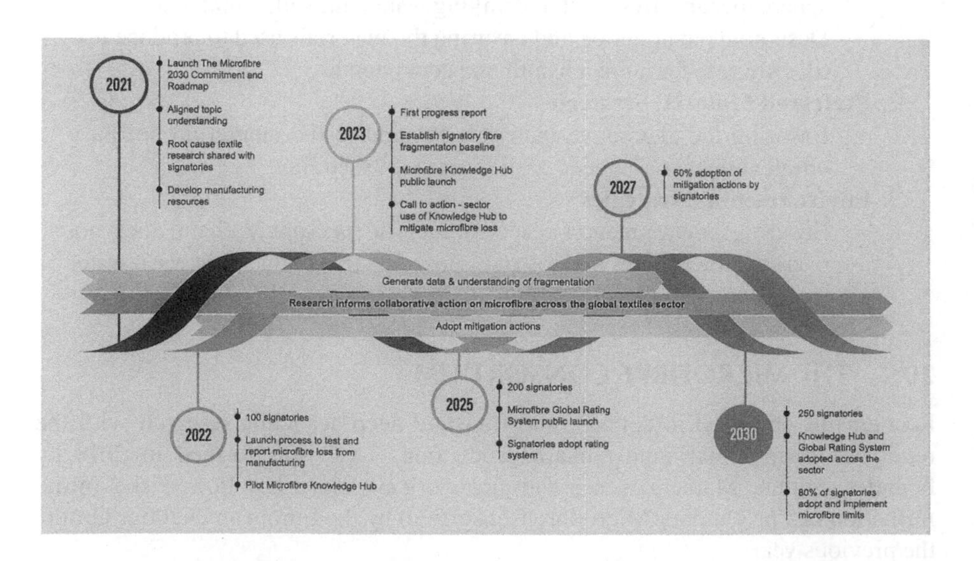

FIGURE 20.2 The Microfibre 2030 Roadmap, version 1.

activity and specific steps required to meet the commitment. It sets out clear milestones against three inter-related pillars:

- **Align** – driving industry commitment to work to one agenda, scale engagement and activity.
- **Understand** – a science-led approach to drive root cause understanding through empirical research.
- **Mitigate** – enabling the sector to take meaningful action at pace whether through root-cause change, or end-of-pipe solutions.

The roadmap timeline is broken down into three stages: activate, implement and scale. The aim of the activation phase is to create the primary research, tools and resources that will enable transition to implementation from 2024 and scaling impact from 2028.

Signatory collaboration in research extends to the voluntary involvement in topic-specific Task Teams. The Task Teams are focussed on a relevant topic of research, working collectively to deliver a science-led, time-relevant output that serves to develop collective knowledge and understanding, driving forward delivery against the roadmap.

20.3 ALIGNING THE SECTOR

The textile industry is not only vast, but diverse, and has many competing priorities when it comes to implementing sustainable practices and reducing detrimental environmental impacts. Critical to the success of developing practical solutions to mitigate microfibre pollution is industry alignment, working to one agenda so that engagement and activity can be easily scaled. In addition to the launch of The Microfibre 2030 Commitment and Roadmap, alignment of language, terminology and tools is critical to its success.

20.3.1 Glossary of Terms

Aligning all stakeholders in the apparel and textiles sector around a common language when speaking of fibre fragmentation is vital to ensure smooth dialogue. Misalignment, misunderstanding and confusion around the key definitions and terms used can cause complications when discussing the topic. To create the clarity and consistency required, the parameters of discussion must be clearly defined. This became particularly pertinent as interest and attention on the subject grew.

The term 'microfibre(s)' is context-dependent, creating challenges for those working to address the issue. Within the environmental sciences and wider public arena, the term microfibre is recognised and commonly used in place of 'textile fibre(s)' or simply 'fibre(s)'. Within the environmental context, microfibres are sometimes classed as 'microplastics' – technically correct if the raw material of the fibre is a fossil fuel, as in the case of synthetic fibres, but incorrect if the microfibre is non-synthetic. Crucially, there is significant confusion of terminology within the textile industry. Technically, a 'microfibre' is the term used to describe a particular type of textile fibre, which is a synthetic fibre with a linear density of less than 1 denier (Ramakrishnan et al., 2009) (and so is exceptionally thin). Thus, 'microfibre pollution' creates bewilderment within the manufacturing industry, whereby a facility that does not create products made from 'microfibres' would consider themselves exempt from being contributors to the problem.

Any solutions to mitigate microfibre pollution, if they are to have sufficient impact, must be developed with the textile industry in mind, and wherever possible through collaboration. To remove unnecessary barriers and align the textile industry, TMC, facilitated discussions with brands, retailers and manufacturers/suppliers across the textile supply chain to agree on a common language for the industry. 'Fibre fragmentation' was identified as the preferred terminology and a more suitable description of microfibre shedding. As such, wherever possible, TMC refers to the loss of microfibres from textiles as 'fibre fragmentation'. The term ensures inclusivity of all fibres – whether synthetic, man-made cellulosic or naturally sourced and applies to all textile manufacturers, irrespective of the types of products they produce.

> Fibre fragmentation is the process of fibre loss from a textile product during its life cycle and / or through its subsequent breakage in the natural environment. This is also referred to as fibre shedding.

20.3.2 The Microfibre Consortium

Once the term fibre fragmentation was agreed, a full glossary of terms was established, taking a further step towards a globally aligned approach. The glossary was developed in consultation with TMC signatories, key industry associations and well-respected bodies, such as American Association of Textile Chemists and Colourists (AATCC) and Textile Exchange (TE), to ensure consistency across the industry. The glossary is publicly available and can be accessed via the TMC website (The Microfibre Consortium, 2023a).

20.3.3 Reporting

Tracking and reporting the progress of TMC in addressing fibre fragmentation and demonstrating impact in the mitigation of microfibre pollution is important to provide transparency. One of the key resources of TMC is the wealth of data collected from measuring the fibre fragmentation of hundreds of different fabrics in use across the industry. The Microfibre Data Portal is the largest dataset of fibre fragmentation of finished fabrics in the world, made possible through industry adoption of a standard testing methodology (see Section 20.4.1). With a sufficient body of data, reference points (or baselines) can be established that enable transparent reporting.

Development and modelling is underway to establish a baseline for fibre fragmentation of finished fabrics. The reference point will be an outcome-focussed quantitative indicator to credibly measure and report on the general status of microfibre loss in tested fabrics over time and demonstrate how the results of TMC's interventions and activities support signatories in mitigating fibre fragmentation of finished fabrics. It will:

- Quantify the average shedding of fabrics tested by The Microfibre 2030 Commitment signatory base and uploaded to The Microfibre Data Portal.
- Provide insights about the data available in The Microfibre Data Portal, the potential gaps and how it evolves over time.
- Enable TMC to quantitatively report on progress over time.

The reference point methodology was published in early 2023, and the first baseline is due to be published at the end of 2023. The reference point will be based upon data from tested finished fabrics housed within The Microfibre Data Portal and is therefore not intended to represent fibre fragmentation in the industry or TMC wider community. Furthermore, it is not intended to be a tool to influence industry actors in their decisions and practices.

20.4 UNDERSTANDING FRAGMENTATION

As a science-led organisation, conducting microfibre research is an integral part of TMC's activities. From closing knowledge gaps to informing product change, to the (eco)toxicological impacts of fibre fragmentation, TMC aims to advance understanding in all aspects of microfibre pollution.

20.4.1 The Microfibre Consortium Test Method: Quantification of Fibre Release from Fabrics during Domestic Laundering (The TMC Test Method)

The majority of TMC signatories are brands and retailers seeking to make positive changes to their products in order to reduce fibre fragmentation. To mitigate microfibre pollution through product change, it is necessary to better understand the key drivers of fibre fragmentation. This is possible only through systematic research that aims to understand why one fabric fragments to a greater (or lesser) extent than another. Such understanding can only be achieved by measuring fibre loss from finished fabrics, using a reliable test method.

Domestic and industrial washing is well established as a major source of microfibre loss and as such has received much research attention (Figure 20.3). Mechanical agitation and the selected washing conditions cause fibres to fragment from clothing, which are subsequently washed away in the wastewater and released into the environment. Multiple studies have since aimed to quantify microfibre release to the environment and determine the factors that affect it; washing factors found to affect loss are wash temperature, wash duration, liquor ratio and the use of products such as fabric softener (Cotton et al., 2020; De Falco et al., 2019; Kelly et al., 2019; Lant et al., 2020). However, the quantification studies often lack detailed technical specifications of the fabrics tested and used an array of diverse methodologies. This made meaningful conclusions difficult to establish and direct comparisons of a fabric's performance impossible (Hartline et al., 2016; Jönsson et al., 2018; Belzagui et al., 2019; De Falco et al., 2020). Consequently, it substantiated the need for a standardised testing methodology and approach to quantify fibre fragmentation of finished fabrics by conducting research to better understand the drivers of fibre fragmentation.

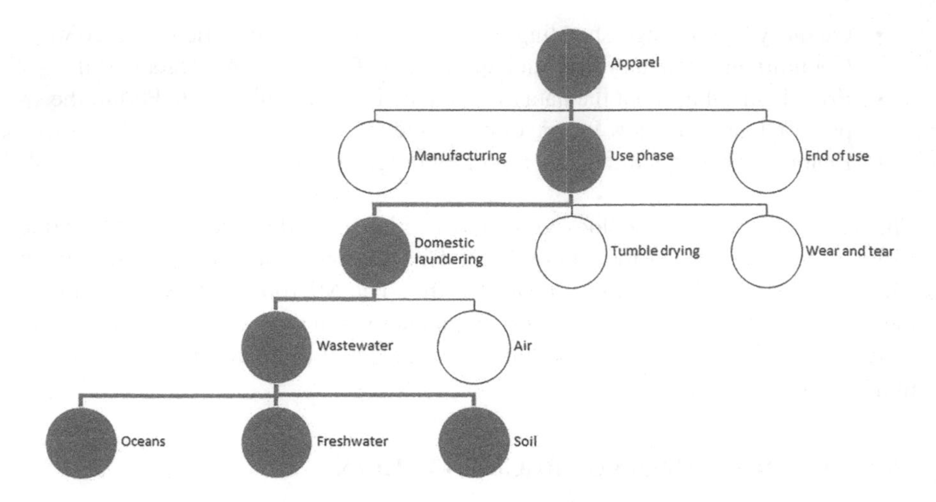

FIGURE 20.3 Loss of fibres from the apparel industry and pathways to their release to the environment through domestic laundering.

Recognising the sector's need, TMC, in collaboration with the European Outdoor Group and the University of Leeds, worked to develop a standardised approach to quantify fibre loss from finished fabrics during simulated domestic laundering. This was the first test method established to reliably measure microfibres released during domestic laundering and is underpinned by detailed fabric specifications (The Microfibre Consortium, 2021c). Prior to public release in 2021, the test method underwent a full validation process involving correlation tests with 10 independent global laboratories and tested 200 fabrics, demonstrating its reliability and reproducibility. Eight fabric specimens, each prepared according to the test protocol, are placed in individual stainless steel cannisters, along with steel balls and a volume of water. Once sealed, the temperature is raised and the cannisters shaken to mimic the movement of a domestic washing machine. Following a fixed duration, the washing effluent is filtered to capture fibre fragments lost from the fabric during laundering, and the fabric is allowed to dry. Gravimetric analysis of pre- and post-washing provides measurement data of both mass loss from the fabric and mass gain of the filter.

The method has since been used as the basis for a new international standard, ISO-4484-1 Textiles and textile products – Microplastics from textile sources – Part 1: Determination of material loss from fabrics during washing (International Organization for Standardization, 2023). There have been some minor adjustments made, mostly for time efficiencies, with a reduction in the number of test specimens from eight to four. It is also closely aligned with AATCC's test methodology, with a key difference being the voluntary addition of detergent (AATCC, 2021).

The TMC Test Method (The Microfibre Consortium, 2021c) can only be performed by TMC-accredited personnel and laboratories. Fabric testing is usually conducted on behalf of brands and retailers by approved commercial laboratories, who are also TMC signatories, or in the case of some brands, by their own internal laboratory technicians. Testing results, along with specification details of the fabric being tested, must be submitted to TMC. The data is collated a confidential and secure repository, The Microfibre Data Portal. Since the launch of The Microfibre 2030 Commitment (The Microfibre Consortium, 2021a), the focus has been on growing such testing to develop a sufficient, reliable body of data from which there is continuous understanding of the subject as patterns emerge. Thus, when joining TMC and in signing up to The Microfibre 2030 Commitment, TMC signatories commit to testing a specified number of fabrics per year, using The TMC Test Method, and together signatories ensure the necessary scaling of data.

20.4.2 The Microfibre Data Portal

Data produced from fabrics tested using The TMC Test Method (The Microfibre Consortium, 2021c) along with detailed fabric specifications are collated and housed in The Microfibre Data Portal. The Microfibre Data Portal is a pivotal tool for scaling research into root causes of fibre fragmentation and supports the ability to analyse the data collected from signatories.

At the time of writing, The Microfibre Data Portal houses over 7,000 data points collated from 653 tested fabrics representing 213 manufacturers from 24 countries. The range of materials from global suppliers provides a snapshot representation of fabrics that the industry is utilising in products today. The Microfibre Data Portal also enables analysis of the diverse processes that are known to contribute to fibre shedding. In submitting technical specification details of the tested fabrics, brands and retailers must answer in the region of 50 questions. These questions span the entire fabric production process, from the raw materials used to make the fabric, to yarn details and fabric structure, dyeing methods and information purporting to the use of any mechanical and chemical finishing processes. It is in these detailed specifications that patterns of the drivers of fragmentation can begin to emerge.

The current dataset, to the best of our knowledge, is the largest global database on fibre fragmentation. It is a solid foundation from which knowledge can be built, but data gaps currently exist that must be filled. For example, the current dataset has a skewed population towards synthetic fibres, with approximately 50% of the tested fabric compositions being polyester-based, and a limited number of tested fabrics of naturally sourced or man-made cellulosic origin. In terms of fabric construction, the data population is knit heavy (jersey construction) and would benefit from the addition of more test data from woven fabrics and warp-knit constructions. In some instances, there is incomplete fabric detail which makes detailed analysis more challenging when looking for insights related to product, materials and finishing types. The skewness in the data may, in part, result from the pro-active nature of outdoor brands who were early in joining TMC and whose products are often dominated by synthetic fabrics. As the signatory base grows and becomes more diverse, so too will the diversity of the fabrics tested, bringing a better balance to the dataset. In the future, the data will support the development of a Global Rating System for finished fabrics.

Up to four times per year, TMC signatories are provided with a detailed report on the performance of the fabrics they have tested, with direct comparisons against The Microfibre Data Portal averages. The report provides details on the effect of key fabric specifications such as fabric structure and yarn type. This information serves to allow signatories to track progress and make product changes, demonstrated by TMC signatory, Patagonia. Following testing with the TMC Test Method (The Microfibre Consortium, 2021c), one of their main fleece products (53% recycled polyester, 42% recycled wool, 3% recycled polyamide, 3% other fibre) was found to shed fibres to a high degree. Areas for improvement were identified in the material design and process steps and through 13 product trials, followed by re-testing, a 44% reduction in fibre fragmentation of the fabric was achieved.

Broad analysis of data within The Microfibre Data Portal has been conducted, which has provided some key learnings, conducive to those evidenced elsewhere (The Microfibre Consortium, 2022a) such as:

- Fabrics of all compositions shed fibres.
- Composition is not the only driver for fibre fragmentation; all fabric specifications influence fragmentation – raw materials and yarn type, fabric structure, colouration and both chemical and mechanical finishing.
- Yarns made from staple fibres release on average 50% more fibre fragments than filament yarns.
- Weft-knit fabrics shed on average twice as many fibre fragments as woven fabrics.
- Mechanical finishing has been found to increase fibre fragmentation.
- There is no clear correlation between fabric weight and fibre loss.

Unsurprisingly, however, there is much inter-dependency of the key drivers of fibre fragmentation which must be systematically untangled. This was evidenced by a detailed investigation of tested fabrics comprised from 100% recycled polyester.

20.4.3 Effect of Recycled Polyester on Fibre Fragmentation

The increase in recycled polyester production has been driven by a desire of the textile industry to reduce its environmental impact, in a move away from using non-renewable petrochemicals (Shen et al., 2010). There has been suggestion that the processing involved in recycling polyester causes weakening of the fibre, which may affect how easily it fragments in comparison to its virgin alternative (Majumdar et al., 2020). Thus, before an environmental issue is purported to be addressed, it is critical that due diligence is carried out. There must be no unintended consequences elsewhere, such as increased fibre fragmentation leading to increased microfibre pollution.

There are a limited number of fibre fragmentation research studies that have included recycled polyester garments in their testing; however, as they were not the focus of the study, the direct effect of recycled polyester on fibre fragmentation is unclear (Belzagui et al., 2019; Hartline et al., 2016; Jönsson et al., 2018). One key study did directly compare recycled and virgin polyester fabrics (as the only variable difference between the two fabrics) and concluded that recycled polyester fabric fragmented to a greater extent. However, the measurement for fibre fragmentation was the number of fragments, rather than a total gravimetric loss, and so it is not known whether the mass loss was the same or not (Özkan and Gündoğdu, 2021).

Whilst isolated studies may indicate the effect of individual variables, the question remains how much of an influence they have more broadly, across a large dataset containing a multitude of variables. For the textile industry, it is important that efforts are focussed on areas which maximise the mitigation of fibre fragmentation. Thus, to investigate the holistic effect of replacing virgin polyester with recycled variants, measuring the fragmentation of a greater number of diverse fabrics was key.

TMC carried out research investigating the effect of recycled polyester on fibre fragmentation. A specific call-to-action to TMC's signatories to test 100% polyester fabrics (recycled or virgin) resulted in testing data from a total of 251 fabrics, across a range of fabric specifications. The data, incorporating 2,977 individual sample data points, validate the collective strength of The Microfibre Data Portal in enabling direct comparison across numerous tested fabrics. The data was analysed in detail, exploring the effect of polyester source, recycling methods and fabric structure (The Microfibre Consortium, 2023b). TMC's research concluded, overall, that fabrics made from mechanically recycled polyester fragmented to the same extent as those made from virgin polyester. Whilst the industry's move towards recycled polyester does not appear to have a detrimental effect on fibre fragmentation, it should remain a concern for all textiles including polyester materials, whether the stock is virgin or recycled. However, as expected, the data highlighted a greater inter-dependency of fabric variables, which served to reinforce the complexity of fibre fragmentation as a topic and the need for a broader and deeper analysis across the entire dataset. This is the current focus of TMC's research activities.

20.4.4 Biodegradability

Many stakeholders regard biodegradable materials as a solution to the fibre fragmentation issue. If microfibres, released into the environment, breakdown into non-harmful substances (in a timely manner), fibre fragmentation may not be considered a concern. A selection of fibres (predominantly naturally sourced and cellulosic) are either inherently biodegradable in their greige unfinished state or are engineered to be biodegradable (e.g. biosynthetics) and, as such, are often presumed to biodegrade in the natural environment. This assumption is often associated with the misconception that the fibres shed from textiles biodegrade before becoming an environmental risk in the context of fibre fragmentation, either from a presence or toxicological perspective (Lykaki et al., 2021). Thus, increasingly, there is a desire to fast-track biodegradability as a solution to the fibre fragmentation issue.

The topic of biodegradation causes confusion among consumers, brands and suppliers about what materials can be considered biodegradable, what their degradation pathways are and how they should be communicated. TMC established a topic-specific Task Team to understand the context in which biodegradation applies to the larger microfibre pollution issue. The Task Team identified numerous issues of concern in relation to the relevance of the testing methodologies used to test biodegradability, including ISO testing standards (The Microfibre Consortium, 2021d). More specifically, methodologies (i) do not usually test the product in its finished state as it would ordinarily be found in the environment; (ii) use ideal testing conditions, not representative of real environmental conditions; (iii) have a narrow focus on the specific type of environmental conditions used for testing; (iv) lack understanding of the toxicological effect of any biodegradable action; and (v) lack standards for testing in freshwater environments.

Following its investigation, TMC does not endorse biodegradability as a 'solution' to fibre fragmentation from textiles, and this sentiment remains at the time of writing. As a result, TMC highlighted the following challenges that the industry must address before biodegradable claims can be accepted:

1. Cross-industry alignment on methodologies (including but not exclusive to determining testing at greige).
2. Methodologies that come as close as possible to represent real-life situations, as this is when the advantages of biodegradation are realised, i.e. low temperatures, limited light or UV, and limited oxygen.
3. Cross-industry alignment on the pathways and therefore the intended receiving environment(s) for testing (i.e. fresh water, marine water, soil, anaerobic and aerobic conditions, etc).
4. Value of testing versus company resources. Identification of which test methods, lab (quicker method) vs field testing (real-life environment) provides cost-effective results and appropriate approaches to be taken for correlation and modelling.
5. A review within the wider sustainability context to ensure no larger issues are raised.
6. Further toxicological work to determine the effect of microfibres as carriers of chemicals into the environment.

20.5 MITIGATING FRAGMENTATION

20.5.1 The Textile Manufacturing Industry

The textile manufacturing supply chain consists of thousands of facilities in dozens of countries. Although the products and processes may vary greatly from one facility to the next, they all have one thing in common: the ability to shed fibres from textiles into the wastewater discharged from each facility (Lim et al., 2022). Thus, the textile and clothing industry is responsible for fibre fragmentation from textiles at both the consumer level and within the manufacturing process.

During the manufacturing and processing of textiles, microfibres can be released from greige fabric during the scouring and cleaning process prior to dyeing and finishing, and from the agitation of the fabric during the dyeing process (Figure 20.4). Microfibres are released from the fibres and yarns within a fabric, whilst particles from the ambient manufacturing environment are entrained within a fabric. TMC advocates a portfolio- scalable approach to the mitigation of microfibre pollution. For example, the capture of microfibre loss through the use of wastewater management at a facility level is a complimentary action to root-cause mitigation that can be done at the textile design and development level to consequential loss.

As the industry-sought best practice guidance to drive mitigation of fibre fragmentation, the need for an in-depth, aligned and globally relevant textile manufacturing

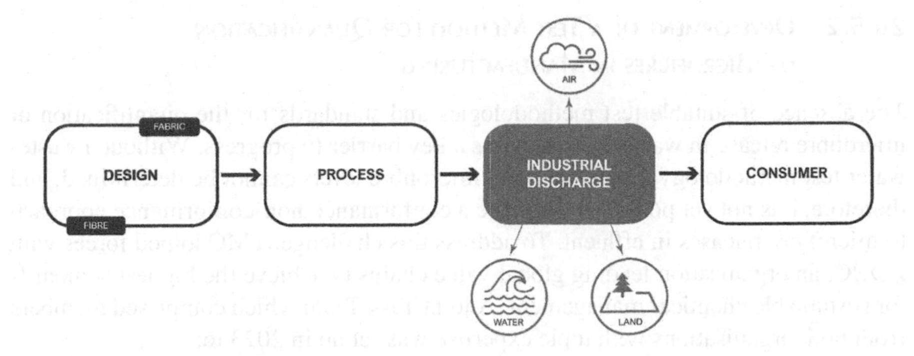

FIGURE 20.4 Industrial discharge of microfibres within the textile manufacturing supply chain.

standpoint became paramount. This was crucial to ensure cross-industry uptake and effective, measurable impact whilst upholding a 'no regrets' attitude within the broader sustainability agenda. Over a 2-year period, a dedicated TMC Task Team convened to develop manufacturing guidelines aimed to reduce microfibre release from facilities (The Microfibre Consortium, 2022b). The objectives were to:

- Define a method of controlling/managing microfibre pollution in manufacturing wastewater.
- Provide information on the technologies which exist to control microfibre pollution from manufacturing.
- Identify and confirm the location within a manufacturing facility at which microfibre pollution can be controlled/measured.
- Provide a vehicle to engage manufacturing partners and identify/landscape where suitable technologies already exist.
- Identify knowledge gaps.

The urgency of measuring microfibre loss in wastewater between production facilities, to manage and ultimately avoid loss, led to the publication of the Preliminary Manufacturing Guidelines, 'Control of Microfibres in Wastewater' in 2022 (The Microfibre Consortium, 2022b), available in Mandarin and simplified Chinese translations. The guidelines have energised engagement within the manufacturing community and established a supply chain network ready to strategise on how to mitigate fibre fragmentation risks to their business and facilities. However, there remains a need to continue to raise awareness in the supply chain and implement TMC guidelines at facility level. To expedite this work, TMC's preliminary guidelines have been developed into a training curriculum, set to be rolled out in 2023. Additionally, a simple snapshot guidance has been developed in conjunction with The Zero Discharge of Hazardous Waste (ZDHC) Foundation (ZDHC), individually tailored to brands/retailers, and suppliers. The guidance details actions that can be taken now to mitigate microfibre release and how to (i) communicate and build awareness about fibre fragmentation, (ii) reduce the formation of fibre fragments and (iii) reduce emissions.

20.5.2 Development of a Test Method for Quantification of Microfibres in Manufacturing

The absence of suitable test methodologies and standards for the quantification of microfibre release in wastewater remains a key barrier to progress. Without a wastewater test methodology, reference point microfibre levels cannot be determined, and therefore, it is not yet possible to operate a conformance/non-conformance approach to microfibre releases in effluent. To address this challenge, TMC joined forces with ZDHC, an organisation leading global value chains to achieve the highest standards for sustainable chemical management. A joint Task Team which comprised members from both organisations with topic expertise was set up in 2023 to:

- Define a test methodology to measure fibre loss within wastewater at a manufacturing level.
- Establish a baseline for microfibre loss from manufacturing facilities.
- Identify a reporting structure that captures the measurement and control of microfibres from manufacturing facilities.
- Set maximum allowable limits for microfibres in discharged effluent.

Early work has identified total suspended solids (TSS) as a potential proxy for the determination of microfibre release in wastewater. For a test methodology to be embraced by the manufacturing supply chain, it must be as frictionless as possible. Key requirements are tests that can be easily implemented across all facilities, simple to perform and low cost. Various systems and technologies exist in manufacturing facilities to target the separation of solids from textile wastewater. Regulatory standards exist for TSS, which cover all types of solids in discharge effluent. The regulations relate to discharge standards of TSS, and failure to meet the standards will almost certainly result in discharge of large quantities of textile fibres (Frehland et al., 2020). A proportion of TSS is made up of microfibres, although the exact proportion is unknown. Thus, the first stage of the research is to determine whether TSS can be used as a medium to reliably quantify microfibre release in wastewater. If so, a suitable testing methodology can be developed and reference points determined, from which standards can be set. This work is planned to be completed in 2023, leading to implementation in 2024. If more stringent standards for discharge of microfibres are introduced in the future, facilities will have to ensure existing processes for removal of solids are optimised. TSS compliance should be followed until specific microfibre test methods and limits are developed and implemented.

20.6 CONCLUSIONS

TMC is a research-led organisation whose mission is to facilitate the development of practical solutions for the textile industry to minimise fibre fragmentation to the environment from textile manufacturing and product life cycle. Through The Microfibre 2030 Commitment and Roadmap, TMC actively engages the textile industry in a number of research-led activities to bring alignment and commitment, develop root cause understanding and deliver sustained action and measurable impact.

The industry uptake of the TMC Test Method to measure microfibre loss from fabrics is now firmly established. Therefore, TMC's focus is shifting towards the analysis of data and closing of knowledge gaps to expedite root-cause understanding. Holistic, in-depth analysis of the data housed within The Microfibre Data Portal is in progress, the outcomes of which will support the development of a vital research strategy. In the long term, the data will allow the development of a global rating system to support the textile industry in making product change.

Significant progress has been made already within the manufacturing community to raise awareness and highlight changes facilities can make today to reduce microfibre pollution. The development of a test methodology for facilities, along with standard limits will see a further step change.

But, challenges remain. There continues to be a lack of understanding of the topic and recognition of the risks posed by fibre fragmentation. We must work to develop stronger, more strategic partnerships with leading sustainability initiatives to drive collective change, faster, ensuring the topic is embedded firmly with the broader sustainability agenda. Organisations must follow suit adopting a 'no-regrets' approach, be active in the development of solutions and commit to working towards zero impact from fibre fragmentation from textiles to the natural environment by 2030.

ACKNOWLEDGEMENTS

We wish to thank Louise Ashdown for her support in enabling this chapter to be written and for her review of the material.

REFERENCES

AATCC. (2021). AATCC TM212-2021, *Test method for fiber fragment release during home laundering*.

Ali, A., Shaker, K., Nawab, Y., Jabbar, M., Hussain, T., Militky, J., & Baheti, V. (2018). Hydrophobic treatment of natural fibers and their composites – A review. *Journal of Industrial Textiles*, 47(8), 2153–2183.

Athey, S. N., Adams, J. K., Erdle, L. M., Jantunen, L. M., Helm, P. A., Finkelstein, S. A., & Diamond, M. L. (2020). The widespread environmental footprint of indigo denim microfibers from blue jeans. *Environmental Science and Technology Letters*, 7(11), 840–847.

Athey, S. N., Carney Almroth, B., Granek, E. F., Hurst, P., Tissot, A. G., & Weis, J. S. (2022). Unraveling physical and chemical effects of textile microfibers. *Water, MDPI*, 14(23), 3797

Athey, S. N., & Erdle, L. M. (2021). Are we underestimating anthropogenic microfiber pollution? A critical review of occurrence, methods, and reporting. *Environmental Toxicology and Chemistry*, 41(4), 822–837.

Atkins, W., Jenkins, P. G., & Warren, F. (1954). The suspended matter in sea water and its seasonal changes as affecting the visual range of the Secchi disc. *Journal of the Marine Biological Association of the United Kingdom*, 33(2), 497–509.

Belzagui, F., Crespi, M., Álvarez, A., Gutiérrez-Bouzán, C., & Vilaseca, M. (2019). Microplastics' emissions: Microfibers' detachment from textile garments. *Environmental Pollution*, 248, 1028–1035.

Belzagui, F., Gutiérrez-Bouzán, C., Álvarez-Sánchez, A., & Vilaseca, M. (2020). Textile microfibers reaching aquatic environments: A new estimation approach. *Environmental Pollution*, 265.

Buchanan, J. B. (1971) Pollution by synthetic fibres. *Marine Pollution Bulletin*, 2(2), 23–23.

Cantrell, S., Roux, C., Maynard, P., & Robertson, J. (2001). A textile fibre survey as an aid to the interpretation of fibre evidence in the Sydney region. *Forensic Science International*, 123, 48–53

Cotton, L., Hayward, A. S., Lant, N. J., & Blackburn, R. S. (2020). Improved garment longevity and reduced microfibre release are important sustainability benefits of laundering in colder and quicker washing machine cycles. *Dyes and Pigments*, 177.

Dalu, T., Banda, T., Mutshekwa, T., Linton, Munyai, F., & Cuthbert, R. N. (2021). Effects of urbanisation and a wastewater treatment plant on microplastic densities along a subtropical river system. *Environmental Science and Pollution Research*, 28(27), 36102–36111.

De Falco, F., Cocca M., Avella M., & Thompson, R. C. (2020). Microfiber release to water, via laundering, and to air, via everyday use: A comparison between polyester clothing with differing textile parameters. *Environmental Science and Technology*, 54(6), 3288–3296.

De Falco, F., Di Pace, E., Cocca, M., & Avella, M. (2019). The contribution of washing processes of synthetic clothes to microplastic pollution. *Scientific Reports*, 9(1), 1–11.

de Sá, L. C., Oliveira, M., Ribeiro, F., Rocha, T. L., & Futter, M. N. (2018). Studies of the effects of microplastics on aquatic organisms: What do we know and where should we focus our efforts in the future? *Science of the Total Environment*, 645, 1029–1039.

Dris, R., Gasperi, J., Mirande, C., Mandin, C., Guerrouache, M., Langlois, V., & Tassin, B. (2017). A first overview of textile fibers, including microplastics, in indoor and outdoor environments. *Environmental Pollution*, 221, 453–458.

Frehland, S., Kaegi, R., Hufenus, R., & Mitrano, D. M. (2020). Long-term assessment of nanoplastic particle and microplastic fiber flux through a pilot wastewater treatment plant using metal-doped plastics. *Water Research*, 182.

Gies, E. A., LeNoble, J. L., Noël, M., Etemadifar, A., Bishay, F., Hall, E. R., & Ross, P. S. (2018). Retention of microplastics in a major secondary wastewater treatment plant in Vancouver, Canada. *Marine Pollution Bulletin*, 133, 553–561.

Grieve, M. C. (2000). Back to the future -40 years of fibre examinations in forensic science, Proceedings of the 40th anniversary meeting of the forensic science society. *Science & Justice*, 40(2), 93–99.

Grieve, M. C., & Biermann, T. (1997). The population of coloured textile fibres on outdoor surfaces. *Science & Justice*, 37(4), 231–239

Grieve, M., Biermann, T., & Schaub, K. (2005). The individuality of fibres used to provide forensic evidence – Not all blue polyesters are the same. *Science & Justice*, 45, 13–28

Hartline, N. L., Bruce, N. J., Karba, S. N., Ruff, E. O., Sonar, S. U., & Holden, P. A. (2016). Microfiber masses recovered from conventional machine washing of new or aged garments. *Environmental Science and Technology*, 50(21), 11532–11538.

Henry, B., Laitala, K., & Klepp, I. G. (2019). Microfibres from apparel and home textiles: Prospects for including microplastics in environmental sustainability assessment. *Science of the Total Environment*, 652, 483–494.

Hu, T., Shen, M., & Tang, W. (2022). Wet wipes and disposable surgical masks are becoming new sources of fiber microplastic pollution during global COVID-19. *Environmental Science and Pollution Research*, 29(1), 284–292.

Ibrahim, Y. S., Tuan Anuar, S., Azmi, A. A., Wan Mohd Khalik, W. M. A., Lehata, S., Hamzah, S. R., Ismail, D., Ma, Z. F., Dzulkarnaen, A., Zakaria, Z., Mustaffa, N., Tuan Sharif, S. E., & Lee, Y. Y. (2021). Detection of microplastics in human colectomy specimens. *JGH Open*, 5(1), 116–121.

International Organization for Standardization. (2023). ISO 4484-1:2023. https://www.iso.org/standard/82238.html accessed 15 August 2023

Jones, J., & Johansson, S. (2023). A population study of textile fibres on the seats at three public venues. *Forensic Science International*, 348, 111604

Jönsson, C., Arturin, O. L., Hanning, A. C., Landin, R., Holmström, E., & Roos, S. (2018). Microplastics shedding from textiles-developing analytical method for measurement of shed material representing release during domestic washing. *Sustainability*, 10(7), 2457.

Kapp, K. J., & Miller, R. Z. (2020). Electric clothes dryers: An underestimated source of microfiber pollution. *PLoS One*, 15.

KeChi-Okafor, C., Khan, F. R., Al-Naimi, U., Béguerie, V., Bowen, L., Gallidabino, M. D., Scott-Harden, S., & Sheridan, K. J. (2023). Prevalence and characterisation of microfibres along the Kenyan and Tanzanian coast. *Frontiers in Ecology and Evolution*, 11.

Kelly, M. R., Lant, N. J., Kurr, M., & Burgess, J. G. (2019). Importance of water-volume on the release of microplastic fibres from laundry. *Environmental Science & Technology*, 53(20), 11735–11744.

Lant, N. J., Hayward, A. S., Peththawadu, M. M. D., Sheridan, K. J., & Dean, J. R. (2020). Microfiber release from real soiled consumer laundry and the impact of fabric care products and washing conditions. *PLoS One*, 15(6), e0233332.

Leslie, H. A., van Velzen, M. J. M., Brandsma, S. H., Vethaak, A. D., Garcia-Vallejo, J. J., & Lamoree, M. H. (2022). Discovery and quantification of plastic particle pollution in human blood. *Environment International*, 107199.

Lim, J., Choi, J., Won, A., Kim, M., Kim, S., & Yun, C. (2022). Cause of microfibers found in the domestic washing process of clothing; focusing on the manufacturing, wearing, and washing processes. *Fashion and Textiles*, 9(1).

Lunstroot, K., Ziernicki, D., & Vanden Driessche, T. (2016). A study of black fleece garments: Can fleece fibres be recognized and how variable are they? *Science & Justice*, 56, 157–164.

Lykaki, M., Markiewicz, M., Stolte, S., & Zhang, Y.-Q. (2021). Findings from water Chemistry: Retention in sewage treatment plants and biodegradability of fibre fragments. *Textile Mission*.

Majumdar, A., Shukla, S., Singh, A. A., & Arora, S. (2020). Circular fashion: Properties of fabrics made from mechanically recycled poly-ethylene terephthalate (PET) bottles. *Resources, Conservation and Recycling*, 161, 104915.

Mbachu, O., Jenkins, G., Pratt, C., & Kaparaju, P. (2020). A new contaminant superhighway? A review of sources, measurement techniques and fate of atmospheric microplastics. *Water, Air, & Soil Pollution*, 231(2), 85.

Naderi Beni, N., Karimifard, S., Gilley, J., Messer, T., Schmidt, A., & Bartelt-Hunt, S. (2023). Higher concentrations of microplastics in runoff from biosolid-amended croplands than manure-amended croplands. *Communications Earth and Environment*, 4(1).

Napper, I. E., Davies, B. F. R., Clifford, H., Elvin, S., Koldewey, H. J., Mayewski, P. A., Miner, K. R., Potocki, M., Elmore, A. C., Gajurel, A. P., & Thompson, R. C. (2020). Reaching new heights in plastic pollution-preliminary findings of microplastics on Mount Everest. *One Earth*, 3(5), 621–630.

Napper, I. E., Parker-Jurd, F. N. F., Wright, S. L., & Thompson, R. C. (2023). Examining the release of synthetic microfibres to the environment via two major pathways: Atmospheric deposition and treated wastewater effluent. *Science of the Total Environment*, 857.

Özkan, İ., & Gündoğdu, S. (2021). Investigation on the microfiber release under controlled washings from the knitted fabrics produced by recycled and virgin polyester yarns. *Journal of the Textile Institute*, 112(2), 264–272.

Pakhomova, S., Berezina, A., Lusher, A. L., Zhdanov, I., Silvestrova, K., Zavialov, P., van Bavel, B., & Yakushev, E. (2022). Microplastic variability in subsurface water from the Arctic to Antarctica. *Environmental Pollution*, 298, 118808.

Palmer, R., & Burch, H. (2009). The population, transfer and persistence of fibres on the skin of living subjects. *Science & Justice*, 49, 259–264.

Palmer, R., Burnett, E., Luff, N., Wagner, C., Stinga, G., Carney, C., & Sheridan, K. (2015). The prevalence of two 'commonly' encountered synthetic target fibres within a large urban environment. *Science & Justice*, 55(2), 103–106.

Pauly, J. L., Stegmeier, S. J., Allaart, H. A., Cheney, R. T., Zhang, P. J., Mayer, A. G., & Streck, R. J. (1998). Inhaled cellulosic and plastic fibers found in human lung tissue. *Cancer Epidemiol Biomarkers Prev.,* 7(5), 419–428.

Pounds, C. A., & Smalldon, K. W. (1975a). The transfer of fibres between clothing materials during simulated contacts and their persistence during wear: Part I-Fibre transference. *Journal of the Forensic Science Society*, 15(1), 17–27.

Pounds, C. A., & Smalldon, K. W. (1975b). The transfer of fibres between clothing materials during simulated contacts and their persistence during wear: Part II-Fibre persistence. *Journal of the Forensic Science Society*, 15(1), 29–37.

Pounds, C. A., & Smalldon, K. W. (1975c). The transfer of fibres between clothing materials during simulated contacts and their persistence during wear: Part III-A preliminary investigation of the mechanisms involved. *Journal of the Forensic Science Society*, 15(3), 197–207.

Prata, J. C. (2018). Airborne microplastics: Consequences to human health? *Environmental Pollution*, 234, 115–126.

Ragusa, A., Svelato, A., Santacroce, C., Catalano, P., Notarstefano, V., Carnevali, O., Papa, F., Rongioletti, M. C. A., Baiocco, F., Draghi, S., D'Amore, E., Rinaldo, D., Matta, M., & Giorgini, E. (2021). Plasticenta: First evidence of microplastics in human placenta. *Environment International*, 146, 106274.

Ramakrishnan, G., Bhaarathidhurai, M., & Mukhopadhyay, S. (2009). An investigation into the properties of knitted fabrics made from viscose microfibers. *Journal of Textile and Apparel Tech*, 6(1).

Rebelein, A., Int-Veen, I., Kammann, U., & Scharsack, J. P. (2021). Microplastic fibers – Underestimated threat to aquatic organisms? *Science of the Total Environment*, 777, 146045.

Shen, L., Worrell, E., & Patel, M. K. (2010). Open-loop recycling: A LCA case study of PET bottle-to-fibre recycling. *Resources, Conservation and Recycling*, 55(1), 34–52.

Stanton, T., Johnson, M., Nathanail, P., MacNaughtan, W., & Gomes, R. L. (2019). Freshwater and airborne textile fibre populations are dominated by 'natural', not microplastic, fibres. *Science of the Total Environment*, 666, 377–389.

The Microfibre Consortium. (2021a). *The microfibre commitment.* https://www.microfibreconsortium.com/2030 Accessed 15 August 2023

The Microfibre Consortium. (2021b). *The microfibre roadmap.* https://www.microfibreconsortium.com/roadmap Accessed 15 August 2023

The Microfibre Consortium. (2021c). *The Microfibre Consortium test method.* https://www.microfibreconsortium.com/the-tmc-test-method Accessed 15 August 2023

The Microfibre Consortium. (2021d). *TMC biodegradability report.* https://www.microfibreconsortium.com/biodegradability-report Accessed 15 August 2023

The Microfibre Consortium. (2022a). *TMC progress report.* https://www.microfibreconsortium.com/tmc-progress-report Accessed 15 August 2023

The Microfibre Consortium. (2022b). *TMC preliminary manufacturing guidelines.* https://www.microfibreconsortium.com/preliminary-manufacturing-guidelines Accessed 15 August 2023

The Microfibre Consortium. (2023a). *Glossary of terms.* https://www.microfibreconsortium.com/tmc-glossary Accessed 15 August 2023

The Microfibre Consortium. (2023b). *Recycled polyester within the context of fibre fragmentation.* https://www.microfibreconsortium.com/rpet-technical-research-report Accessed 15 August 2023

Thompson, R., Olsen, Y., Mitchell, R., Davis, A., Rowland, S., John, A., Mcgonigle, D., & Russell, A. (2004). Lost at Sea: Where is all the Plastic? *Science*, 304(5672), 838.

UNESCO World Water Assessment Programme (2018). *The United Nations world water development report 2018: Nature-based solutions for water.*

Watt, R., Roux, C., & Robertson, J. (2005). The population of coloured textile fibres in domestic washing machines. *Science & Justice*, 45, 75–83.

Weis, J. S., & De Falco, F. (2022). Microfibers: Environmental problems and textile solutions. *Microplastics*, 1(4), 626–639.

Wu, H., Hou, J., & Wang, X. (2023). A review of microplastic pollution in aquaculture: Sources, effects, removal strategies and prospects. *Ecotoxicology and Environmental Safety*, 252, 114567.

Zambrano, M. C., Pawlak, J. J., Daystar, J., Ankeny, M., Cheng, J. J., & Venditti, R. A. (2019). Microfibers generated from the laundering of cotton, rayon and polyester based fabrics and their aquatic biodegradation. *Marine Pollution Bulletin*, 142, 394–407.

Index

Printed in the United States
by Baker & Taylor Publisher Services

Printed in the United States
by Baker & Taylor Publisher Services